INSECT TRANSGENESIS
Methods and Applications

COVER PHOTOGRAPHS

Front Cover
(counterclockwise from upper left-hand corner to center)

Ceratitis capitata adult transformed with *piggyBac/white+* and *white eye* mutant host (S. McCombs and A. Handler); *Aedes aegypti* larvae infected with TE/3'2J/GFP Sindbis virus (bottom) and uninfected (top) (S. Higgs); *Ae. aegypti* late larva transformed with *Hermes*/GFP (A. Pinkerton and P. Atkinson); *Ae. aegypti kynurenine hydroxylase-white* (khw) adult mutant host and *Hermes/cinnabar+* transformant (A. James and F. Collins); *C. capitata* adult transformed with *piggyBac*/GFP (A. Handler and S. McCombs); *Anastrepha suspensa* embryos transformed with *piggyBac*/GFP (left) and untransformed (right) (R. Harrell and A. Handler).

Back Cover
(top to bottom)

Aedes aegypti adult transformed with *Hermes*/GFP (A. Pinkerton and P. Atkinson); *Anopheles albimanus* pupae transformed with *piggyBac*/GFP (bottom) and untransformed (top) (R. Harrell, O. Perera, and A. Handler); *Drosophila melanogaster* transformed with *piggyBac/white+* showing complete (top) and variegated (bottom) *white+* expression (A. Handler and R. Harrell); *A. suspensa* larva transformed with *piggyBac*/GFP (R. Harrell and A. Handler); *Ae. aegypti* larva transformed with *Hermes*/GFP (A. Pinkerton and P. Atkinson).

INSECT TRANSGENESIS
Methods and Applications

EDITED BY

Alfred M. Handler

Anthony A. James

CRC Press
Boca Raton   London   New York   Washington, D.C.

Library of Congress Cataloging-in-Publication Data
Insect transgenesis : methods and applications / edited by Alfred M. Handler, Anthony A. James
      p. cm.
  Includes bibliographical references (p.).
  ISBN 0-8493-2028-3
    1. Insect cell biotechnology. 2. Insects—Genetic engineering. I. Handler, Alfred M. II. James, Anthony A.
TP248.27.I56 I577 2000
638—dc21
                            00-023660
                                CIP

This work was created in the performance of a Cooperative Research and Development Agreement with the U.S. Department of Agriculture.

This book contains information obtained from authentic and highly regarded sources. Reprinted material is quoted with permission, and sources are indicated. A wide variety of references are listed. Reasonable efforts have been made to publish reliable data and information, but the author and the publisher cannot assume responsibility for the validity of all materials or for the consequences of their use.

Neither this book nor any part may be reproduced or transmitted in any form or by any means, electronic or mechanical, including photocopying, microfilming, and recording, or by any information storage or retrieval system, without prior permission in writing from the publisher.

All rights reserved. Authorization to photocopy items for internal or personal use, or the personal or internal use of specific clients, may be granted by CRC Press LLC, provided that $.50 per page photocopied is paid directly to Copyright Clearance Center, 222 Rosewood Drive, Danvers, MA 01923 USA. The fee code for users of the Transactional Reporting Service is ISBN 0-8493-2028-3/00/$0.00+$.50. The fee is subject to change without notice. For organizations that have been granted a photocopy license by the CCC, a separate system of payment has been arranged.

The consent of CRC Press LLC does not extend to copying for general distribution, for promotion, for creating new works, or for resale. Specific permission must be obtained in writing from CRC Press LLC for such copying.

Direct all inquiries to CRC Press LLC, 2000 N.W. Corporate Blvd., Boca Raton, Florida 33431.

Trademark Notice: Product or corporate names may be trademarks or registered trademarks, and are used only for identification and explanation, without intent to infringe.

© 2000 by CRC Press LLC

No claim to original U.S. Government works
International Standard Book Number 0-8493-2028-3
Library of Congress Card Number 00-023660
Printed in the United States of America 1 2 3 4 5 6 7 8 9 0
Printed on acid-free paper

Preface

The transfer of genetic material into the genome of insects has been a goal of geneticists and entomologists for more than 30 years, and with the successful transposon-mediated transformation of *Drosophila melanogaster* reported by Rubin and Spradling in 1982, efforts have been focused on duplicating this methodology in insects of agricultural and medical importance. The transfer of *Drosophila* transformation systems to other insects, however, has not been straightforward, and only in the past several years have notable successes been reported. These recent successes have resulted from a maturation over the past decade of all aspects of gene transfer technology. As with *Drosophila*, transformation in tephritid fruit flies, mosquitoes, and moths has been achieved with transposon-based vectors, although most have taken advantage of newly discovered elements. Viral transformation vectors are beginning to meet with success, and symbiont-mediated transformation is already finding practical application. The development of reliable marker gene systems has been important to the progress achieved in transgenesis. Genes encoding the green fluorescent protein or its variants appear to have broad utility as marker genes for screening transformants, and these genes may be used as reporter genes in transformation-based analyses of promoter function. Furthermore, genes encoding enzymes important in establishing eye color in insects continue to be useful when appropriate mutant recipient strains are available. Work continues on selectable marker genes such as those that encode enzymes involved in insecticide or antibiotic resistance, and these may be available soon for analyzing large numbers of potential transformants.

The high level of interest in insect transgenesis has resulted in two major forums for the discussion of this research: the Keystone Symposium, Toward the Genetic Manipulation of Insects, held in 1995 and 1998, and the International Workshop Series on Transgenesis of Invertebrate Organisms, held every 2 years since 1995. The idea for this book arose, in part, from discussions at these meetings that indicated that the field was at a turning point where rapid progress was being made and a wider dissemination of ideas was necessary for these results and techniques to reach the scientific mainstream. This was especially important since many of the researchers worldwide who might take greatest advantage of the methodology have remained unaware of the current state of the art. It also has become clear that despite recent breakthroughs, none of the successful systems or specific techniques will be useful for all insects. Thus, it seemed important to present the existing methodologies so that others might understand the potentials and pitfalls relevant to their species of interest (which also includes noninsect invertebrate systems). While some of the available systems might be directly utilized, it is also a hope that the information provided here will serve as a foundation and guide for independent investigation leading to new methods and strategies. Only in this way will the field continue to advance.

As insect transgenesis becomes more routine and widespread, more applications will depend on the release, if not mass release, of transgenic strains. This raises important ecological concerns that undoubtedly will be challenging, and the need for biological risk assessment to address these concerns in a rational and comprehensive manner cannot be understated. Many of the chapters in this volume address these questions relevant to particular systems, with a more in-depth consideration given in the final two chapters that discuss risk assessment from a scientific and regulatory standpoint. It is clear that these issues will differ for each release in terms of the specific transgene, the transgenic host, and the particular ecological niches into which the insects are released or which they must invade. Each investigator interested in creating transgenic strains for release must be highly aware of these issues, and take them into consideration in the planning stages of vector development and strategies for use of the transgenic strain. We also hope that the information provided here will be a starting point for what will be an ongoing discussion.

The contributors to this book include many of the invited speakers to the Keystone Symposia and Transgenesis Workshops, all of whom are held in high regard in their fields of expertise. While we have tried to include all of the major areas of importance to insect transgensis, some existing or potential vector and marker systems may have been given only limited attention. In a rapidly developing field such as this, new systems appearing close to the time of publication may have been omitted, although we have tried as much as possible to anticipate these possibilities. In terms of strategies for the use of transgenic insects, there are numerous possibilities for both basic and applied purposes, and here we provide only a sampling of strategies for insects that are plant predators and vectors of disease. Again, the purpose of this book is to serve as a foundation and guide to this emerging field, and as with most scientific disciplines, it is important if not critical to be continually kept up to date by frequent literature reviews. Comprehensive reference lists, appendices, and Web-site listings have been included in this volume to help with this process.

In many ways, and possibly due to the many roadblocks encountered early on, the field of insect transgenesis has become a large, collaborative effort. We therefore very sincerely thank the contributors to this book who have been on the vanguard of this effort, as well as numerous other colleagues who have helped lay the foundation for this technology in past years and those who have more recently provided data and ideas that have contributed to successes. We especially acknowledge the contribution of one of our mentors, Howard A. Schneiderman, to whom this book is dedicated. Schneiderman was the founder and first director of the Developmental Biology Center (DBC) at the University of California, Irvine, where we did research as a student (AAJ) and post-doc (AMH). Schneiderman was a leading insect physiologist, who early on appreciated the power of genetics to understand all phases of insect biology, and this led him to become one of the first insect biologists to turn a large portion of his research efforts toward the use of *Drosophila* as a model system. This was not only to understand genetics, but also to understand insects. It is not surprising that some of the first attempts at *Drosophila* transformation were undertaken at the DBC. Later on in his career, Schneiderman become Vice President for Research and Development at Monsanto where he led this company's pioneering endeavors into plant transgenics. Beyond ourselves, others who were trained as *Drosophila* geneticists at the DBC have gone on to become leaders in genetic analysis and transgenesis of non-drosophilid insects and other organisms and, by doing so, are hopefully carrying forward and helping complete the circle of Schneiderman's original thoughts and inspiration.

We are grateful to those who provided enormous assistance throughout the development and production of this book including Lynn Olson at the University of California, Irvine, and members of CRC Press editorial and production staff including Christine Andreasen, Pat Roberson, and John Sulzycki.

Alfred M. Handler
Gainesville, Florida

Anthony A. James
Irvine, California

Editors

Alfred M. Handler, Ph.D., is a Research Geneticist at the USDA, ARS Center for Medical, Agricultural and Veterinary Entomology in Gainesville, Florida.

Dr. Handler received his B.S. in biology from the State University of New York at Stony Brook in 1972, and his Ph.D. in biology (developmental genetics) from the University of Oregon in 1977. He held a postdoctoral fellowship in genetics at the Division of Biology, California Institute of Technology, Pasadena, from 1977 to 1979, and then joined the Developmental Biology Center at the University of California, Irvine as a research biologist from 1979 to 1985. During this time he was a visiting research associate at the Zoological Institute, University of Zürich, Switzerland and the Technische Hochschule, Darmstadt, Germany. In 1985 he joined the Agricultural Research Service in Gainesville, Florida, and in 1995 he was a visiting scientist at the Laboratory of Comparative Pathology, University of Montpellier II, France.

Dr. Handler is the U.S. scientific coordinator for a cooperative scientific program between the USDA–ARS and the French Centre National Recherche Scientifiques (CNRS) on "Transgenesis of Invertebrate Organisms of Economic and Medical Importance." As part of this role he has served as conference organizer for the "International Workshops on Transgenesis of Invertebrate Organisms." Dr. Handler has served as a consultant and expert panel member for several international organizations in the field of insect genetics and transgenesis.

Dr. Handler's research at the USDA–ARS has centered on the use of transgenic insects for biological control programs. Most of his efforts have focused on the development of efficient gene transfer vector and marker systems.

Anthony A. James, Ph.D., is Professor of Molecular Biology and Biochemistry at the University of California, Irvine (UCI).

Dr. James received his B.S. in biology from UCI in 1973, and his Ph.D. in developmental biology from UCI in 1979. He held postdoctoral positions in the Department of Biological Chemistry, Harvard Medical School, and the Department of Biology, Brandeis University, before joining the faculty of the Department of Tropical Public Health at the Harvard School of Public Health in 1985. Dr. James returned to his alma mater in 1989, where he has remained.

Dr. James was a principal investigator with the Network on the Biology of Parasite Vectors funded by the John D. and Catherine T. MacArthur Foundation and was a recipient of the Molecular Parasitology Award from the Burroughs–Wellcome Fund. He is a fellow of the American Association for the Advancement of Science and the Royal Entomological Society of London. He is a founding editor of *Insect Molecular Biology*, and is on the editorial board of *Experimental Parasitology*.

Dr. James' areas of interest include vector–parasite interactions, mosquito molecular biology, and other problems in insect developmental biology. His current work focuses on using genetic and molecular genetic tools to interrupt parasite transmission by mosquitoes.

Contributors

Boris Afanasiev
Arthropod-Borne Infectious Disease Lab
Department of Microbiology
Colorado State University
Fort Collins, CO 80523

Peter W. Atkinson
Department of Entomology
University of California, Riverside
Riverside, CA 92521

Charles B. Beard
Division of Parasitic Diseases
Centers for Disease Control and Prevention
4770 Buford Highway
Chamblee, GA 30341

Mark Q. Benedict
Division of Parasitic Diseases
Centers for Disease Control and Prevention
4770 Buford Highway
Chamblee, GA 30341

Alain Bucheton
Institut de Génétique Humaine
141 rue de la Cardonille
34396 Montpellier
France

Jane C. Burns
Department of Pediatrics
School of Medicine
University of California, San Diego
9500 Gilman Drive
La Jolla, CA 92093

Jonathan Carlson
Arthropod-Borne Infectious Disease Lab
Department of Microbiology
Colorado State University
Fort Collins, CO 80523

Frank H. Collins
Department of Biological Sciences
University of Notre Dame
Notre Dame, IN 46556

Ravi V. Durvasula
Department of Internal Medicine
Yale University School of Medicine
60 College Street
New Haven, CT 06520

Paul Eggleston
School of Life Sciences
Keele University
Huxley Building
Staffordshire ST5 5BG
United Kingdom

Richard H. ffrench-Constant
Department of Biology and Biochemistry
University of Bath
Bath BA2 7AY
United Kingdom

Arnold S. Foudin
Animal and Plant Health Inspection Service
U.S. Department of Agriculture
4700 River Road
Riverdale, MD 20737

Gerald Franz
IAEA Laboratories
Agriculture and Biotechnology Laboratory
Entomology Unit
A-2444 Seibersdorf
Austria

Malcolm J. Fraser, Jr.
Department of Biological Sciences
University of Notre Dame
Notre Dame, IN 46556

Kent G. Golic
Department of Biology
University of Utah
Salt Lake City, UT 84112

Alfred M. Handler
Center for Medical, Agricultural,
 and Veterinary Entomology
USDA/ARS
1700 SW 23rd Drive
Gainesville, FL 32608

Stephen Higgs
Department of Pathology
University of Texas Medical Branch
301 University Boulevard
Galveston, TX 77550

Marjorie A. Hoy
Department of Entomology and Nematology
P. O. Box 110620
University of Florida
Gainesville, FL 32611

Shirley P. Ingebritsen
Animal and Plant Health Inspection Service
U. S. Department of Agriculture
4700 River Road
Riverdale, MD 20737

Anthony A. James
Department of Molecular Biology-
 Biochemistry
3205 Bio Sci II
University of California, Irvine
Irvine, CA 92697

David J. Lampe
Department of Biological Sciences
Duquesne University
913 Bluff Street
Pittsburgh, PA 15219

David L. Lewis
Laboratory of Molecular Biology
University of Wisconsin
1525 Linden Drive
Madison, WI 53706

David A. O'Brochta
Center for Agricultural Biotechnology
University of Maryland Biotechnology Institute
College Park, MD 20742

Kenneth E. Olson
Arthropod-Borne Infectious Disease Lab
Department of Microbiology
Foothills Campus
Colorado State University
Fort Collins, CO 80523

Scott L. O'Neill
Department of Epidemiology and Public Health
Yale University School of Medicine
60 College Street
New Haven, CT 06520

Alain Pélisson
Institut de Génétique Humaine
141 rue de la Cardonille
34396 Montpellier
France

Frank F. Richards
Department of Internal Medicine
Yale University School of Medicine
60 College Street
New Haven, CT 06520

Hugh M. Robertson
Department of Entomology
University of Illinois
505 S. Goodwin Avenue
Urbana, IL 61801

Alan S. Robinson
IAEA Laboratories
Agriculture and Biotechnology Laboratory
Entomology Unit
A-2444 Seibersdorf
Austria

Yikang S. Rong
Department of Biology
University of Utah
Salt Lake City, UT 84112

Abhimanyu Sarkar
Department of Biological Sciences
University of Notre Dame
Notre Dame, IN 46556

John M. Sherwood
Department of Entomology
University of Illinois
505 S. Goodwin Avenue
Urbana, IL 61801

Steven P. Sinkins
Liverpool School of Tropical Medicine
Pembroke Place
Liverpool L3 5QA
United Kingdom

Erica Suchman
Arthropod-Borne Infectious Disease Lab
Department of Microbiology
Colorado State University
Fort Collins, CO 80523

Christophe Terzian
Institut de Génétique Humaine
141 rue de la Cardonille
34396 Montpellier
France

Kimberly K. O. Walden
Department of Entomology
University of Illinois
505 S. Goodwin Avenue
Urbana, IL 61801

Bruce A. Webb
Department of Entomology
University of Kentucky
Lexington, KY 40546

Orrey P. Young
Animal and Plant Health Inspection Service
U. S. Department of Agriculture
4700 River Road
Riverdale, MD 20737

Yuguang Zhao
NERC Institute of Virology
  and Environmental Microbiology
Mansfield Road
Oxford OX1 3SR
United Kingdom

*Dedicated to the memory of
Howard A. Schneiderman*

Contents

SECTION I — INTRODUCTION

Chapter 1
An Introduction to the History and Methodology of Insect Gene Transfer 3
Alfred M. Handler

SECTION II — GENE TARGETING

Chapter 2
Targeted Transformation of the Insect Genome 29
Paul Eggleston and Yuguang Zhao

Chapter 3
Site-Specific Recombination for the Genetic Manipulation of Transgenic Insects 53
Yikang S. Rong and Kent G. Golic

SECTION III — TRANSGENIC SELECTION

Chapter 4
Eye Color Genes for Selection of Transgenic Insects 79
Abhimanyu Sarkar and Frank H. Collins

Chapter 5
Green Fluorescent Protein (GFP) as a Marker for Transgenic Insects 93
Stephen Higgs and David L. Lewis

Chapter 6
Resistance Genes as Candidates for Insect Transgenesis 109
Richard H. ffrench-Constant and Mark Q. Benedict

SECTION IV — VIRAL VECTORS

Chapter 7
Pantropic Retroviral Vectors for Insect Gene Transfer 125
Jane C. Burns

Chapter 8
Densonucleosis Viruses as Transducing Vectors for Insects 139
Jonathan Carlson, Boris Afanasiev, and Erica Suchman

Chapter 9
Sindbis Virus Expression Systems in Mosquitoes: Background, Methods, and Applications 161
Ken E. Olson

Chapter 10
Retrotransposons and Retroviruses in Insect Genomes 191
Christophe Terzian, Alain Pélisson, and Alain Bucheton

Chapter 11
Polydnaviruses and Insect Transgenic Research 203
Bruce A. Webb

SECTION V — TRANSPOSABLE ELEMENT VECTORS

Chapter 12
Hermes and Other *hAT* Elements as Gene Vectors in Insects 219
Peter W. Atkinson and David A. O'Brochta

Chapter 13
Genetic Engineering of Insects with *mariner* Transposons 237
David J. Lampe, Kimberly K. O. Walden, John M. Sherwood, and Hugh M. Robertson

Chapter 14
The TTAA-Specific Family of Transposable Elements: Identification, Functional
Characterization, and Utility for Transformation of Insects 249
Malcolm J. Fraser, Jr.

SECTION VI — SYMBIONT VECTORS

Chapter 15
Wolbachia as a Vehicle to Modify Insect Populations 271
Steven P. Sinkins and Scott L. O'Neill

Chapter 16
Bacterial Symbiont Transformation in Chagas Disease Vectors 289
Charles B. Beard, Ravi V. Durvasula, and Frank F. Richards

SECTION VII — STRATEGIES, RISK ASSESSMENT, AND REGULATION

Chapter 17
The Application of Transgenic Insect Technology in the Sterile Insect Technique 307
Alan S. Robinson and Gerald Franz

Chapter 18
Control of Disease Transmission through Genetic Modification of Mosquitoes 319
Anthony A. James

Chapter 19
Deploying Transgenic Arthropods in Pest Management Programs: Risks and Realities 335
Marjorie A. Hoy

Chapter 20
Regulation of Transgenic Arthropods and Other Invertebrates in the United States 369
Orrey P. Young, Shirley P. Ingebritsen, and Arnold S. Foudin

Index 381

Section I

Introduction

1 An Introduction to the History and Methodology of Insect Gene Transfer

Alfred M. Handler

CONTENTS

1.1 Historical Perspective on Insect Gene Transfer 3
    1.1.1 *P*-Element Transformation 4
        1.1.1.1 *P* Vectors and Markers 6
        1.1.1.2 Use of *P* as a Genetic Tool 7
    1.1.2 Use of *P* for Non-Drosophilid Gene Transfer 8
1.2 New Transformation Vectors 9
    1.2.1 New Transposon Vectors 9
    1.2.2 Viral and Symbiont Vectors 11
1.3 Transformation methodology 12
    1.3.1 DNA Preparation 12
    1.3.2 Dechorionation 13
    1.3.3 Preparation for Embryo Injection 13
        1.3.3.1 Needles 14
    1.3.4 Microinjection Apparatus and Procedures 14
    1.3.5 Methods for DNA Delivery 16
        1.3.5.1 Lipofection 16
        1.3.5.2 Biolistics 17
        1.3.5.3 Electroporation 17
1.4 Identification of Transgenic Insects and Transgenes 18
1.5 Perspectives on the Use of Insect Gene Transfer 19
1.6 Appendix 20
    Equipment 20
    Addresses 21
Acknowledgments 22
References 22

1.1 HISTORICAL PERSPECTIVE ON INSECT GENE TRANSFER

The use of genetic material as recombinant DNA and the ability to integrate it into a host genome has proved to be a powerful method for genetic analysis and manipulation, providing a major new era in the field of genetics. Procaryotic gene transformation was actually realized early on, and in fact the pivotal bacterial transformation studies by Avery et al. (1944) gave definitive proof to DNA being the inherited genetic material. Continued procaryotic genetic transformation studies, indeed, helped lay the foundation for modern molecular genetics. It is thus not surprising that geneticists attempted to duplicate this methodology in eucaryotes as well, long before eucaryotic DNA could be isolated as recombinant molecules and analyzed in a meaningful way.

The genetic transformation of insects was first attempted in *Ephestia* nearly 35 years ago, when mutant larvae were injected with wild-type DNA, with some developing into adults with wild-type wing scales (Caspari and Nawa, 1965). In subsequent studies with *Ephestia* (Nawa and Yamada, 1968) and *Bombyx mori* (Nawa et al., 1971), complementation of eye color mutations was observed after treatment with wild-type DNA. While these experiments yielded wild-type adults and at least limited non-Mendelian inheritance of the normal phenotype, it is likely that these initial insect transformations were somatic with inheritance occurring extrachromosomally. Shortly after the initial studies in moths, transformation of *Drosophila melanogaster* was similarly attempted, although delivery of wild-type DNA was achieved by soaking embryos in genomic DNA within ringers or sucrose solutions. As with the moth studies, somatic mosaics resulted, but inheritance of the reverted phenotyes was not clearly Mendelian and it was concluded that genetic transformation had occurred extrachromosomally, with episomal transmission and not chromosomal integration (Fox and Yoon, 1966; 1970; Fox et al., 1970).

More recent approaches to insect transformation began with studies in *Drosophila* that relied on the direct injection of wild-type DNA into embryos. These attempts to revert the *vermilion* (*v*) mutant line met with some success (Germeraad, 1976), although integrations were not verified beyond the genetic mapping of the complementing gene outside of the *v* locus (suggesting that a direct *v* reversion had not occurred), but the transformed lines were subsequently lost without further genetic or biochemical verification.

1.1.1 *P*-Element Transformation

Concurrent with the *vermilion* studies, the role of P factors in *Drosophila* hybrid dysgenesis was being eludicated (Kidwell et al., 1977), culminating in the identification and isolation of the *P* transposable element as the responsible agent. In now classical experiments by Rubin et al. (1982) and Rubin and Spradling (1982), *P* was first isolated from a *P*-induced mutation of *white* in *D. melanogaster*, and then developed into the first transposon-based system to transform the germline of *D. melanogaster* efficiently and stably (see Engels, 1989, for a comprehensive review of the discovery and early analysis of *P*).

P was found to be 2.9 kb in length with 31 bp inverted terminal repeats (O'Hare and Rubin, 1983), similar in general structure to *Activator*, the first transposable element to be discovered in maize by Barbara McClintock (see Federoff, 1989). Both of these elements, as well as all the subsequently discovered transposons used for insect germline transformation, belong to a general group of transposable elements known as Class II short inverted terminal repeat transposons (see Finnegan, 1989). These elements transpose via a DNA-intermediate and generally utilize a cut-and-paste mechanism that creates a duplication of the insertion site. Within the terminal repeats of these elements is a transcriptional unit that encodes a transposase molecule that acts at or near the termini to catalyze excision and transposition of the complete element. As first described by Rubin and Spradling (1982), the ability of the transposase to act in *trans* has allowed the development of binary vector-helper systems (Figure 1.1). Typically the vector plasmid includes the mobile terminal repeats of the element and requisite proximal internal sequences that surround a marker gene. The vector is made nonautonomous by having the transposase gene either deleted or disrupted by insertion of the marker gene, and thus it is unable to move by itself. The transposase is provided on a separate helper plasmid, and, after introduction into germ cell nuclei, the helper mediates transposition of the vector into the genome. The original helper was an autonomous *P* element (pπ25.1) that had the ability to integrate as well, and its presence could cause instability of the vector in subsequent generations (if not earlier). This problem was ameliorated somewhat by having much higher vector-to-helper ratios, but was solved more

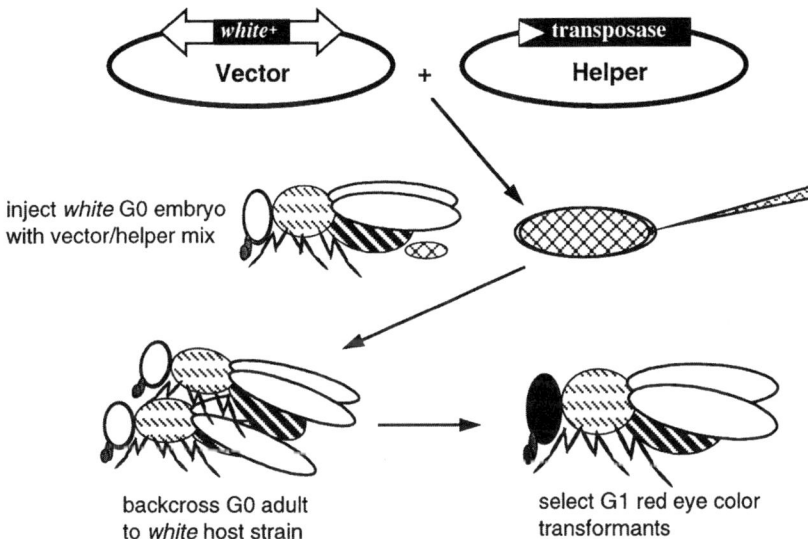

FIGURE 1.1 Schematic diagram of typical *Drosophila* transformation using a *white+* marker selection in a white eye host strain. Vector and helper plasmids are mixed at appropriate concentrations in injection buffer and microinjected into preblastoderm G0 embryos. G0 adults are backcrossed to white eye host strain flies with resulting G1 progeny screened for eye pigmentation.

definitively by the creation of a helper (pπ25.7wc) rendered immobile by deletion of most of the 3′ terminal sequence (Karess and Rubin, 1984). Without the ability to integrate, the helper plasmid is lost after subsequent cell divisions, allowing the vector to remain stably integrated.

Although the efficiency of this system relied on the high mobility properties of the *P* element, several factors were involved that allowed successful *Drosophila* transformation with this system at that time. For the *P* system in particular, the fortuitous existence of M strains devoid of *P* elements was important. These included wild and mutant strains maintained in laboratories previous to 1950, a time before *P* invaded *D. melanogaster* in the wild (Anxolabéhère et al., 1988). Most wild P strains collected after this time contain mostly defective *P* elements that provide a repressive cellular environment for *P* movement, and initial transformations in these strains would have proved frustrating, if not impossible. Curiously, some other vector systems are not, or are less repressed by host strains containing the transposon, as was found with *piggyBac* in *Bactrocera dorsalis* (A. Handler and R. Harrell, unpublished), although this may be a function of the number and structure of elements present. Given the variability of transposon regulation in any given host environment, the safest strategy for the use of a vector in an insect species for the first time is to assess its function by transient assays in the host strain before attempting transformation experiments (see Chapter 12 by Atkinson and O'Brochta).

Another important factor in the success of the *P* experiments was the ability to use a cloned wild-type gene in high molar concentrations as a transgenic marker. This is in contrast to the use of wild-type genes diluted by the rest of the genome in the original experiments. As described further on in this chapter and book, several critical factors are involved in every gene transfer system in any particular organism. These include a functional vector system, unambiguous and easily detectable markers, and simple methods for DNA delivery. All of these factors must be functional at a minimal level for successful gene transfer of any particular insect species, yet increased efficiency for any one may offset inefficiencies for the others. For example, an inefficient vector may remain useful if DNA can be delivered easily to many thousands of eggs. Obviously, optimization of all these parameters will be required for the most efficient and widespread use of the technology.

1.1.1.1 *P* Vectors and Markers

While specific differences exist between *P* and other vector systems currently used for non-drosophilid insects, optimization of these and prospective transposon and viral vectors will clearly benefit from the knowledge gained after two decades of experimentation with the *P* transposon and its use in transformation studies. Of particular importance to all researchers contemplating insect transformation is a practical knowledge of *Drosophila* transformation, and arguably it is quite worthwhile to attempt transformation of this species before any other. Presented here is a general overview of *P* transformation of *Drosophila*, and how it relates to non-drosophilid gene transfer. Several excellent in-depth reviews are available for more specific details on the structure, behavior, and use of *P*, which may be related to other vector systems, as well as transformation methodology (see Karess, 1985; Spradling, 1986; Engels, 1989; Handler and O'Brochta, 1991). Of particular interest is the book and methods manual by Ashburner (1989a,b) that review the various vectors, markers, and methodologies used for *Drosophila* transformation, as well as early techniques used to manipulate *Drosophila* embryos. We have found this information highly relevant to other insect systems.

After its description in 1982, *P* transformation was rapidly utilized by hundreds of *Drosophila* laboratories, and modifications and refinement of the *P* system occurred quickly in terms of vectors and markers (Pirrotta, 1988). This was also aided by basic investigations of transposable element structure, function, and evolution (see Berg and Howe, 1989). As noted above, the original helpers under *P* promoter regulation were autonomous elements that worked effectively, but could integrate along with the vector resulting in vector instability. A nonautonomous "wings-clipped" helper was created having its 3′ terminus deleted (Karess and Rubin, 1984), but many laboratories found it to be somewhat less effective. Subsequently, a new wings-clipped helper was created having transposase under highly active *hsp70* regulation that resulted in routine transformation (Steller and Pirrotta, 1986). It was soon realized that these frequencies, if not the ability to transform at all, was influenced by vector structure. Critical factors were vector size, placement of markers, and amounts of subterminal DNA left in the vector. Sequences within 138 bp of its 5′ end and 216 bp of its 3′ end were found necessary for *P* movement, and while the terminal repeats were identical, the adjacent sequences were not interchangeable (Mullins et al., 1989). An interesting finding was that *P* transposase has strongest binding affinity for sequences internal (~50 bp) to the inverted terminal repeats (Kaufman et al., 1989). For other vectors such as *mariner* (*Mos*1), functionality can be drastically influenced by discrete base-pair changes (Hartl et al., 1997). Vectors having requisite sequences are otherwise limited by the amount of marker DNA inserted, with the frequency of transformation inversely influenced by the size of the vector. Early tests with *P* marked with *rosy* showed that 8 kb vectors could yield transformation frequencies of 50% per fertile G_0 or greater, while the frequency with 15 kb vectors decreased to about 20% (see Spradling, 1986). Vectors as large as 20 kb could integrate, but at frequencies of 1% or less. Of course, actual frequencies depend greatly on technical ability, and various parameters including DNA purity and concentration, method of microinjection, ambient conditions, among others. We can expect other systems to be similarly affected by vector size and requisite sequences, in addition to how particular vectors and markers function in specific insect host strains.

The success of the *P* system in *Drosophila* was due in large part to the availability of easily detectable visible marker systems that rely on the complementation of a mutant allele in the host strain by a cloned wild-type gene in the vector (see Chapter 4 by Sarkar and Collins and Color Figure 1*). The original marker for "mutant-rescue" was the *rosy* gene, although this had the drawback of being cloned within a relatively large DNA fragment of ~7 to 8 kb. The *white* gene was subsequently used, but this was part of an even larger genomic fragment, and transformation frequencies were relatively low. New mini-*white* constructs (approximately 4 kb) having the large first intron deleted, with and without *hsp70* promoters, were much more effective. While these

---

* Color Figure 1 follows p. 108.

white markers are routinely used, they do suffer from genomic position effect variegation that acts to suppress gene expression (Hazelrigg et al., 1984; Pirrotta et al., 1985; Klemenz et al., 1987), and thus some transformants may be difficult if not impossible to detect by the visible phenotype. The same effect has been observed in tephritid fruit fly (Loukeris et al., 1995a; Handler et al., 1998) and mosquito (Coates ct al., 1998; Jasinskiene et al., 1998) transformants. In one experiment comparing *white* to a green fluorescent protein (GFP) marker in *Drosophila*, we found that less than 40% of transformant G_1 flies could be detected by *white*⁺ (pigmented eyes) expression alone (Handler and Harrell, 1999).

Additional markers based upon chemical selections or enzymatic activity were also developed for *Drosophila*, and these included alcohol dehydrogenase (*Adh*) (Goldberg et al., 1983) and dopa decarboxylase (*Ddc*) (Scholnick et al., 1983) that complemented existing mutations, and neomycin phosphotransferase (NPT or *neo*) (Steller and Pirrotta, 1985), β-galactosidase (Lis et al., 1983), organophosphorus dehydrogenase (*opd*) (Benedict et al., 1995), and dieldrin-resistance (*Rdl*) (ffrench-Constant et al., 1991) which are dominant selections not requiring preexisting mutations. These are reviewed in more detail in Ashburner (1989a) and in Chapter 6 by ffrench-Constant and Benedict. While these markers have the advantage of utilizing various mutant or wild-type host strains, and in some cases selection can be done early in G_1 development, none of them compares in the ease of use and reliability to the eye color markers. GFP markers, however, not only share the benefit of not depending on preexisting mutations, but they can actually be more effective than *white* (Handler and Harrell, 1999; see Chapter 5 by Higgs and Lewis). The primary advantage of chemical selections is that they allow selection *en masse*, which should be useful for all transgenic selections, and possibly critical for insects that cannot be transformed at a high efficiency. Now it is possible to refine chemical selections by linking the requisite genes to a GFP marker so that *bona fide* transformants can be selected and tested to increase the reliability and efficiency of the selection. Another significant benefit of GFP is that when regulated by a promoter active throughout development (e.g., polyubiquitin), transgenics can be selected early in their life cycle, saving time and resources. A caveat, however, is that expression in G_1 insects may not be detectable until late or postembryonic stages, though transgenic embryos can be detected in subsequent generations owing to a maternal contribution of GFP (see Davis et al., 1995). One drawback of using GFP markers is that a somewhat costly dedicated ultraviolet optical system is required for their detection, which may prohibit preliminary studies for many laboratories.

1.1.1.2 Use of *P* as a Genetic Tool

The *P*-element transposon has found its greatest use as a vector to import genes into the *Drosophila* genome, but has also been used (1) as a mutagenic agent to transposon-tag genes simplifying their cloning (Searles et al., 1982; Sentry and Kaiser, 1992); (2) in enhancer-trap and GAL4/UAS studies to identify and analyze temporally and spatially distinct regulatory elements (Bellen et al., 1989; Wilson et al., 1989; Smith et al., 1993; Brand et al., 1994; Gustafson and Boulianne, 1996); and (3) for targeted transposition (or site-directed gene conversion) which allows gene replacement for the creation of specific mutant phenotypes (Gloor et al., 1991; see Chapter 2 by Eggleston and Zhao). Targeting also controls for chromosomal position effects that vary when transposition is random. These methods have revolutionized genetic analysis in *Drosophila* and have the potential to expand enormously the genetic analysis and manipulation of non-drosophilid insects as well, for both basic and applied purposes. Indeed, a primary benefit of developing transposon-based vectors has been this versatility for genetic manipulation beyond simple gene transfer.

As in other systems, both transposon tagging and enhancer trapping in insects should allow the efficient identification and isolation of genes and genetic systems involved in their development, behavior, and reproduction. While a primary goal will be to use these genes or genetic elements in transgenic strains for biological control, the analysis of these genes and their use in transgenic strains for laboratory experimentation should also provide new targets for insect control using highly specific conventional methods in addition to molecular methods. For example, a genetic dissection of neuronal pathways may reveal targets that do not exist in vertebrates, or are specific for particular insects, allowing the development of pesticides innocuous to nontargeted organisms. Genes expressed in reproductive tissues may be manipulated to produce sterility in the laboratory (e.g., for sterile insect technique (SIT); see Chapter 17 by Robinson and Franz) or their gene products may be targeted to induce sterility in the field. Manipulation of endocrine gland genes may disrupt development and reproduction, and genes expressed in antennal lobes may disrupt chemosensory reception necessary for foraging or mating. These are only a few of the potential conventional applications of information generated by basic studies of transgenic insects.

While this information will certainly improve conventional insect control strategies, the greatest potential for insect control using transgenics lies in their direct use. Many transgenic strains will be simply maintained as inbred lines; yet, some of these strains may include gene constructs that result in sterility or lethality making them difficult, if not impossible, to rear. A clever method to create strains that would normally be inviable is to utilize the yeast transcriptional activator GAL4/UAS system, which is used widely in *Drosophila* and other organisms for developmental analyses and control of gene expression (Brand et al., 1994). Essentially, GAL4 is a transcriptional activator that works by binding to upstream activating sequences (UAS) to promote transcription of downstream reading frames. It is thus possible to create a transgenic strain having a selectable or "lethal" gene linked to UAS with no negative effect in the absence of the GAL4 protein. Another strain may be created having GAL4 production regulated by a conditional, tissue-specific, or sex-specific promoter. Parental strains homozygous for the transgenes would remain unaffected, but, upon mating, their progeny will produce the UAS-linked gene product in response to the developmentally or conditionally regulated GAL4 product. For example, female-lethality may be achieved by having a strain with UAS linked to a toxin or cell-death gene mated to a strain having a female-specific promoter linked to GAL4. The UAS/GAL4 system is highly versatile as well, since libraries of GAL4 and UAS strains can be maintained with specific types of gene expression possible by mating the appropriate strains. Thus, the same UAS-cell-death gene strain used for female-lethality could be mated to a testis-promoter-GAL4 strain resulting in male sterile progeny. The use of enhancer traps and GAL4/UAS studies will help identify and test a wide array of transcriptional enhancers that can be immediately used for the sophisticated manipulation of insect populations. The power and potential of these manipulations may be vast.

1.1.2 Use of *P* for Non-Drosophilid Gene Transfer

The success with *P* transformation in *Drosophila* did not go unnoticed by those working on non-drosophilid species, and especially those interested in their genetic manipulation. The potential use of *P* in these insects was encouraged by the use of the vector to transform two drosophilid species that did not contain the *P* transposon, *D. simulans* (Scavarda and Hartl, 1984) and *D. hawaiiensis* (Brennan et al., 1984). The latter experiment was of particular interest due to the distant relationship between the Hawaiian drosophilids and *D. melanogaster*, increasing the possibility that it might function as well in non-drosophilids. Another critical factor that allowed the testing of *P* was the development of a chemical resistance marker based on the bacterial NPT gene, that allowed selection of *Drosophila* transformants by resistance to a neomycin analogue, G418 (or Geneticin®; Steller and Pirrotta, 1985). In the absence of visible marking systems, it would be otherwise impossible to select transgenics in non-drosophilid insects. Unfortunately, several years of fruitless effort ensued by several laboratories working with *P* vectors marked with NPT in mosquitoes (Miller et al., 1987; McGrane et al., 1988; Morris et al., 1989), tephritid fruit flies (McInnis et al., 1990; A. Handler and S. Gomez, unpublished), locusts (Walker, 1990), and possibly several other species, although many of these results are anecdotal due to the failure to select transposon-mediated transformants. Interesting results were obtained by several of the mosquito laboratories, however, which selected G418-resistant flies that were transformed, but apparently all occurred by low-frequency random integrations or recombination events and not *P* transposition. These results take on added significance in light of recent results indicating transposase-dependent recombinations for other vector systems in mosquitoes (Jasinskiene et al., 1999). These observations highlight the need to consider vector function in the context of the individual cellular and genomic environment of each species.

Despite the interesting results in mosquitoes, use of the *P* vector did not result in transposon-mediated events in any of the non-drosophilid insects tested, and it appeared that neomycin-resistance selections resulted in a significant number of false positives. These and other variables involved in transformation methodology did not allow a straightforward assessment of the possible limiting factor(s). Considering vector function to be most critical, O'Brochta developed a series of *in vivo* transient excision assays that could simply and rapidly assess vector function in the insect embryo (O'Brochta and Handler, 1988; O'Brochta et al., 1991). These assays initially took advantage of *in vitro* assays for *P* function (Rio et al., 1986), and the concurrent observation of transient expression of plasmid-encoded genes injected into *Drosophila* embryos (Martin et al., 1986). These, and much more sophisticated transposition assays developed more recently (O'Brochta et al., 1994; Sarkar et al., 1997) are now used as a standard procedure to determine vector function in most insects of interest, and are discussed more fully in Chapter 12 by Atkinson and O'Brochta. The *P* excision assays quickly indicated to us that mobility of the element decreases as a function of relatedness to *D. melanogaster*, with a lack of mobility outside of the Drosophilidae (Handler et al., 1993). This provided a turning point in efforts to transform non-drosophilids, as it was realized that new vector systems were critical.

1.2 NEW TRANSFORMATION VECTORS

With the utility of *P* unlikely for gene transfer in non-drosophilids, two choices remained for vector-mediated gene transfer. One was to continue testing existing vectors or transposable elements from *Drosophila* or other organisms, or attempting to isolate new transposons or viral systems for vector development. As discussed in detail in this book, both approaches were taken with various levels of success. While *P* and other transposon-based vectors have proved their usefulness in specific species, their analysis in a variety of insects by transient assays and transformation experiments indicate that their function differs in different cellular environments, and perhaps in response to differing genomic organization. Thus, vectors successful for some species may act differently or not at all in others, and new types of vectors may be more effective for particular organisms or for particular applications. This suggests that long-term strategies for gene transfer of non-drosophilid insects will require broad-based approaches, including viral- and symbiont-based systems, as well as recombination systems and methods to improve gene targeting.

1.2.1 NEW TRANSPOSON VECTORS

Early consideration was given to the only other transformation vector tested in *D. melanogaster*, the *hobo* element. *hobo* was discovered by its association with mutant alleles in *D. melanogaster* about the time *P* transformation was first reported (McGinnis et al., 1983; see Blackman and Gelbart, 1989), and as another short inverted terminal repeat transposon; within several years it also was developed into an efficient gene transfer system (Blackman et al., 1989). Nevertheless, there was no reason, *a priori*, to believe that the *hobo* range of function would be any greater than that of *P*. This possibility was reconsidered when amino acid sequence alignments showed that *hobo* shared related motifs with plant transposons *Ac* from maize and *Tam*3 from the snapdragon (Calvi et al., 1991). Unlike *P*, whose related elements were all apparently limited to the drosophilids and closely related species (Lansman et al., 1985; Daniels and Strausbaugh, 1986; Anxolabéhère et al., 1988), *hobo* was part of a wide-ranging transposable element family. This suggested that it might have a broad range of function like *Ac* (Baker et al., 1986), and related elements might be discovered in other species that could be used as vectors in these and related insects. Although transient assays suggested that *hobo* may have low-frequency vector function in non-drosophilids (O'Brochta et al., 1991), transformation has yet to be demonstrated for such insects. However, *hobo* has transformed *D. virilis*, which is distantly related to *D. melanogaster*, at about a 1% frequency (Lozovskaya et al., 1995; Gomez and Handler, 1997).

Importantly, homologies between *hobo* and *Ac* have allowed the identification of new elements such as *Hermes* from *Musca domestica* (Warren et al., 1994), which has been shown to be a highly active vector in *D. melanogaster* (O'Brochta et al., 1995), and it is one of the few useful vectors for non-drosophilids (Jasinskiene et al., 1998; Pinkerton et al., 2000). Notably, *Hermes* and other new insect transposon systems within the *hobo, Ac, Tam3* (*hAT*) family (see O'Brochta and Atkinson, 1996; Handler and Gomez, 1996) have been specifically isolated for their potential use as vectors, and are discussed further in Chapter 12 by Atkinson and O'Brochta. All of the other non-drosophilid transposon vector systems were found fortuitously and were not specifically isolated for potential vector function.

Minos was the first transposon vector to successfully transform a non-drosophilid, the medfly *Ceratitis capitata* (Loukeris et al., 1995b), and it was originally discovered in *D. hydei* as part of a ribosomal RNA transcriptional unit (Franz and Savakis, 1991). Several *Minos* elements were subsequently isolated and the functional element was found to have 254-bp inverted terminal repeats and a transposase encoded by two exons separated by a 60-bp intron sequence (Franz et al., 1994). *Minos* appears to be a member of the *Tc* family, sharing more than 40% coding sequence identity with *Tc1*, and, like other *Tc* elements, it causes a TA duplication of its insertion site. The ability of *Minos* to function as a gene vector was first demonstrated in *D. melanogaster* using a *white* marked vector and a *hsp70* regulated helper (Loukeris et al., 1995a). Transformation frequencies were relatively low, in the range of about 5% per fertile G_0, although transformants were consistently produced in several experiments and *Minos* integrations were verified by Southern hybridization and sequencing of several insertion sites. Notably, chromosomal insertions were verified by PCR sequencing across the insertion sites in the parental host strain. In one of the transgenic lines, the *Minos* integration was remobilized by crossing to a transposase carrier line, that reaffirmed a *Minos*-mediated integration and provided support for the possible use of *Minos* for enhancer trapping.

The availability of the cloned *white* gene cDNA from the medfly (Zwiebel et al., 1995) made it possible to test *Minos* vector function in a medfly *white eye* host strain using *Drosophila* protocols (Loukeris et al., 1995b). Similar to the *Drosophila* experiments, several transformant lines were generated at an overall frequency of less than 5%, although this was more difficult to assess due to group matings of the G_1 flies. Nevertheless, this was the first *bona fide* transposon-mediated transformation of a non-drosophilid, and at a frequency useful for routine experiments. Although there have not been subsequent published reports of *Minos* transformation, it has been used repeatedly in the medfly (C. Savakis, personal communication), in *D. virilis* (L. Megna and T. Cline, personal communication), and *Minos* is likely to have a broad range of vector function in insects.

The *piggyBac* element was used as the second transformation agent in medflies (Handler et al., 1998) and, at the time of this writing, is the most successful vector for non-drosophilids that include dipterans, lepidopterans, and a coleopteran (although molecular verification for some is in progress). *piggybac* was originally discovered in *Trichoplusia ni* cell lines as a result of its insertion into infecting baculoviruses resulting in few polyhedra (FP) mutations (Fraser et al., 1983). Since the medfly *white* gene already had been isolated and tested as a transgenic marker in the *Minos* transformations, it was possible to test *piggyBac* transformation directly in a non-drosophilid species (Handler et al., 1998). Medfly transformation with a self-regulated transposase helper (p3E1.2 having its 5′ terminus deleted) occurred at relatively low frequencies (~2 to 5% per fertile G_0), but did indicate autonomous function for the transposon in a dipteran. Vector function was subsequently proved for *D. melanogaster* (Handler and Harrell, 1999), with frequencies elevated to 26% using a *hsp70* regulated helper, and subsequently two other tephritid fruit fly species have been transformed, *Anastrepha suspensa* (A. Handler and R. Harrell, unpublished) and *Bactrocera dorsalis* (A. Handler and S. McCombs, unpublished). A broad range of vector function for *piggyBac* is further supported by recent transformation experiments in the silkmoth, *Bombyx mori* (Tamura et al., 2000), as well as the red flour beetle, *Tribolium castaneum* (Berghammer et al., 1999), and this work is discussed in more detail in Chapter 14 by Fraser.

The *mariner* element is another Class II transposon, originally discovered in *D. mauritiana* in association with a mutant *white* allele (Haymer and Marsh, 1986; Medhora et al., 1988). While it probably has been the most intensively studied element beyond *P*, it has taken more than a decade since its discovery for it to be used as a vector in a non-drosophilid. The *mariner* story is discussed in detail in Chapter 13 by Lampe et al., but, suffice it to say, it is part of one of the largest known transposon families traversing many orders of animals by horizontal transmission (Robertson and Lampe, 1995), and it functions accurately *in vitro* in the absence of any cofactors. Nevertheless, it has thus far demonstrated only relatively weak vector function in several drosophilids, with related active elements thus far failing to transform *D. melanogaster* after extensive testing. Only recently has *mariner* been shown to have vector function in a mosquito species (Coates et al., 1998), yet is has exhibited vector function in vertebrate species including chickens (Sherman et al., 1998) and zebrafish (Fadool et al., 1998). At present, *mariner* appears to have great potential as a broadly active vector, although it remains to be seen how useful and widely functional it will be in insects.

It is notable that the discovery of vector function for all of these transposons depended upon first developing reliable marking systems. Both of the medfly transformations depended on isolation of the medfly *white* gene cDNA (Zwiebel et al., 1995), and the testing of *Hermes* and *mariner* in *Aedes aegypti* relied on the finding that the *D. melanogaster cinnabar* gene could complement a mosquito white-eye mutation (Cornel et al., 1997). Thus, while much effort and interest have been focused on vector systems (see O'Brochta and Atkinson, 1996), the lack of suitable transgenic marker systems has been an equal, if not greater, bottleneck for successful insect transformation.

1.2.2 VIRAL AND SYMBIONT VECTORS

While transposon-based vectors are preferable for the creation of stable transgenic strains at this time, and especially for basic studies of gene expression and genetic manipulation, it is realized that different types of systems will be necessary for particular field or experimental applications, and for particular insect species not amenable to transposon function or the typical methodology used for germline transformation. For example, many insect species have long generation times or complex life cycles that make untenable the testing of germline transformation, or the creation of mass-reared transgenic strains. For some purposes, the need to create stable transgenic strains is not essential, and the development of extrachromosomal transient expression systems that might be much simpler to create, could be a higher priority if not preferable. Such transient systems might have the added benefit of acting as carriers for inefficient transposon vectors or gene-targeting systems allowing them to perdure within cells, increasing their chance for integration. Thus, while the development of germline transformation systems for non-drosophilid insects has been a priority, the need for other types of systems for particular purposes and particular species is clearly of equal importance. For these situations infectious agents such as viruses or gene expression from endosymbionts may be more efficacious, and several of these systems are discussed in this book.

Several viral systems are under consideration such as the Sindbis RNA virus that is highly effective as an expression system both *in vitro* and *in vivo* (see Chapter 9 by Olson), and DNA viruses such as densoviruses that have been used as transduction agents in mosquito larvae (see Chapter 8 by Carlson et al.). Retroviruses have long been used as vehicles for gene transfer into mammalian cells (Eglitis and Anderson, 1988), and recently the host range of a mammalian retroviral vector was increased considerably by pseudotyping to include vertebrate and invertebrate systems including several insect species (see Chapter 7 by Burns). With the recent elucidation of retroviruses in insects such as *gypsy*, these systems also have potential for insect vector function with minimal modification (see Chapter 10 by Terzian et al.).

A bit different from the typical concept of germline or somatic (transient) transformation is the use of symbiotic organisms to express genes of choice in a host organism, a type of gene transfer referred to as paratransgenesis. Several systems are under development that have great potential for particular applications. For example, several prospective field applications will require driving transgenes into an insect population and bacterial endosymbionts, such as *Wolbachia,* that can spread through populations very rapidly have considerable potential for achieving this (see Chapter 15 by Sinkins and O'Neill). Of course, the genes to be spread must be transferred into the endosymbiont first, and this may take advantage of viral or transposon vectors. Bacterial symbionts of *Rhodnius prolixus* have already been transformed with genes lethal to parasitic trypanosomes that cohabit in the insect host, and the use of such paratransgenic insects for disease prevention may be implemented shortly (see Chapter 16 by Beard et al.).

As the development of more sophisticated transgenic strains becomes necessary for applied uses, we will find the need for gene transfer of multiple transgenes that are highly stable and noninteracting. The greatest assurance for having a widely applicable genetic toolbox for many different species will only come from the development of vector systems that are mechanistically distinct and varied.

1.3 TRANSFORMATION METHODOLOGY

Since *Drosophila* is the only routinely transformed insect, it is the best model system for transformation methodology, although, undoubtedly modifications will be necessary if not critical for other types of vectors and insect systems. Our studies with tephritid fruit flies have generally followed the standard *Drosophila* procedures (see Spradling, 1986), and most other studies with other insects have done so as well. Although several of these studies have been successful, it is quite possible that modifications in injection buffer, DNA concentrations, and DNA delivery might greatly improve gene transfer efficiency. Some specific modifications developed for mosquito transformation are discussed by Morris (1997). While primary concerns for non-drosophilid transformation usually centered on functional vectors and selectable marker systems, the ability to deliver DNA into the germline or soma of the host organism has been equally daunting, and, at present, it is probably one of the greatest general roadblocks.

1.3.1 DNA Preparation

The preparation and amount of DNA injected is critical to successful transformation. Early *Drosophila* experiments utilized only cesium chloride purified plasmid DNA, usually requiring double ultracentrifugation with rigorous removal of ethidium bromide and any organic solvents (see Sambrook et al., 1989). The recent availability of plasmid preparation kits provides much simpler methods. While we and others have successfully transformed with plasmid prepared with such kits, for some plasmids grown in some particular bacterial hosts, contaminating toxic proteins may not be easily removed and the cesium method is generally the most fool-proof. Qiagen now provides an endotoxin-free kit that may alleviate some of these concerns (see Appendix to this chapter for location and contact information for this and other companies mentioned).

Plasmid DNA for injection is usually a mixture of the desired amounts of vector and helper (or vector alone) that is ethanol precipitated and resuspended in injection buffer. Most insect transformations have used the *Drosophila* injection buffer (5 mM KCl, 0.1 M sodium phosphate pH 6.8), although it is likely that this is not ideal for all insects. Resuspended DNA should be used within a few days, and kept frozen until use, with an aliquot run on an agarose gel to ensure DNA integrity and the general concentration desired. Previous to injection, the DNA should be centrifuged at 12,000 × *1212* for 5 min to eliminate any particles that might clog needles or contaminate the eggs.

Generally, total DNA concentrations have not exceeded 1 mg/ml with vector/ helper ratios ranging from 2:1 to 9:1, with actual molar ratios being a consideration. Excessive amounts of DNA are considered to be harmful if not lethal to embryos, probably because of the nucleic acids themselves and unavoidable contaminants in the solution. It is likely, however, that many embryos larger than *Drosophila* can withstand higher concentrations or amounts injected. Other considerations are the size of the plasmid (usually the vector), since as plasmid size increases the number of molecules injected will decrease for a standard DNA concentration, but increasing the concentration may result in shearing the molecules during injection. Shearing may be alleviated by larger-bore needles, although embryo survival may be compromised. These and other trade-offs are pervasive in gene transfer methodology, and the best system for any particular insect must be determined empirically by extensive control experiments.

1.3.2 Dechorionation

Drosophila benefits from having an easily removable chorion, either manually or chemically, and can be quickly desiccated for microinjection of DNA. Manual dechorionation is perfomed by gently rolling freshly laid eggs on double-stick tape with forceps, with the egg generally popping out from the chorion. Chemical dechorionation uses a diluted hypochlorite solution (liquid bleach) at a final concentration of 1.5 to 2.0% (note that hypochlorite concentrations in bleach vary around the world). The bleach should also be fresh — open for no more than 2 to 3 weeks, with diluted solutions being no more than 2 days old. The time and concentration of hypochlorite treatment must be determined empirically. If a 1.5% solution does not dechorionate within 2 to 3 min, slightly higher concentrations should be tried. After dechorionation, eggs must be repeatedly washed in distilled water, which is usually aided by the addition of Triton-X 100 or NP-40 (0.02% final concentration). After three washes the eggs can be prepared for desiccation.

Depending on the number of eggs collected, hypochlorite dechorionation can be done in a watch glass or small culture dish using a drawn out Pasteur pipette or syringe to remove fluid and a brush to swirl and move the eggs. Larger numbers of eggs are more easily (and quickly) manipulated by using a small Buchner funnel (~45 mm inner diameter) and a filtering flask. Eggs collected (or washed) onto white filter paper are placed in the funnel and rinsed in water which is gently drawn off with a water vacuum (use very low vacuum and control by keeping flask stopper loose). Diluted bleach is added to the funnel, swirled around, and gently drawn off after the apppropriate time. Wash solution is quickly added with a squirt wash bottle, swirled, and drawn off. The filter paper with eggs is taken out of the funnel, and the eggs washed onto black filter paper (eggs are more easily observed on black filter paper, which is bleached by the hypochlorite). Washing is repeated two to three times with very gentle aspiration (filter should always be wet and not sucked dry or eggs will be crushed).

1.3.3 Preparation for Embryo Injection

Dechorionated *Drosophila* eggs are kept moist on damp filter paper (from the last wash) and their development may be slowed by keeping them in a cool incubator or ice bucket (although not directly on ice). By using a fine brush (000) or forceps, eggs are placed on a thin strip (1 to 2 mm) of double-stick tape placed on a slide, often at the edge. We have found it more convenient to have the tape on a rectangular coverslip (30 × 22 mm) which is then placed on top of a slide on the mechanical stage. *Drosophila* eggs are typically injected under halocarbon oil, although recently some laboratories have found it possible to inject in open air, using food coloring in the DNA mixture to visualize better the injected DNA. We continue to inject under oil, and therefore place the tape inside a thick rectangle drawn with a wax pencil which holds the oil around the eggs (we can fit two rectangles with tape on each coverslip; wax lines should be ~5 mm from the edge of the tape). Eggs are placed on the tape with with the posterior end facing the edge of the coverslip so that injection may be made near the pole plasm.

Most dipteran eggs we have injected must be desiccated to reduce the internal volume and pressure so that injected DNA can be accepted without extrusion. Proper desiccation may be the most critical factor for good injection and embryo survival, and optimal desiccation times must be determined empirically for each experiment, and often change within a several-hour period. Over-desiccation often results in death, and underdesiccation often results in extrusion of the DNA and yolk, resulting in death, sterility, or a lower gene transfer frequency for those that survive. Although embryos can sometimes survive considerable loss of material, these embryos are likely to be sterile due to loss of pole plasm, or nontransformed due to loss of injected DNA. We typically desiccate in open air at 22°C and 50% relative humidity or lower for 8 to 12 min. Desiccation times are certainly affected by ambient temperature and humidity, and may be better controlled by placing the eggs in a closed chamber with a drying agent, with or without a gentle vacuum. Obviously manual dechorionation requires less desiccation time than bleach dechorionation.

1.3.3.1 Needles

Close in importance to proper desiccation for microinjection are properly prepared needles. To a certain extent there is a trade-off between desiccation and needles in that very fine sharp needles can inject well into underdesiccated eggs, with the benefit of better viability. On the other hand, large-bore needles do not clog as easily and may be necessary to prevent shearing of large plasmids, but eggs must be desiccated as much as possible to accept the injection. For insects such as moth and mosquito species where dechorionation is not possible, large-bore needles are required to "cut through" the chorion and not get clogged. Typically, needles are drawn out on needle pullers usually used for creating glass electrodes for neurophysiology and can be found at World Precision Instruments (WPI) and Narishige, among other suppliers. The same glass capillaries for electrodes may be used (some including microfilaments allowing easy fluid flow) or more simple micropipettes (such as 25 μl Drummond microcaps). Previous to drawing out the needle, the capillaries should be siliconized or silanized by one of a variety of methods (e.g., Sigmacote from Sigma), and thoroughly washed and dried. For embryos that have a soft chorion or can be dechorionated, a finely drawn-out needle with the tip scraped off along the edge of the glass slide (that supports the coverslip) before injection works quite well. For the many embryos that cannot be dechorionated, such as mosquitoes (see Jasinskiene et al., 1998), beveling the tips of the needles will probably be required, and this is helpful for all injections and especially those requiring large-bore needles (Morris, 1997). We use a beveler available from Sutter Instruments (BV-10), and WPI and Narishige also provide similar types of needle bevelers.

DNA may be backfilled into the needle or sucked up prior to injection. Backfilling may result in less shearing, and needles for backfilling can be created from siliconized 100 μl microcaps that are drawn out over a flame and broken in the middle. A few microliters of DNA sucked into these needles by capillarity can then be backfilled into several injection needles.

1.3.4 MICROINJECTION APPARATUS AND PROCEDURES

The typical apparatus for *Drosophila* microinjection requires an inverted or stereozoom microscope having a 10 to 80× magnification range and transmitted light, a mechanical stage, micromanipulator, and a mechanism for DNA injection (Figure 1.2). Direct illumination (instead of or in addition to transmitted) is required for embryos that are pigmented or cannot be dechorionated. The general procedure involves aligning the needle at the injection site with the micromanipulator, while the actual injection is done by using the mechanical stage to move the egg into the needle. Once inside the egg, DNA is injected manually or with an electronic air-pulse system. *Drosophila* is typically injected under oil, although early injection experiments were done in open air, and several methods

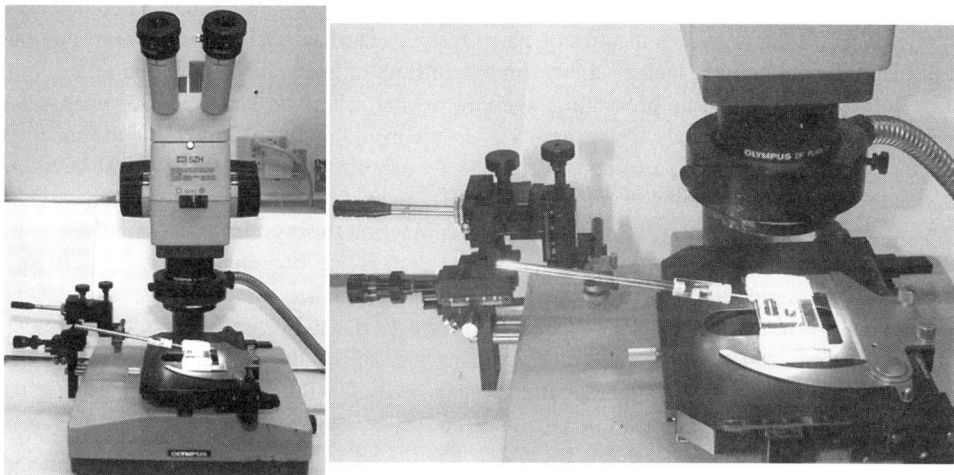

FIGURE 1.2 Microscope setup for insect embryo injections. The setup includes an Olympus SZH stereozoom microscope with a transmitted light base and BH2-SH mechanical stage (with SZH-STAD1 adaptor). A Narishige MN-151 micromanipulator with a B-8B ball joint pipette holder is mounted on the microscope base, with an attached WPI MPH-1 micropipette holder. The air-injection system is not shown, but it connects perpendicularly into a female luer fitting within the plastic cylinder portion of the needle holder. Embryos prepared for injection on a coverslip are placed on a slide taped to a U-shaped plastic carrier for elevation.

were used to seal the puncture wound including a dissolved gum damar and a 1.2% sulfosalysilic acid solution (see Ashburner, 1989a). While *Drosophila* is again injected in open air by some laboratories, typically the injection site is not sealed, although for other insect embryos that cannot be submerged in oil and require large-bore needles, sealing the puncture should limit leakage and increase survival. An alternative sealant is any one of a number of "super" or "crazy" glues, although the effect on viability of any of these sealants should be tested on uninjected and injected eggs. We continue to inject several dipteran species under oil without sealing.

Drosophila injections were originally performed with an inverted microscope because they had or could be fitted with a mechanical stage, there was working space for a micromanipulator, and, in all likelihood, they were generally available. We found more recently that the Olympus SZH stereozoom microscope (now SZX) with a transmitted light base could be fitted with a flat Olympus stage (model BH2-SH) which is almost ideal for easy handling. The mechanical stage and adaptor (SZH-STAD1) can actually be fitted on most stereozoom dissecting microscopes from other manufacturers after tapping screw holes for the adaptor into the base; however, a longer carrier pole for the microscope body may be needed for the additional height (400 mm pole used on Olympus). This setup has the advantage of zoom optics and easy handling of eggs and needle while under view, and for non-tissue-culture laboratories, the scope has utility when the injection apparatus is removed (e.g., scoring transformants).

The micromanipulator should be adjustable in three dimensions (X, Y, and Z axes), and finely adjustable in the axis used for proper needle height (axis can vary depending on how the manipulator is mounted). We use two manipulators, which work well: a Leitz manipulator that is free-standing (current Leica manipulators may differ) and a Narishige MN-151 manipulator (with B-8B ball joint pipette holder) that can be mounted with adaptors onto the microscope base. The Leitz manipulator was originally suggested for *Drosophila* injections, but we prefer the Narishige for its simplicity and it is considerably less expensive (and the lower cost models MN-152 and 153 also may be used).

Apparatus for DNA delivery varies greatly — from simple connections using a 50-cent syringe and tubing, to air-pulse controller systems costing up to $20,000. As might be expected, the cheaper apparatus probably works best for those who routinely inject and have a refined technique. We use something in between which is relatively inexpensive and efficient, and modest upgrades are also possible that mimic the very elegant and expensive systems. Our system is also based on components used in neurophysiology that include a pressurized air-tank or house air-supply connected to high-pressure tubing with an in-line pressure gauge and regulator, and an electric valve operated by a simple switch and battery supply. These components are available from Clippard who sells through regional suppliers, at a total cost of less than $120. A male luer fitting or cutoff 1-ml luer syringe is fitted at the end, and this connects into a WPI MPH-1 microneedle holder. The switch and battery holder are available from any local electronics store. We find this system to be quite inexpensive, easy to use, and efficient. An upgraded air-pulse system is the pneumatic picopump from WPI (PV830), which has pressure and vacuum controls and a foot switch for approximately $2000. The vacuum capability allows the uptake of DNA through the needle and is useful in unclogging needles. Several laboratories use a somewhat more elegant version of the picopump from Eppendorf, currently marketed as the Transjector 5246 (that can be used with their micromanipulator called InjectMan), which is quite expensive. Both WPI and Eppendorf sell premade microneedles, but these seem prohibitively expensive ($3 to $5 per needle) for routine injection experiments where many needles may be required.

1.3.5 METHODS FOR DNA DELIVERY

Although microinjection is currently the method of choice to deliver large quantities of DNA into embryos, as noted above, different methods of DNA delivery have been tried ever since insect gene transfer experiments were first attempted. Some of the first were variations on soaking embryos in DNA, that attempted to increase efficiency of DNA entry by trying first to permeabilize the eggs with organic solvents (Limbourg and Zalokar, 1973). The more recent development of methodologies and equipment for introducing DNA into plants, cell lines, and bacteria have offered a wide range of possibilities. The most well known of these include lipofectin reagent, biolistics (originally ballistics), electroporation, and modifications of microinjection. Most of these have been attempted with *Drosophila* or other insects for at least transient DNA delivery and expression, and while none have proved efficient yet for routine germline gene transfer, further experimentation is certainly warranted for those species not amenable to efficient microinjection. Certainly for those systems that allow DNA delivery *en masse*, routine gene transfer may be possible even with high mortality and inefficient vector systems. Modifications of microinjection also have been attempted, such as injection into ovarian egg follicles previous to oviposition (Presnail and Hoy, 1992), or injection into the female hemocoel for DNA uptake into egg follicles along with vitellin. For some of these methods, information has not been published, is found in abstracts, or is anecdotal. The following brief overview should provide some useful information that may be used as a basis for further experimentation.

1.3.5.1 Lipofection

The development of cationic lipids that form liposomes encapsulating DNA has provided a routine method for DNA delivery into cultured cells, that occurs after liposome fusion with cell membranes. Modified forms of the cationic lipid have been developed and are commercially available for the procedure known as lipofection. Although used routinely for transfection of cell lines, lipofection also has allowed DNA delivery into cells *in vivo* in a variety of animal systems (Nicolau et al., 1983; Felgner et al., 1987), and transient expression has been reported for cultured mosquito salivary glands (Morris et al., 1995) and heart tissue from the Pacific oyster (Boulo et al., 1996). Possible advantages of using lipofectin would be the uptake of DNA into egg follicles from maternal hemolymph, or the uptake of DNA into the germline after pole cell formation, allowing a considerably longer time for injection during blastoderm formation. A systematic analysis of lipfectin–DNA mixtures with functional vectors in *Drosophila* or other species should be a high priority.

1.3.5.2 Biolistics

Two DNA delivery systems using mechanical means are biolistics and electroporation. Biolistics involves the coating of microscopic pellets (usually gold or tungsten microparticles) with DNA by ethanol precipitation, which are bombarded into cells or tissues. It was derived from a ballistic method, so-called since it actually used a modified form of a shotgun to shoot the particles into plant tissue (Klein et al., 1987). A highly interesting study used the ballistic method to deliver DNA into *Drosophila* embryos successfully, resulting in high frequencies of transient expression, and in a *P* transformation experiment a single transformant from several thousand eggs was recovered and verified (Baldarelli and Lengyel, 1990). Although this report met with great optimism, repetition of the experiment has not been reported. Ballistics was subsequently modified to a biolistics method using pressurized helium within a chamber and is commercially available from BioRad. Biolistics is routinely used for plant transformation, and a more recent modified apparatus is the Helios Gene Gun System® that can be used for subcutaneous injection. Further details on these current systems are available from the company (see catalog and Web site). An apparent problem using the biolistics PDS-1000/He chamber for insects has been disintegration or dispersal of eggs after the high-pressure bombardment. Miahle and Miller (1994) addressed this problem and successfully achieved transient expression in mosquito eggs by having the DNA-coated particles in an aqueous suspension. They increased levels of reporter gene expression (luciferase) significantly using this suspension under high or low pressure, although transformation has not been reported. These results indicate that a ballistic or biolistic method can effectively deliver DNA into insect embryos and, at some level, achieve transformation.

1.3.5.3 Electroporation

DNA delivery by electroporation has also received considerable attention in recent years. Essentially, cells in solution are given an electric shock which serves to increase pore size in cell membranes allowing the passive introduction of molecules from the solution. It is used routinely for bacterial transformations, and modifications in the shock parameters, including waveform, resistance, and voltage, have optimized DNA delivery for a variety of procaryotic and eucaryotic systems. Multicellular organisms have been more challenging, although great interest was generated by the relatively high frequency of xanthine dehydrogenase transient expression in *Drosophila* larvae that had been electroporated as embryos under fairly standard conditions (Kamdar et al., 1992). Additional challenges exist for insect eggs that exceed the size of, or cannot be submerged in, standard cuvettes (usually 1 to 4 mm wide). Leopold and colleagues (1996) have successfully electroporated eggs under uniform conditions from lepidopteran and dipteran species using a slot cuvette design with electrodes placed on a microscope slide. While the reports for successful DNA delivery, as concluded from transient reporter gene expression, are highly encouraging, germline transformation in *Drosophila* has not been reported. Nevertheless, the benefits of such a procedure and success with transient expression from these initial reports indicate that continued high-priority efforts in this area of DNA delivery are also warranted.

Outlined above are the primary systems of DNA delivery currently being used or considered, although the range of size, structure, and habitat of insect eggs suggests that any system successful for a particular species (or related species) will require modifications for others. Microinjection of oviposited eggs has already been modified so that DNA may be delivered *in situ* in mites and wasps (Presnail and Hoy, 1992; see Chapter 19 by Hoy). Preliminary experiments in our laboratory indicate that DNA injected into female abdomens is taken up by maturing egg follicles and can be recovered in oviposited eggs, although recovered plasmid amounts have been low (<100 molecules per egg) (S. Gomez and A. Handler, unpublished). Another variation on injection involves the puncturing of eggs with a tungsten needle coated with DNA, and a combination of methods such as the use of lipofectin-reagent with any of the various mechanical methods may help increase their efficiency.

Continued testing for all these methods is clearly warranted now given the new vectors and markers that have become available since most of the methods were originally considered.

1.4 IDENTIFICATION OF TRANSGENIC INSECTS AND TRANSGENES

A significant portion of this book is dedicated to markers for transgenic selection. These markers are gene products that restore a normal phenotype in a mutant host, confer a new visible phenotype, or confer resistance to a chemical or drug. Some, and possibly most of these markers, provide reasonably convincing evidence that a marked G_1 insect is indeed a transformant. Yet, there are a variety of reasons to remain cautious, and it is now realized that transgenics must be identified unambiguously by definitive molecular tests. This is especially so for a proof of gene transfer in an insect species with a new vector system for the first time. For "mutant-rescue" selections resulting in a wild-type phenotype, nontransformants may be erroneously selected due to strain contamination and, more rarely, reversions of the mutation. As noted above for G418 selections, non-vector-related chemical resistant organisms may be selected, and this may occur in the host strain itself, or in host strain symbionts. Beyond the simple determination of gene transfer, an assessment of vector activity and the potential release of a transgenic strain depend on knowing the number and possibly the chromosomal location of the vector integrations. In addition, an assessment of vector stability and the potential use of a vector as a genetic tool depends on knowing whether an integration was vector mediated, as opposed to a fortuitous or illegitimate recombination. Non-vector-mediated integrations cannot, typically, be remobilized, which may add to their stability, although this might hinder the use of the vector for enhancer-trap studies. Finally, there may be safety issues or practical concerns related to exactly what part of the vector entered the host genome. Most vector plasmids have at least one or more antibiotic resistance genes, bacterial reporter genes, and an origin of replication. If vector integration occurs by recombination and not transposon or viral transposition, then extraneous DNA may inadvertently enter the genome as well. It should be noted that *P* vector insertions, which normally integrate individually, have been detected as multimeric concatamers. This and other types of rearrangements may occur previous to integration, or afterward if the vector transposase or cross-mobilizing system exists in the genome.

The determination of a chromosomal integration, and the number of integrations (if fewer than 10), is most easily achieved by Southern DNA hybridization; however, extrachromosomal DNA (e.g., nonintegrated injected plasmid) may be detected if in large enough quantities. Nevertheless, proper diagnostics with carefully chosen restriction site digestions and probes should normally yield unambiguous results (see Sambrook et al., 1989). Some studies have utilized direct PCR for molecular detection, but contamination may be problematic since vector plasmids are probably widespread within the same laboratory, and at the very least control reactions are necessary using several primer sets to plasmid sequence outside of the transposon (although this will detect non-integrated plasmid as well). Chromosomal integrations and their number can also be determined by *in situ* hybridization to chromosomes (see Ashburner 1989a, b), and *Drosophila* transformants have been typically analyzed in this way, but this is most easily performed in the relatively few species with large polytenized chromosomes.

The most definitive proof of vector-mediated chromosome insertions is by sequencing the insertion site junctions, since most transposons create an insertion site duplication that should be diagnostic for that vector. Sequencing an insertion site is most simply achieved by isolating the junction sequences by inverse PCR (Ochman et al., 1988), but this is most straightforward for genomes having a single integration, and, again, plasmid contamination must be avoided. For multiple integrations within a transgenic strain, it may be necessary to create and screen small genomic libraries, and this was found necessary for mosquito strains where only one of the junction fragments could be isolated by inverse PCR (Jasinskiene et al., 1999). An alternative strategy useful in *Drosophila* is plasmid rescue (Steller and Pirrotta, 1986), which is based on having a pUC vector backbone within the transposon vector termini. Thus, transformed genomic DNA can be digested, religated to form circles, and transformed into bacteria. Recovered plasmid should contain the transgene with adjacent insertion site DNA, and this is highly advantageous as well for transposon-tagging and enhancer-trap screens, allowing easy isolation of the tagged genomic DNA. Drawbacks of this method are that it will add 2 to 3 kb of additional DNA to the vector, and, as mentioned above, antibiotic resistance genes may raise safety concerns for released transgenic insects.

Sequences adjacent to the insertion site should be different from those in the vector plasmid, and can be verified as such by alignment to the plasmid and using them to create primers for PCR. Genomic insertions can then be definitively verified by direct PCR of nontransformed host and transformed DNA, that should yield products of known length and sequence depending on the presence or absence of the transgene. It is also worthwhile to subject the chromosomal sequence to BLAST analysis (Altschul et al., 1990), since this will quickly determine if the sequence is nonplasmid (which is not always evident if it is rearranged), and may possibly identify interesting sequences within the genome by homology to sequences within the database (with the sequence thus transposon tagged).

1.5 PERSPECTIVES ON THE USE OF INSECT GENE TRANSFER

As noted in the beginning of this chapter, and as will become more apparent throughout this book, the genetic transformation of insects will allow enormous strides in further understanding the genetic and biochemical basis of insect biology, and will present many new and efficient strategies to control the population and behavior of beneficial and pest insects. One only needs to review the incredible advancement in the knowledge of basic model systems such as *D. melanogaster, Caenorhabditis elegans*, mice, and *Arabidopsis thaliana*, made possible in large part by the analysis of transgenics. The primary motivation, however, for creating transgenic insects is for applied purposes, and in recent years we have already seen the revolutionary influence of gene transfer technology on the commercial uses of plant and animal systems. A major difference in the use of transgenic insects, however, is that unlike other transgenic organisms most of the applications for insects will require their release into the environment where their future and that of their descendants will not be under direct regulation. The only exceptions to this will be the release of sterile transgenics, created to optimize biological control programs such as the sterile insect technique (Handler, 1992; see Chapter 17 by Robinson and Franz), or the release of conditional lethals that will die in response to changing environmental conditions (Fryxell and Miller, 1994). Several other strategies discussed in this book rely on the release of genetically modified insects that are not sterile or lethal, and, as the technology advances, we can expect new and more sophisticated strategies requiring the release of fertile transgenics. This raises important questions relating to the ecological and environmental impact of these new insect systems, and it is clearly in our interest to consider these issues carefully as we develop strategies for the use of transgenic insects, and the specific design of vectors to create these strains.

Biological risk assessment and current regulatory practices for transgenic arthropod transport and release are addressed in this book (see Chapter 19 by Hoy and Chapter 20 by Young et al.), and it is realized that individual consideration must be given to each insect species modified by specific recombinant DNA molecules carried within particular vectors. Some of the greatest concerns, however, may be minimized by careful analysis of vector and marker function and by the use of available technology to manipulate the vector postintegration. For example, a major consideration for risk assessment is vector stability and the potential for horizontal transmission of transgenes into nontarget organisms. Use of bipartite vector systems and testing with transient assays should generally ensure vector stability in specific target host organisms, but other organisms in nature may have the same transposon or a related cross-mobilizing system (see Sundararajan et al., 1999), and in some contexts the transgene may have the opportunity to move into the genome of such organisms if they possess a functional transposase, or integrase for viral systems. A method to prevent such movement is to create "suicide vectors" that allow the deletion or rearrangement of vector sequences, such as terminal and subterminal sequences, postintegration to immobilize the transgene effectively. Mechanisms by which this is possible are recombination systems such as FLP/*FRT* (Golic and Lindquist, 1989) and Cre/*lox*P (Siegal and Hartl, 1996), which are discussed in Chapter 3 by Rong and Golic. The use of transgenic markers such the green fluorescent protein (Prasher et al., 1992), which should be functional in almost all organisms (Chalfie et al., 1994; see Chapter 5 by Higgs and Lewis), also provides a means to effectively monitor released transgenics, as well as possible transgene movement outside the host species under experimental and field conditions.

From a more practical standpoint, one of major attendant problems with the ability to create transgenic strains is the required funds and facilities to rear them. A critical component in the use of transgenic organisms is to have many varied strains, some of which can be interbred for desired offspring, and it is not always obvious if or when a strain may be critical for a specific purpose. *Drosophila* was chosen as an organism for genetic research because its rearing was simple and took relatively little space (for a eucaryote at least), although at this point, where many thousands of wild, mutant, and transgenic strains exist, stock collections are routinely culled due to space and cost considerations. Almost all other insect systems are more difficult and costly to rear, and this has been a clear constraint on their use for genetic studies. Indeed, many important insect strains have been discarded due to lack of funds or the loss of a primary investigator. It is clear that for transgenic technology and genetic manipulation to advance for any insect species, the ability to store germ plasm must be a high priority. Possibilities include cryopreservation of insect embryos by chemical and mechanical means, which is already possible for some dipteran species (Leopold et al., 1998), as well as the potential for cold storage enhanced by expression of antifreeze proteins (AFP) that are produced by some insects for overwintering (Tyshenko et al., 1997). For the latter possibility, a marked chromosome with an appropriate AFP construct could be crossed into a strain for storage, and crossed out when necessary for normal rearing.

The creation of transgenic insects will provide many challenges for their safe, efficient, and effective use in laboratory and field studies, with vector stability and strain maintenance among them. It is our hope that the same thoughtfullness and creativity that is making this technology a reality will be used to meet these challenges effectively. Only in this way will the continued and productive use of transgenic insects go forward, allowing us to learn more about the biology of insects, as well as the ability to control their population size and behavior.

1.6 APPENDIX*

Equipment

1. Inverted or stereozoom microscope
   — Magnification range of ~8 to 80×
   — Transmitted and/or direct illumination
   — Mechanical stage (Olympus BH2-SH with SZH-STAD1 adaptor can be mounted on most stereozoom microscopes)
2. Micromanipulators
   — Narishige MN-151 (MN-152 or MN-153)
   — Leica
   — Eppendorf Injectman
3. Air-pulse injection systems

---

* Mention of a proprietary product does not constitute an endorsement or the recommendation for its use by USDA.

a. Clippard components (assembly needed):
      — Vinyl hose tubing, EV3-3 electronic valve, MAR-1 air regulator, PG-100 pressure gauge, MAF-1 air-filter (connect to air supply), L-fittings, T-fittings, hose-fittings, and gaskets
      — Male Luer fitting or cutoff 1-ml syringe (connect to hose fitting and needle holder)
      — Two 1.5-V C batteries and battery holder
      — Switch
      — WPI MPH-1 needle holder
      — House or tank air or nitrogen supply
   b. WPI pneumatic picopump with foot switch
   c. Eppendorf Transjector 5246
4. Needle preparation
   — WPI capillaries or needles
   — Drummond microcapillaries — 25 and 100 μl
   — Narishige needle puller
   — WPI needle puller
   — Sutter Instruments microbeveler BV-10
   — WPI micobeveler
   — Sigmacote
5. Egg dechorionation
   — Liquid bleach
   — Triton-X 100 or NP-40 nonionic detergents
   — 45-mm I.D. Buchner funnel
   — 500-ml Erlenmeyer flask
   — 43-mm-diameter white and black filter circles
   — Small vacuum chamber with drierite
   — Billups–Rothenberg tissue culture chamber MIC-101
   — Oxygen supply

ADDRESSES

Billups–Rothenberg, Inc., P.O.Box 977, Del Mar, CA 92014
Bio-Rad Laboratories, Life Sciences Group, 2000 Alfred Noble Drive, Hercules, CA 94547-1804; (800) 4BIORAD; *www.bio-rad.com*
Clippard Instrument Laboratory, Inc., 7390 Colerain Rd., Cincinnati, OH 45239; (513) 521-4261; *www.clippard.com*
Eppendorf Scientific, Inc., Cantiague Lane, Westbury, NY 11590-2852; (516) 876-6800; *www.eppendorfsi.com*
Leica Microsystems, Inc., 111 Deer Lake Rd., Deerfield, IL 60015-4986; (847) 405-0123; *www.leica-microsystems.com*
Narishige USA, Inc., 1 Plaza Rd., Greenvale, NY 11548-1027; (516) 621-4588; *www.narishige.co.jp/main.htm*
Olympus America, Inc., 2 Corporate Center Drive, Melville, NY 11747-3157; *www.olympus.com*
Qiagen, Inc., 28159 Stanford Ave., Valencia, CA 91355; (800) 362-7737; *www.qiagen.com*
Sigma Chemical Co., 3050 Spruce St., St. Louis, MO 63103-2564; (800) 325-3010; *www.sigma-aldrich.com*
Sutter Instruments Co., 51 Digital Drive. Novato, CA 94949; (415) 883-0128; *www.sutter.com*
World Precision Instruments Inc., 175 Sarasota Center Blvd., Sarasota, FL 34240-8750; (941) 371-1003; *www.wpiinc.com*

ACKNOWLEDGMENTS

Grateful appreciation is extended to Herbert Oberlander and Steven Valles for comments on the manuscript, and to those who shared unpublished results. The USDA–NRI Competitive Grants Program is acknowledged for support of the author's research discussed in this chapter.

REFERENCES

Altschul SF, Gish W, Miller W, Myers EW, Lipman DJ (1990) Basic local alignments search tool. J Mol Biol 215:403–410
Anxolabéhère D, Kidwell M, Periquet G (1988) Molecular characteristics of diverse populations are consistent with the hypothesis of a recent invasion of *Drosophila melanogaster* by mobile *P* elements. Mol Biol Evol 5:252–269
Ashburner M (1989a) *Drosophila*: A Laboratory Handbook. Cold Spring Harbor Laboratory Press, Cold Spring Harbor, NY
Ashburner M (1989b) *Drosophila*: A Laboratory Manual. Cold Spring Harbor Laboratory Press, Cold Spring Harbor, NY
Avery OT, Macleod CM, McCarty M (1944) Studies on the chemical nature of the substance inducing transformation of pneumococcal types. I. Induction of transformation by a desoxyribonucleic acid fraction isolated from pneumococcus type III. J Exp Med 79:137–158
Baker B, Schell J, Loerz H, Fedoroff N (1986) Transposition of the maize controlling element Activator in tobacco. Proc Natl Acad Sci USA 83:4844–4848
Baldarelli RM, Lengyel JA (1990) Transient expression of DNA after ballistic introduction into *Drosophila* embryos. Nucl Acids Res 18:5903–5904
Bellen HJ, O'Kane CJ, Wilson C, Grossniklaus U, Pearson RK, Gehring WJ (1989) *P*-element-mediated enhancer detection: a versatile method to study development in *Drosophila*. Development 3:1288–1300
Benedict MQ, Salazar CE, Collins FH (1995) A new dominant selectable marker for genetic transformation: Hsp70-opd. Insect Biochem Mol Biol 25:1061–1065
Berg DE, Howe MM (1989) Mobile DNA. American Society for Microbiology, Washington, D.C.
Berghammer AJ, Klingler M, Wimmer EA (1999) A universal marker for transgenic insects. Nature 402:370–371
Blackman RK, Gelbart WM (1989) The transposable element *hobo* of *Drosophila melanogaster*. In Berg DE, Howe MM (eds) Mobile DNA. American Society for Microbiology, Washington, D.C., pp. 523–525
Blackman RK, Macy M, Koehler D, Grimaila R, Gelbart WM (1989) Identification of a fully functional *hobo* transposable element and its use for germ-line transformation of *Drosophila*. EMBO J 8:211–217
Boulo V, Cadoret JP, Le Marrec F, Dorange G, Miahle E (1996) Transient expression of luciferase reporter gene after lipofection in oyster (*Crassostrea gigas*) primary cell cultures. Mol Mar Biol Biotechnol 5:167–174
Brand AH, Manoukian AS, Perrimon N (1994) Ectopic expression in *Drosophila*. Methods Cell Biol 44:635–654
Brennan MD, Rowan RG, Dickinson WJ (1984) Introduction of a functional *P* element into the germ line of *Drosophila hawaiiensis*. Cell 38:147–151
Calvi BR, Hong TJ, Findley SD, Gelbart WM (1991) Evidence for a common evolutionary origin of inverted repeat transposons in *Drosophila* and plants: *hobo*, *Activator*, and *Tam3*. Cell 66:465–471
Caspari E, Nawa S (1965) A method to demonstrate transformation in *Ephestia*. Z Naturforsch 206:281–284
Chalfie M, Tu Y, Euskirchen G, Ward W, Prasher DC (1994) Green fluorescent protein as a marker for gene expression. Science 263:802–805
Coates CJ, Jasinskiene N, Miyashiro L, James AA (1998) *Mariner* transposition and transformation of the yellow fever mosquito, *Aedes aegypti*. Proc Natl Acad Sci USA 95:3742–3751
Cornel AJ, Benedict MQ, Rafferty CS, Howells AJ, Collins FH (1997) Transient expression of the *Drosophila melanogaster cinnabar* gene rescues eye color in the white eye (WE) strain of *Aedes aegypti*. Insect Biochem Mol Biol 27:993–997
Daniels SB, Strausbaugh LD (1986) The distribution of *P* element sequences in *Drosophila*: the *willistoni* and *saltans* species groups. J Mol Evol 23:138–148

Davis I, Girdham CH, O'Farrell PH (1995) A nuclear GFP that marks nuclei in living *Drosophila* embryos; maternal supply overcomes a delay in the appearance of zygotic fluorescence. Dev Biol 170:726–729

Eglitis MA, Anderson WF (1988) Retroviral vectors for introduction of genes into mammalian cells. BioTechniques 6:608–615

Engels WR (1989) P elements in *Drosophila melanogaster*. In Berg DE and Howe MM (eds) Mobile DNA. American Society of Microbiology, Washington, D.C., pp. 439–484.

Fadool JM, Hartl DL, Dowling JE (1998) Transposition of the *mariner* element from *Drosophila mauritiana* in zebrafish. Proc Natl Acad Sci USA 95:5182–5186

Federoff N (1989) Maize transposable elements. In Berg, DE and Howe, MM (eds) Mobile DNA. American Society for Microbiology, Washington, D.C., pp. 375–411.

Felgner PL, Gadek TR, Holm M, Roman R, Chan HW, Wenz M, Northrop JP, Ringold GM, Danielsen M (1987) Lipofection: a highly efficient, lipid-mediated DNA-transfection procedure. Proc Natl Acad Sci USA 84:7413–7417

ffrench-Constant RH, Mortlock DP, Shaffer CD, MacIntyre RJ, Roush RT (1991) Molecular cloning and transformation of cyclodiene resistance in *Drosophila*: an invertebrate gamma-aminobutyric acid subtype A receptor locus. Proc Natl Acad Sci USA 88:7209–7213

Finnegan DJ (1989) Eucaryotic transposable elements and genome evolution. Trends Genet 5:103–107

Fox AS, Yoon SB (1966) Specific genetic effects of DNA in *Drosophila melanogaster*. Genetics 53:897–911

Fox AS, Yoon SB (1970) DNA-induced transformation in *Drosophila*: locus specificity and the establishment of transformed stocks. Proc Natl Acad Sci USA 67:1608–1615

Fox AS, Duggleby WF, Gelbart WM, Yoon SB (1970) DNA-induced transformation in *Drosophila*: evidence for transmission without integration. Proc Natl Acad Sci USA 67:1834–1838

Franz G, Savakis C (1991) *Minos*, a new transposable element from *Drosophila hydei*, is a member of the Tc-1-like family of transposons. Nucl Acids Res 19:6646

Franz G, Loukeris TG, Dialektaki G, Thompson CRL, Savakis C (1994) Mobile *Minos* elements from *Drosophila hydei* encode a two-exon transposase with similarity to the paired DNA-binding domain. Proc Natl Acad Sci USA 91:4746–4750

Fraser MJ, Smith GE, Summers MD (1983) Acquisition of host-cell DNA-sequences by baculoviruses — relationship between host DNA insertions and FP mutants of *Autographa californica* and *Galleria mellonella* nuclear polyhedrosis viruses. J Virol 47:287–300

Fryxell KJ, Miller TA (1994) Autocidal biological control: a general strategy for insect control based on genetic transformation with a highly conserved gene. J Econ Entomol 85:1240–1245

Germeraad S (1976) Genetic transformation in *Drosophila* by microinjection of DNA. Nature 262:229–231

Gloor GB, Nassif NA, Johnson-Schlitz DM, Preston CR, Engels WR (1991) Targeted gene replacement in *Drosophila* via P element-induced gap repair. Science 253:1110–1117

Goldberg DA, Posakony JW, Maniatis T (1983) Correct developmental expression of a cloned alcohol dehydrogenase gene transduced into the *Drosophila* germ line. Cell 34:59–73

Golic KG, Lindquist SL (1989) The FLP recombinase of yeast catalyzes site-specific recombination in the *Drosophila* genome. Cell 59:499–509

Gomez, SP, Handler AM (1997) A *Drosophila melanogaster hobo-white*+ vector mediates low frequency gene transfer in *D. virilis* with full interspecific *white*+ complementation. Insect Mol Biol 6:1–8

Gustafson K, Boulianne GL (1996) Distinct expression patterns detected within individual tissues by the GAL4 enhancer trap technique. Genome 1:174–182

Handler AM (1992) Molecular genetic mechanisms for sex-specific selection. In Anderson TE and Leppla NC (eds), Advances in Insect Rearing for Research and Pest Management, Westview Press, Boulder, CO, pp 11–32

Handler AM, Gomez SP (1996) The *hobo* transposable element excises and has related elements in tephritid species. Genetics 143:1339–1347

Handler AM, Harrell RA (1999) Germline transformation of *Drosophila melanogaster* with the *piggyBac* transposon vector. Insect Mol Biol 8:449–457

Handler AM, O'Brochta DA (1991) Prospects for gene transformation in insects. Annu Rev Entomol 36:159–183

Handler AM, Gomez SP, O'Brochta DA (1993) A functional analysis of the *P*-element gene-transfer vector in insects. Arch Insect Biochem Physiol 22: 373–384

Handler AM, McCombs SD, Fraser MJ, Saul SH (1998) The lepidopteran transposon vector, *piggyBac*, mediates germline transformation in the Mediterranean fruitfly. Proc Natl Acad Sci USA 95:7520–7525

Hartl DL, Lohe AR, Lozovskaya ER (1997) Modern thoughts on an ancyent marinere: function, evolution, regulation. Annu Rev Genet 31:337–358

Haymer DS, Marsh JL (1986) Germ line and somatic instability of a *white* mutation in *Drosophila mauritiana* due to a transposable element. Dev Genet 6:281–291

Hazelrigg T, Levis R, Rubin GM (1984) Transformation of *white* locus DNA in *Drosophila*: dosage compensation, zeste interaction, and position effects. Cell 64:1083–1092

Jasinskiene N, Coates CJ, Benedict MQ, Cornel AJ, Rafferty CS, James AA, Collins FH (1998) Stable, transposon mediated transformation of the yellow fever mosquito, *Aedes aegypti*, using the *Hermes* element from the housefly. Proc Natl Acad Sci USA 95:3743–3747

Jasinskiene N, Coates CJ, James AA (2000) Structure of *Hermes* integrations in the germline of the yellow fever mosquito, *Aedes aegypti*. Insect Mol Biol 9:11–18

Kamdar P, Von Allmen G, Finnerty V (1992) Transient expression of DNA in *Drosophila* via electroporation. Nucl Acids Res 11:3526

Karess RE (1985) *P* element mediated germ line transformation of Drosophila. In Glover DM (ed) DNA Cloning Vol. II: A Practical Approach. IRL Press, Oxford, pp 121–141

Karess RE, Rubin GR (1984) Analysis of *P* transposable element functions in *Drosophila*. Cell 38:135–146

Kaufman PK, Doll RF, Rio DC (1989) *Drosophila P* element transposase recognizes internal *P* element DNA sequences. Cell 59:359–371

Kidwell MG, Kidwell JF, Sved JA (1977) Hybrid dysgenesis in *Drosophila melanogaster*: a syndrome of aberrant traits including mutation, sterility, and male recombination. Genetics 86:813–833

Klein TM, Wolf ED, Wu R, Sanford JC (1987) High-velocity microprojectiles for delivering nucleic acids into living cells. Nature 327:70–73

Klemenz R, Weber U, Gehring WJ (1987) The white gene as a marker in a new *P*-element vector for gene transfer in *Drosophila*. Nucl Acids Res 15:3947–3959

Lansman RA, Stacey SN, Grigliatti TA, Brock HW (1985) Sequences homologous to the P mobile element of *Drosophila melanogaster* are widely distributed in the subgenus *Sophophora*. Nature 318:561–563

Leopold RA, Hughes KJ, DeVault JD (1996) Using electroporation and a slot cuvette to deliver plasmid DNA to insect embryos. Genet Anal 12:197–200

Leopold RA, Rojas RR, Atkinson PW (1998) Post pupariation cold storage of three species of flies: increasing chilling tolerance by acclimation and recurrent recovery periods. Cryobiology 36:213–224

Limbourg B, Zalokar M (1973) Permeabilization of *Drosophila* eggs. Dev Biol 35:382–387

Lis JT, Simon JA, Sutton CA (1983) New heat shock puffs and β-galactosidase activity resulting from transformation of *Drosophila* with an *hsp70-LacZ* hybrid gene. Cell 35:403–410

Loukeris TG, Arca B, Livadras I, Dialektaki G, Savakis C (1995a) Introduction of the transposable element *Minos* into the germ line of *Drosophila melanogaster*. Proc Natl Acad Sci USA 92:9485–9489

Loukeris TG, Livadaras I, Arca B, Zabalou S, Savakis C (1995b) Gene transfer into the medfly, *Ceratitis capitata*, with a *Drosophila hydei* transposable element. Science 270:2002–2005

Lozovskaya ER, Nurminsky DI, Hartl DL, Sullivan DT (1995) Germline transformation of *Drosophila virilis* mediated by the transposable element *hobo*. Genetics 142:173–177

Martin P, Martin A, Osmani A, Sofer W (1986) A transient expression assay for tissue-specific gene expression of alcohol dehydrogenase in *Drosophila*. Dev Biol 117:574–580

McGinnis W, Shermoen AW, Beckendorf SK (1983) A transposable element inserted just 5′ to a *Drosophila* glue protein gene alters gene expression and chromatin structure. Cell 34:75–84

McGrane V, Carlson JO, Miller BR, Beaty BJ (1988) Microinjection of DNA into *Aedes triseriatus* ova and detection of integration. Am J Trop Med Hyg 39:502–510

McInnis DO, Haymer DS, Tam SYT, Thanaphum S (1990) *Ceratitis capitata* (Diptera: Tephritidae): transient expression of a heterologous gene for resistance to the antibiotic geneticin. Ann Entomol Soc Am 83:982–986

Medhora MM, MacPeek AH, Hartl DL (1988) Excision of the *Drosophila* transposable element *mariner*: identification and characterization of the *Mos* factor. EMBO J 7:2185–2189

Mialhe E, Miller LH (1994) Biolistic techniques for transfection of mosquito embryos (*Anopheles gambiae*). BioTechniques 16:924–931

Miller LH, Sakai RK, Romans P, Gwadz RW, Kantoff P, Coon HG (1987) Stable integration and expression of a bacterial gene in the mosquito *Anopheles gambiae*. Science 237:779–781

Morris AC (1997) Microinjection of mosquito embryos. In Crampton, JM, Beard, CB, and Louis, C (Eds), Molecular Biology of Insect Disease Vectors: A Methods Manual. Chapman & Hall, London

Morris AC, Eggelston P, Crampton JM (1989) Genetic transformation of the mosquito *Aedes aegypti* by microinjection of DNA. Med Vet Entomol 3:1–7

Morris AC, Pott GB, Chen J, James AA (1995) Transient expression of a promoter-reporter construct in differentiated adult salivary glands and embryos of the mosquito *Aedes aegypti*. Am J Trop Med Hyg 52:456–460

Mullins MC, Rio DC, Rubin GM (1989) *Cis*-acting DNA sequence requirements for *P*-element transposition. Cell 3:729–738

Nawa S, Yamada S (1968) Hereditary change in *Ephestia* after treatment with DNA. Genetics 58:573–584

Nawa S, Sakaguchi B, Yamada MA, Tsujita M (1971) Hereditary change in *Bombyx* after treatment with DNA. Genetics 67:221–234

Nicolau C, LePape A, Soriano P, Fargette F, Juhel M (1983) *In vivo* expression of rat insulin after intravenous administration of the liposome-entrapped gene for rat insulin I. Proc Natl Acad Sci USA 80:1068–1072

O'Brochta DA, Atkinson PW (1996) Transposable elements and gene transformation in non-drosophilids. Insect Biochem Mol Biol 26:739–753

O'Brochta DA, Handler AM (1988) Mobility of *P* elements in drosophilids and nondrosophilids. Proc Natl Acad Sci USA 85:6052–6056

O'Brochta DA, Gomez SP, Handler AM (1991) *P* element excision in *Drosophila melanogaster* and related drosophilids. Mol Gen Genet 225:387–394

O'Brochta DA, Warren WD, Saville KJ, Atkinson PW (1994) Interplasmid transposition of *Drosophila hobo* elements in non-drosophilid insects. Mol Gen Genet 244:9–14

O'Brochta DA, Warren WD, Saville KJ, Atkinson PW (1995) *Hermes*, a functional non-drosophilid insect gene vector. Genetics 142:907–914

Ochman H, Gerber AS, Hartl DL (1988) Genetic implications of an inverse polymerase chain reaction. Genetics 120:621–623

O'Hare K, Rubin GM (1983) Structures of *P* transposable elements and their sites of insertion and excision in the *Drosophila melanogaster* genome. Cell 34:25–35

Pinkerton AC, Michel K, O'Brochta DA, Atkinson PW (2000) Green fluorescent protein as a genetic marker in transgenic *Aedes aegypti*. Insect Mol Biol 9:1–10

Pirrotta V (1988) Vectors for *P*-mediated transformation in Drosophila. In Rodriguez, RL and Denhardt, DT (eds) Vectors — A Survey of Molecular Cloning Vectors and Their Uses. Butterworths, Boston, pp 437–456

Pirrotta V, Steller H, Bozzetti MP (1985) Multiple upstream regulatory elements control the expression of the *Drosophila white* gene. EMBO J 4:3501–3508

Prasher DC, Eckenrode VK, Ward WW, Prendergast FG, Cormier MJ (1992) Primary structure of the *Aequorea victoria* green fluorescent protein. Gene 111:229–233

Presnail JK, Hoy MA (1992) Stable genetic transformation of a beneficial arthropod, *Metaseiulus occidentalis* (Acari: Phytoseiidae), by a microinjection technique. Proc Natl Acad Sci USA 89:7732–7736

Rio DC, Laski FA, Rubin GM (1986) Identification and immunochemical analysis of biologically active *Drosophila P* element transposase. Cell 44:21–32

Robertson HM, Lampe DJ (1995) Distribution of transposable elements in arthropods. Annu Rev Entomol 40:333–357

Rubin GM, Spradling AC (1982) Genetic transformation of *Drosophila* with transposable element vectors. Science 218:348–353

Rubin GM, Spradling AC (1983) Vectors for *P*-element-mediated gene transfer in *Drosophila*. Nucl Acid Res 11:6341–6351

Rubin GM, Kidwell MG, Bingham PM (1982) The molecular basis of P-M hybrid dysgenesis: the nature of induced mutations. Cell 29:987–994

Sambrook J, Fritsh EF, Maniatis T (1989) Molecular Cloning: A Laboratory Manual. Cold Spring Harbor Laboratory Press, Cold Springs Harbor, NY

Sarkar A, Coates CJ, Whyard S, Willhoeft U, Atkinson PW, O'Brochta DA (1997) The *Hermes* element from *Musca domestica* can transpose when introduced into divergent insect hosts. Genetica 99:15–29

Scavarda NJ, Hartl DL (1984) Interspecific DNA transformation in *Drosophila*. Proc Natl Acad Sci USA 81:7615–7619

Scholnick SB, Morgan BA, Hirsh J (1983) The cloned *Dopa* decarboxylase gene is developmentally regulated when reintegrated into the *Drosophila* genome. Cell 34:37–45

Searles LL, Jokerst RS, Bingham PM, Voelker RA, Greenleaf AL (1982) Molecular cloning of sequences from a *Drosophila* polymerase II locus by *P* element transposon tagging. Cell 31:585–592

Sentry JW, Kaiser K (1992) *P* element transposition and targeted manipulation of the *Drosophila* genome. Trends Genet 8:329–331

Sherman A, Dawson A, Mather C, Gilhooley H, Li Y, Mitchell R, Finnegan D, Sang H (1998) Transposition of the *Drosophila* element *mariner* into the chicken germ line. Nat Biotechnol 16:1050–1053

Siegal ML, Hartl DL (1996) Transgene coplacement and high efficiency site-specific recombination with the Cre/loxP system in *Drosophila*. Genetics 144:715–726

Smith D, Wohlgemuth J, Calvi BR, Franklin I, Gelbart WM (1993) *Hobo* enhancer trapping mutagenesis in *Drosophila* reveals an insertion specificity different from *P* elements. Genetics 135:1063–1076

Spradling AC (1986) *P* element-mediated transformation. In Roberts, DB (Ed), *Drosophila*: A Practical Approach. IRL Press, Oxford, pp 175–197

Steller H, Pirrotta V (1985). A transposable P vector that confers selectable G418 resistance to *Drosophila* larvae. EMBO J 4:167–171

Steller H, Pirrotta V (1986) P transposons controlled by the heat shock promoter. Mol Cell Biol 6:1640–1649

Sundararajan P, Atkinson PW, O'Brochta DA (1999) Transposable element interactions in insects: crossmobilization of *hobo* and *Hermes*. Insect Mol Biol 8:359–368

Tamura T, Thibert T, Royer C, Kanda T, Eappen A, Kamba M, Kômoto N, Thomas J-L, Mauchamp B, Chavancy G, Shirk P, Fraser M, Prudhomme J-C, Couble P (2000) A *piggybac* element-derived vector efficiently promotes germ-line transformation in the silkworm *Bombyx mori* L. Nat Biotechnol 18:81–84

Tyshenko MG, Doucet D, Davies PL, Walker VK (1997) The antifreeze potential of the spruce budworm thermal hysteresis protein. Nat Biotech 15:887–890

Walker VK (1990) Gene transfer in insects. In Maramorosch, K (Ed), Advances in Cell Culture, Vol. 7, Academic Press, New York pp 87–124.

Warren WD, Atkinson PW, O'Brochta DA (1994) The *Hermes* transposable element from the house fly, *Musca domestica*, is a short inverted repeat-type element of the *hobo*, *Ac*, and *Tam3* (*h*AT) element family. Genet Res Camb 64:87–97

Wilson C, Pearson RK, Bellen HJ, O'Kane CJ, Grossniklaus U, Gehring WJ (1989) P-element-mediated enhancer detection: an efficient method for isolating and characterizing developmentally regulated genes in *Drosophila*. Genes Dev 3:1301–1313

Zwiebel LJ, Saccone G, Zacharapoulou A, Besansky NJ, Favia G, Collins FH, Louis C, Kafatos FC (1995) The *white* gene of *Ceratitis capitata*: a phenotypic marker for germline transformation. Science 270:2005–2008

Section II

Gene Targeting

2 Targeted Transformation of the Insect Genome

Paul Eggleston and Yuguang Zhao

CONTENTS

2.1 Introduction 29
2.2 Homologous Recombination...30
    2.2.1 Biological Significance of Homologous Recombination 30
    2.2.2 Extrachromosomal Homologous Recombination..30
    2.2.3 Case Study: Extrachromosomal Homologous Recombination in Mosquito Cells....31
    2.2.4 Genome Manipulation through Homologous Recombination 34
        2.2.4.1 Targeting Vector Design..35
        2.2.4.2 Replacement and Insertion Vectors for Gene Targeting.................36
        2.2.4.3 Enrichment for Targeted Integration..38
    2.2.5 Case Study: Negative Selection with HSV-*tk* in Mosquito Cells 38
    2.2.6 Case Study: Targeted Gene Knockout in the Mosquito Genome 42
    2.2.7 Prospects for *in Vivo* Gene Targeting in the Insect Genome 43
2.3. Double-Strand Gap Repair..44
    2.3.1 Transposon-Mediated Gap Repair in *Drosophila* 45
    2.3.2 Transposon-Mediated Gap Repair in Other Insects 46
References 48

2.1 INTRODUCTION

The genetic manipulation of insect genomes may herald novel strategies for the control of insect-borne disease and could provide the means both to limit economic damage by crop pests and to increase productivity in commercially important insects. Such manipulation is now considered routine in the fruit fly, *Drosophila melanogaster*, and is based on the exploitation of transposable genetic elements such as *P*. Current attempts at the transformation of non-drosophilid insects have also focused on this approach, but phylogenetic restriction in mobility of the *P* element has necessitated a search for alternative functional transposons (Handler et al., 1993). As a result, the *Minos* element from *D. hydei* has since been shown to transpose into the genome of the medfly, *Ceratitis capitata* (Loukeris et al., 1995), the *hobo* element from *D. melanogaster* into *D. virilis* (Lozovskaya et al., 1996), the *mariner* element from *D. mauritiana* into *D. virilis* (Lohe and Hartl, 1996) and *Aedes aegypti* (Coates et al., 1998), the *Hermes* element from *Musca domestica* into *D. melanogaster* (O'Brochta et al., 1996), *A. aegypti* (Jasinskiene et al., 1998), and *Anopheles gambiae* cells (Zhao and Eggleston, 1998), and the *piggyBac* element from the lepidopteran, *Trichoplusia ni* into *C. capitata* (Handler et al., 1998).

    Apart from this focus on transposable elements, other approaches to the generation of transgenic insects include the use of viral vectors, such as Sindbis virus (Seabaugh et al., 1998) and *Aedes* densovirus (Afanasiev et al., 1994). Such vectors have proved effective at transducing genes into mosquitoes, but they are limited as to the size of transgene they can incorporate and they have so far proved unable to mediate stable germline transformation. Similarly, pantropic retroviruses have been used to mediate stable gene transfer and gene expression in somatic cells from a variety of insect species (Matsubara et al., 1996; Teysset et al., 1998).

Despite these recent successes, both transposon and viral-mediated strategies are constrained to some extent by the quasi-random nature of the integration sites. This can give rise to insertional inactivation of essential genes and all transgenes introduced in this way can suffer dramatically from position effects on expression (Wilson et al., 1990). For example, the transgene may not be expressed (or may be expressed suboptimally) if integration occurs in a transcriptionally inactive region of the genome.

As an alternative approach, we have been investigating the potential for targeted transformation of the insect genome. In any broad view of this strategy, it is possible to focus on three distinct approaches. The first of these is gene targeting in the conventional sense, which exploits the mechanism of homologous recombination between donor plasmid and genomic sequences. This will be the primary consideration in this chapter and we will review our experimental data using cultured mosquito cells as a model system. The second strategy involves the integration of donor sequences by double-strand gap repair following the excision of transposable genetic elements or cleavage at a rare or introduced endonuclease recognition sequence. Much progress has been made with this strategy in *D. melanogaster*, although it has yet to be extended to other insects. This chapter will summarize the approach and review the prospects for its wider application. Finally, there is the prospect of exploiting site-specific recombination systems such as FLP-*FRT* from yeast and *Cre/loxP* from bacteriophage P1. These provide a direct link between transposon-mediated transformation and site-specific integration since the initial receptor sites (*FRT* and *loxP*, respectively) must first be introduced as stable transgenes into the genome. Indeed, the integration of *FRT* sites into the *Drosophila* genome by *P*-element transformation and the subsequent integration of transgenes at these sites has already been demonstrated (Golic et al., 1997). The site-specific recombinase systems are reviewed elsewhere in this book and will not be covered in further detail here (see Chapter 3 by Rong and Golic).

2.2 HOMOLOGOUS RECOMBINATION

2.2.1 BIOLOGICAL SIGNIFICANCE OF HOMOLOGOUS RECOMBINATION

Recombination between homologous DNA sequences plays a central role in the life and evolution of organisms. During meiosis, homologous recombination is responsible for the reassortment of genetic information, and, during mitosis, homologous recombination (among other mechanisms) plays a role in the repair of DNA damage that occurs spontaneously or is induced by external agents (Kucherlapati and Smith, 1988; Low, 1988). It can produce genome rearrangements, including deletions and duplications, when it involves dispersed homologous sequences (Fitch et al., 1991; Onda et al., 1993) and can also be involved in contraction or expansion of tandem repeated sequences (Bishop and Smith, 1989). Homologous recombination is also involved in the generation of multigene families and the amplification of genes in response to various stimuli (Kafatos et al., 1985).

2.2.2 EXTRACHROMOSOMAL HOMOLOGOUS RECOMBINATION

Much useful information concerning the requirements and efficiency of homologous recombination can be gleaned from studies of recombination between DNA molecules introduced into cultured cells. Indeed, this is a typical approach to establishing that particular tissues have the necessary cellular machinery for homologous recombination. Extrachromosomal recombination can occur at high frequency, making it possible to obtain data quickly from many independent events, and it is relatively easy to study the effects of modifying the sequences undergoing recombination (Bollag et al., 1989). One particularly useful strategy for demonstrating and quantifying homologous recombination is the regeneration of selectable marker genes from inactive substrates. In such experiments two plasmids, each carrying the same selectable marker (e.g., neomycin resistance) but deficient at unique sites, are transfected into cultured cells (Small and Scangos, 1983; Song et al., 1985; Baur et al., 1990). Cells that survive selection must have regenerated a functional copy of the marker gene through homologous recombination. Such experiments have been used in mammalian cells to optimize many of the parameters that affect gene targeting frequencies. For example, recombination frequency has been found to increase with length of homology and isogenicity of homologous sequence (Rubnitz and Subramani, 1984; Ayares et al., 1986; Waldman and Liskay, 1988). Poor sequence homology appears to disrupt initiation rather than propagation of the recombination events, and recombination efficiency seems to be determined by the maximum length of perfect homology rather than the overall percentage homology (Waldman and Liskay, 1988). The introduction of double-strand breaks into the recombining molecules has also been found to stimulate recombination efficiency. Such breaks are only effective if they occur within, or near, the region of homology and it is thought that they work primarily because the broken strands are able to act as recipients in nonreciprocal exchanges. These effects are cumulative such that a double-strand break in one substrate might yield a 10-fold improvement in efficiency, whereas double-strand breaks in both substrates might increase efficiency by 100-fold (Bollag et al., 1989). One difficulty in carrying out such experiments with conventional selectable markers (e.g., neomycin resistance) is that the transfected DNA must remain in the cells for sufficient time to allow nonresistant cells to be killed and resistant cell clones to become established. Recently, variations of this technique have been developed that involve studies of reporter gene activity. For example, restoration of luciferase reporter gene activity from truncated substrates can be used to monitor the efficiency of extrachromosomal homologous recombination. Measurements of reporter gene activity are not only rapid and convenient, but also give more precise quantitative data that help to reveal small differences in recombination efficiency.

2.2.3 CASE STUDY: EXTRACHROMOSOMAL HOMOLOGOUS RECOMBINATION IN MOSQUITO CELLS

We have used the restoration of luciferase activity from a pair of truncated, but overlapping, extrachromosomal luciferase substrates to study homologous recombination in cultured mosquito cells (Y. Zhao and P. Eggleston, unpublished). The recombination substrates carried the firefly luciferase reporter gene driven by the *actin 5C* promoter from *D. melanogaster*. One substrate (DL) was designed with a deletion at the 5′ or left-hand end of the luciferase coding sequence. The second (DR) carried a nonoverlapping deletion at the 3′ or right-hand end of the luciferase coding sequence. Thus, the DL and DR substrates, although individually defective, shared a region of homology within the luciferase gene that provided the opportunity for restoration of luciferase activity through homologous recombination (Figure 2.1).

Various combinations of linear and circular substrates were cotransfected into a cell line derived from the malaria vector mosquito, *Anopheles gambiae*. After a period of time to allow for extrachromosomal recombination, the luciferase activity generated by the cells was determined. In all cases, this involved standard luciferase assays alongside both positive (intact luciferase gene) and negative controls. By controlling both substrate topology (circular vs. linear) and the site of linearization, it was envisaged that the substrate structures would help to determine the nature of the recombinational events taking place. As expected, high luciferase activity was seen in cells transfected with the intact luciferase gene, but transfection with the deletion constructs alone gave no detectable luciferase activity above background. However, all cotransfections involving combinations of the two deletion constructs gave rise to significant recovery of luciferase activity. This was true even when both deletion constructs were present as circular molecules, which are generally regarded as poor substrates for homologous recombination. Much higher luciferase activities were anticipated following

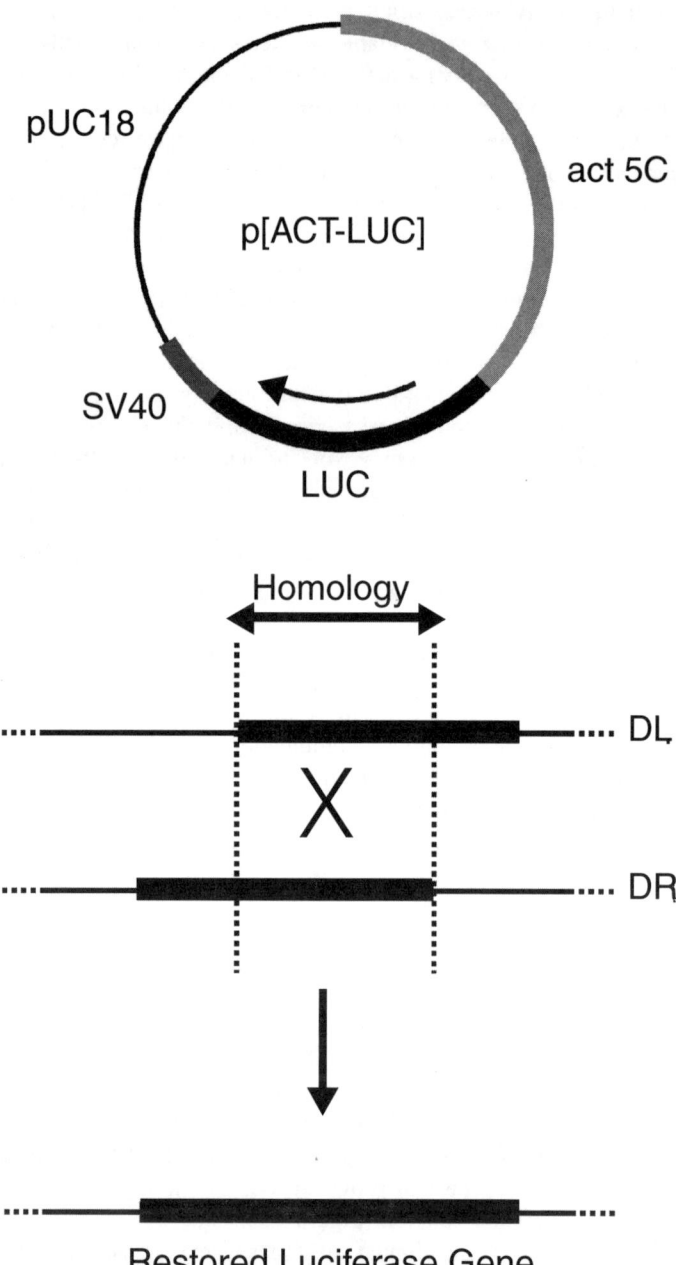

FIGURE 2.1 Extrachromosomal homologous recombination in mosquito cells. The upper panel shows the parental plasmid p[ACT-LUC] from which the deletion substrates were created. This carries the firefly luciferase coding sequence (black) driven by the *actin 5C* promoter from *Drosophila* (light gray) with transcription terminated by the SV40 polyadenylation sequence (mid-gray). The central panel shows the deletion substrates transfected into the mosquito cells. DL carries a deletion at the left-hand or 5' end of the luciferase coding sequence. DR carries a nonoverlapping deletion at the right-hand or 3' end of the luciferase coding sequence. Both DL and DR are individually inactive, but they share a region of homology within the luciferase gene that provides the opportunity for restoration of luciferase activity through homologous recombination. The lower panel shows the restored luciferase coding sequence resulting from either reciprocal exchange or gene conversion.

the introduction of double-strand breaks into the luciferase substrates. To this end, DL and DR were linearized by restriction digests at different locations and cotransfected (either both in linear form or in various combinations with circular substrates) into the mosquito cells. Such topological modifications to the substrates can be used to investigate the mechanism(s) through which homologous recombination occurs in mosquito cells. Some of these are believed to involve reciprocal exchange of material, others the one-way transfer of sequences through gene conversion.

For example, linearization of both substrates can be used to test the single-strand-annealing model (Figure 2.2) proposed by Lin et al. (1984) in which DNA ends at a double-strand break are rendered single-stranded by a strand-specific exonuclease or as a result of unwinding (Wake et al., 1985). When homologous sequences are present, this process ends with complementary single strands that are capable of reannealing. Repair synthesis and ligation then complete the formation of the nonreciprocal homologous junction. In the case of the substrates employed here, linearization near to the region of homology would favor the single-strand-annealing model. In fact, this modification did provide the greatest restoration of luciferase activity with levels around 70-fold higher than for cotransfections with the circular substrates. This result was consistent with previous research suggesting that the single-strand-annealing model was the most efficient pathway for extrachromosomal homologous recombination in mammalian cells (Lin et al., 1984) and plant cells (De-Groot et al., 1992; Bilang et al., 1992; Puchta et al., 1994).

Restoration of luciferase activity was less pronounced when double-strand breaks were introduced outside the region of homology. In fact, luciferase activities recovered from such substrates were around fivefold less than those from circular substrates, despite the fact that linear molecules are considered a more efficient substrate for homologous recombination. Similar results have been documented for mammalian cells where double-strand breaks near the deletion sites enhanced the recombination frequency, whereas breaks outside the region of homology had little effect (Song et al., 1985).

Although the single-strand-annealing model describes an efficient recombinant pathway, in the context of gene targeting it is not generally feasible to introduce double-strand breaks into the target sequence. That is to say, the target molecules are, in most cases, unbroken strands of DNA. Perhaps the best model of such a situation, with respect to the parameters affecting gene targeting, is to use a circular plasmid to mimic the chromosomal target site and a linear molecule as an analogue of the targeting vector (Segal and Carroll, 1995). Our data in mosquito cells suggest that restoration of luciferase activity following cotransfection with one linear and one circular substrate is effective but not as efficient as that seen when both substrates are linearized near the region of homology. Once again, this would appear to be in broad agreement with similar studies in other organisms.

Taken as a whole, these experiments suggest that homologous recombination may be an effective mechanism in mosquito cells and encourage the further exploration of gene targeting strategies for the generation of transgenic mosquitoes. The data indicate that linear targeting vectors are better substrates for homologous recombination than circular ones and that the introduction of double-strand breaks into or near the region of homology can stimulate homologous recombination frequencies. Predictably, it would appear that a number of different recombination pathways might be involved and there is evidence that particular pathways are favored by specific substrate topologies. For example, where two linear substrates are involved, there is the potential for nonhomologous end-joining of the two plasmids. This phenomenon has been highlighted recently (Hagmann et al., 1998), where it was described as the predominant mechanism in zygotes and early embryos of the zebrafish (*Danio rerio*) and *Drosophila melanogaster*. In the context of the experiments described here, end-joining (unlike homologous recombination) would not restore a functional luciferase gene and would therefore be overlooked by the luciferase assay. However, the larger products resulting from end-joining of linearized substrates would be detectable by Southern blot. Indeed, such products were seen in all cases where cotransfections involved two linearized substrates but not where one or both substrates were in circular form. Thus, end-joining clearly takes place between linearized substrates in mosquito cells. However, in the context of targeted genome

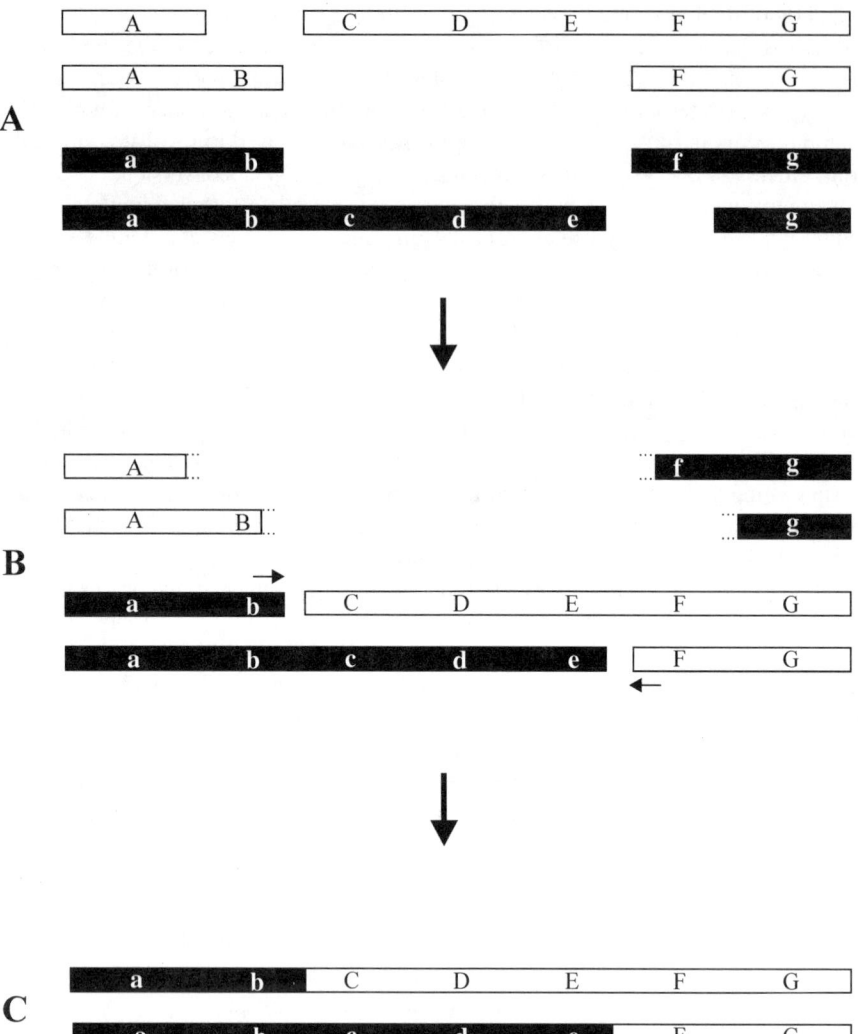

FIGURE 2.2 The single-strand-annealing model of extrachromosomal homologous recombination (Lin et al., 1984). White and black bars represent homologous double-stranded DNA molecules (A-G and a-g, respectively). Panel A: Double-strand breaks in each molecule serve as entry points for single-strand exonucleases or for unwinding, resulting in overhanging complementary single strands. Panel B: Recombination is promoted by complementary base pairing of single-stranded DNA within the region of homology and the ends of each molecule not involved in the exchange are degraded (dotted lines). Panel C: Resolution of the recombination event by repair synthesis and ligation, using the corresponding ends of the double-stranded DNA as primers (shown as arrows in Panel B). The process is nonconservative with one molecule being restored and the other degraded. However, in half of such events the other two DNA ends would be involved. Note that the flanking sequences are always exchanged in this model.

manipulation, it should be noted that end-joining reactions can only arise if a double-strand break is first introduced into the target genome and this is considered in Section 2.3.

2.2.4 Genome Manipulation through Homologous Recombination

Gene targeting is the term used to describe precise genome manipulation mediated by homologous recombination. It involves recombination between a "donor" sequence (typically a plasmid construct) and the corresponding homologous region of the "recipient" genome. A number of alternative recombination pathways might be involved in determining the outcome of such exchanges. For example, such events may involve true reciprocal exchange (crossover) between the double-stranded DNA molecules of the donor and target (Figure 2.3A). This will lead to the genome accepting part of the targeting vector plasmid which, in turn, will pick up the genomic sequence involved in the exchange. However, not all homology-driven recombination involves such reciprocal exchange. Many events arise through gene conversion, which is the nonreciprocal transfer of sequence information from donor to target molecules. Gene conversion can occur either independently (Figure 2.3B) or in association with a crossover (Figure 2.3C) and these two possibilities appear to be equally likely (Hooper, 1992).

In principle, gene targeting can be used to direct specific modifications to any genomic region that has previously been cloned. It has the potential to facilitate the precise introduction of transgenes into predetermined chromosomal sites of demonstrated transcriptional activity, with copy numbers that are determined by the available number of targeting sites in the genome. It also offers the potential to engineer precise modifications in a target gene, either to correct defects or to inactivate gene activity through precise disruptions (knockouts) of coding sequences or promoters. Gene-targeted modification of both alleles within a recessive homozygote, leading to an alteration of phenotype, has also been observed (Metsaeranta and Vuorio, 1992). Moreover, with appropriate construction of the gene-targeting vector, it is possible to introduce specific mutations into a target gene and study the resulting phenotype (Rubinstein et al., 1993) or to revert mutant to wild-type alleles (Shesely et al., 1991).

Gene targeting has been exploited in yeasts (Hinnen et al., 1978), mammalian cells (Capecchi, 1989a, b; Rossant and Nagy, 1995; Templeton et al., 1997), protozoans (Cruz and Beverly, 1990; Blundell et al., 1996; Wu et al., 1996; Papadopoulou and Dumas, 1997), slime mould (Morrison et al., 1997), plant cells (Puchta et al., 1994), and intact plants such as the moss, *Physcomitrella patens* (Schaefer and Zryd, 1997) and *Arabidopsis thaliana* (Kempin et al., 1997). Although little information is available for insects, the investigation and optimization of targeting strategies has been greatly facilitated by using cultured somatic cells as a model system. Through such approaches, the machinery of homologous recombination has been demonstrated in both mosquitoes (Baldridge and Fallon, 1996) and *Drosophila* (Cherbas and Cherbas, 1997). Gene targeting therefore provides a valuable tool for genome manipulation that offers several advantages over alternative transformation strategies. However, as with most valuable tools there are significant problems to be overcome. These include the need for detailed sequence information at the proposed targeting site and the relatively low frequencies of homologous recombination encountered in higher eukaryotic systems.

2.2.4.1 Targeting Vector Design

The most significant progress in gene targeting has involved the use of mouse embryonic stem (ES) cells and these studies have revealed the importance of vector design (Galli-Taliadoros et al., 1995). In particular, factors such as overall length of homology, isogenicity between donor and target sequences, and vector topology play an important role. On the other hand, increasing the copy number of the targeting vector or the target site does not appear to improve targeting efficiency (Zheng and Wilson, 1990). Thus, it would appear that the initial juxtaposition of the homologous sequences in the vector and target is not rate limiting, and, indeed, lower copy numbers of the targeting vector may help to avoid high frequencies of random integration. In any attempt at gene targeting, the design of the targeting vector itself is of paramount importance. The vector must carry a region of sequence homologous to that of a previously cloned section of genomic DNA. This region of homology thus fixes the site of genomic modification and, in simple terms, the resulting efficiency of targeted integration will increase with length of homology and with the degree of genetic identity between the donor and target sequences (isogenicity). By altering the basic structure of the targeting vector, it is possible to achieve either the replacement of a section

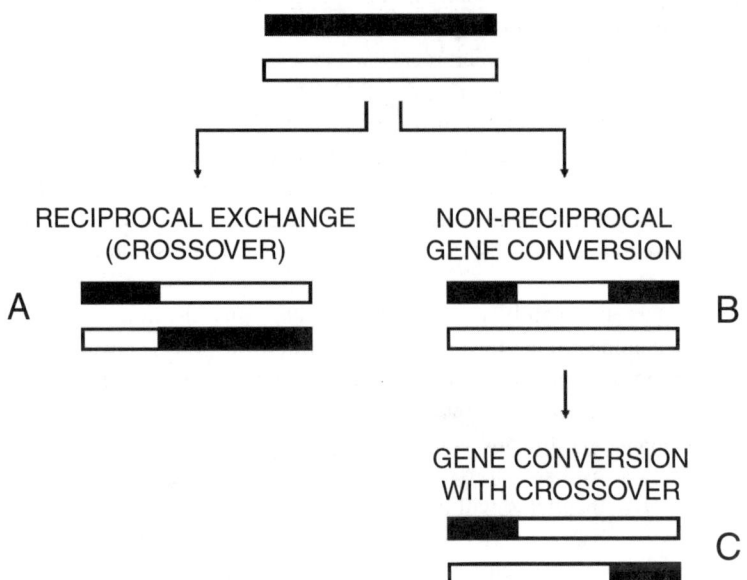

FIGURE 2.3 Potential homologous recombination pathways between two double-stranded DNA molecules, represented as either black or white bars. Panel A shows the reciprocal exchange between double-stranded DNA molecules that occurs during a conventional crossover. Panel B shows a nonreciprocal gene conversion process during which information is transferred from one double-stranded molecule to the other. Panel C shows a nonreciprocal gene conversion event in association with a crossover.

of chromosome with homologous sequence or to insert additional sequences at the site of homology. With both types of vector the overall frequency of homologous recombination is likely to be low, and therefore appropriate methods for the selection and isolation of targeted integration events must be employed. Moreover, these selection methods must be capable of distinguishing between homologous and nonhomologous integration events, particularly since the latter appear to be more common in higher eukaryotic organisms.

2.2.4.2 Replacement and Insertion Vectors for Gene Targeting

In a replacement vector (Figure 2.4A), the region of homology is interrupted by nonhomologous sequence, usually a selectable marker or reporter gene that will subsequently allow identification of the targeted integration event. Typically, the vector is linearized within the region of homology, and, following juxtaposition of the homologous sequences, recombination will occur between the donor and target. This results in replacement of part of the genomic sequence by the corresponding section of the targeting vector, taking with it the intervening nonhomologous sequence. If the nonhomologous sequence disrupts an essential exon in the target gene, then the result will be gene inactivation (knockout). However, with appropriate design more subtle modifications can be introduced, including the correction of mutations. The simplest interpretation of replacement vector function implies that two separate recombination events are required on either side of the nonhomologous sequence. This is not, however, an absolute prerequisite and it has been suggested that one recombination event may suffice (Ellis and Bernstein, 1989). In this model a single-strand exchange at one of the flanking regions of homology would be followed by branch migration through the nonhomologous sequence and resolution at the second region of homology.

In an insertion vector (Figure 2.4B) the nonhomologous sequence (e.g., marker gene cassette) is organized so that it flanks the region of homology, rather than disrupting it. The vector is linearized within the region of homology and a single recombination event, incorporating both free ends, results

Targeted Transformation of the Insect Genome

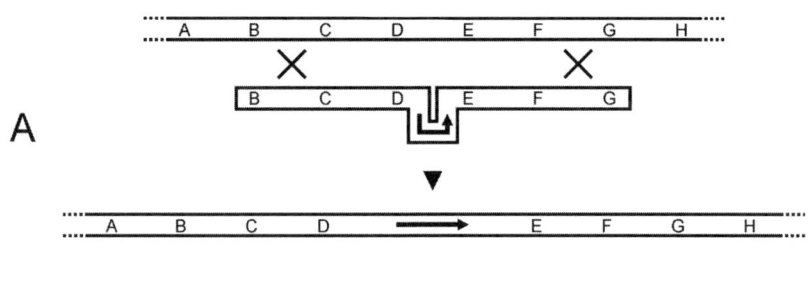

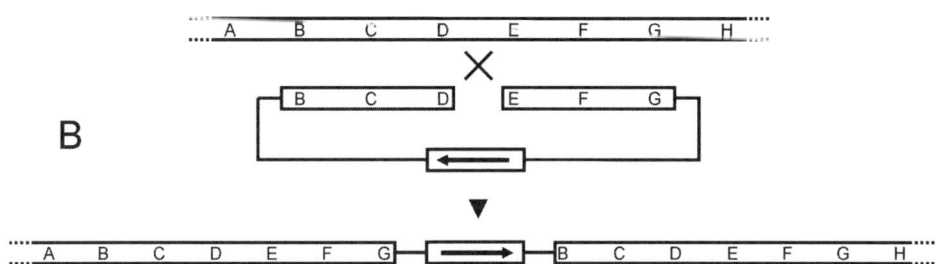

FIGURE 2.4 The design of replacement and insertion vectors for the gene-targeted modification of eukaryotic genomes. Genomic sequences at the targeting site (denoted by the letters A to H) are represented as bars flanked by dotted lines. In the replacement strategy (Panel A) the targeting vector carries a region of homology to the targeting site (B to G) that is disrupted by an intervening nonhomologous sequence, typically a selectable marker cassette (black arrow). Homologous recombination between vector and target sites (indicated by crosses) leads to replacement of part of the genomic sequence (located between the crossover sites) by the corresponding section of the targeting vector. The end result is the integration of the nonhomologous sequence into the targeting site. In the insertion strategy (Panel B), the targeting vector also carries a region of genomic homology (B to G), but in this case the nonhomologous sequence (black arrow) is organized so that it flanks, rather than disrupts, the region of homology. The vector is linearized within the region of homology, and a single recombination event (indicated by a cross) leads to integration of the entire targeting vector including both homologous and nonhomologous sequences. The end result is a duplication of part of the genomic sequence (B to G) as well as transgene integration into the targeting site.

in the insertion of the entire vector into the targeted sequence. In some instances, the insertion of multiple copies of the targeting vector can be seen (Hasty et al., 1991). A direct consequence of this mechanism is the duplication of homologous sequence at the target site. Depending on the vector design, this could result in gene knockout either because transcription is terminated at the polyadenylation signal of the marker gene or because read-through transcription yields a spliced product with duplicated exons. Although unlikely, it is possible for a read-through transcript to generate a spliced product excluding the duplicated exons. The product would be indistinguishable from that of the wild-type gene (Moens et al., 1992), and this may be important if absolute gene knockout must be achieved.

The relative efficiency of each type of vector has not been clearly determined and work in mammalian cells provides conflicting views. Thomas and Capecchi (1987) reported that both insertion and replacement vectors (carrying the same homologous sequence) yielded comparable targeting efficiencies in mouse ES cells. On the other hand, Hasty et al. (1991) found that insertion vectors were much more efficient and that replacement vectors commonly gave rise to integrations of recircularized or concatenated vector. It therefore seems likely that the specific outcome may be influenced by differences in the target genome or homologous sequence, or indeed by vector design.

2.2.4.3 Enrichment for Targeted Integration

In higher eukaryotic systems, it appears that overall homologous recombination frequencies are relatively low and that nonhomologous recombination predominates. This is in contrast to yeast, for example, where homologous recombination appears to be the preferential pathway (Orr-Weaver and Szostak, 1985). Thus, there is a need to enrich for targeted events in any transformation strategy mediated by homologous recombination and several mechanisms have been suggested. First, the targeting vector can incorporate a promoterless marker that is only expressed from an endogenous promoter in the context of targeted integration. Such "promoter-trap" strategies can incorporate conventional selectable markers, such as neomycin resistance, or reporter genes such as green fluorescent protein (GFP). The approach relies on the active expression of the targeted gene since the promoter of the targeted gene is subsequently "hijacked" by the incoming marker. Following precise (gene-targeted) integration, the marker becomes incorporated into the targeting site such that its expression is driven by the endogenous promoter. Targeted integrations can therefore be identified both by expression of the incoming marker and by the loss of activity (knockout) of the targeted gene. One potentially exciting development of the promoter-trap approach would be to employ large-scale automated isolation of targeted cells. For example, cells expressing GFP following the targeted integration of a promoterless construct could be enriched by fluorescence activated cell sorting (FACS).

A second strategy that has been used to great effect in mouse ES cells is known as positive–negative selection. Here, the targeting vector incorporates a positive selection marker to identify transformed cells irrespective of whether the integration is mediated by homologous or nonhomologous recombination. It also incorporates a negative selectable marker such that simultaneous selection against non-homologous recombination can be applied. Negative selection markers kill the cells in which they are expressed. They do this either because their expression *per se* is toxic, e.g., diphtheria toxin A (Yagi et al., 1993) or because the gene product catalyzes chemicals leading to toxic metabolites that ultimately kill the cells, e.g., Herpes simplex virus Type I thymidine-kinase — HSV-*tk* (Mansour et al., 1988). The negative selection marker is incorporated into the targeting vector in such a way that it is lost following a precise homologous recombination event (Figure 2.5). Cells transformed in this way will therefore lose the negative marker and survive selection whereas cells transformed by nonhomologous integration will die as a result of negative marker expression. In this way, positive–negative selection has typically generated enrichment factors of around 20-fold in mammalian cells (Mombaerts et al., 1991).

Finally, mechanisms that do not rely on the expression of a selectable marker or a reporter gene have also been described. One such strategy relies on PCR amplification between unique primers located in the targeting vector and the recipient genome (Wilmut et al., 1991). In the case of targeted integration the two primers are brought into close proximity in the genome allowing the possible exponential amplification of the intervening sequence. In the absence of integration, no PCR product will be amplified and, in the case of nonhomologous integration, linear amplification will yield a product that is not readily detectable by staining after gel electrophoresis. Wilmut et al. (1991) described how this technique could be used for clonal isolation of targeted cells. Potentially targeted cells are divided into pools and subjected to PCR. Positive pools are then divided into a number of low-density "sib" cultures such that only a few might be expected to carry targeted cells. These sib cultures are again subjected to PCR and the process repeated until targeted cells are sufficiently enriched to allow clonal isolation.

2.2.5 CASE STUDY: NEGATIVE SELECTION WITH HSV-*TK* IN MOSQUITO CELLS

We have tested the efficiency of negative selection mediated by HSV-*tk* in *Anopheles gambiae* (Mos55) cells. HSV-*tk* has previously been used extensively as a negative selection marker in mouse ES cells (Mansour et al., 1988) and, more recently, in experiments directed at human gene therapy

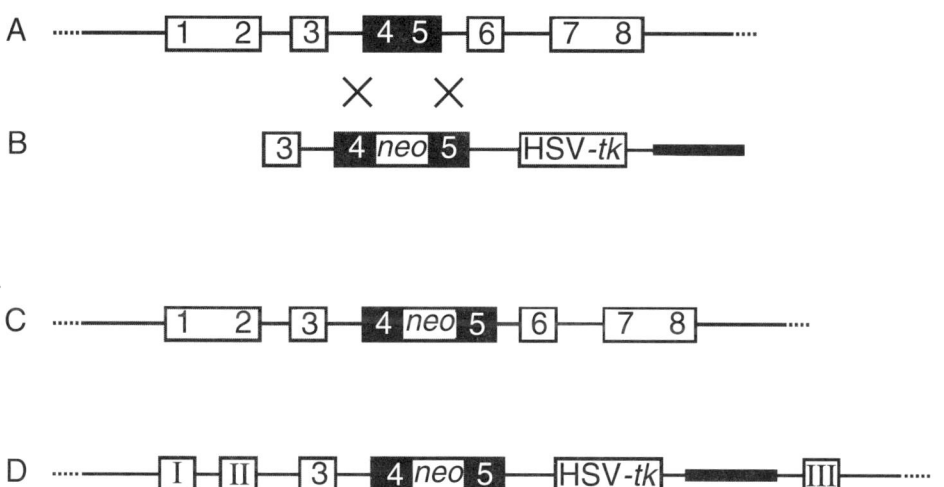

FIGURE 2.5 Positive–negative selection for the enrichment of gene-targeted integration events. Panel A represents genomic sequences (1 to 8) at the targeting site, flanked by dotted lines and organized into five exons (boxes) separated by introns (thin lines). The exon (4–5) targeted for modification is shown in black. Panel B shows the gene targeting replacement vector that carries a region of homology to the genome (3–5) and includes both a positive selection marker (*neo*) and a negative selection marker (HSV-*tk*). The vector is constructed in a suitable plasmid (thick line) and organized such that the targeted exon (4–5) is disrupted by the *neo* gene while the HSV-*tk* gene flanks the region of homology. Panel C shows the anticipated outcome following targeted (homologous) integration of replacement vector sequences. The genomic copy of the targeted exon (4–5) is replaced by the disrupted exon from the targeting vector (4–*neo*–5) and the flanking HSV-*tk* gene (which lies outside the region of homology) is lost. Such events are likely to involve inactivation of the targeted gene (knockout) and can be identified both by positive selection with neomycin and by the loss of the HSV-*tk* gene. Panel D shows the end result of nonhomologous integration where the entire targeting vector plasmid, including both positive and negative selection markers, is integrated into a novel genomic site, here represented by the sequence I–III flanked by dotted lines. Such events can also be identified by positive selection with neomycin but the presence of the HSV-*tk* gene means that cells transformed in this way will be sensitive to the cytotoxic effects of nucleoside analogues such as gancyclovir. Thus, negative selection can be used to remove cells expressing HSV-*tk* preferentially, thereby enriching for gene-targeted integration events that exclude the negative selection marker.

for cancer (Freeman et al., 1996). It has a relaxed substrate specificity that enables it to phosphorylate a variety of nucleoside analogues, such as gancyclovir (Boehme, 1984). Cellular *tk* genes, on the other hand, have more stringent substrate specificity and cannot phosphorylate nucleoside analogues. It is this difference in activity that is exploited in negative selection. Expression of HSV-*tk* phosphorylates the nucleoside analogues and these are converted by cellular kinases to nucleoside triphosphates that inhibit host cell DNA replication by chain termination (Furman et al., 1984; Reardon and Spector, 1989). The HSV-*tk* gene can be used as a complete selection cassette and does not need to be combined with a promoterless positive selection marker. As a result, there is greater flexibility in the choice of targeting site, which need not be transcriptionally active (Johnson et al., 1989).

The successful application of HSV-*tk* as a negative selection marker in insect tissue requires an understanding of the activity of the cellular *tk* gene. In lepidopteran insect cells, negative selection has been used to select recombinant baculoviruses that have lost HSV-*tk* as a result of homologous recombination (Godeau et al., 1992). In these experiments, the lepidopteran cellular *tk* gene appeared to behave in a similar way to the mammalian homologue although the *Sf*-9 insect cells needed much higher concentrations of nucleoside analogues to inhibit cell growth. There may also be other differences since certain nucleoside analogues, such as acyclovir or edoxudine, which are

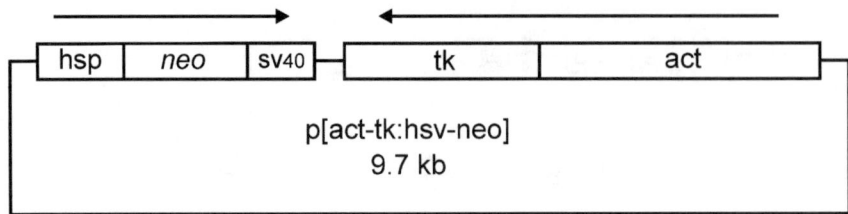

FIGURE 2.6 Positive–negative selection in mosquito cells. The plasmid construct p[act-tk:hsp-neo] carries a neomycin resistance gene (*neo*) as a positive selection marker and an HSV-*tk* gene as a negative selection marker (tk). The *neo* coding sequence is driven from the *hsp*70 *Drosophila* promoter with transcription terminated by the SV40 polyadenylation sequence. The HSV-*tk* coding sequence has its own polyadenylation sequence and is driven from the *actin 5C Drosophila* promoter. The *neo* and HSV-*tk* genes are transcribed in opposite orientations.

effective in mammalian cells, appear not to be so effective in *Aedes* mosquito cells (Mazzacano and Fallon, 1995).

We designed the construct p[act-tk:hsp-neo] to carry both positive and negative selection markers (Figure 2.6). The positive selection cassette comprised a neomycin resistance gene (*neo*) driven by the *hsp*70 promoter from *Drosophila* and the negative selection cassette comprised the HSV-*tk* gene driven by the *actin 5C* promoter from *Drosophila*. Both promoters were chosen because of their demonstrated effectiveness in mosquito cells (Zhao and Eggleston, 1999). Mos55 cells were transfected with p[act-tk:hsp-neo] and positively selected with the neomycin derivative G418 at a concentration of 800 μg/ml. This concentration was sufficient to kill untransformed cells within 2 weeks and, under these conditions, there was no evidence of colony formation from untransfected cells. G418-resistant colonies were, however, isolated from transfected cells and one clone (*neo-tk*-3) was chosen for detailed analysis.

To confirm that the G418-resistant cell clone *neo-tk*-3 contained both the *neo* gene and the HSV-*tk* gene Southern blot hybridizations were performed (Figure 2.7). High-molecular-weight genomic DNA was isolated from *neo-tk*-3 cells and from untransfected Mos55 cells and digested with various restriction enzymes. Following transfer to a membrane, the DNA was probed with the *neo* gene and washed at high stringency. No signals were detected in untransfected control DNA (Figure 2.7A; Lane C) indicating the absence of *neo* in the Mos55 genome. Signals were, however, detected in DNA from the transfected *neo-tk*-3 cells. Two pieces of evidence strongly suggested that the transfected DNA was integrated into the Mos55 genome, rather than existing as free plasmid or as unintegrated concatemers. First, the undigested DNA (Figure 2.7A; Lane U) revealed very high molecular weight signals that did not conform to the low-molecular-weight expectation for free plasmid. Second, the undigested DNA yielded signals of higher molecular weight than those derived from *Sal*I digestion (Figure 2.7A; Lane S). *Sal*I does not cut the original transformation vector and so, if the transfected DNA had existed as unintegrated concatemers, the signals in undigested and *Sal*I digested DNA would have been of the same size. *Eco*RI digestion of *neo-tk*-3 DNA (Figure 2.7A; Lane E) gave rise to a major band of ~9.7 kb, corresponding to linearized p[act-tk:hsp-neo]. Double digestion of *neo-tk*-3 DNA with *Xba*I and *Bam*HI (Figure 2.7A; Lane XB) revealed major bands of ~2.2 and ~5.8 kb corresponding to those predicted from digestion of circular p[act-tk:hsp-neo]. These data showed that the transformation vector plasmid was integrated into the *neo-tk*-3 genome in the form of tandem head-to-tail arrays. Such tandem arrays appear to be a common feature of nonhomologous integration in insect cells and similar results have been observed elsewhere (Rio and Rubin, 1985; Monroe et al., 1992; Lycett and Crampton, 1993; Shotkoski and Fallon, 1993; Vulsteke et al., 1993).

Probing the same blot with the HSV-*tk* gene supported this interpretation. No signals were detected in the DNA from untransfected control cells (Figure 2.7B; Lane C) indicating the absence of endogenous *tk* homology. The major signals derived from *Eco*RI (Figure 2.7B; Lane E) and *Sal*I

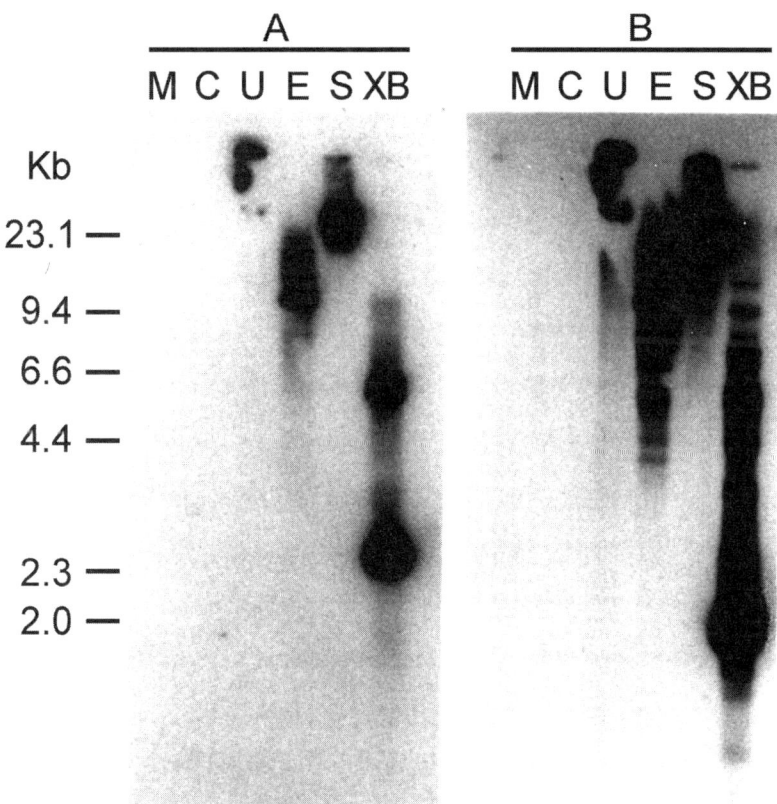

FIGURE 2.7 Southern analysis of p[act-tk:hsp-neo] transformed *Anopheles gambiae* cell clone *neo-tk*-3 DNA. High-molecular-weight genomic DNA was isolated from untransfected Mos55 (control) cells and digested to completion with *Eco*RI (Lane C). High-molecular-weight genomic DNA from transformed *neo-tk*-3 cells was either left uncut (Lane U) or digested to completion with *Eco*RI (Lane E); *Sal*I (Lane S) or *Xba*I and *Bam*HI (Lane XB). Following electrophoretic separation of the DNA (~5 μg per lane) alongside λ *Hin*dIII markers (Lane M) the reaction products were transferred to a nylon membrane and hybridized sequentially to a *neo* probe (Panel A) and an HSV-*tk* probe (Panel B), ensuring that the membrane was completely stripped between hybridizations. In both cases, the filter was washed in 0.1 × SSC; 0.1% SDS for 30 min at 55°C, although the autoradiograph in Panel A was exposed for 2 h and that in Panel B for 5 h at –70°C.

(Figure 2.7B; Lane S) digestion corresponded to those identified by the *neo* probe. Digestion with *Xba*I and *Bam*HI (Fig. 7B; Lane XB) identified a major signal at 1.7 kb corresponding to the HSV-*tk* gene in p[act-tk:hsp-neo].

Northern analysis was conducted to verify the transcription of the HSV-*tk* gene and to link this activity directly to cell sensitivity to gancyclovir. Total cellular RNA was isolated from *neo-tk*-3 cells and from untransfected Mos55 cells and hybridized with an HSV-*tk* specific probe. The resulting autoradiograph revealed a single band of around 1.7 kb specifically in RNA derived from *neo-tk*-3 cells (data not shown) and this corresponded to the anticipated size of the HSV-*tk* transcription unit. Together with the Southern blot data, these results suggested that the HSV-*tk* gene was not only stably integrated into the mosquito genome, but also actively transcribed.

The efficacy of negative selection was tested by treating *neo-tk*-3 cells with gancyclovir, which is the nucleoside analogue most commonly employed in mammalian cells. Preliminary experiments had shown that untransfected Mos55 cells were resistant to gancyclovir concentrations up to 2 m*M*. However, growth inhibition was evident in *neo-tk*-3 cells at gancyclovir concentrations in excess of 5 μ*M* and at a concentration of 100 μ*M* granular particles could be seen inside the cells, indicating cytotoxicity. Despite this growth inhibition, *neo-tk*-3 cell death was only apparent after extended periods of high negative selective pressure (1 m*M*). This is in contrast to mammalian cells, where around 2 μ*M* gancyclovir is usually sufficient to kill HSV-*tk* transfected cells (Mansour et al., 1988; Beck et al., 1995). However, much higher concentrations (100 μ*M*) were found necessary to kill HSV-*tk* transformed cells of the lepidopteran insect cell line, *Sf*-9 (Godeau et al., 1992). The relative insensitivity of *neo-tk*-3 cells to gancyclovir apparently does not result from poor expression of the HSV-*tk* gene, since the Southern and Northern data indicated that HSV-*tk* was both stably integrated and efficiently transcribed. It is more likely that individual cell lines simply vary in their inherent gancyclovir sensitivity. This is certainly true of mammalian cells (Beck et al., 1995) and there is even evidence that some mammalian cell lines carrying the HSV-*tk* gene can survive long periods of gancyclovir selection (Boviatsis et al., 1994). Taken together, these results indicate that the HSV-*tk* gene might serve as an efficient negative selection marker in insect tissue but that this might require high nucleoside analogue concentrations to achieve complete negative selection (i.e., cell death).

2.2.6 CASE STUDY: TARGETED GENE KNOCKOUT IN THE MOSQUITO GENOME

The design of gene-targeting vectors requires extensive data on genomic sequence at the proposed targeting site. Apart from a few loci, such data are not widely available among non-drosophilid insects. One way around this difficulty would be to target a well-characterized transgene that had previously been introduced into the genome by nonhomologous recombination. We have demonstrated the feasibility of gene targeting in mosquitoes by targeting a hygromycin resistance transgene previously introduced into an *Aedes aegypti* cell line genome and stably incorporated as multicopy head-to-tail tandem arrays (Y. Zhao and P. Eggleston, unpublished). To do this, we employed a replacement vector in which the region of homology (hygromycin resistance gene and SV40 terminator) was disrupted by a promoterless neomycin resistance gene (*neo*). As described in Section 2.2.4.3 this design utilizes the *neo* gene as a promoter trap for the enrichment of targeted integration events. In this way, neomycin resistance can only be expressed from the promoter of the hygromycin resistance gene in the event of precise gene targeting. Targeted integration events would therefore be detectable both by efficient expression of neomycin resistance and by inactivation (knockout) of hygromycin resistance.

Aedes aegypti cells were first stably transformed with a hygromycin resistance cassette in which the bacterial *hph* gene was driven by the *Drosophila actin 5C* promoter with transcription terminated by the SV40 polyadenylation signal. Southern analysis was used to show that the hygromycin resistance transgenes were integrated into the recipient genome as multicopy head-to-tail tandem arrays. In such situations, Southern blots give rise to banding patterns consistent with digests of the circular transformation vector. Similar patterns have been observed elsewhere following the stable transformation of mosquito cell lines (Monroe et al., 1992; Lycett and Crampton, 1993; Shotkoski and Fallon, 1993), *Drosophila* cell lines (Bourouis and Jarry, 1983; Rio and Rubin, 1985; Vulsteke et al., 1993), and mammalian cell lines (Pomerantz et al., 1983). It is believed the head-to-tail tandem arrangement occurs independently of the replicative capacity of the transforming DNA and is facilitated by the use of large quantities of DNA during transfection. It has been suggested that such arrays might be formed by homologous recombination between transfected plasmids either before or after integration into the chromosomes (Pomerantz et al., 1983).

The targeting vector directed at the resident hygromycin resistance sequences was of the replacement type, where the region of homology is disrupted by an intervening sequence. In this case the isogenic region of homology comprised the hygromycin resistance gene and SV40 termination sequence. The intervening sequence was a promoterless neomycin resistance gene that in the event of targeted integration would be driven from the heterologous *actin 5C* promoter. It is known that homologous recombination efficiency is maximized by the introduction of double-strand breaks near the region of homology (Lin et al., 1984). To this end, the targeting vector was linearized at the 5' end of the *hph* gene and transfected into the hygromycin resistant mosquito cells. Following selection with G418, neomycin resistant clones were established and this suggested that resistance was being expressed from the *actin 5C* promoter following targeted integration. This resistance remained stable even when the cells were maintained for over 6 months in the absence of neomycin selection pressure, strongly indicating that the *neo* gene within the targeting vector had become stably integrated into the recipient genome. Further analysis revealed that these neomycin resistant cells had lost all hygromycin resistance, as might be expected following targeted disruption of the *hph* gene. Subsequently, both PCR and sequencing analysis were used to confirm the targeted integration of the *neo* coding sequence into the hygromycin resistance transgenes.

2.2.7 PROSPECTS FOR *IN VIVO* GENE TARGETING IN THE INSECT GENOME

Experiments such as those described above clearly demonstrate that the cellular mechanisms necessary for homologous recombination are present in insect cells. They also demonstrate how aspects of targeting vector design might be optimized to improve efficiency and to enrich for homologous integration events. The major difficulty in applying this technology to intact insects would appear to be the relatively low frequency of homologous recombination encountered in higher eukaryotic systems. The tried and trusted method for introducing DNA into insects is still microinjection of preblastoderm embryos and it is not clear whether this technique will allow sufficient opportunity for homologous recombination events to be isolated and characterized. Of course, many aspects of the procedure remain to be optimized and there are numerous possibilities for additional modifications that could increase homologous recombination efficiency. A number of proteins are known to be involved in homologous recombination pathways, including the products of the bacterial *recA* gene and its homologue in yeast, *Rad51*. These genes are thought to catalyze pairing and strand exchange between homologous DNA molecules. Isolation of related genes from insects and their overexpression in target tissues could be used to increase homologous recombination frequencies (Bezzubova et al., 1997). That this is feasible has been shown by the recent isolation of the *Rad51dm* gene from *Drosophila* using degenerate PCR based on conserved regions of the yeast *Rad51* gene (McKee et al., 1996).

Another protein known to be involved is the product of the poly-ADP-ribose polymerase (PARP) gene (Morrison et al., 1997). The knockout of PARP activity appears to lead to greatly elevated levels of homologous recombination (Wang et al., 1997), and this involvement could explain a number of recent findings. For example, it has been shown that homologous recombination frequencies in *Xenopus* oocytes are much higher than those in mature eggs or embryos (Hagmann et al., 1996). Interestingly, this transition stage from oocyte to egg is also the first developmental stage when PARP activity can be detected (Aoufouchi and Shall, 1997). Thus, homologous recombination appears to be the preferential pathway for DNA repair in early *Xenopus* oocytes. If this turns out to be generally true of germ cells, then it could be argued that gene targeting might be much more efficient if it were directed at gametes. Whether this tissue difference reflects PARP activity, the haploid status of the gametes *per se*, or some other germline-specific environmental factor is not clear. However, a similar conclusion was reached in a recent study of gene targeting in the moss, *Physcomitrella patens* (Schaefer and Zryd, 1997). Unusually high frequencies of homologous recombination were seen in this plant, with targeting efficiencies of 90% that are comparable with those found in yeast. The authors suggested that this might be associated with the predominantly haploid gametophytic life cycle and went on to speculate that efficient gene targeting might generally be correlated with the haploid (gametic) phase in higher eukaryotes. Thus, gene-targeting vectors directed at insect oocytes (or perhaps more easily, sperm) may be able to generate high frequencies of homologous recombination and targeted transgene integration. In principle at least, such gene-targeted gametes (whether modified *in vivo* or *in vitro*) could subsequently be fused to generate transgenic insects. This is an intriguing prospect, but one that might require considerable preliminary work to facilitate the isolation, modification, and survival of the targeted gametes.

More recently, a novel strategy for gene targeting in the silk worm, *Bombyx mori*, using a modified baculovirus has been described (Yamao et al., 1999). In these experiments, the *Bombyx* fibroin *light (L) chain* gene was cloned and a GFP gene inserted into exon 7 of the genomic sequence. This chimeric targeting sequence was used to replace the polyhedrin gene of the *Autographica californica* nucleopolyhedrovirus (AcNPV) to generate a baculovirus-targeting vector. This strategy therefore benefits from highly efficient viral-mediated delivery of the targeting sequence utilizing a vector that is capable of replication within *B. mori* but which does not give rise to symptoms of nuclear polyhedrosis. In practice, female silk moths were infected with the recombinant virus and mated to normal males. F_1 embryos were screened by PCR for the presence of the GFP transgene and siblings of the PCR-positive embryos were reared to adulthood and mated *inter-se*. F_2 embryos from these matings were screened by PCR for the presence of a novel DNA junction that could only be generated following targeted integration of the chimeric transgene. The overall efficiency of gene targeting in these experiments was around 0.16%, thereby providing an exciting opportunity to explore the potential of precise genomic modification in this species. Whether viral-mediated gene targeting can be more universally applied in insect transgenesis remains to be seen but experiments such as these clearly support the exploration of such technology.

2.3. DOUBLE-STRAND GAP REPAIR

Transformation of the *Drosophila* genome was pioneered by work with the *P* transposable element and is now a routine part of the armory of the fruitfly geneticist. It could be argued that the very success of this procedure prevented the search for alternative transformation methods. For example, in contrast to work with the mouse, until very recently there had been no attempt to pursue targeted genome modification in *Drosophila*. This was compounded by the lack of a cell culture system comparable with the ES cells of the mouse, where virtually all of the pioneering work in gene targeting has been conducted. Now, as more is being discovered about recombinational DNA repair in higher eukaryotes, there is a growing interest in targeted genome modification. Recombinational DNA repair is involved in the resolution of double-strand breaks in the DNA duplex. It requires the exchange of single DNA strands between homologous duplexes (although the exchange need only be temporary) and at least one of the strands needs to be intact. It is this transfer of information between the single strands that leads to double-strand break repair. Several examples have now been described in *Drosophila* where double-strand breaks have been introduced into the genome and used as "docking sites" for the targeted introduction of transgenes. The first of these mechanisms directly exploits the *P*-transposable element which leaves a double-strand break upon excision from the chromosome (Gloor et al., 1991; Nassif et al., 1994; Keeler et al., 1996; Dray and Gloor, 1997). This "transposon-induced gap repair" is known to occur in both somatic and germline cells and in both male and female flies. The double-strand break or gap can be repaired using information from a variety of templates, including genomic sequences located *in cis* or *in trans* and plasmid constructs introduced into the targeted cells or microinjected into insect embryos. It should be noted that transposon-induced gap repair is not a peculiarity of *P*-element transposition. Indeed, any mechanism leading to a double-strand break will serve to create the necessary docking site. For example, rare-cutting endonucleases have been used in mouse cells to induce double-strand breaks (Choulika et al., 1994; Rouet et al., 1994; Jasin, 1996) and site-specific cleavage sites have been introduced as transgenes into the *Drosophila* genome to generate similar gaps. In other words, the *P*-element excision simply acts as a tool to create the double-strand break rather than constituting a necessary part of the repair.

There is little doubt that this technique provides a powerful new tool for *Drosophila* genetics but what is particularly exciting is the potential for more widespread exploitation among insects.

The recent successful transformation of *Aedes aegypti* with the *Hermes* (Jasinskiene et al., 1998) and *mariner* (Coates et al., 1998) elements and of an *Anopheles gambiae* cell line genome with the *Hermes* element (Zhao and Eggleston, 1998) provides the raw material to begin to test this hypothesis. However, since none of these experiments has yet been carried out in insects other than *Drosophila*, this chapter will be confined to a simple description of the technique and some of the parameters known to affect targeting efficiency.

2.3.1 TRANSPOSON-MEDIATED GAP REPAIR IN *DROSOPHILA*

The *P*-transposable element in *Drosophila* is one of the most widely studied eukaryotic transposons. It is described as a Class II mobile element in that transposition takes place directly from DNA to DNA. This distinguishes it from the Class I elements (retrotransposons) that require the generation of an RNA intermediate. Many other Class II element families have been described, including the *hAT* family (defined in terms of the *hobo, Ac,* and *Tam* elements but also containing *Hermes;* see Chapter 12 by Atkinson and O'Brochta), the *Tc1-mariner* family (which also contains *Minos;* see Chapter 13 by Lampe et al.) and the TTAA family of elements that includes *piggyBac*, (see Chapter 14 by Fraser). The biological characteristics of several of these elements are described elsewhere in this book and need not concern us here. What does concern us is the structural biology that unifies these diverse elements. Typically, they comprise a central region encoding a transposase enzyme that catalyzes the transposition reaction. This is flanked by inverted terminal repeat sequences (ITRs) that are a fundamental requirement for transposition. In simple terms, the transposase is believed to bind to the ITRs, bringing them into close association and facilitating the excision reaction. For some elements, such as *P*, this interaction is known to involve other gene products (e.g., IRBP, inverted repeat binding protein; Beall and Rio, 1996). On the other hand, elements of the *Tc1-mariner* family appear to be completely autonomous in that transposition can be mediated by purified transposase within cell-free systems. The outcome of the interaction is the excision of the element and the creation of a double-strand break. The break is then repaired by copying sequence information from a suitable template. Work with the *P* element has shown that the repair template is often the sister DNA strand or homologous chromosome and that these typically carry the same *P*-element sequence. This is one explanation of the replicative nature of transposition encountered for such mobile elements. The double-strand break repair copies *P*-element sequences from the homologous template into the vacant site. This recreates the original site while leaving the excised element free to integrate at an additional site (Lankenau, 1995; Lankenau et al., 1996). This aspect of their biology also explains the efficiency with which the *P* element was able to invade the worldwide population of *D. melanogaster*.

Repair templates, however, are not restricted to *P*-element sequences on the sister strand or homologous chromosome. It is believed that the broken DNA strands following element excision initiate a homology search to facilitate repair. Once such homology is found, a template is established and the repair mechanism will copy the available sequence into the broken site. However, certain aspects of the biology of this repair differ from the comparable double-strand break repair mechanisms of yeast. In yeast, the conversion process is also initiated by homology-dependent template establishment, but this is understood to be followed by the creation and resolution of Holliday junctions, leading to the frequent occurrence of heteroduplex DNA and the crossing-over of flanking markers (Stahl, 1996). By way of contrast, sequence information in *Drosophila* appears to be copied from the template as a continuous tract with the result that heteroduplex formation and crossover of flanking markers is rarely seen. This has resulted in the proposal of a modified model to account for transposon-mediated gap repair, which is known as SDSA or synthesis-dependent strand annealing (Figure 2.8). SDSA seems to be widespread in higher eukaryotes (Rubin and Levy, 1997) and may provide evidence for an evolutionary strategy that maintains genome stability through the repair of double-strand breaks without the chromosomal rearrangements that would result from frequent crossing-over between the template and the site of the break.

As described previously, the template that is established for transposon-mediated gap repair need not be resident on the sister strand or homologous chromosome. It can quite easily be located at nonhomologous genomic sites (ectopic localization) or indeed on an introduced plasmid construct. This provides the facility to target heterologous sequence into the break site and therefore bring about transgene integration. In fact, large heterologous sequences can be introduced into the genome in this way. Nassif et al. (1994) showed that an 8-kb sequence could be introduced at frequencies comparable with that for a point mutation and Keeler et al. (1996) reported the integration of a 13-kb heterologous sequence.

Several studies have looked at the homology requirements for transposon-mediated gap repair. Dray and Gloor (1997) showed that the presence of 3' homology was an absolute requirement for targeting. In the presence of fixed 5' homology of around 2.5 kb, targeting frequency was abolished in the absence of 3' homology but increased with 3' homology up to around 500 bp and then became saturated. This is comparable with the situation in mammalian cells (Hasty et al., 1991), although here the homology saturation point is around 10 kb. Dray and Gloor (1997) also indicated that longer heterologous inserts may require longer tracts of homology to achieve efficient targeting. In common with work in mammalian cells (te Riele et al., 1992) there is also evidence that targeting efficiency is maximized with isogenic stretches of homology. For example, Nassif and Engels (1993) showed that conversion efficiency in transposon-mediated gap repair decreased linearly with increasing numbers of single-base heterologies. However, it seems likely that the actual homology requirements may depend on the searching ability of specific DNA ends and that this may need to be optimized in individual situations.

2.3.2 TRANSPOSON-MEDIATED GAP REPAIR IN OTHER INSECTS

As far as we are aware, there have so far been no attempts to utilize transposon-mediated gap repair for targeted transformation outside of the genus *Drosophila*. This is, perhaps, not surprising given the very recent nature of transposon-mediated transformation in non-drosophilid insects. In fact, the first reported event of this kind was the use of the *Minos* element from *D. hydei* to transform the genome of the medfly, *Ceratitis capitata* (Loukeris et al., 1995). This major breakthrough probably arose because of the availability of an effective selectable marker namely, the cloned medfly *white* gene, to facilitate the unambiguous detection of transformed flies. Progress with other non-drosophilid insects was compromised by poor selectable markers, and it was to be a further 3 years before additional successes were reported. Significantly, these included the transformation of the yellow fever mosquito, *Aedes aegypti*, with both the *Hermes* element from *Musca domestica* (Jasinskiene et al., 1998) and the *mariner* element from *D. mauritiana* (Coates et al., 1998) elements. A further breakthrough came with the transformation of the medfly, *C. capitata*, by the *piggyBac* element from the lepidopteran, *Trichoplusia ni* (Handler et al., 1998). In both cases, progress was undoubtedly linked to the availability of unambiguous selectable markers. The successful application of the medfly *white* gene had been demonstrated previously, but progress with *A. aegypti* came from an unexpected direction. One of the *white* eye mutations in *Aedes aegypti* arises because of a defect in the kynurenine hydroxylase gene that catalyzes metabolic synthesis of ommochrome eye pigments (Bhalla, 1968). It was found that the wild-type *cinnabar* gene from *D. melanogaster* (encoding kynurenine hydroxylase) can complement this mutation and serve as a transformation marker by rescuing the mutant strain (Cornel et al., 1997). In this way, transformed mosquitoes of the white-eyed kh^w strain could be detected by the restoration of eye pigment in a manner similar to that routinely employed in *Drosophila* transformation. At the same time, progress was being made with the malaria vector mosquito, *Anopheles gambiae*, through the successful transformation of a cell line genome with the *Hermes* element (Zhao and Eggleston, 1998). In this case, however, further progress toward the transformation of intact insects still awaits the establishment of an

Targeted Transformation of the Insect Genome

INVASION

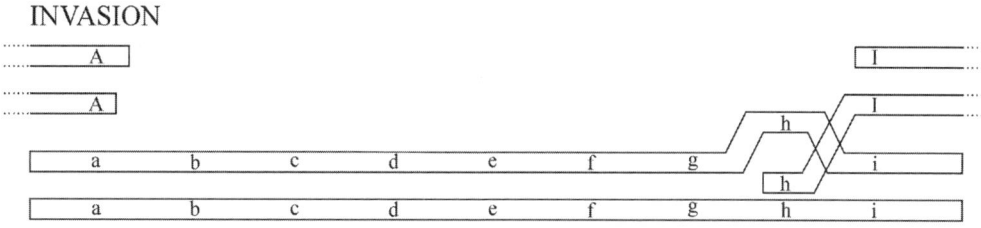

BUBBLE MIGRATION

ANNEALING

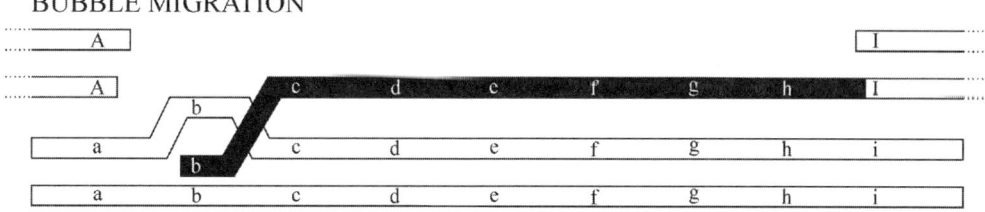

FILL-IN SYNTHESIS

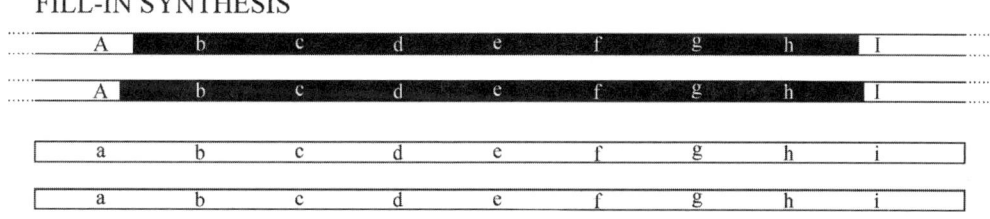

FIGURE 2.8 The SDSA model for double-strand break repair (Nassif et al., 1994). The four panels represent successive stages of the interaction between double-stranded chromosomal DNA (here represented by the sequence A to I flanked by dotted lines) and the homologous repair template (here represented by the sequence a to i). Repair begins with INVASION, where homology is established between the broken chromosome end and a homologous repair template. This is followed by BUBBLE MIGRATION as the newly synthesized DNA (shown in black) is displaced from the template and then ANNEALING of the newly synthesized DNA to the opposite side of the break. Subsequently, a process of FILL-IN SYNTHESIS completes the chromosomal repair. The process is essentially one of gene conversion with repair converting the broken chromosomal sequence to that of the homologous repair template.

effective protocol for *Anopheles*. At the time of writing, there are preliminary and unpublished reports of the successful transformation of *Anopheles stephensi* using the *Minos* element (Catteruccia et al., unpublished), clearly representing an exciting development in this area.

These recent developments all highlight the emerging use of transposable elements in the transformation of non-drosophilid insects. In all cases, the successful transformations were mediated by Class II mobile genetic elements (*Minos, Hermes, mariner,* and *piggyBac*) which, like the *P* element, are either known or believed to transpose via a cut-and-paste mechanism. These events therefore provide the raw material to begin to test the feasibility of targeted transformation through transposon-mediated gap repair. Although much preliminary work remains to be done in these non-drosophilid systems, targeted transformation would provide a number of real benefits. Foremost among these would be freedom from the position effects commonly associated with random transgene integration. It is generally accepted that transgene expression can vary widely in response to integration site (Wilson et al., 1990; Koller and Smithies, 1992). These position effects arise either because of the transcriptional status of the integration site or because of the effects of enhancers or suppressors of gene expression located *in cis* or *in trans*. For example, the recent transformation of the white-eyed kh^w strain of *Aedes aegypti* with *Hermes* and *mariner* vectors carrying the *Drosophila* wild-type *cinnabar* gene gave rise to considerable variation in the amount of restored eye color. Thus, identical transgenes commonly give rise to different expression patterns depending on the site of genomic integration. These position effects will necessarily compromise any comparative study of transgene activity since expression cannot be linked directly to the structure and organization of the introduced sequence. These problems would not exist, however, if alternative transgenes could be introduced at precisely the same site by targeted transformation as described above.

The most obvious way forward would be to identify transposon integration sites giving good transgene expression, namely, those showing high levels of marker gene expression. These could then serve as recipients for additional transgenes through transposon-mediated gap repair in the knowledge that the integration site was transcriptionally active and expression competent. Once the necessary homology requirements had been empirically determined, these additional transgenes could be introduced as heterologous sequences on appropriate plasmid constructs. These would be coinjected into the relevant insect embryos alongside an appropriate helper construct to catalyze excision of the resident element. In principle at least, double-strand break repair would then have the potential to bring about transgene integration by copying of sequence information from the plasmid template. It will be of interest to see if the enormous potential benefits of this approach can be demonstrated outside of *D. melanogaster*.

REFERENCES

Afanasiev BN, Kozlov YV, Carlson JO, Beaty BJ (1994) Densovirus of *Aedes aegypti* as an expression vector in mosquito cells. Exp Parasitol 79:322–339

Allis CD, Waring GL, Mahowald AP (1977) Mass isolation of pole cells from *Drosophila melanogaster*. Dev Biol 56:372–381

Allis CD, Underwood EM, Caulton JH, Mahowald AP (1979) Pole cells of *Drosophila melanogaster* in culture. Dev Biol 69:451–465

Aoufouchi S, Shall S (1997) Regulation by phosphorylation of *Xenopus laevis* poly(ADP-ribose) polymerase enzyme activity during oocyte maturation. Biochem J 325:543–551

Ayares D, Chekuri L, Song K-Y, Kucherlapati R (1986) Sequence homology requirements for intermolecular recombination in mammalian cells. Proc Natl Acad Sci USA 83:5199–5203

Baldridge GD, Fallon AM (1996) Evidence for a DNA homologous pairing activity in nuclear extracts from mosquito cells. Insect Biochem Mol Biol 26:667–676

Baur M, Potrykus I, Paszkowski J (1990) Intermolecular homologous recombination in plants. Mol Cell Biol 10:492–500

Beall EL, Rio DC (1996) *Drosophila* IRBP/Ku p70 corresponds to the mutagen-sensitive *mus309* gene and is involved in *P*-element excision *in vivo*. Genes Dev 10:921–933

Beck C, Cayeux S, Lupton SD, Doerken B, Blankenstein T (1995) The thymidine kinase/gancyclovir mediated "suicide" effect is variable in different tumor cells. Hum Gene Ther 6:1525–1530

Bezzubova O, Silbergleit A, Yamaguchi-Iwai Y, Takeda S, Buerstedde JM (1997) Reduced X-ray resistance and homologous recombination frequencies in a RAD54$^{(-/-)}$ mutant of the chicken DT40 cell line. Cell 89:185–193

Bhalla SC (1968) Genetic aspects of pteridines in mosquitoes. Genetics 58:249–258

Bilang R, Peterhans A, Bogucki A, Paszkowski J (1992) Single-stranded DNA as a recombination substrate in plants as assessed by stable and transient recombination assays. Mol Cell Biol 12:329–336

Bishop JO, Smith P (1989) Mechanism of chromosomal integration of microinjected DNA. Mol Biol Med 6:283–289

Blundell PA, Rudenko G, Borst P (1996) Targeting of exogenous DNA into *Trypanosoma brucei* requires a high degree of homology between donor and target DNA. Mol Biochem Parasitol 76:215–229

Boehme RE (1984) Phosphorylation of the antiviral precursor 9-(1,3-dihydrox-2-propoxymethyl) guanylate kinase isozymes. J Biol Chem 259:12346–12349

Bollag RJ, Waldman AS, Liskay RM (1989) Homologous recombination in mammalian cells. Annu Rev Genet 23:199–225

Bourouis M, Jarry B (1983) Vectors containing a prokaryotic dihydrofolate reductase gene transform *Drosophila* cell to methotrexate-resistance. EMBO J 2:1099–1104

Boviatsis EJ, Park JS, Sena-Esteves M, Kramm CM, Chase M, Efird JT, Wei MX, Breakefield XO, Chiocca EA (1994) Long-term survival of rats harboring brain neoplasms treated with gancyclovir and a herpes simplex virus vector that retains an intact thymidine kinase gene. Cancer Res 54:5745–5751

Capecchi MR (1989a) The new mouse genetics: altering the genome by gene targeting. Trends Genet 5:70–76

Capecchi MR (1989b) Altering the genome by homologous recombination. Science 244:1288–1292

Cherbas L, Cherbas P (1997) "Parahomologous" gene targeting in *Drosophila* cells: an efficient, homology-dependent pathway of illegitimate recombination near a target site. Genetics 145:349–358

Choulika A, Perrin A, Dujon B, Nicolas J-F (1994) Induction of homologous recombination in mammalian chromosomes by using the I-SceI system of *Saccharomyces cerevisiae*. Mol Cell Biol 15:1968–1973

Coates CJ, Jasinskiene N, Miyashiro L, James AA (1998) *Mariner* transposition and transformation of the yellow fever mosquito, *Aedes aegypti*. Proc Natl Acad Sci USA 95:3748–3751

Cornel AJ, Benedict MQ, Rafferty CS, Howells AJ, Collins FH (1997) Transient expression of the *Drosophila melanogaster cinnabar* gene rescues eye color in the white-eye (WE) strain of *Aedes aegypti*. Insect Biochem Mol Biol 27:993–997

Cruz A, Beverley SM (1990) Gene replacement in parasitic protozoa. Nature 348:171–173

De-Groot MJA, Offringa R, Does MP, Hooykaas PJJ, van-den-Elzen PJM (1992) Mechanisms of intermolecular homologous recombination in plants as studied with single- and double-stranded DNA molecules. Nucl Acids Res 20:2785–2794

Dray T, Gloor GB (1997) Homology requirements for targeting heterologous sequences during *P*-induced gap repair in *Drosophila melanogaster*. Genetics 147:689–699

Ellis J, Bernstein A (1989) Gene targeting with retroviral vectors:recombination by gene conversion into regions of non-homology. Mol Cell Biol 9:1621–1627

Evans MJ, Kaufman MH (1981) Establishment in culture of pluripotential cells from mouse embryos. Nature 292:154–156

Fitch DHA, Bailey WJ, Tagle DA, Goodman M, Sieu L, Slightom JL (1991) Duplication of the gamma-globin gene mediated by L1 long interspersed repetitive elements in an early ancestor of simian primates. Proc Natl Acad Sci USA 88:7396–7400

Freeman SM, Whartenby KA, Freeman JL, Abboud CN, Marrogi AJ (1996) *In situ* use of suicide genes for cancer therapy. Semin Oncol 23:31–45

Furman PA, St. Clair MH, Spector T (1984) Acyclovir triphosphate is a suicide inactivator of the herpes simplex virus DNA polymerase. J Biol Chem 259:9575–9579

Galli-Taliadoros LA, Sedgwick JD, Wood SA, Koerner H (1995) Gene knockout technology:a methodological overview for the interested novice. J Immunol Methods 181:1–15

Gloor GB, Nassif NA, Johnson-Schlitz DM, Preston CR, Engels WR (1991) Targeted gene replacement in *Drosophila* via *P*-element-induced gap repair. Science 253:1110–1117

Godeau F, Saucier C, Kourilsky P (1992) Replication inhibition by nucleoside analogues of a recombinant *Autographa californica* multicapsid nuclear polyhedrosis virus harboring the herpes thymidine kinase gene driven by the IE-1(0) promoter:A new way to select recombinant baculoviruses. Nucl Acids Res 20:6239–6246

Golic MM, Rong YS, Peterson RB, Lindquist SL, Golic KG (1997) FLP-mediated DNA mobilisation to specific target sites in *Drosophila* chromosomes. Nucl Acids Res 25:3665–3671

Hagmann M, Adlkofer K, Pfeiffer P, Bruggmann R, Georgiev O, Rungger D, Schaffner W (1996) Dramatic changes in the ratio of homologous recombination to non-homologous DNA-end joining in oocytes and early embryos of *Xenopus laevis*. Biol Chem Hoppe-Seyler 377:239–250

Hagmann M, Bruggmann R, Xue L, Georgiev O, Schaffner W, Rungger D, Spaniol P, Gerster T (1998) Homologous recombination and DNA-end joining reactions in zygotes and early embryos of zebrafish (*Danio rerio*) and *Drosophila melanogaster*. Biol Chem 379:673–681

Handler AM, Gomez SP, O'Brochta DA (1993) A functional analysis of the *P*-element gene transfer vector in insects. Arch Insect Biochem Physiol 22:373–384

Handler AM, McCombs SD, Fraser MJ, Saul SH (1998) The lepidopteran transposon vector, *piggyBac*, mediates germline transformation in the Mediterranean fruitfly. Proc Natl Acad Sci USA 95:7520–7525

Hanson KD, Sedivy JM (1995) Analysis of biological selections for high-efficiency gene targeting. Mol Cell Biol 15:45–51

Hasty P, Rivera-Pérez J, Chang C, Bradley A (1991) Target frequency and integration pattern for insertion and replacement vectors in embryonic stem cells. Mol Cell Biol 11:4509–4517

Hinnen A, Hicks JB, Fink GR (1978) Transformation of yeast. Proc Natl Acad Sci USA 75:1929–1933

Hooper ML (1992) Embryonal Stem Cells: Introducing Planned Changes into the Animal Germline. Harwood Academic Publishers, Chur, Switzerland

Jasin M (1996) Genetic manipulation of genomes with rare-cutting endonucleases. Trends Genet 12:224–228

Jasinskiene N, Coates CJ, Benedict MQ, Cornel AJ, Rafferty CS, James AA, Collins FH (1998) Stable transformation of the yellow fever mosquito, *Aedes aegypti*, with the *Hermes* element from the housefly. Proc Natl Acad Sci USA 95:3743–3747

Johnson RS, Sheng M, Greenberg ME, Kolodner RD, Papaioannou VE, Spiegelman BM (1989) Targeting of non-expressed genes in embryonic stem cells *via* homologous recombination. Science 245:1234–1236

Kafatos FC, Orr W, Delidakis C (1985) Developmentally regulated gene amplification. Trends Genet 1:301–305

Keeler KJ, Dray T, Penney JE, Gloor GB (1996) Gene targeting of a plasmid-borne sequence to a double-strand DNA break in *Drosophila melanogaster*. Mol Cell Biol 16:522–528

Kempin SA, Liljegren SJ, Block LM, Rounsley SD, Yanofsky MF, Lam E (1997) Targeted disruption in *Arabidopsis*. Nature 389:802–803

Koller BH, Smithies O (1992) Altering genes in animals by gene targeting. Annu Rev Immunol 10:705–730

Kucherlapati R, Smith GR, (eds) (1988) Genetic Recombination. American Society for Microbiology, Washington, D.C.

Lankenau DH (1995) Genetics of genetics in *Drosophila*: *P*-elements serving the study of homologous recombination, gene conversion and targeting. Chromosoma 103:659–668

Lankenau DH, Corces VG, Engels WR (1996) Comparison of targeted-gene replacement frequencies in *Drosophila melanogaster* at the *forked* and *white* loci. Mol Cell Biol 16:3535–3544

Lin FL, Sperle K, Sternberg N (1984) Model for homologous recombination during transfer of DNA into mouse L cells:role for DNA ends in the recombination process. Mol Cell Biol 4:1020–1034

Lohe AR, Hartl DL (1996) Germline transformation of *Drosophila virilis* with the transposable element *mariner*. Genetics 143:365–374

Loukeris TG, Livadaras I, Arca B, Zabalou S, Savakis C (1995) Gene transfer into the medfly, *Ceratitis capitata*, with a *Drosophila hydei* transposable element. Science 270:2002–2005

Low KB (ed) (1988) The Recombination of Genetic Material. Academic Press, New York

Lozovskaya ER, Nurminsky DI, Hartl DL, Sullivan DT (1996) Germline transformation of *Drosophila virilis* mediated by the transposable element *hobo*. Genetics 142:173–177

Lycett GJ, Crampton JM (1993) Stable transformation of mosquito cell lines using a hsp70::neo fusion gene. Gene 136:129–136

Mansour SL, Thomas KR, Capecchi MR (1988) Disruption of the proto-oncogene *int-2* in mouse embryo-derived stem cells: a general strategy for targeting mutations to non-selectable genes. Nature 336:348–352

Matsubara T, Beeman RW, Shike H, Besansky NJ, Mukabayire O, Higgs S, James AA, Burns JC (1996) Pantropic retroviral vectors integrate and express in cells of the malaria mosquito, *Anopheles gambiae*. Proc Natl Acad Sci USA 93:6181–6185

Mazzacano CA, Fallon AM (1995) Evaluation of a viral thymidine kinase gene for suicide selection in transfected mosquito cells. Insect Mol Biol 4:125–134

McKee BD, Ren XJ, Hong CS (1996) A *recA*-like gene in *Drosophila melanogaster* that is expressed at high-levels in female but not male meiotic tissues. Chromosoma 104:479–488

Metsaeranta M, Vuorio E (1992) Transgenic mice as models for heritable diseases. Ann Med 24:117–120

Moens CB, Auerbach AB, Conlon RA, Joyner AL, Rossant J (1992) A targeted mutation reveals a role for *N-myc* in branching morphogenesis in the embryonic mouse lung. Genes Dev 6:691–704

Mombaerts P, Clarke AR, Hooper ML, Tonegawa S (1991) Creation of a large genomic deletion at the T-cell antigen receptor ϑ-subunit locus in mouse embryonic stem cells by gene targeting. Proc Natl Acad Sci USA 88:3084–3087

Monroe TJ, Muhlmann-Diaz MC, Kovach MJ, Carlson JO, Bedford JS, Beaty BJ (1992) Stable transformation of a mosquito cell line results in extraordinarily high copy numbers of the plasmid. Proc Natl Acad Sci USA 89:5725–5729

Morrison A, Marschalek R, Dingermann T, Harwood AJ (1997) A novel, negative selectable marker for gene disruption in *Dictyostelium*. Gene 202:171–176

Nassif N, Engels W (1993) DNA homology requirements for mitotic gap repair in *Drosophila*. Proc Natl Acad Sci USA 90:1262–1266

Nassif N, Penney J, Pal S, Engels WR, Gloor GB (1994) Efficient copying of non-homologous sequences from ectopic sites *via* P-element-induced gap repair. Mol Cell Biol 14:1613–1625

O'Brochta DA, Warren WD, Saville KJ, Atkinson PW (1996) *Hermes*, a functional non-drosophilid insect gene vector from *Musca domestica*. Genetics 142:907–914

Onda M, Kudo S, Rearden A, Mattei M-G, Fukuda M (1993) Identification of a precursor genomic segment that provided a sequence unique to glycophorin B and E genes. Proc Natl Acad Sci USA 90:7220–7224

Orr-Weaver TL, Szostak JW (1985) Fungal recombination. Microbiol Rev 49:33–58

Papadopoulou B, Dumas C (1997) Parameters controlling the rate of gene targeting frequency in the protozoan parasite *Leishmania*. Nucl Acids Res 25:4278–4286

Pieper FR, De-Wit ICM, Pronk ACJ, Kooiman PM, Strijker R, Krimpenfort PJA, Nuyens JH, De-Boer HA (1992) Efficient generation of functional transgenes by homologous recombination in murine zygotes. Nucl Acids Res 20:1259–1264

Pomerantz BJ, Naujokas M, Hassell JA (1983) Homologous recombination between transfected DNAs. Mol Cell Biol 3:1680–1685

Puchta H, Swoboda P, Hohn B (1994) Homologous recombination in plants. Experientia 50:277–284

Reardon JE, Spector T (1989) Herpes simplex virus type 1 DNA polymerase: mechanism of inhibition by acyclovir triphosphate. J Biol Chem 264:7405–7411

Rio DC, Rubin GM (1985) Transformation of cultured *Drosophila melanogaster* cells with a dominant selectable marker. Mol Cell Biol 5:1833–1838

Rossant J, Nagy A (1995) Genome engineering:the new mouse genetics. Nat Med 1:592–597

Rouet P, Smih F, Jasin M (1994) Introduction of double-strand breaks into the genome of mouse cells by expression of a rare-cutting endonuclease. Mol Cell Biol 14:8096–8106

Rubin E, Levy AA (1997) Abortive gap repair: underlying mechanism for DS element formation. Mol Cell Biol 17:6294–6302

Rubinstein M, Mortrud M, Liu B, Low MJ (1993) Rat and mouse proopiomelanocortin gene sequences target tissue-specific expression to the pituitary gland but not to the hypothalamus of transgenic mice. Neuroendocrinology 58:373–380

Rubnitz J, Subramani S (1984) The minimum amount of homology required for homologous recombination in mammalian cells. Mol Cell Biol 4:2253–2258

Schaefer DG, Zryd JP (1997) Efficient gene targeting in the moss, *Physcomitrella patens*. Plant J 11:1195–1206

Seabaugh RC, Olsen KE, Higgs S, Carlson JO, Beaty BJ (1998) Development of a chimeric Sindbis virus with enhanced *per Os* infection of *Aedes aegypti*. Virology 243:99–112

Segal DJ, Carroll D (1995) Endonuclease-induced, targeted homologous extrachromosomal recombination in *Xenopus* oocytes. Proc Natl Acad Sci USA 92:806–810; erratum 3632

Shesely EG, Kim HS, Shehee WR, Papayannopoulou T, Smithies O, Popovich BW (1991) Correction of a human beta super(S)-globin gene by gene targeting. Proc Natl Acad Sci USA 88:4294–4298

Shotkoski FA, Fallon AM (1993) An amplified mosquito dihydrofolate reductase gene:amplicon size and chromosomal distribution. Insect Mol Biol 2:155–161

Small J, Scangos G (1983) Recombination during gene transfer into mouse cells can restore the function of deleted genes. Science 219:174–176

Song K-Y, Chekuri L, Rauth S, Ehrlich S, Kucherlapati RS (1985) Effect of double-strand breaks on homologous recombination in mammalian cells and extracts. Mol Cell Biol 5:3331–3336

Stahl F (1996) Meiotic recombination in yeast: coronation of the double-strand-break repair pathway. Cell 87:965–968

Templeton NS, Roberts DD, Safer B (1997) Efficient gene targeting in mouse embryonic stem cells. Gene Ther 4:700–709

te Riele H, Robanus Maandag E, Berns A (1992) Highly efficient gene targeting in embryonic stem cells via homologous recombination with isogenic DNA constructs. Proc Natl Acad Sci USA 89:5128–5132

Teysset L, Burns JC, Shike H, Sullivan BL, Bucheton A, Terzian C (1998) A Moloney murine leukaemia virus-based retroviral vector pseudotyped by the insect retroviral *gypsy* envelope can infect *Drosophila* cells. J Virol 72:853–856

Thomas KR, Capecchi MR (1987) Site-directed mutagenesis by gene targeting in mouse embryo-derived stem cells. Cell 51:503–512

Vulsteke V, Janssen I, Vanden-Broeck J, De-Loof A, Huybrechts R (1993) Baculovirus immediate early gene promoter based expression vectors for transient and stable transformation of insect cells. Insect Mol Biol 2:195–204

Wake CT, Vernaleone F, Wilson JH (1985) Topological requirement for homologous recombination among DNA molecules transfected into mammalian cells. Mol Cell Biol 5:2080–2089

Waldman AS, Liskay RM (1988) Dependence of intrachromosomal recombination in mammalian cells on uninterrupted homology. Mol Cell Biol 8:5350–5357

Wang Z-Q, Stingl L, Morrison C, Jantsch M, Los M, Schulze-Osthoff K, Wagner E (1997) PARP is important for genomic stability but dispensable for apoptosis. Genes Dev 11:2347–2358

Wheeler MB (1994) Development and validation of swine embryonic stem cells:a review. Reprod Fertil Dev 6:563–568

Wilmut I, Hooper ML, Simons JP (1991) Genetic manipulation of mammals and its application in reproductive biology. J Reprod Fertuk 92:245–279

Wilson C, Bellen HJ, Gehring WJ (1990) Position effects on eukaryotic gene expression. Annu Rev Cell Biol 6:679–714

Wu Y, Kirkman LA, Wellems TE (1996) Transformation of *Plasmodium falciparum* malaria parasites by homologous integration of plasmids that confer resistance to pyrimethamine. Proc Natl Acad Sci USA 93:1130–1134

Yagi T, Nada S, Watanabe N, Tamemoto H, Kohmura N, Ikawa Y, Aizawa S (1993) A novel negative selection for homologous recombinants using diphtheria toxin A fragment gene. Anal Biochem 214:77–86

Yamao M, Katayama N, Nakazawa H, Yamakawa M, Hayashi Y, Hara S, Kamei K, Mori H (1999) Gene targeting in the silkworm by use of a baculovirus. Genes Dev 13:511–516

Zhao Y-G, Eggleston P (1998) Stable transformation of an *Anopheles gambiae* cell line mediated by the *Hermes* mobile genetic element. Insect Biochem Mol Biol 28:213–219

Zhao Y-G, Eggleston P (1999) Comparative analysis of promoters for transient gene expression in cultured mosquito cells. Insect Mol Biol 8:31–38

Zheng H, Wilson JH (1990) Gene targeting in normal and amplified cell lines. Nature 344:170–173

Zheng Z, Hayashimoto A, Li Z, Murai N (1991) Hygromycin resistance gene cassettes for vector construction and selection of transformed rice protoplasts. Plant Physiol 97:832–835

3 Site-Specific Recombination for the Genetic Manipulation of Transgenic Insects

Yikang S. Rong and Kent G. Golic

CONTENTS

3.1 Introduction 53
3.2 Site-Specific Recombination Systems 54
    3.2.1 The FLP Recombination Target...54
    3.2.2 FLP Recombinase 56
3.3 Applications in Insect Transgenesis 57
    3.3.1 Manipulating Transgenes 57
        3.3.1.1 Marker Genes..57
        3.3.1.2 Controlling Gene Expression and Analyzing Gene Function.....................59
    3.3.2 Chromosome Rearrangements 60
    3.3.3 Site-Specific Recombination in Transformation 61
        3.3.3.1 Techniques for Integrating DNA at an *FRT*...62
        3.3.3.2 Efficiency of Recombinase-Mediated Integration 63
            3.3.3.2.1 Donor: Target Pairing...63
            3.3.3.2.2 Stabilizing the Integrated DNA 63
            3.3.3.2.3 Position Effects 65
3.4 Dominant Marker Genes for Insect Transformation 66
    3.4.1 *pug*^{D} 66
    3.4.2 Other Dominant Transgenes 67
3.5 Summary 71
Acknowledgments 71
References 71

3.1 INTRODUCTION

The investigator who sets out to make a transgenic organism must consider two issues. First, a method of transformation must be chosen, or invented; and second, the transgene construct should be designed so that the transgenic organisms are maximally useful. The relative import of these two issues depends largely on the ease of making the transformants. If an easy, well-established transformation procedure is available, then most emphasis will fall on designing the construct so that the informative experiments can be performed with the transgenic organisms. On the other hand, if transformation is difficult, or has not yet been achieved, then the construct may be primarily designed merely to achieve transformation. This chapter will discuss how site-specific recombination may be employed to aid in both aspects of transgenesis. The majority of this chapter discusses methods of site-specific recombination that have been used in the model insect *Drosophila mela-*

nogaster; however, relevant results from other systems will also be introduced. Applications of site-specific recombination in gene targeting and transgene manipulation will be considered. Finally, a recently discovered mutation of *D. melanogaster* will be considered as a potential dominant marker gene for general insect transformation.

3.2 SITE-SPECIFIC RECOMBINATION SYSTEMS

Although many site-specific recombination systems are known, there are two that predominate in transgenic analysis. They are the FLP recombinase–*FRT* (FLP Recombination Target) target site system derived from the yeast 2μ plasmid and the Cre (Causes recombination) recombinase–*lox* (locus of crossover) target site system from the bacteriophage P1. The two systems both consist of a single protein recombinase and a short DNA sequence that the recombinase recognizes. Both recombinases mediate reciprocal physical exchange between two copies of the target sequence, they require no other proteins to function, and they have both been transferred to a number of different organisms where they carry out recombination with high efficiency, including the model insect *D. melanogaster* (Golic and Lindquist, 1989; Siegal and Hartl, 1996). This chapter will focus mainly on the 2μ system, which has successfully been used in bacteria, plants, mammals, and other insects. The Cre–*lox* system has also been applied in many different organisms and functions almost identically (Sauer, 1998). The biochemistry of FLP site-specific recombination has been covered in a recent review (Sadowski, 1995). Here we provide only those details that are pertinent to a consideration of making and analyzing transgenic organisms.

3.2.1 THE *FLP* RECOMBINATION TARGET

The *FRT* is 34 bp long, consisting of two 13 bp inverted repeats flanking an 8 bp central spacer (Figure 3.1), although as little as 28 bp serves to provide full recombinational efficiency (Senecoff

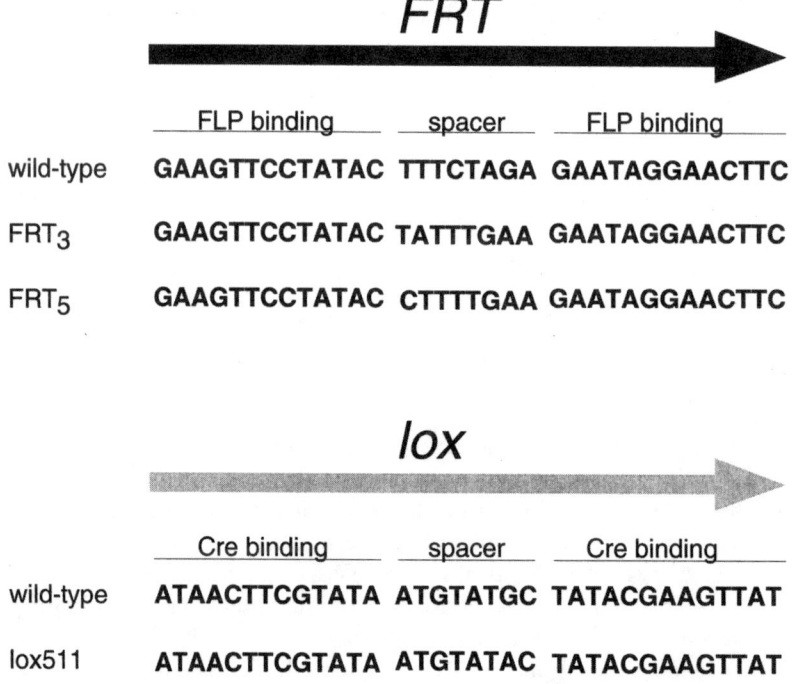

FIGURE 3.1 The sequences of the wild-type FLP–*FRT* and Cre target site (*lox*) are shown. The sequences of recombination-competent mutant target sites that do not recombine with the wild-type targets, because of mutations in the spacer sequences, are shown beneath the wild-type sequences (see Section 3.3.3.2.2).

et al., 1985). The 13 bp repeats serve as FLP binding sites (Senecoff et al., 1988). The 8 bp wild-type spacer sequence possesses an asymmetry which confers a directionality to the *FRT* that determines the outcome of recombination between two *FRT*s (Senecoff and Cox, 1986). The *lox* site is quite similar (Figure 3.1).

The products of recombination between two *FRT*s can be easily predicted in a schematic diagram by aligning those *FRT*s in a parallel direction, one on top of the other, and then drawing a simple reciprocal crossover between the two aligned *FRT*s (Figure 3.2). When considering intramolecular recombination, if two *FRT*s lie in the same orientation, the outcome of recombination will be removal of the *FRT*-flanked DNA from the parent molecule as a circle carrying a single *FRT*. A single *FRT* is also left behind in the parent molecule (Figure 3.2A). This is the most efficient reaction carried out by FLP *in vivo*. In *Drosophila*, this was first demonstrated by placing the *white*⁺ (w^+) eye color gene between direct repeats of the *FRT*, and transforming this via P-element transformation into a *white* mutant strain, to generate flies with pigmented eyes (Golic and Lindquist, 1989). These were crossed to flies that carried an inducible *FLP* transgene. After induction, FLP caused excision of the w^+ gene, and it was then lost when cells divided since it lacked a centromere. In cells of the eye this loss produced white cells in a pigmented background. Excision and loss can be frequent enough to produce almost entirely white-eyed flies. Test crosses also showed that excision occurred efficiently in the male and female germlines.

Intramolecular recombination between inverted *FRT*s results in the inversion of the DNA that lies between the *FRT*s (Figure 3.2B). The results of intermolecular reactions can be similarly predicted, with the outcome depending on whether the *FRT*s are on linear or circular DNA molecules. If at least one of the two recombining *FRT*s is located on a circular molecule, then the

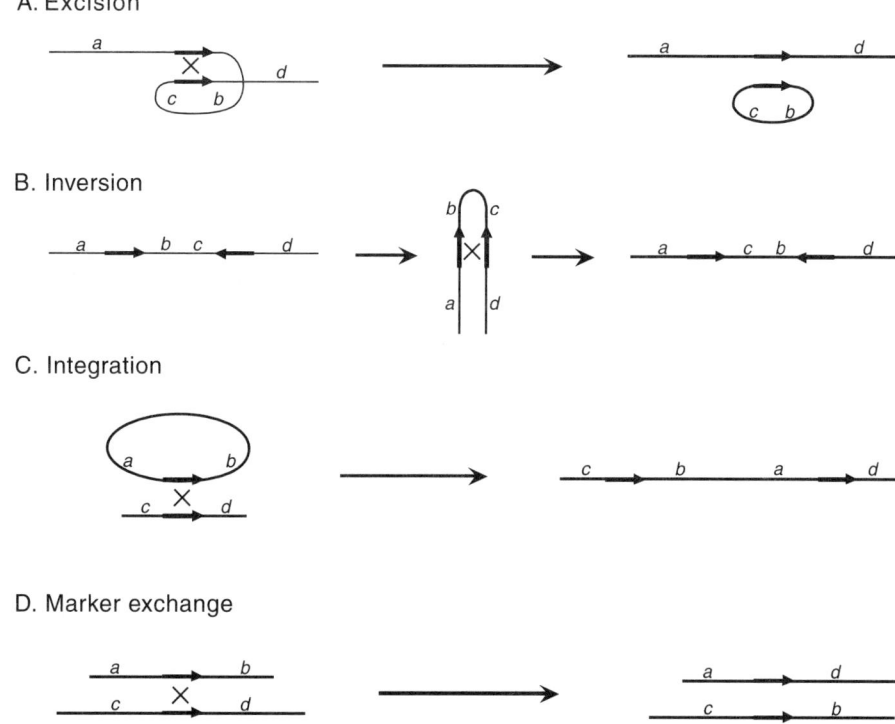

FIGURE 3.2 Reactions catalyzed by recombinases. The basic intramolecular (A, B) and intermolecular (C, D) site-specific exchange events are diagrammed. Target sites are indicated as arrows embedded in the DNA strands (thin lines). Arbitrary genetic or molecular markers are indicated by lowercase letters.

result of recombination will be integration of that circle into the other molecule, where it will lie flanked by *FRT*s (Figure 3.2C — essentially a reversal of the reaction diagrammed in Figure 3.2A). If both *FRT*s are on linear molecules, FLP catalyzes an exchange of flanking DNA segments between the two molecules to generate two linear recombinant DNAs (Figure 3.2D). These reactions have also been demonstrated *in vivo* in *D. melanogaster* and many other organisms. Specific examples will be cited in following sections.

The precise sequence of the spacer is not critical to the recombination reaction and may be altered to generate novel *FRT*s that can recombine with another *FRT* of like sequence, but not with the original wild-type *FRT* or with other differently altered *FRT*s (Senecoff and Cox, 1986). In addition, *FRT*s with symmetrical, but identical, spacers recombine with each other but without regard to directionality, yielding alternate products from the same molecules. The ability to vary the spacer sequence provides an added dimension to the design of experiments and, as will be discussed later, can significantly improve the ability to recover the products of intermolecular recombination *in vivo*.

3.2.2 FLP Recombinase

The *FLP* coding sequence is 1272 bp long, contains no introns, and thus encodes a 423 amino acid protein. This recombinase is capable of catalyzing both intra- and intermolecular recombination between *FRT*s *in vivo* and *in vitro*. Bacterial strains are available that synthesize FLP and that may be used for testing constructs, or possibly as an aid in some cloning steps (Snaith et al., 1996; Bucholz et al., 1996).

In *Drosophila*, hybrid transgenes have been the most frequently used source of FLP protein. The promoter of the *Drosophila hsp70* gene has been used to great effect by eliciting heat-shock-induced expression of FLP (Golic and Lindquist, 1989; Golic et al., 1996). In other insects, the *hsp70-FLP* fusion gene is also likely to prove useful, and has already been used successfully in mosquitoes (Morris et al., 1991). In *Drosophila*, the promoter of the testis-specific β2-*tubulin* gene has been used for male germline-specific expression (Struhl et al., 1993; Golic et al., 1997), and the GAL4-UAS expression system has been used to provide tissue-specific expression (K. Beuemer and K. Golic, unpublished results; Frise et al., 1996). The Cre recombinase has also been expressed in *Drosophila* using a hybrid promoter composed of sequences from the *Drosophila hsp70* gene and the *Mos1* transposon. This construct exhibited constitutive and maternal expression (Siegal and Hartl, 1996).

An alternative method of expression that holds great promise in insect transgenesis is the use of FLP mRNA injection. Konsolaki et al. (1992) injected synthetic FLP mRNA into *Drosophila* embryos, along with *FRT*-bearing plasmds. They found that the injected mRNA directed FLP synthesis, as detected by recombination between the coinjected plasmids. Baek and Ambrosio (1994) have also shown that the injection of artificial mRNA can be used as a method of gene expression in early *Drosophila* embryos.

We wished to determine whether FLP synthesized from injected mRNA would be capable of carrying out efficient recombination between chromosomal *FRT*s. FLP mRNA was synthesized *in vitro* from a construct in which the FLP coding sequence was cloned downstream of a T7 promoter (the construct is available from the authors upon request) using the mMESSAGE mMACHINE kit from Ambion (Austin, TX). The FLP mRNA was injected into precellular blastoderm stage embryos in 0.1× PBS, following essentially the same injection procedure that is used for *P*-element transformation (Spradling, 1986). FLP activity in the injected embryos was detected by two methods. First, by X-Gal staining after injection into a strain that carried a β-*galactosidase* gene that could be activated by FLP (Buenzow and Holmgren, 1995). In approximately half of the injected embryos β-galactosidase activity was detected by blue staining, indicating that FLP was synthesized and carried out recombination, while embryos injected with buffer alone showed no staining. A second method allowed easy quantitation of germline recombination that resulted from the mRNA injection.

Embryos that carried a homozygous insertion of the *FRT*-flanked *white*+ gene, in a *white* null mutant background, were injected with the synthetic FLP mRNA in the posterior end of the embryo. Somatic excision was observed among the adults that survived, but rarely, probably because this would require FLP synthesis near the anterior of the embryo. However, when the injected flies were test-crossed, we observed germline excision at rates of 70 to 90% in approximately three quarters of the tested adults, including males and females. Thus, the injection of FLP mRNA into syncitial stage embryos can efficiently cause recombination between chromosomal *FRT*s in the germline. Because this method requires no further manipulations of the embryos beyond the injections, and does not require generating a strain of insects carrying an *FLP* transgene, it may ultimately prove the most useful method for purposes of transgenesis. mRNA injection has also been used as a method to express Cre recombinase (de Wit et al., 1998).

A novel post-translational method for regulating FLP activity was devised by Logie and Stewart (1995). They constructed a gene that encodes a fusion protein consisting of FLP and the ligand-binding domain of a steroid hormone receptor. This fusion protein shows little recombinase activity in the absence of hormone, but becomes active upon treatment with the hormone ligand.

3.3 APPLICATIONS IN INSECT TRANSGENESIS

Site-specific recombination has proved to be an efficient and versatile tool for manipulation of eukaryotic genomes. FLP–*FRT* and Cre–*lox* recombination have been useful for a variety of manipulations at the level of single genes, as well as for larger-scale events that have led to the production of a variety of chromosome rearrangements.

3.3.1 Manipulating Transgenes

3.3.1.1 Marker Genes

One of the major problems facing any investigator who wishes to make transgenic insects is the paucity of useful marker genes. In a particular species there may be but a single useful marker gene, if indeed there are any at all. This can present a problem if, subsequent to an initial transformation event, it is desired to transform a second construct into the same strain. As discussed by Dale and Ow (1991), in such cases site-specific recombination can be used to remove the marker gene from the first transgenic construct so that it may be reused in a second round of transformation. If the transformation construct is built with directly repeated *FRT*s flanking the marker gene, then, subsequent to transformation, FLP can remove the marker gene from the integrated construct. As mentioned above, deleting a gene from the chromosome by FLP-mediated recombination is typically very efficient using either an *hsp70-FLP* transgene or by injection of FLP mRNA.

This could also be a useful procedure in cases where the marker gene interferes with the subsequent utility or analysis of the transgenic organism. For instance, in some cases a chemical resistance gene may be used as a transformation marker. If the ultimate goal of a particular program of research is some form of biological control that will use the transformed insect, then it is probably not desirable to allow the insects that carry this resistance gene to be released. In such a case site-specific recombination may be used to remove the resistance gene after the transformed strain is established.

Site-specific recombination can also be useful when the transformation method results in the integration of multiple end-to-end repeats of the transgenes, as is often the case with transformation technologies based on transfection of DNA into cultured cells. Such multicopy transformants can sometimes lead to silencing of the genes carried within the construct (Linn et al., 1990; Dorer and Henikoff, 1994; Garrick et al., 1998; reviewed by Selker, 1999). If the transformed construct carries an *FRT*, FLP may be used to carve down the repeat array to just one copy if the outermost elements are oriented in the same direction (Figure 3.3A). If the outermost

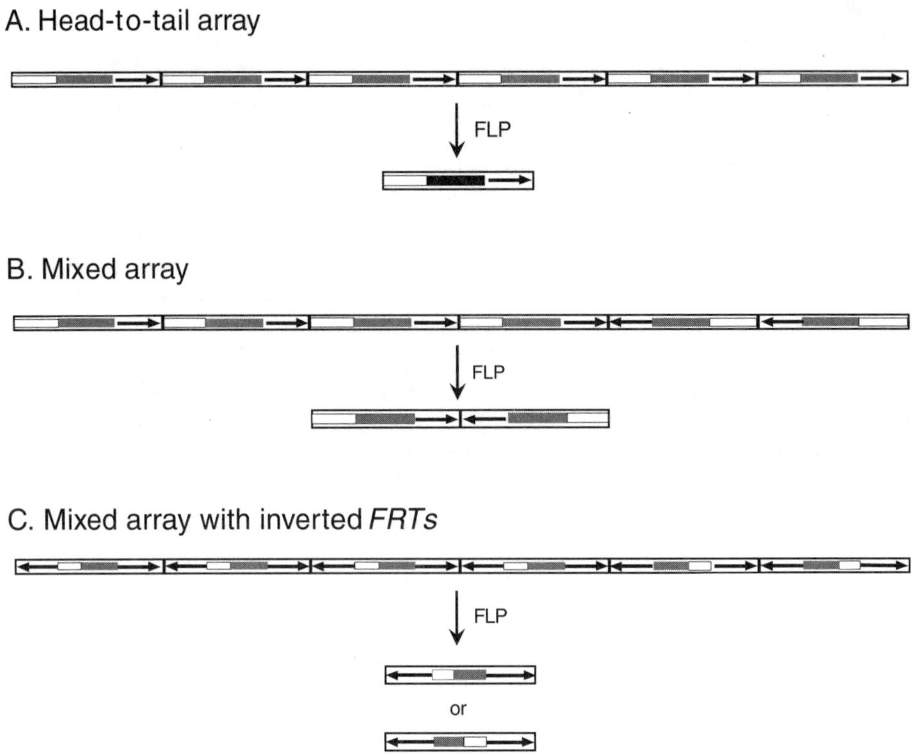

A. Head-to-tail array

B. Mixed array

C. Mixed array with inverted FRTs

FIGURE 3.3 Resolution of tandem repeat by site-specific recombination. Site-specific recombination can reduce multimeric arrays to monomers or dimers, depending upon the disposition of the individual repeats. A single repeat unit is indicated by the large rectangle; components within each repeat are indicated as smaller rectangles; *FRT*s as arrows.

elements are inverted, then two elements will remain (Figure 3.3C). If two inverted *FRT*s are placed in the transformation construct, then it should be possible to arrive at a single copy of the transgene regardless of the orientation of the outside elements (Figure 3.3C). However, *in vivo*, this situation presents an additional complication.

When inverted *FRT*s lie near one another, a very frequent type of FLP-mediated recombination is unequal sister-chromatid exchange in G_2 of the cell cycle (Golic, 1994). This exchange produces dicentric chromosomes and acentric chromosome fragments (Figure 3.4A), although this does not necessarily prevent recovery of the desired product. For example, we have used FLP-mediated inversion of a portion of the *white* gene as a means of switching its expression on or off (Ahmad and Golic, 1996). In most cases though, the propensity for inverted *FRT* (or *lox*) repeats to form dicentric chromosomes should be taken into consideration when planning the experiment.

Finally, if a transformed gene is flanked by directly repeated *FRT*s, then it is also possible to use site-specific recombination to generate a dosage series of the transformed gene. In *Drosophila*, such amplified insertions are not necessarily silenced, and an increase in effective dosage of the amplified genes may be observed (Welte et al., 1993). When unequal sister-chromatid exchange occurs between directly repeated *FRT*s, one product of the exchange is a tandem duplication of the material between the *FRT*s (Figure 3.4B). This event is much less frequent than excision of the material between *FRT*s, making up perhaps 0.5% of the progeny following *hsp70-FLP* induction. These duplications can still be recovered with relative ease if there is some phenotypic screen for their occurrence, for instance, if a gene between the *FRT*s produces a distinctly different phenotype when it is duplicated (Golic and Lindquist, 1989), or if the novel junction produced at the site of

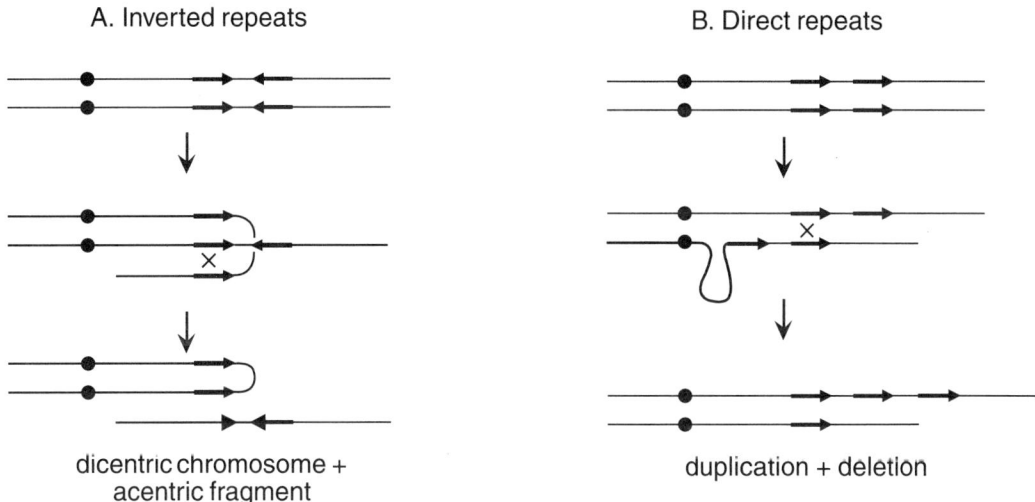

FIGURE 3.4 Unequal sister-chromatid exchange. The products of recombinase-mediated unequal sister-chromatid exchange between inverted or directly repeated target sites are diagrammed. Chromosomes are indicated as thin lines, centromeres as filled circles.

exchange brings together two parts of a gene to generate a functional copy of that gene (Ahmad and Golic, 1996; Aladjem et al., 1997).

3.3.1.2 Controlling Gene Expression and Analyzing Gene Function

Site-specific recombination has proved to be an extremely useful aid in the analysis of gene function. One of the most frequent areas of application has been in controlling gene expression to generate functional mosaics. Genes can be activated or inactivated by site-specific recombination. Constructs can be designed so that one gene is activated while another is simultaneously inactivated, providing an elegant method for generating a mutant cell and simultaneously marking it with a gratuitous phenotype so that it can be easily visualized (Struhl and Basler, 1993). If these gene-switching events do not occur in 100% of cells, then the animal will become a genetic mosaic. These mosaics have been extremely valuable in answering questions such as where and when a gene functions, and whether that action is cell autonomous. In addition, mosaic analysis can greatly facilitate the analysis of recessive lethal mutations, because in many cases clones of cells with a lethal genotype may survive even though an entirely mutant animal does not. Site-specific recombination even allows for the analysis of dominant lethal genotypes through the route of building constructs *in vitro* that would normally be lethal (for example by misexpression of a developmental regulator), but that are expressed only when activated by site-specific recombination *in vivo*. To generate mosaics, *FRT*s have been successfully placed in nontranslated leader regions of genes, in introns, and even within coding sequences.

These methods rely upon having a cloned copy of a gene whose expression can be manipulated in a mutant background. But site-specific recombination can also be used to make mosaics for mutations in genes that have not been cloned. FLP can efficiently catalyze recombination between *FRT*s located at the same site on homologous chromosomes. When such an exchange occurs in G_2 of the cell cycle, the recombinant chromatids typically segregate to opposite poles at the following mitosis, and produce daughter cells that are homozygous for alleles that had originally been heterozygous, giving rise to a genetic mosaic (Figure 3.5; Golic, 1991; Beumer et al., 1998). FLP-catalyzed mitotic recombination is efficient in *Drosophila* — a single heat-shock induction of FLP typically produces many clones in a single individual, and even in a single tissue.

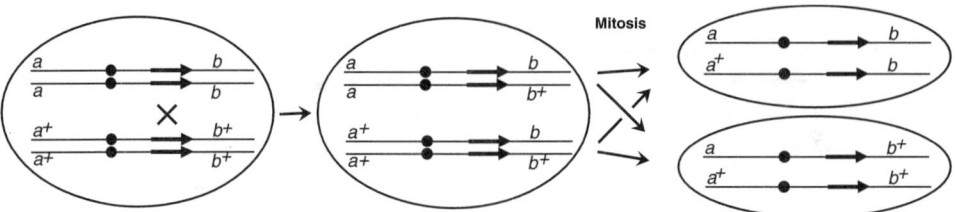

FIGURE 3.5 Mosaicism produced by mitotic recombination in G2. When recombinase-mediated exchange between homozygous target sites occurs in G2 of the cell cycle, it is typically followed by segregation of the recombinant chromatids. This produces two daughter cells that are homozygous for the alleles (previously heterozygous) that lie distal to the site of exchange. The diagram shows the chromosomal composition of a cell in G2 prior to the exchange (at left), after the exchange (middle), and the two daughter cells after mitosis (right).

This high frequency of mitotic recombination may be especially useful in a species where there are few or no pre-existing mutations because it can be adapted to provide a rapid and efficient genetic screen for newly induced recessive mutations. Any mutation lying distal to a homozygous insertion of *FRT*s can be made homozygous in a fraction of cells at an early stage of development. By inducing mitotic recombination in the F_1 progeny of a mutagenized parent, individuals that carry a mutation causing a phenotype that is visible in adults may be recognized (Xu and Rubin, 1993). The major benefit of this screen is that it saves two generations of laborious crosses when compared with traditional screens for autosomal recessive mutations.

It is not the purpose here to discuss in any detail the generation or analysis of genetic mosaics. This topic has been covered in reviews that should be consulted if more detail is desired (Golic, 1993; Xu and Harrison, 1994; Theodosiou and Xu, 1998). We simply wish to point out the possibilities that site-specific recombination offers for the manipulation of gene expression.

3.3.2 CHROMOSOME REARRANGEMENTS

Site-specific recombination has also been used to generate large- and small-scale chromosome rearrangements (Matsuzaki et al., 1990). A similar approach might be used to generate rapidly, in other insect species, at least some of the genetic tools that have long been available in *D. melanogaster*. For instance, inversions might be easily generated to provide a basic set of balancer chromosomes. Small deficiencies and duplications could be produced to facilitate screening for mutations in a defined region, or for analyzing genes located within the deleted region.

The construction of chromosome rearrangements by site-specific recombination is considerably less efficient than the type of short-range events that have been discussed to this point. Recombination can occur between *FRT*s that are separated by large distances, but at least in *Drosophila*, the frequency of recombination between two *FRT*s is partially governed by the distance between those *FRT*s. Using a standard 1-h heat shock to induce an *hsp70-FLP* gene, recombination between closely spaced *FRT*s in *cis*, for instance ~5 kb apart, occurs with almost 100% efficiency. This high frequency of recombination extends until *FRT*s are nearly 200 kb apart (by extrapolation of data obtained from generating large and small chromosome rearrangements). Then, as *FRT*s become separated by greater distances, the frequency of recombination decreases until, when *FRT*s are approximately 30 Mb apart, they recombine at a rate of approximately 0.03%. Additionally, recombination between *FRT*s in *trans* is even less frequent than recombination between *FRT*s in *cis* (Golic and Golic, 1996a).

Recombinants that arose at this low frequency were recovered with relative ease by virtue of a simple screening system. Two *P*-element vectors that each carry half of the *Drosophila white* gene were used. One element carries the 5' half of the gene, ending at an *FRT* located within the first intron. The second element carries the 3' half of *white*, starting at an *FRT* located at the same

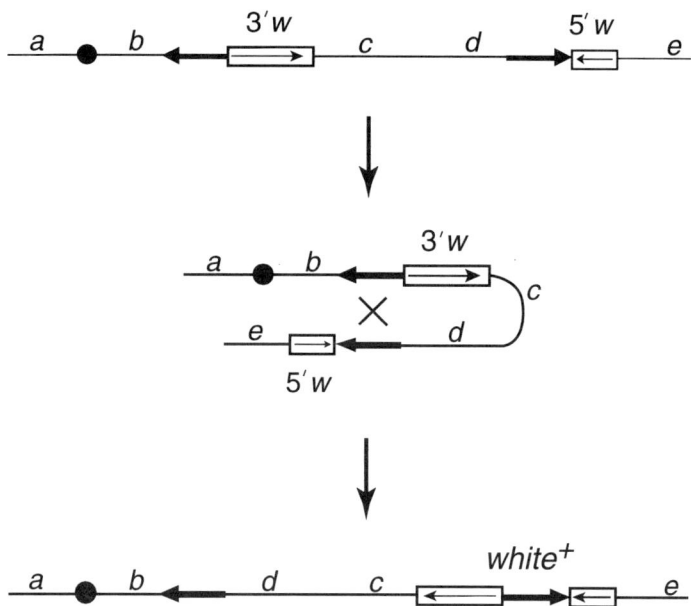

FIGURE 3.6 Positive marking of a recombinase-mediated inversion. Chromosome rearrangements generated by site-specific recombination can be positively marked by using a pair of transgene constructs that, upon recombination, join two nonfunctional gene halves to generate a functional dominant marker. In the example diagrammed here, the 5′ and 3′ halves of the *Drosophila white* gene are used. Rectangles indicate the halves of the gene, with small arrows indicating the direction of transcription.

site and in the same orientation in the first intron. Neither half is functional by itself, but when the two elements recombine at the *FRT*s, a functional *white* gene is reconstructed by the recombination event (Figure 3.6). Thus, in a *white* mutant background, chromosome rearrangements are easily recovered by heat-shocking flies that carry insertions of an *hsp70-FLP* gene and these two *P* elements, mating them *inter se*, and screening their offspring for individuals with pigmented eyes (Golic and Golic, 1996a). Similarly, in plants and mammals chromosome rearrangements were easily recovered by using chemical selections for genes constructed at the rearrangement breakpoint by the action of site-specific recombination (Medberry et al., 1995; Osborne et al., 1995; Ramirez-Solis et al., 1995).

The type of rearrangement produced depends on the relative orientation of the two elements, and their locations within the genome. The products are simply a large-scale equivalent of those diagrammed in Figure 3.2. In the two simplest cases, chromosomal inversions can be produced by intramolecular recombination between inverted *FRT*s, and deficiencies can be generated by intramolecular recombination between directly repeated *FRT*s. Reciprocal translocations have also been produced by site-specific recombination (Qin et al., 1994; Smith et al., 1995; Beumer et al., 1998).

3.3.3 SITE-SPECIFIC RECOMBINATION IN TRANSFORMATION

When a transformation method is developed for a species, further rounds of transformation may be accomplished by that same method. Potential techniques are discussed elsewhere in this book. However, there may be significant advantages in the consideration of site-specific recombination as a transformation tool. When the initial transformation step is difficult or inefficient it may be advantageous to use site-specific recombination as an alternative method of producing transgenic animals. Moreover, a transformation method based on site-specific recombination might also be suitable for the integration of large DNA fragments that cannot be transformed with available methods.

If the initial transformation construct includes a target site for site-specific recombination, then subsequent rounds of transformation may take advantage of this chromosomal target site by directing integration to that site with the site-specific recombinase. In some cases it may be worth investigating whether sequences that resemble *FRT*s or *lox* sites already exist in the genome, since it is possible that these can be used as target sites for integration (Sauer, 1996).

In its simplest form recombinase-mediated integration may be achieved with a single *FRT* in the chromosome and a matching *FRT* on a circular DNA molecule that is to be integrated. In such a case, site-specific recombination may integrate the circular molecule, as diagrammed in Figure 3.2C. With organisms that are amenable to chemical selection in single cells, there are many examples of using site-specific recombination for this approach to transformation (Falco et al., 1982; Sauer and Henderson, 1990; O'Gorman et al., 1991; Huang et al., 1991). Here, the experimental evidence from *Drosophila* that such an approach may work in insects is reviewed, and a potentially useful elaboration on this method discussed.

3.3.3.1 Techniques for Integrating DNA at an *FRT*

With insects, no methods are currently available that allow *in vitro* manipulation of single cells followed by the return of those cells to intact animals. Until such techniques are developed, experiments designed to produce genetically altered or transgenic individuals must be carried out using the intact organism. Typically, some form of genetic screen is used to recover transgenic animals. Genetic screens in *Drosophila* were used to show that site-specific recombination can be used to integrate an extrachromosomal *FRT*-bearing circle at a chromosomal *FRT* target site (Golic et al., 1997). Crosses were used to generate flies that carried three transgene components that had previously been transformed using *P* elements. These were (1) a heat-shock-inducible *FLP* gene; (2) a target element carrying a single *FRT* target site; and (3) a donor element carrying, between directly repeated *FRT*s, the DNA that was to be integrated at the target site. Flies that carried all three elements were heat-shocked to induce FLP synthesis. Recombination between the *FRT*s of the donor construct excised an *FRT*-bearing circle and, in the same cell cycle, a second round of recombination between this extrachromosomal circle and the target *FRT* integrated the circle at the target site. In other words, FLP carried out the reaction diagrammed in Figure 3.2A (donor excision), then the reaction shown in Figure 3.2C (integration at the target) at a different chromosomal *FRT* insertion. The result was FLP-mediated mobilization of the *FRT*-flanked donor DNA to the target.

The products of this targeted mobilization were recovered by two different schemes. In the first, the donor was an *FRT*-flanked *white$^+$* gene located on the *X* chromosome or on an autosome carrying a gratuitous dominant genetic marker. Heat-shocked males were test-crossed to *white* females and their progeny were screened either for w^+ sons (when an *X*-linked donor was used) or for w^+ progeny without the gratuitous marker (when the autosomal donor was used). In these cases w^+ segregated away from the chromosome that carried it originally. Such progeny must arise from an event in the parental germline in which site-specific recombination mobilized the donor DNA to the target *FRT*. Further genetic and molecular tests confirmed that almost all such progeny did carry the expected w^+ gene inserted at the target site.

In a second screen, donor and target elements were constructed such that each carried half of a functional *Drosophila white* gene. After donor excision and reintegration at the target site a full w^+ gene was constructed at the target site and provided the phenotypic signal for integration, similar to the screen that was used for recovering chromosome rearrangements. The heat-shocked males and females, which carried a null mutation of the normal *white* gene, were mated *inter se* and their offspring were screened for pigmented eyes. Crosses were typically set up with ratios of two to three males for each female per vial. From these crosses w^+ progeny were recovered with relative ease, arising in approximately 2% of the vials. In experiments that used the β2-*tubulin* promoter to express FLP, site-specific mobilization of the donor to the target was detected in nearly 5% of all vials examined. Further tests showed that virtually all resulted from FLP-mediated site-specific recombination at the target site.

These experiments demonstrate that FLP–*FRT* site-specific recombination can integrate extrachromosomal DNA at a chromosomal target site. Because efficient methods of transformation have already been developed for *Drosophila,* this technique is at present most useful for overcoming the problem of variable chromosomal position effects. Several transgene constructs may be placed at precisely the same chromosomal site, eliminating variable position effects on gene expression. However, in the context of this book, the more interesting possibility is that site-specific recombination may be used as a direct transformation tool. In our experiments a single extrachromosomal circle was able to find and recombine with a single chromosomal target site (two copies of each if the cells are in G_2). Thus, it does not seem unreasonable to suppose that if an *FRT*-bearing plasmid were injected into an insect strain that already carried a chromosomal *FRT*, then FLP might catalyze exchange between the chromosomal target site and the injected DNA resulting in integration of that DNA. Morris et al. (1991) and Konsolaki et al. (1992) have shown that intermolecular FLP-mediated recombination does occur when two different *FRT*-bearing plasmids are coinjected into insect embryos along with an FLP expression plasmid or FLP mRNA, lending further credence to the notion that an injected *FRT*-bearing plasmid may be able to recombine with an *FRT* located in a chromosome.

3.3.3.2 Efficiency of Recombinase-Mediated Integration

3.3.3.2.1 Donor: Target Pairing

The results of our FLP-mediated mobilization experiments suggest that the degree of sequence homology between the donor episome and the chromosomal target site may influence integration efficiency. We measured integration frequencies with donor/target combinations that differed in the extent of sequence homology they shared (Golic et al., 1997). When using the heat-inducible *FLP* gene, with donor/target pairs sharing ~4.1 kb of homology, we recovered 52 site-specific integration events from 2844 crosses (1.8%). For donor/target pairs with ~1.1 kb homology, only two site-specific integration events were recovered from 904 crosses (0.2%). With ~200 to 600 bp of homology between donor and target, only two site-specific integration events were found from more than 2000 crosses (<0.1%).

This apparent dependence on donor/target homology may be based on the fact that in *Drosophila,* and in the Diptera generally, homologous chromosomes pair in mitotic cells. Cytological and genetic evidence demonstrates that homologues are associated with each other during the cell cycle in *Drosophila* (discussed by Golic and Golic, 1996b). Stevens (1908) and Metz (1916) reported frequent association between homologous chromosomes in metaphase spreads in ~80 dipteran species, including *Drosophila.* Many insect species also possess giant polytene chromosomes in which the endoreduplicated homologues are paired along their entire lengths. The existence of this mechanism for homologous pairing may underlie the apparent pairing dependence of site-specific integration. The *FRT*s of the donor and the target could be brought to relative proximity by pairing interactions between homologous sequences. This proximity would, in turn, promote the direct interaction between the *FRT*s that is required for recombination. Although further experiments are needed to confirm these initial observations, it is certainly possible that the same effect will also apply in other Diptera. Therefore, when attempting to use site-specific recombination as a method of DNA integration, at least in Diptera, it would probably be wise to include a reasonable extent of homology (>4 kb) between the donor and target constructs.

3.3.3.2.2 Stabilizing the Integrated DNA

One of the main problems limiting the efficiency of site-specific recombination as a tool for DNA integration is the tendency for an integrated circle to reexcise by another round of recombination between the *FRT*s that now flank the integrated DNA. This intramolecular excision reaction is kinetically favored over the intermolecular integration reaction. A particularly useful variant of FLP-mediated DNA integration was developed by Schlake and Bode (1994) to deal with this problem. Their method was tested in cultured cells and showed substantially greater efficiency than was provided by simple reciprocal exchange between an extrachromosomal *FRT* and a chromosomal *FRT*. At the target site, they used two *FRT*s with different spacer sequences (one wild-type *FRT* and one mutant, either *FRT₃* or *FRT₅*; see Figure 3.1). Because exchange partners are determined by the sequence of the 8 bp spacer, these *FRT*s do not recombine with each other, and no recombination was observed between them even in the long-term presence of constitutive FLP expression (Seibler and Bode, 1997). The donor construct carried the same two *FRT*s, allowing the donor *FRT*s to recombine with those of the target. A double reciprocal exchange replaces the sequence at the target that is flanked by *FRT*s with the donor sequence that is flanked by *FRT*s (Figure 3.7). The integrated donor sequence cannot be reexcised by simple recombination between the flanking *FRT*s. Thus, the material between *FRT*s acts as a cassette that can be swapped by a double exchange (or by two successive single exchanges). Highly efficient cassette exchange was observed in cultured cells, in both transient and stably transfected lines. An added advantage of this method is that two types of selection may be used. Selection may be applied for acquisition of the donor cassette, or for loss of the target cassette. Because such loss requires double exchange with the incoming cassette, these loss events should coincide with gain of the donor cassette.

Bethke and Sauer (1997) used the cassette exchange strategy in yeast with the Cre-*lox* site-specific recombination system and saw a 20-fold increase in integration frequency compared with single reciprocal exchange, with an absolute frequency of integration as high as 1 of 20 cells (see Figure 3.1 for *lox* mutant sequence).

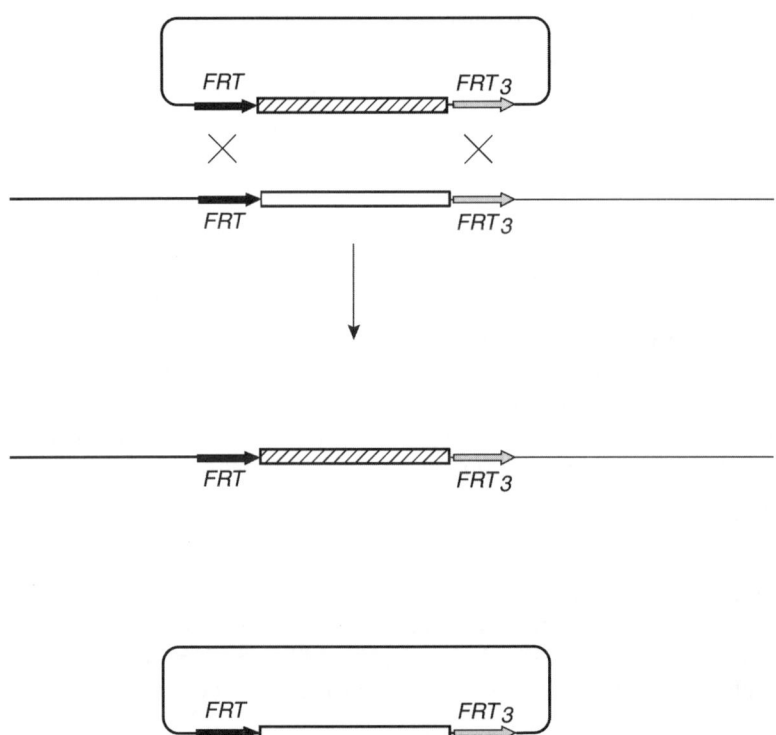

FIGURE 3.7 Recombinase-mediated cassette exchange. Diagram of cassette exchange directed by using incompatible target sites (one wild-type and one mutant) in *cis*. Matching target sites are used on the donor molecule. A double reciprocal exchange swaps the DNA that is flanked by the target sites.

A similar method that may be considered is the use of a target site and donor that each carry two wild-type *FRT*s in inverted, rather than direct orientation. This setup would similarly allow for a double reciprocal exchange to place donor DNA at the target site. The donor DNA could not reexcise by exchange between the flanking *FRT*s, although it could invert. However, because there are added complications that arise from the use of inverted repeats (see Section 3.3.1.1) the method developed by Bode and co-workers is preferred.

Another method that has been tried is the use of *FRT*s that are mutated in the FLP binding sites. The 46A mutant *FRT* has a mutation in one of the FLP binding sites and the 46T mutant *FRT* has a mutation in the other binding site. Because recombination occurs within the spacer region of the *FRT* the two recombinant *FRT*s are actually hybrids, consisting of one 13-bp inverted repeat from each of the original *FRT*s. Recombination between these two mutant *FRT*s generates one wild-type *FRT* and one *FRT* with mutations in both of the inverted repeats (46A/46T double mutant). *In vitro* studies suggested that this recombination proceeds with an efficiency that is only slightly reduced relative to wild-type *FRT*s. However, the reversal of this reaction (recombination between the 46A/46T double mutant and a wild-type *FRT*) proceeds at a much lower efficiency (Senecoff et al., 1988). Therefore, if the donor and the target each carry one of the two mutant *FRT*s, recombination should "lock" the integrated DNA fragment in the chromosome. This pair of mutant *FRT*s has been tested in several systems for the ability to prevent the reversal of an FLP-catalyzed recombination (Huang et al., 1991; Morris et al., 1991; Y. Rong and K. Golic, unpublished results). However, this method has not proved as useful *in vivo* as the *in vitro* results suggested. It seems that the efficiency of recombination between these two *FRT*s has been compromised *in vivo* by the mutations, and no further progress with this method has been made. However, a similar double-mutant design has been proved productive for site-specific integration by the Cre–*lox* system. Araki et al. (1997) observed up to a 32-fold increase of integration frequency using a pair of mutant *lox* sites that prevent reexcision *in vivo*. Therefore, it may still be worthwhile to test other pairs of mutant *FRT*s.

Finally, the stability of DNA integrated via site-specific recombination may be stabilized by controlling the synthesis of the recombinase. If recombinase is no longer present after integration, then the integrated DNA cannot excise. Experiments *in vivo* suggest that there is an optimal FLP concentration that allows integration, but is low enough so that excision does not inevitably follow (Huang et al., 1991; Golic et al., 1997). To prevent reexcision, recombinase may be synthesized transiently. Use of the *hsp70* promoter allows this — as cells recover from heat shock they turn off the *hsp70-FLP* gene. Varying the severity of heat shock changes the extent of *hsp70-FLP* induction, and the rapidity of recovery. FLP-expressing plasmids have been introduced into cells so that subsequent loss of the plasmids by nuclease degradation or loss at cell division assures transient expression of the recombinase (Morris et al., 1991; O'Gorman et al., 1991). Purified Cre recombinase has been introduced into cells to achieve high rates of integration (Baubonis and Sauer, 1993), and injection of FLP mRNA should exhibit a similar property of transient expression.

3.3.3.2.3 Position Effects

Another factor that can influence targeting efficiency is the position of the target site (Huang et al., 1991; Baubonis and Sauer, 1993; Golic et al., 1997; Y. Rong and K. Golic, unpublished). Chromatin configuration at the target site may affect the interaction among the donor, the target, and the recombinase. An "open" region might permit easy access of the recombinase and more intimate interaction between the *FRT*s. In fact, it has been shown that *FRT*s lying adjacent to centric heterochromatin, and thus subject to the variegated silencing that is called position effect variegation, are recombined less efficiently by FLP than are *FRT*s at typical euchromatic sites (Ahmad and Golic, 1996). Thus, if possible, it can be worthwhile to use more than a single target site when attempting site-specific integration because the efficiency of integration at different target sites may differ considerably.

3.4 DOMINANT MARKER GENES FOR INSECT TRANSFORMATION

A successful germline transformation system requires two components: a transformation system that can introduce foreign DNA into the host genome and a dominant genetic marker that allows the recognition of individuals with successful integration of foreign DNA. For a variety of insect species transposable elements serve well as transformation vectors (for review, see O'Brochta and Atkinson, 1996). However, suitable transformation markers have not been developed for most insect species. To produce a marker gene for transformation, typically a screen is undertaken to find a recessive mutant with an easily visible phenotype. The mutation is then mapped and the gene cloned. The wild-type gene is then used as a dominant marker to transform recessive mutant animals. There are a few examples in which insect species other than *D. melanogaster* have been successfully transformed by following this general approach (e.g., Loukeris et al., 1995; Lozovskaya et al., 1996; Jasinskiene et al., 1998; see Chapter 4 by Sarkar and Collins). However, this process is tedious, has to be repeated for every species, and is often hampered by the lack of developed molecular and genetic tools in those species.

To eliminate this tedious process of marker development, the idea that a common marker might be developed for transformation of wild-type animals of a variety of insect species is very attractive. Such a common marker must have three essential properties. First, it must be dominant to the wild-type allele so that a recessive mutant host line is not necessary. Second, it should affect a nonessential biological trait that is shared by many species, for example, eye coloration. Third, the marker gene must be properly expressed, and its product functional in heterologous organisms. This third property cannot be easily verified except by the recovery of a transgenic animal bearing the marker gene and expressing its phenotype. However, if a marker gene were developed from a dominant mutation in a gene that is highly conserved among different species, one might reasonably suspect that the mutant allele would be functional in many species.

3.4.1 PUG^D

We have identified a dominant mutation in *D. melanogaster* that has the potential to serve as a common transformation marker for other insect species. The *pugilist^{Dominant}* (pug^D) mutation causes dominant defects in *Drosophila* eye pigmentation (Rong and Golic, 1998). As in many other insect species, the color of the *Drosophila* eye results from the synthesis of two families of pigment molecules, ommochromes and pteridines (for reviews, see Phillips and Forrest, 1980, Summers et al., 1982; see Chapter 4 by Sarkar and Collins). Ommochromes are synthesized from tryptophan. Pteridines are made from GTP which can be synthesized from purine precursors. The pug^D mutation reduces ommochromes and virtually eliminates pteridines in the middle of the eye, while preserving normal pigmentation at the eye margin. The mutant phenotype is undiminished by the presence of two copies of the wild-type gene.

Ommochromes and pteridines are present in many insect species besides *Drosophila*. Ommochromes were identified in the following orders: Blattaria, Coleoptera, Diptera, Heteroptera, Hymenoptera, Lepidoptera, Neuroptera, Odonata, Phasmida, and Saltatoria (Phillips and Forrest, 1980). Pterins, which include pteridines, were found in the following orders: Coleoptera, Dictyoptera, Diptera, Hemiptera, Hymenoptera, Lepidoptera, Neuroptera, Orthoptera, and Phasmida (Summers et al., 1982). In other insects these molecules contribute to the pigmentation of many body parts, including the eye.

We have cloned the wild-type and mutant alleles of *pug*. The pug^D mutation lies in a gene that encodes the trifunctional enzyme methylenetetrahydrofolate dehydrogenase (MTHFD). The mutant allele consists of the N-terminal one fifth of the coding region of MTHFD fused to 1 kb of simple repetitive DNA. This piece of repetitive DNA consists of about 140 near-perfect units of the sequence AGAGAGA. A 3.4-kb DNA fragment from pug^D regenerates the dominant phenotype when transformed into wild-type *Drosophila*. The dominance of the mutant almost certainly requires translation of the repetitive stretch (Rong and Golic, 1998).

The MTHFD enzyme provides essential tetrahydrofolate derivatives for purine *de novo* synthesis. Genes that encode MTHFD have been found in a large variety of organisms (for reviews, see Benkovic, 1980; Appling, 1991), and their amino acid sequences are highly conserved from yeast to human. In fact, when expressed in yeast, a mammalian form and an insect form of this enzyme can rescue the auxotrophy that is caused by mutations in *ade3*, the corresponding yeast gene (Thigpen et al., 1990; Tremblay and MacKenzie, 1995). In light of this high degree of conservation, it is very likely that other insect species have highly homologous versions of the same gene, and it certainly seems possible that *pug^D* could have a dominant effect on pigment formation in other insects.

The mechanism by which *pug^D* causes reduction in eye pigmentation is not known, but it probably acts by interfering with purine *de novo* synthesis. This is consistent with the fact that pteridines are derived from purine precursors, and that other *Drosophila* mutations in purine synthesis often cause defects in eye pigmentation (Johnstone et al., 1985; Henikoff et al., 1986; Tiong et al., 1989). Since purine *de novo* synthesis is a highly conserved process, we expect that *pug^D* would act similarly in other insect species.

Before *pug^D* can be developed into a common transformation marker, one needs to consider some practical points. The first concerns expression of the *pug^D* gene. The 3.4-kb genomic fragment from *pug^D* flies may be usable in other species without modification. There are many examples in which *Drosophila* genes are expressed normally in other insects, and vice versa (e.g., Loukeris et al., 1995; Skavdis et al., 1996; Jasinkiene et al., 1998). Alternatively, we constructed a *pug^D* cDNA clone driven by the *Drosophila hsp70* promoter, which functions in other insect species (e.g., Loukeris et al., 1995). However, this *hspug^D* gene only produced a mild pigment defect, even when present in multiple copies. If the *Drosophila pug^D* gene is not expressed in another species, a better solution would be to isolate the species-specific regulatory elements of the respective *pug$^+$* gene and use them to direct trans-specific expression of the *Drosophila pug^D* gene.

Another practical point concerns cloning with *pug^D*. The repetitive sequences of *pug^D* are unstable in regular bacterial cloning strains. DNA clones with shorter repeats are spontaneously generated during overnight culturing. Since translation of a certain minimum number of these repeats seems to be necessary to produce the phenotype (Rong and Golic, 1998), it is important to maintain these repeats during the cloning process. We find that the repeats are fairly stable during bacterial growth at 30°C, rather than 37°C in the TOP10 strain (Invitrogen, Carlsbad, CA; other strains may also work). It is, nonetheless, a good idea to check repeat length at each step of the cloning process, as well as checking the final DNA preparation that is to be used for transformation, to make sure that the full-length *pug^D* gene has been maintained.

Finally, we have to consider the effect of *pug^D* on viability. In some cases, the ultimate goal of a program of transgenesis is to release genetically engineered insects into the wild. The transgenic animals must have good overall fitness so that they can compete effectively in the wild. In the laboratory, we did not observe any obvious deterioration in viability of flies carrying a single copy of the *pug^D* transgene. However, it seems likely that sensitive measurements would reveal some loss of fitness if *pug^D* does interfere with purine synthesis. This problem can be overcome by using site-specific recombination to remove the *pug^D* marker gene after transformation (Section 3.3.1.1).

The *pug^D* mutation appears to possess the essential properties of a common transformation marker. It is a mutation in a highly conserved gene, it is dominant over the wild-type gene, and it causes visible defects in the deposition of pigments that are also found in other insects.

3.4.2 OTHER DOMINANT TRANSGENES

It should be noted that *pug^D* may be but one of many dominant mutations from *D. melanogaster* that have the potential to serve as markers for insect transformation. Even dominant mutations that produce strong developmental defects might serve as transformation markers if site-specific recombination is used to excise that marker gene after transformation. Thus, mining the genetic resources

TABLE 3.1
Trangenes That Produce Dominant Phenotypes in *D. melanogaster*

Gene	Phenotypes	Mechanism	Regulatory Elements	Notes	Refs.
Antennapedia (*Antp*)	Antenna to leg transformation	Ectopic expression of the wild-type gene	*heat shock protein 70 (hsp 70)*	Most animals failed to survive but might be useful with more limited espression; most effective stages of heat shock are the second and third instar; ~50% penetrance	Schneuwly et al., 1987; Gibson and Gehring, 1988
Mouse *Hox 2.2* (homologous to *Antp*)	Antenna to leg transformation	Ectopci expression of the wild-type gene	*hsp70*	Animals failed to survive to adulthood	Malicki et al., 1990
argos	Eye and wing defects	Ectopic overexpression of the wild-type gene	*hsp70*	—	Sawamoto et al., 1994
corkscrew	Rough and reduced eye	Production of dominant negative protein	2× *sev/hsp70*[a]	—	Allard et al., 1996
Human *cyclin-dependent kinase inhibitor*	Rough eye	Ectopic expression of the wild-type gene	*GMR*[b]	Requires a very high level of expression	do Nooij and Hariharan, 1995
dacapo (cyclin-dependent kinase inhibitor)	Rough eye	Overexpression of the wild-type gene	*GMR*	Requires a very high level of expression	de Nooij et al., 1996
Deformed	Head defects	Ectopic expression of the wild-type gene	*hsp70*	Most animals failed to survive to adulthood	Chadwick et al., 1990
Human *Hox 4.2* (homologous to *Deformed*)	Head defects	Ectopic expression of the wild-type gene	*hsp70*	Low penetrance, often leads to lethality	McGinnis et al., 1990
Delta	Eye or wing defects	Expression of various truncated genes producing dominant negative proteins	*GRM, Gal4/UAS*,[c] *sevenless*	—	Sun and Artavanis-Tsakonas, 1996, 1997
Ellipse (EGFR)	Eye ablation	Overproduction of dominant negative protein	*Gal4/UAS*	—	Freeman, 1996
eyes absent	Ectopic eye formation	Ectopic expression of the wild-type gene	*Gal4/UAS*	Requires a high level of expression	Bonini et al., 1997
eyeless	Ectopic eye formation	Ectopic expression of the wild-type gene	*Gal4/UAS*	Many Gal4 insertions causes lethality when combined with the UAS-eyeless transgene	Halder et al., 1995

Gene	Phenotype	Purpose	Promoter	Comments	References
head involution defective (hid)	Eye ablation	Ectopic expression of the wild-type gene	GMR	Flies have normal viability and fertility even when homozygous; might be a good system to ablate other nonessential tissues	Grether et al., 1995
myosin heavy chain (mhc)	Flightless, reduced jumping ability	Production of the embryonic isoform throughout development	mhc	Normal mating ability in laboratory	Wells et al., 1996
Notch	Missing bristles	Production of constitutively active protein	Actin 5C, hsp70	Only mosaic animals were generated	Struhl et al., 1993
Notch	Defects in wing, eye and/or bristle formation	Various partial deletions of the gene producing antimorphic or neomorphic mutations	hsp70	Requires a high level of expression	Rebay et al., 1993
P35 from baculovirus	Rough eye	Ectopic expression of the wild-type gene	GMR	—	Hay et al., 1994
pugilist[D]	Eye pigment defects	Dominant mutant allele	pugilist, hsp70	—	Rong and Golic, 1998
ras1	Rough eye	Production of constitutively active protein	sevenless	Very dosage sensitive	Fortini et al., 1992
ras2	Wing, eye defects	Production of constitutively active protein	hsp70, ras2	Uninduced hsp70-driven expression led to low viability	Bishop and Corces, 1988
ras2	Wing, eye defects	Production of constitutively active protein	Gal4/UAS	—	Brand and Perrimon, 1993
reaper	Eye ablation	Ectopic expression of the wild-type gene	GMR	Sensitive to dosage and requires a high level of expression	White et al., 1995; Hay et al., 1995
Human SCA3/MJD	Rough eye	Expression of glutamine-repeat disease gene	Gal4/UAS	—	Warrick et al., 1998
Serrate	Scalloped wing	Ectopic expression of the wild-type gene	Gal4/UAS	—	Thomas et al., 1995
Serrate	Eye or wing defects	Expression of various truncated genes producing dominant negative proteins	GMR, Gal4/UAS, sevenless	—	Sun and Artavanis-Tsakonas, 1996, 1997
Serrate	Wing defects	Overproduction of dominant negative protein	Gal4/UAS	Sensitive to dosage	Hukriede et al., 1997
sevenless	Rough eye	Production of constitutively active protein	2× sev/hsp70	—	Basler et al., 1991
Sex combs reduced (Scr)	homeotic transformation	Ectopic overexpression of the wild-type gene	hsp70	Most adults failed to survive	Gibson et al., 1990; Zhao et al., 1993

continued

TABLE 3.1 (continued)
Trangenes That Produce Dominant Phenotypes in *D. melanogaster*

Gene	Phenotypes	Mechanism	Regulatory Elements	Notes	Refs.
Mouse *Hox 1.3* (homogous to *Scr*)	Homeotic transformation	Ectopic overexpression of the wild-type gene	*hsp70*	Most adults failed to survive	Zhao et al., 1993
Ultrabithorax	Homeotic transformation	Ectopic overexpression of the wild-type gene	*hsp70*	A mild heat shock produced mutant phenotypes in 50 to 80% of the viable animals	Mann and Hogness, 1990
wingless	Rough and reduced eye	Ectopic expression of the wild-type gene	2× *sev/hsp70*, *Gal4/UAS*	Requires a high level of expression	Brunner et al., 1999
yan	Rough and reduced eye	Production of constitutively active protein	*GMR, seven less*	—	Rebay and Rubin, 1995

[a] 2× *sev/hsp70*: two copies of the *seven less* enhancer plus the *hsp70* promoter (Basler et al., 1991).
[b] *GMR*: *glass* multimer reporter (Ellis et al., 1993; Hay et al., 1994).
[c] *Gal4/UAS*: transcriptional activation systems from yeast (Fischer et al., 1988; Brand and Perrimon, 1993).

of *D. melanogaster* might greatly aid the discovery of markers for transformation of other insects. Table 3.1 lists several genes that have produced dominant phenotypes within transgenic constructs in *Drosophila* and that may also be suitable for use in other insects. Although this is certainly only a partial listing, it does show that many possibilities exist. In many instances, the biological processes that are affected, such as cell–cell signaling or intracellular signal transduction, are highly conserved. One problem will be to find promoters that can be used in other species. Many of the constructs are (or are likely to be) dominantly lethal if expressed ubiquitously. However when their expression is limited to a specific tissue, such as the eye, they produce easily visualized phenotypes. The promoters that have been useful in *Drosophila* are given in Table 3.1.

3.5 SUMMARY

Site-specific recombination has proved its value as a tool for the directed manipulation of genes and genomes in several organisms. Recombinases have been used to manipulate the expression of transgenes, to delete genes from transgenic constructs, to insert transgenes at specific chromosomal target sites, and to rearrange chromosomes to produce deletions, duplications, or balancer chromosomes. Many of the applications of site-specific recombination are best realized if the use of site-specific recombination is planned from the outset. This may involve as little as including a token recombinase target site in a construct, on the chance that it may later prove to be valuable, possibly as a target site for further rounds of DNA integration. When the purpose is clear, more elaborate constructs may be employed. One area that may be creatively explored is the combined use of more than one recombinase. For instance, if site-specific recombination is used to integrate DNA at a chromosomal target site, one may still wish to manipulate the inserted DNA by site-specific recombination. This might be easily accomplished by using a second site-specific recombinase that recognizes target sites distinct from those employed in the integration step.

Site-specific recombination gives a further advantage to attempts at transforming new insect species — it considerably broadens the possible choices for transformation marker genes. Because a marker gene can be removed after a transformant has been recovered, genes that have serious fitness consequences can be used as markers to detect transformants. Many such gene constructs with dominant phenotypes have been transformed into *Drosophila*; some of these genes may be suited for use in other insects.

ACKNOWLEDGMENTS

The authors thank Kami Ahmad and Mary Golic for technical assistance. Work in our laboratory is supported by grants from the National Science Foundation (MCB-9728070) and from the National Institutes of Health (HD28694).

REFERENCES

Ahmad K, Golic KG (1996) Somatic reversion of chromosomal position effects in *Drosophila melanogaster*. Genetics 144:657–670

Aladjem MI, Brody LL, O'Gorman S, Wahl GM (1997) Positive selection of FLP-mediated unequal sister chromatid exchange products in mammalian cells. Mol Cell Biol 17:857–861

Allard JD, Chang HC, Herbst R, McNeill H, Simon MA (1996) The SH2-containing tyrosine phosphatase corkscrew is required during signaling by sevenless, Ras1 and Raf. Development 122:1137–1146

Appling DR (1991) Compartmentation of folate-mediated one carbon metabolism in eukaryotes. FASEB J 5:2645–2651

Araki K, Araki M, Yamamura K (1997) Targeted integration of DNA using mutant *lox* sites in embryonic stem cells. Nucl Acids Res 25:868–872

Baek K-H, Ambrosio L (1994) An efficient method for microinjection of mRNA into *Drosophila* embryos. BioTechniques 17:1024–1026

Basler K, Christen B, Hafen E (1991) Ligand-independent activation of the sevenless receptor tyrosine kinase changes the fate of cells in the developing *Drosophila* eye. Cell 64:1069–1081

Baubonis W, Sauer B (1993) Genomic targeting with purified Cre recombinase. Nucl Acids Res 21:2025–2029

Benkovic SJ (1980) On the mechanism of action of folate- and biopterin-requiring enzymes. Annu Rev Biochem 49:227–251

Bethke B, Sauer B (1997) Segmental genomic replacement by Cre-mediated recombination: genotoxic stress activation of the p53 promoter in single-copy transformants. Nucl Acids Res 25:2828–2834

Beumer K, Pimpinelli S, Golic KG (1998) Induced chromosomal exchange directs the segregation of recombinant chromatids in mitosis of *Drosophila*. Genetics 150:173–188

Bishop JG, Corces VG (1988) Expression of an activated *ras* gene causes developmental abnormalities in transgenic *Drosophila melanogaster*. Genes Dev 2:567–577

Bonini NM, Bui QT, Gray-Board GL, Warrick JM (1997) The *Drosophila eyes absent* gene directs ectopic eye formation in a pathway conserved between flies and vertebrates. Development 124:4819–4826

Brand AH, Perrimon N (1993) Targeted gene expression as a means of altering cell fates and generating dominant phenotypes. Development 118:401–415

Brunner E, Brunner D, Fu W, Hafen E, Basler K (1999) The dominant mutation *glazed* is a gain-of-function allele of *wingless* that, similar to loss of APC, interferes with normal eye development. Dev Biol 206:178–188

Buchholz F, Angrand PO, Stewart AF (1996) A simple assay to determine the functionality of Cre or FLP recombination targets in genomic manipulation constructs. Nucl Acids Res 24:3118–3119

Buenzow DE, Holmgren R (1995) Expression of the *Drosophila* gooseberry locus defines a subset of neuroblast lineages in the central nervous system. Dev Biol 170:338–349

Chadwick R, Jones B, Jack T, McGinnis W (1990) Ectopic expression from the *Deformed* gene triggers a dominant defect in *Drosophila* adult head development. Dev Biol 141:130–140

Dale EC, Ow DW (1991) Gene transfer with subsequent removal of the selection gene from the host genome. Proc Natl Acad Sci USA 88:10558–10562

de Nooij JC, Hariharan IK (1995) Uncoupling cell fate determination from patterned cell division in the *Drosophila* eye. Science 270:983–985

de Nooij JC, Letendre MA, Hariharan IK (1996) A cyclin-dependent kinase inhibitor, *dacapo*, is necessary for timely exit from the cell cycle during *Drosophila* embryogenesis. Cell 87:1237–1247

de Wit T, Drabek D, Grosveld F (1998) Microinjection of Cre recombinase RNA induces site-specific recombination of a transgene in mouse oocytes. Nucl Acids Res 26:676–678

Dorer DR, Henikoff S (1994) Expansions of transgene repeats cause heterochromatin formation and gene silencing in *Drosophila*. Cell 77:993–1002

Ellis MC, O'Neill EM, Rubin GM (1993) Expression of *Drosophila* glass protein and evidence for negative regulation of its activity in non-neuronal cells by another DNA-binding protein. Development 119:855–865

Falco SC, Li Y, Broach JR, Botstein D (1982) Genetic properties of chromosomally integrated 2 mu plasmid DNA in yeast. Cell 1982 29:573–584

Fischer JA, Giniger E, Maniatis T, Ptashne M (1988) Gal4 activates transcription in *Drosophila*. Nature 332:853–856

Fortini ME, Simon MA, Rubin GM (1992) Signaling by the sevenless protein tyrosine kinase is mimicked by Ras1 activation. Nature 355:559–561

Freeman M (1996) Reiterative use of the EGF receptor triggers differentiation of all cell types in the *Drosophila* eye. Cell 87:651–660

Frise E, Knoblich JA, Younger-Shepherd S, Jan LY, Jan YN (1996) The *Drosophila* Numb protein inhibits signaling of the Notch receptor during cell-cell interaction in sensory organ lineage. Proc Natl Acad Sci USA 93:11925–11932

Garrick D, Fiering S, Martin DI, Whitelaw E (1998) Repeat-induced gene silencing in mammals. Nat Genet 18:56–59

Gibson G, Gehring WJ (1988) Head and thoracic transformations caused by ectopic expression of *Antennapedia* during *Drosophila* development. Development 102:657–675

Gibson G, Schier A, LeMotte P, Gehring WJ (1990) The specificities of Sex combs reduced and Antennapedia are defined by a distinct portion of each protein that includes the homeodomain. Cell 62:1087–1103

Golic KG (1991) Site-specific recombination between homologous chromosomes in *Drosophila*. Science 252: 958–961

Golic KG (1993) Generating mosaics by site-specific recombination. In Hartley D (ed) Cellular Interactions in Development: A Practical Approach. Oxford University Press, New York pp. 1–31

Golic, KG (1994) Local transposition of *P* elements in *Drosophila melanogaster* and recombination between duplicated elements using a site-specific recombinase. Genetics 137: 551–563

Golic KG, Golic MM (1996a) Engineering the *Drosophila* genome: chromosome rearrangements by design. Genetics 144:1693–1711

Golic MM, Golic KG (1996b) A quantitative measure of the mitotic pairing of alleles in *Drosophila melanogaster* and the influence of structural heterozygosity. Genetics 143:385–400

Golic KG, Lindquist S (1989) The FLP recombinase of yeast catalyzes site-specific recombination in the *Drosophila* genome. Cell 59:499–509

Golic MM, Rong YS, Petersen RB, Lindquist SL, Golic KG (1997) FLP-mediated DNA mobilization to specific target sites in *Drosophila* chromosome. Nucl Acids Res 25:3665–3671

Grether ME, Abrams JM, Agapite J, White K, Steller H (1995) The *head involution defective* gene of *Drosophila melanogaster* functions in programmed cell death. Genes Dev 9:1694–1708

Halder G, Callaerts P, Gehring WJ (1995) Induction of ectopic eyes by targeted expression of the *eyeless* gene in *Drosophila*. Science 267:1788–1792

Hay BA, Wolff T, Rubin GM (1994) Expression of baculovirus P35 prevents cell death in *Drosophila*. Development 20:2121–2129

Hay BA, Wassarman DA, Rubin GM (1995) *Drosophila* homologs of baculovirus inhibitor of apoptosis proteins function to block cell death. Cell 83:1253–1262

Henikoff S, Nash D, Hards R, Bleskan J, Woolford JF, Naguib F, Patterson D (1986) Two *Drosophila melanogaster* mutations block successive steps of *de novo* purine synthesis. Proc Natl Acad Sci USA 83:3919–3923

Huang L-C, Wood EA, Cox MM (1991) A bacterial model system for chromosomal targeting. Nucl Acids Res 19:443–448

Hukriede NA, Gu Y, Fleming RJ (1997) A dominant-negative form of Serrate acts as a general antagonist of Notch activation. Development 124:3427–3437

Jasinskiene N, Coates CJ, Benedict MQ, Cornel AJ, Rafferty CS, James AA, Collins FH (1998) Stable transformation of the yellow fever mosquito, *Aedes aegypti*, with the *Hermes* element from the housefly. Proc Natl Acad Sci USA 95:3743–3747

Johnstone ME, Nash D, Naguib FNM (1985) Three purine auxotrophic loci on the second chromosome of *Drosophila melanogaster*. Biochem Genet 23:539–555

Konsolaki M, Sanicola M, Kozlova T, Liu V, Arca B, Savakis C, Gelbart WM, Kafatos FC (1992) FLP-mediated intermolecular recombination in the cytoplasm of *Drosophila* embryos. New Biol 4:551–557

Linn F, Heidmann I, Saedler H, Meyer P (1990) Epigenetic changes in the expression of the maize A1 gene in Petunia hybrida: role of numbers of integrated gene copies and state of methylation. Mol Gen Genet 222:329–336

Logie C, Stewart AF (1995) Ligand-regulated site-specific recombination. Proc Natl Acad Sci USA 92:5940–5944

Loukeris TG, Livadaras I, Arca B, Zabalou S, Savakis C (1995) Gene transfer into the medfly, *Ceratitis capitata*, with a *Drosophila hydei* transposable element. Science 270:2002–2005

Lozovskaya ER, Nurminsky DI, Hartl DL, Sullivan DT (1996) Germline transformation of *Drosophila virilis* mediated by the transposable element *hobo*. Genetics 142:173–177

Malicki J, Schughart K, McGinnis W (1990) Mouse *Hox-2.2* specifies thoracic segmental identity in *Drosophila* embryos and larvae. Cell 63:961–967

Mann RS, Hogness DS (1990) Functional dissection of Ultrabithorax proteins in *D. melanogaster*. Cell 60:597–610

Matsuzaki H, Nakajima R, Nishiyama J, Araki H, Oshima Y (1990) Chromosome engineering in *Saccharomyces cerevisiae* by using a site-specific recombination system of a yeast plasmid. J Bactiol 172:610–618

McGinnis N, Kuziora MA, McGinnis W (1990) Human *Hox-4.2* and *Drosophila deformed* encode similar regulatory specificities in *Drosophila* embryos and larvae. Cell 63:969–976

Medberry SL, Dale E, Qin M, Ow DW (1995) Intra-chromosomal rearrangements generated by Cre-*lox* site-specific recombination. Nucl Acids Res 23:485–490

Metz CW (1916) Chromosome studies on the Diptera II: the paired association of chromosomes in Diptera, and its significance. J Exp Zool 21:213–279

Morris AC, Schaub TL, James AA (1991) FLP-mediated recombination in the vector mosquito, *Aedes aegypti*. Nucl Acids Res 19:5895–5900

O'Brochta DA, Atkinson PW (1996) Transposable elements and gene transformation in non-drosophilid insects. Insect Biochem Mol Biol 26:739–753

O'Gorman S, Fox DT, Wahl GM (1991) Recombinase-mediated gene activation and site-specific integration in mammalian cells. Science 251:1351–1355

Osborne BI, Wirtz U, Baker B (1995) A system for insertional mutagenesis and chromosomal rearrangement using the *Ds* transposon and Cre-*lox*. Plant J 7:687–701

Phillips JP, Forrest HS (1980) Ommochromes and pteridines. In Ashburner M, Wright TRF (eds) The Genetics and Biology of *Drosophila*, Vol 2d. Academic Press, New York, pp 541–623

Qin M, Bayley C, Stockton T, Ow DW (1994) Cre recombinase-mediated site-specific recombination between plant chromosomes. Proc Natl Acad Sci USA 91:1706–1710

Ramirez-Solis R, Liu P, Bradley A (1995) Chromosome engineering in mice. Nature 378:720–724

Rebay I, Rubin GM (1995) Yan functions as a general inhibitor of differentiation and is negatively regulated by activation of the Ras1/MAPK pathway. Cell 81:857–866

Rebay I, Fehon RG, Artavanis-Tsakonas S (1993) Specific truncations of *Drosophila* Notch define dominant activated and dominant negative forms of the receptor. Cell 74:319–329

Rong YS, Golic KG (1998) Dominant defects in *Drosophila* eye pigmentation resulting from a euchromatin–heterochromatin fusion gene. Genetics 150:1551–1566

Sadowski PD (1995) The Flp recombinase of the 2-μm plasmid of *Saccharomyces cerevisiae*. Prog Nucl Acid Res Mol Biol 51:53–91

Sauer B (1996) Multiplex Cre/*lox* recombination permits selective site-specific DNA targeting to both a natural and an engineered site in the yeast genome. Nucl Acids Res 24:4608–4613

Sauer B (1998) Inducible gene targeting in mice using the Cre/*lox* system. Methods 14:381–392

Sauer B, Henderson N (1990) Targeted insertion of exogenous DNA into the eukaryotic genome by the Cre recombinase. New Biol 2:441–449

Sawamoto K, Okano H, Kobayakawa Y, Hayashi S, Mikoshiba K, Tanimura T (1994) The function of argos in regulating cell fate decisions during *Drosophila* eye and wing vein development. Dev Biol 164:267–276

Schlake T, Bode J (1994) Use of mutated FLP recognition target (*FRT*) sites for the exchange of expression cassettes at defined chromosomal loci. Biochemistry 33:12746–12751

Schneuwly S, Klemenz R, Gehring WJ (1987) Redesigning the body plan of *Drosophila* by ectopic expression of the homeotic gene *Antennapedia*. Nature 325:816–818

Selker EU (1999) Gene silencing: repeats that count. Cell 97:157–160

Senecoff JF, Cox MM (1986) Directionality in FLP protein-promoted site-specific recombination is mediated by DNA–DNA pairing. J Biol Chem 261:7380–7386

Senecoff JF, Bruckner RC, Cox MM (1985) The FLP recombinase of the yeast 2-micron plasmid: characterization of its recombination site. Proc Natl Acad Sci USA 82:7270–7274

Senecoff JF, Rossmeissl PJ, Cox MM (1988) DNA recognition by the FLP recombinase of the yeast 2 μ plasmid: a mutational analysis of the FLP binding site. J Mol Biol 201:405–421

Seibler J, Bode J (1997) Double-reciprocal crossover mediated by FLP-recombinase: a concept and an assay. Biochemistry 36:1740–1747

Siegal ML, Hartl DL (1996) Transgene coplacement and high efficiency site-specific recombination with the Cre/*loxP* system in *Drosophila*. Genetics 144:715–726

Skavdis G, Siden-Kiamos I, Muller H-M, Crisanti A, Louis C (1996) Conserved function of *Anopheles gambiae* midgut-specific promoters in the fruitfly. EMBO J 15:344–350

Smith AJ, De Sousa MA, Kwabi-Addo B, Heppell-Parton A, Impey H, Rabbitts PA (1995) A site-directed chromosomal translocation induced in embryonic stem cells by Cre-*loxP* recombination. Nat Genet 9:376–385

Snaith MR, Kilby NJ, Murray JA (1996) An *Escherichia coli* system for assay of Flp site-specific recombination on substrate plasmids. Gene 180:225–227

Spradling AC (1986) *P*-element-mediated transformation. In Roberts DB (ed) *Drosophila*: A Practical Approach. IRL Press, Oxford, pp 175–198

Stevens NM (1908) A study of the germ cells of certain Diptera, with reference to the heterochromosomes and the phenomena of synapsis. J Exp Zool 5:359–383

Struhl G, Basler K (1993) Organizing activity of wingless protein in *Drosophila*. Cell 72:527–540

Struhl G, Fitzgerald K, Greenwald I (1993) Intrinsic activity of the Lin-12 and Notch intracellular domains *in vivo*. Cell 74:331–345

Summers KM, Howells AJ, Pyliotis NA (1982) Biology of eye pigmentation in insects. In Berridge MJ, Treherne JE, Wigglesworth VB (eds) Advances in Insect Physiology. Academic Press, London, pp 119–166

Sun X, Artavanis-Tsakonas S (1996) The intracellular deletions of DELTA and SERRATE define dominant negative forms of the *Drosophila* Notch ligands. Development 122:2465–2474

Sun X, Artavanis-Tsakonas S (1997) Secreted forms of DELTA and SERRATE define antagonists of Notch signaling in *Drosophila*. Development 124:3439–3448

Theodosiou NA, Xu T (1998) Use of FLP/FRT system to study *Drosophila* development. Methods 14:355–365

Thigpen AE, West MG, Appling DR (1990) Rat C1-tetrahydrofolate synthase. cDNA isolation, tissue-specific levels of the mRNA, and expression of the protein in yeast. J Biol Chem 265:7907–7913

Thomas U, Jonsson F, Spiecher SA, Knust E (1995) Phenotypic and molecular characterization of Ser^D, a dominant allele of the *Drosophila* gene *Serrate*. Genetics 139:203–213

Tiong SYK, Keizer C, Nash D, Bleskan J, Patterson D (1989) *Drosophila* purine auxotrophy: new alleles of *adenosine 2* exhibiting a complex visible phenotype. Biochem Genet 27:333–348

Tremblay GB, MacKenzie RE (1995) Primary structure of a folate-dependent trifunctional enzyme from *Spodoptera frugiperda*. Biochim Biophys Acta 1261:129–133

Warrick JM, Paulson HL, Gray-Board GL, Bui QT, Fischbeck KH, Pittman RN, Bonini NM (1998) Expanded polyglutamine protein forms nuclear inclusions and causes neural degeneration in *Drosophila*. Cell 93:939–949

Wells L, Edwards KA, Bernstein SI (1996) Myosin heavy chain isoforms regulate muscle function but not myofibril assembly. EMBO J 15:4454–4459

Welte MA, Tetrault JM, Dellavalle RP, Lindquist SL (1993) A new method for manipulating transgenes: engineering heat tolerance in a complex multicellular organism. Curr Biol 3:842–853

White K, Tahaoglu E, Steller H (1996) Cell killing by the *Drosophila* gene *reaper*. Science 271:805–807

Xu T, Harrison SD (1994) Mosaic analysis using FLP recombinase. Methods Cell Biol 44:655–681

Xu T, Rubin GM (1993) Analysis of genetic mosaics in developing and adult *Drosophila* tissues. Development 117:1223–1237

Zhao JJ, Lazzarini RA, Pick L (1993) The mouse *Hox-1.3* gene is functionally equivalent to the *Drosophila Sex combs reduced* gene. Genes Dev 7:343–354

Section III

Transgenic Selection

4 Eye Color Genes for Selection of Transgenic Insects

Abhimanyu Sarkar and Frank H. Collins

CONTENTS

4.1 Introduction 79
4.2 Structure and Functional Organization of the Insect Compound Eye 80
4.3 Pigments in the Insect Compound Eye 81
    4.3.1 Ommochrome Pigments 81
    4.3.2 Pteridine Pigments 82
4.4 Mutations Affecting Pigmentation in the Insect Eye 82
    4.4.1 Transport Mutations 84
    4.4.2 Granule Group Mutations 85
    4.4.3 Ommochrome Biosynthesis Pathway Mutations 85
    4.4.4 Pteridine Biosynthesis Pathway Mutations 86
    4.4.5 Uncharacterized and Lost Mutations in Insects 86
4.5 Use of Eye Color Genes as Markers for Transgenic Insects 87
4.6 Conclusions 88
Appendix 88
Acknowledgments 89
References 89

4.1 INTRODUCTION

One of the most important tools for studying the biology of organisms at the molecular level is the ability to manipulate their genomes by transformation. A critical component of a transformation system is the choice of markers for the detection of the transgenic organism. Markers can be either selectable or scorable. Selectable markers, such as antibiotic resistance, are useful in detecting transformants in prokaryotic and unicellular eukaryotic systems. However, in complex multicellular organisms, the use of selectable markers to detect transformants is often confounded by the existence of inherent, variable levels of resistance to the selection agent in the population of the organism targeted for transformation. In addition, selection itself can often be biased toward the identification of transformed individuals with multiple copies of the resistance marker. This results from multiple integrations of the gene vector in the initial transformation event or because of subsequent selection in transformed lines for individuals in which the selectable marker is amplified. Scorable markers, usually representing the presence or absence of an easily detectable phenotype, are thus preferred as markers for the detection of transformation in complex organisms such as insects. There is no selection pressure that can result in changes in the number of marker copies in established transformed lines, and such markers often produce phenotypes that can be used to distinguish individuals that are heterozygous from those that are homozygous for the marker.

This chapter discusses the use of eye color genes as scorable markers for detection of transformation in insects. A general overview of the structure and function of the insect compound eye and the pigment biosynthesis pathway is presented to help the reader understand the biology of the genes that affect insect eye color and also to assist in identifying novel genes that may be useful as markers for the detection of transgenic insects. Although various genes that confer obvious, visible phenotypic differences that readily distinguish transformed from untransformed individuals without any significant fitness cost can be used, genes that cause changes in eye color are especially valuable for insects. Most insects have large compound eyes that are easily observed with little or no magnification. The most commonly used strategy to select insect transformants has involved introduction of wild-type eye color alleles to rescue color in a strain in which the target gene is mutated. Although eye color pigments are doubtless important in insect vision, many different insect species have been identified with natural or induced mutations in genes affecting eye color. In general, most such mutant strains are highly fit as laboratory colonies and pose no specific challenge to maintenance. Moreover, genes involved in eye color pigment synthesis and deposition in the insect eye have been well studied in many insects, especially *Drosophila melanogaster*, and cloning of homologous genes from other insects is relatively straightforward. The task of identifying and cloning these genes will also be facilitated by the information available from the Berkeley Drosophila Genome Project and the Celera Genomics Corporation cooperative *D. melanogaster* genome sequencing effort.

4.2 STRUCTURE AND FUNCTIONAL ORGANIZATION OF THE INSECT COMPOUND EYE

The typical insect compound eye consists of the peripheral retina or the ommatidial layer, which is separated from the three underlying optic neuropile masses by the basement membrane (reviewed by Summers et al., 1982).

The peripheral retina consists of units called ommatidia or facets and is responsible for the reception of visual information. An ommatidium is made up of the peripheral dioptric apparatus, the photoreceptor cells or retinula, and the pigment cells. (See Figure 4.1 for a detailed diagram of the structure of the general insect ommatidium.) The outermost part of the dioptric apparatus is the corneal lens or lenslet, a transparent modification of the exoskeleton that admits light into the ommatidium; below the lenslet is the crystalline cone, resting on four Semper or cone cells. The cone cells connect the dioptric apparatus to the photoreceptor cells. Each ommatidium has eight photoreceptor cells and the plasma membranes of these cells are modified to form a microvillar structure called the rhabdomere. The photosensitive visual pigment, rhodopsin, is associated with the rhabdomere. The axons of the photoreceptor cells connect to the outermost optic neuropile mass, called the lamina, which serves as the site of integration of visual information and is connected to the middle optic neuropile, the medulla, by the optic nerve. The medulla is connected to the innermost optic neuropile, the lobula, by the optic chiasma. The lobula in turn transmits the visual information to the protocerebrum of the insect brain.

Pigment cells are modified sheath cells and are associated with both the cells of the dioptric apparatus and the photoreceptor cells. Two primary or corneal pigment cells surround the dioptric apparatus. At least six secondary or retinal pigment cells extend from the corneal lens to the basement membrane, surrounding the photoreceptor cells and the corneal pigment cells. Certain cells in the basement membranes, called basal pigment cells, contain pigment granules, and these pigment cells of the insect compound eye are not directly involved in photoreception but serve to screen and optically isolate each ommatidium. This enables the insect compound eye to process visual information in a spatial manner and also increases contrast perception.

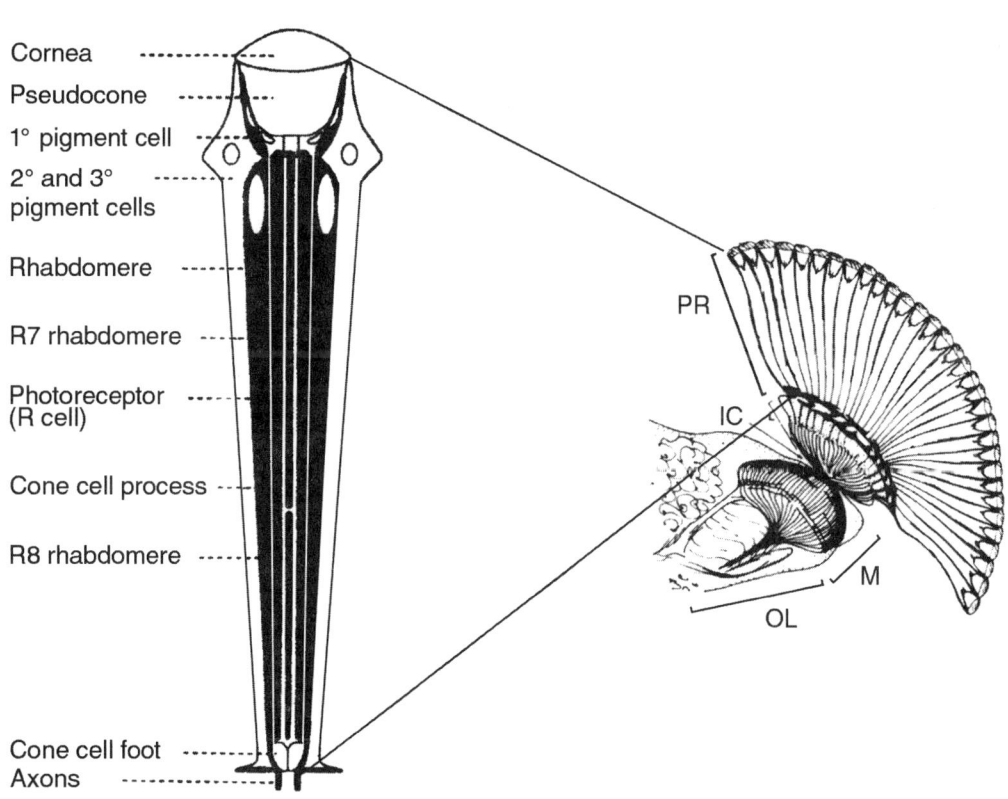

FIGURE 4.1 The insect (dipteran) compound eye. The compound eye (right) consists of the peripheral retina (PR), the lamina (L), the intermediate chiasma (IC), the medula (M), and the optic lobe (OL). A detailed diagram of a single ommatidium is shown on the left. (Modified and reprinted from Lloyd, V., et al., *Trends Cell Biol.*, 8(7):257–259, 1998; Summers, K.M., et al., in Berridge M.J., Treherue, J.E., Wigglesworth V.B. (eds), *Advances in Insect Physiology*, 16, Academic Press, Orlando, FL, 1982, 119–166. With permission.)

4.3 PIGMENTS IN THE INSECT COMPOUND EYE

Two major classes of screening pigments are found in the insect compound eye, the ommochromes or brown pigments and the pteridines or the red and red-yellow pigments (reviewed by Summers et al., 1982). The pigments in the pigment cells are localized in membrane-bound granules that are specialized lysosomes. In *D. melanogaster*, the ommochrome pigment (xanthommatin)–containing granules are called Type I granules, while the pteridine pigment (drosopterin)–containing granules are called Type II granules (Shoup, 1966). In some insects, pteridines are found in solution in the cytoplasm (Langer, 1975). It has been suggested that the formation of the ommochrome and the pteridine pigment-containing granules are interdependent (Ferre et al., 1986).

4.3.1 Ommochrome Pigments

The ommochrome pigments are present in the compound eyes of all insects studied thus far. Apart from xanthommatin, ommins and ommidins are also found in the eyes of insect species of the order Orthoptera while in the hymenopteran *Apis mellifera* (honeybee), xanthommatin and ommins are present (reviewed by Summers et al., 1982). In *D. melanogaster*, both xanthommatin and its derivative dihydroxanthommatin are found, with the latter form predominating (Ferre et al., 1986).

The first step in the biosynthesis of xanthommatin, the major ommochrome in dipterans and most other insects, is the conversion of the amino acid tryptophan to formylkynurenine by the enzyme tryptophan oxygenase (EC 1.13.11.11; tryptophan 2,3-dioxygenase; see Figure 4.2 for the ommochrome synthetic pathway). Formylkynurenine is converted to kynurenine by kynurenine formamidase. The activity of kynurenine formamidase is typically 50 to 100 times more than that of tryptophan oxygenase, and this results in the rapid conversion of formylkynurenine to kynurenine in insect cells. One of the major roles of these two enzymes is to regulate the levels of free tryptophan in the insect. The conversion of kynurenine to 3-hydroxykynurenine is catalyzed by kynurenine hydroxylase, also called kynurenine monoxygenase. This step appears to be specific to the synthesis of the ommochrome pigments. Finally, 3-hydroxykynurenine is converted to xanthommatin by phenoxazinone synthase.

4.3.2 Pteridine Pigments

Different insects appear to have different kinds of pteridine pigments in the compound eye, and individual species may have more than one kind of pteridine pigment. For example, in *D. melanogaster*, important pteridines include isoxanthopterin, pterin, biopterin, sepiapterin, and the drosopterins, a class of at least six closely related compounds. In contrast, the mosquito *Anopheles gambiae* apparently has no visible or fluorescing pteridine pigments in its eyes (Beard et al., 1995). The ultraviolet fluorescing pteridines sepiapterin and biopterin have been reported from *Aedes aegypti* (Bhalla, 1966, 1968), but these pigments are probably found in the body (as they are in *An. gambiae*; Beard et al., 1995) rather than the eyes. *Aedes aegypti* that are incapable of producing ommochrome pigments have entirely white eyes, indicating that the ultraviolet fluorescing pteridine pigments biopterin and sepiaterin do not contribute to eye color in daylight (Cornel et al., 1997).

The biosynthesis of pteridines utilizes a branched pathway (Figure 4.3). The common first step in this pathway is the conversion of guanosine triphosphate (GTP) to dihydroneopterin triphosphate by the enzyme GTP cyclohydrolase (EC 3.5.4.16; GTP 7,8-8,9-dihydrolase; Mackay and O'Donnell, 1983). The pathway branches at this point, with one of the branches leading to the formation of pterin and subsequently isoxanthopterin. In this pathway, dihydroneopterin triphosphate is converted to dihydroneopterin. Dihydroneopterin is then converted to dihydropterin 6-CH_2OH, which is then converted to pterin by dihydropterin oxidase (Silva et al., 1991). Pterin is converted to isoxanthopterin by the enzyme xanthine dehydrogenase (EC 1.1.1.204; xanthine:NAD^+ oxidoreductase). Xanthine dehydrogenase also plays an important role in the purine metabolism of insects, catalyzing the conversion of xanthine to hypoxanthine and subsequently uric acid. Another branch of the pteridine biosynthesis pathway leads to the formation of sepiapterin and biopterin. Dihydroneopterin triphosphate is converted to sepiapterin by sepiapterin synthase in the presence of NADPH and Mg^{2+} ions. Sepiapterin is converted to dihydrobiopterin while the oxidized form of sepiapterin is converted to biopterin, with both of these reactions being catalyzed by biopterin synthase. There are two families of drosopterins, each having three isomeric members. One of the families includes drosopterin, isodrosopterin, and neodrosopterin, and the other family consists of the related aurodrosopterins. All the drosopterins are dipteridine compounds, with a carbon atom bridge between two ring systems. The first committed step in the synthesis of drosopterins is the conversion of dihydroneopterin triphosphate to an unstable intermediate, 6-pyruvoyltetrahydropterin, by the removal of the phosphate groups. This unstable intermediate is then converted into pyrimidodiazepine (PDA) by the enzyme PDA synthase (Wiederrecht and Brown, 1984). This is followed by a multistep pathway that involves the condensation of two pteridine ring systems and the removal of a three-carbon side chain.

4.4 MUTATIONS AFFECTING PIGMENTATION IN THE INSECT EYE

Many kinds of mutations affect the pigmentation of the insect compound eye, including those that affect the ommochrome and the pteridine biosynthesis pathways, the transport of pigments and their precursors, as well as those that affect pigment granule biogenesis.

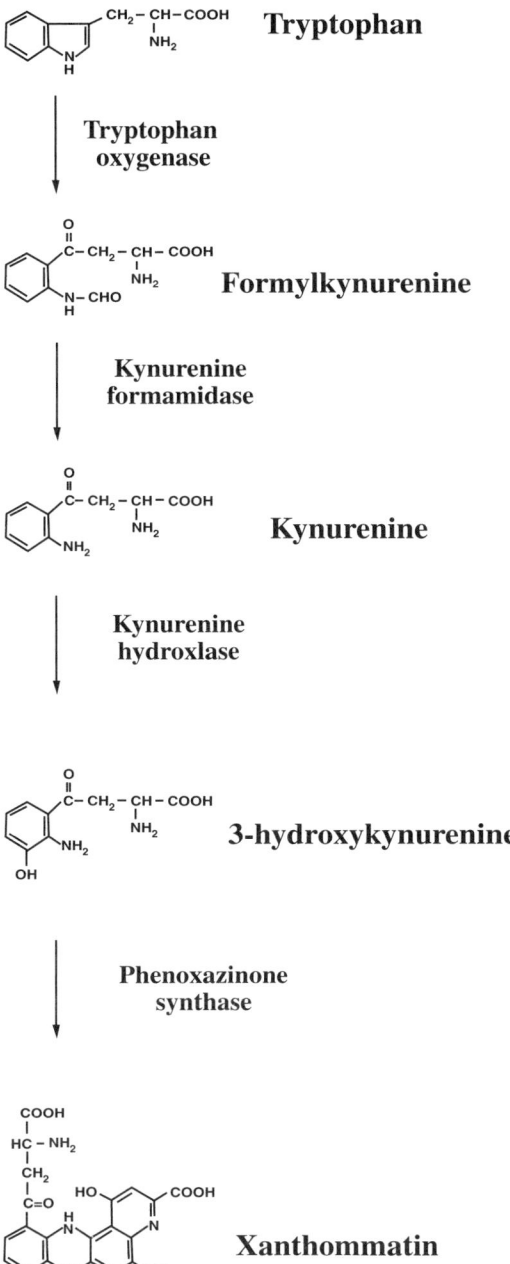

FIGURE 4.2 The ommochrome biosynthesis pathway. The ommochrome biosynthesis pathway, with chemical structures of the intermediates and enzymes, which catalyzes the reactions. (Modified and reprinted from Summers, K.M., et al., in Berridge, M.J., Treherue, J.E., Wigglesworth, V.B. (eds.), *Advances in Insect Physiology*, 16, Academic Press, Orlando, FL, 1982, 119–166. With permission.)

Insects in which mutations ablate either the synthesis of these screening pigments or the importation of these pigments and their proper assembly into granules may not be completely blind in the strict sense of total absence of visual information reaching the brain. A large number of insect species with null mutations in either the synthesis or pigment importation pathways are able

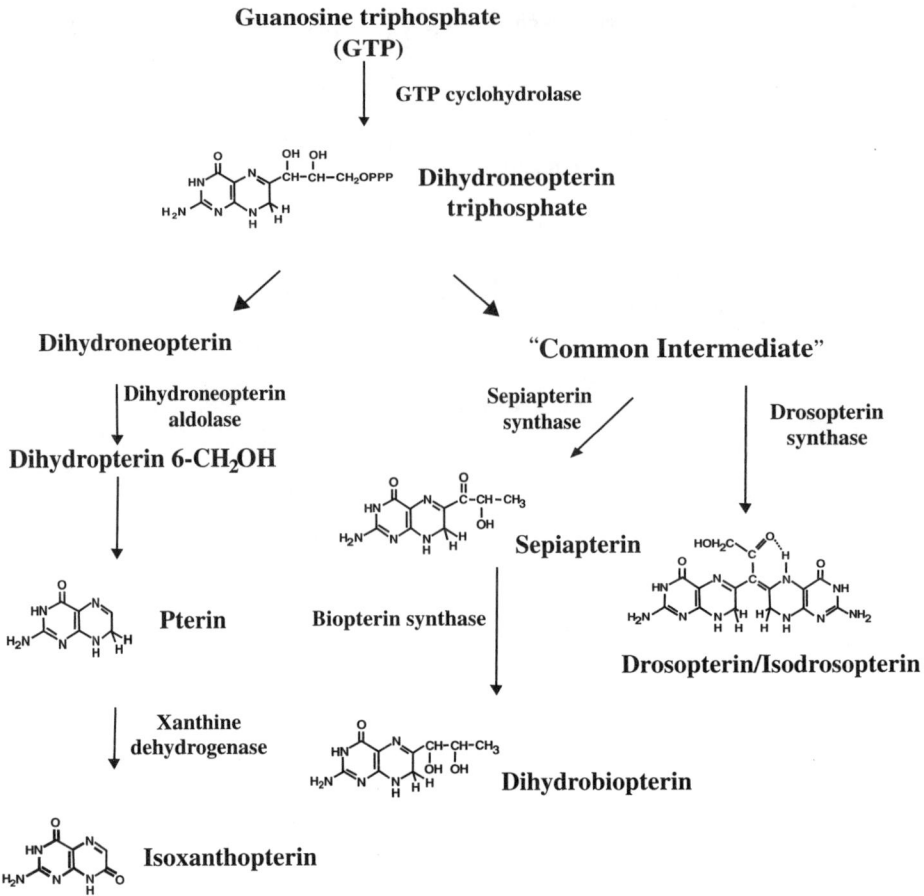

FIGURE 4.3 The pteridine biosynthesis pathway. The pteridine biosynthesis pathway, with chemical structures of the intermediates and enzymes, which catalyzes the reactions. (Modified and reprinted from Summers, K.M., et al., in Berridge, M.J., Treherue, J.E., Wigglesworth, V.B. (eds.), *Advances in Insect Physiology*, 16, Academic Press, Orlando, FL, 1982, 119–166. With permission.)

to feed, mate, and even fly. Interestingly, however, mosquitoes and some other insects with such null mutations are not able to adjust the degree of their cuticle pigmentation in response to the shade of the background environment. The inability of mosquitoes with eye color mutations to make this adjustment (homochromy) has actually been used to screen for eye color mutations in *Anopheles* species (Benedict and Chang, 1996). Larvae of most *Anopheles* species with wild-type eye color genes, when reared in black pans, will develop a dark pigment in their cuticle, while larvae with null or even leaky eye color mutations will not make this adjustment and will have a relatively pale cuticle pigmentation. Such larvae are readily distinguishable from their wild-type peers even without the use of a microscope.

4.4.1 Transport Mutations

Many genes affect the transport of pigments and pigment precursors in the insect compound eye. The *white*, *brown*, and *scarlet* genes of *D. melanogaster* encode three different proteins that belong to the Traffic ATPase superfamily of transmembrane permeases (also called ABC or *A*TP *B*inding *C*assette superfamily; Ewart et al., 1994). These are responsible for the transport of precursors for the synthesis of the ommochrome and pteridine pigments, and the protein products of these three genes may act as a transport complex. Null mutations in the *white* gene block transport and assembly of both ommochrome and pteridine pigments in the granules, and thus insects with null mutants in this gene have completely white or pigmentless eyes. The *white* gene encodes an approximately 2.5-kb transcript and the complete gene or *white* gene fragments have been cloned from a number of different insects for the potential value of the wild-type gene as a transformation reporter and for phylogenetic analysis (e.g., Besansky and Fahey, 1997). In a number of non-drosophilid insects such as the Mediterranean fruit fly, *Ceratitis capitata*, *An. gambiae*, and *An. albimanus*, the complete gene has been cloned; also, strains carrying mutations in the *white* gene have been identified (Besansky et al., 1995; Benedict et al., 1996; Zwiebel et al., 1995). The cloning of the *white* gene from these insects has been facilitated by the conservation of certain sequence motifs in this gene between the various species, including the *Lucilia cuprina* (Elizur et al., 1990) and *D. melanogaster white* homologue (reviewed by Hazelrigg, 1987). The *white* gene product contains both ABC and hydrophobic membrane spanning domains. The ABC is present in the amino terminal of the protein and contains three highly conserved motifs, the nucleotide binding Walker A motif, the Walker B motif, and the consensus motif present in members of the ABC transporter superfamily of proteins (Walker et al., 1982; Elizur et al., 1990; Besansky et al., 1995; Zwiebel, 1995; Coates et al., 1997). The hydrophobic membrane spanning domain is also conserved (Zwiebel, 1995).

The *scarlet* gene of *D. melanogaster* encodes a 2.3-kb transcript whose conceptual protein product shows similarities to the *white* gene product (Tearle et al., 1989). Mutations in this gene prevent the formation of the brown xanthommatin pigment granules by disrupting the transport and storage of the pigment or pigment precursors. Formation of the red pteridine pigment granules is unaffected by this mutation; thus lines of *Drosophila* mutant for *scarlet* have bright red (scarlet-colored) eyes. While both the *white* and *scarlet* genes are required for the formation of the ommochrome pigment granules, they do not appear to affect the transcription levels of each other. The *scarlet* gene has been cloned not only from *D. melanogaster* but also from *L. cuprina*, the Australian sheep blowfly, where it is called the *topaz* gene (Elizur et al., 1990). Null mutations of the *brown* gene eliminate the red pteridine pigments from the *Drosophila* eye, again by disrupting the transport and storage of the pteridine pigment precursors (Ferre et al., 1986). Ommochrome pigment import is unaffected; thus flies with null mutations in the *brown* gene have brown-colored eyes.

4.4.2 GRANULE GROUP MUTATIONS

Another set of genes affects insect eye color because its members are involved in protein trafficking to the pigment granules, which are essentially specialized lysosomes (reviewed by Lloyd et al., 1998). The *garnet* gene encodes the *Drosophila* homologue of the delta-adaptin subunit of clathrin coat adaptor. Mutations at this locus are defective for the intracellular transport processes required for the formation of the pigment granules (Ooi et al., 1997). Also, abolition of *white* gene activity causes the elimination of the transport of both ommochrome and pteridine pigment precursors and disrupts the formation of the pigment granules in the insect eye (Montell et al., 1992).

4.4.3 OMMOCHROME BIOSYNTHESIS PATHWAY MUTATIONS

The ommochrome biosynthesis pathway is relatively simple and well characterized. The *D. melanogaster vermilion* (*v*) strain has a mutation in the gene that encodes the first enzyme in this pathway, tryptophan oxygenase (Green, 1949; Baglioni, 1959). (*Note:* This spelling was used in the original description of the mutant and has been preserved because of the gene-naming precedence policy in the *Drosophila* field.) This leads to the buildup of high levels of tryptophan and also the absence of subsequent intermediates in the ommochrome biosynthesis pathway. Strains with mutations in the tryptophan oxygenase gene in other insects include *snow* in *A. mellifera* (Dustman, 1968; 1975), *a* in the meal moth, *Ephestia kühniella* (Caspari and Gottlieb, 1975), *green* in the housefly, *Musca domestica* (Milani, 1975), *yellowish* in *L. cuprina* (Summers and Howells, 1978), and *ivory* in the flesh fly, *Sacrophaga barbata* (Trepte, 1978). The product of the *tryptophan oxygenase* gene is not cell autonomous and can be transferred across cell membranes and even tissue boundaries. The *tryptophan oxygenase* gene has been cloned from many insects including *D. melanogaster* (Searles and Voelker, 1986; Searles et al., 1990) and *An. gambiae* (Mukabayire et al., 1996) and has been demonstrated to be useful as a marker in heterologous species (White et al., 1996).

The *carnation* mutation in *M. domestica* has reduced activity for the next enzyme in the ommochrome biosynthesis pathway, kynurenine formamidase, while the *M. domestica green* mutation lacks this enzyme activity entirely (Grigolo, 1969). However, mutant animals deficient in this enzyme activity are not known in other dipteran species.

Kynurenine hydroxylase or kynurenine monoxygenase is encoded by the *cinnabar* locus in *D. melanogaster* (Warren et al., 1996). Strains carrying mutations in the gene for this enzyme in other insects include *ivory* of *A. mellifera* (Dustman, 1968; 1975), *ocra* of *M. domestica* (Milani, 1975), *yellow* of *L. cuprina* (Summers and Howells, 1978), and *kynurenine hydroxylase^{white}* (kh^w) of *Ae. aegypti* (Bhalla, 1968; Cornel et al., 1997). The product of the *kynurenine hydroxylase* gene is also not cell autonomous.

The enzyme phenoxazinone synthase is involved the final step of the ommochrome biosynthesis pathway. Strains of various insects that lack this enzyme include *tangerine* in *L. cuprina* (Summers and Howells, 1978), *chartreuse* in *A. mellifera* (Dustman, 1968; 1975), and *alb* in *E. kühniella* (Collin and Klett, 1978). The *cardinal* mutant of *D. melanogaster* (Howells et al., 1977) has reduced activity of this enzyme while *karmoisin* appears to be totally blocked (as null mutation) in this final step of the ommochrome synthesis pathway.

4.4.4 PTERIDINE BIOSYNTHESIS PATHWAY MUTATIONS

The pteridine biosynthesis pathway is more complex in terms of the number of steps involved, their regulation, and the interrelationships between the various steps of the pathway. Mutations in the *raspberry* and *prune* loci of *D. melanogaster* have altered levels of GTP cyclohydrolase activity, the first enzyme of the pteridine biosynthetic pathway. This enzyme is encoded by the *Punch* locus (Mackay and O'Donnell, 1983). Xanthine dehydrogenase, which converts pterin to isoxanthopterin, is encoded by the *rosy* locus in *D. melanogaster* (Reaume et al., 1991). The *purple* locus is involved in the regulation of pteridine biosynthesis in one of the steps converting dihydroneopterin triphosphate to sepiapterin (Wilson and Jacobson, 1977). Mutations in this gene cause reduction in the enzyme 6-pyruvoyl tetrahydropterin synthase (previously called sepiapterin synthase A) or PTP synthase. This 83-kDa enzyme catalyzes the conversion of dihydroneopterin triphosphate to 6-pyruvoyl tetrahydropterin phosphate and also is involved in the synthesis of biopterin. The *little isoxanthopterin* (*lix*) gene is the structural gene for dihydropterin oxidase in *D. melanogaster* (Silva et al., 1991).

The *pugilist* gene of *D. melanogaster* encodes a trifunctional enzyme methylenetetrahydrofolate dehydrogenase (MTHFD; E.C.1.5.1.5, E.C.3.5.4.9, E.C.6.3.4.3) and is involved in purine biosynthesis (Rong and Golic, 1998; see Chapter 3 by Rong and Golic). The *pugilist^{Dominant}* (pug^D) mutation reduces both ommochrome and pteridine pigmentation in the compound eye. The primary effect of this mutation appears to disrupt purine biosysnthesis. Since pteridines are synthesized from the precursor GTP, this causes a drastic reduction in the pteridine pigmentation. Since pteridines are required for normal ommochrome pigment granule formation in *Drosophila* (Summers et al., 1982; Reaume et al., 1991), this leads to a reduction in the amount of ommochrome pigments. However, it is not known if this dominant mutation would affect eye color in insects such as *Anopheles*, that do not have pteridine eye pigments.

4.4.5 UNCHARACTERIZED AND LOST MUTATIONS IN INSECTS

Eye color mutations have been frequently observed in various insects; however, the genetic lesions that cause these mutations is often unknown. A factor that often leads to confusion is the convention of naming the mutations in various species according to their phenotype. Thus, we find a plethora of "white" mutations for different species in the literature, none of which may be actually a mutation in the ABC transporter *white* gene, and that refer to different genetic lesions in different species. An example of this would be the strain of *Ae. aegypti* having a white eye phenotype originally described as *white* (Bhalla, 1968), but was later found to be due to a null mutation in the kynurenine hydroxylase gene and renamed *kynurenine hydroxylase^{white}* (or *kh^w*; Cornel et al., 1997). The *white* gene of *Ae. aegypti* has been cloned (Coates et al., 1997), but none of the three *Ae. aegypti* mutations that produce white eyes — *kh^w*, *cream* (L. Munstermann, unpublished), and another mutant in the *cream* locus identified by S. Higgs, unpublished — is a mutation in the "*white gene*" responsible for pigment transport. Mutations resulting in white or other non-wild-type eye colors have been described in mosquitoes, including *An. gambiae*, *An. albimanus*, *An. quadrimaculatus* (see Benedict et al., 1996), *An. stephensi* (e.g., Akhtar and Sakai, 1985), *Ae. aegypti*, *Ae. mascarensis* (Hartberg and Craig, 1974), the *Culex pipiens* complex (Sakai et al., 1980; Narang and Seawright, 1982), *Cx. tritaeniorhynchus* (Dubash et al., 1980), and many others. Most of these mosquitoes were well studied genetically, and often as many as three different loci were found that affected eye color. Unfortunately, with the notable exceptions of *An. gambiae*, *An. albimanus*, *An. quadrimaculatus*, *An. stephensi*, and *Ae. aegypti*, most of these mutations have since been lost.

4.5 USE OF EYE COLOR GENES AS MARKERS FOR TRANSGENIC INSECTS

Various genes affecting the color of the compound eye have been successfully used as scorable markers in the creation of transgenic insects (see Color Figure 1).* The first reported germline transformation of an insect was that of a *D. melanogaster* strain carrying the *rosy* mutation. A *P*-transposable element marked with the wild-type *rosy* (xanthine dehydrogenase) gene was used to transform stably and rescue the wild-type eye color of the fruitfly (Rubin and Spradling, 1982). The *white* gene was subsequently cloned and used as a marker of choice for routine *P*-element-mediated transformation of *D. melanogaster* (Klemenz et al., 1987). The *white* gene of *D. melanogaster* has also been used as a marker for the transformation of *D. melanogaster* using the *mariner* (Lidholm et al., 1993), *Hermes* (O'Brochta et al., 1996) and *piggyBac* (Handler and Harrell, 1999) transposable elements. Earlier, *D. virilis* was transformed using the *hobo* transposable element, with the *D. melanogaster white* gene as the marker (Lozovskaya et al., 1996; Gomez and Handler, 1997), serving as an example of the use of eye color gene complementation in heterologous species. The *white* gene of *C. capitata*, was cloned (Zwiebel et al., 1995) and used as a transformation marker for a strain of that species with a null mutation at the *white eye* locus. The transformation vectors used were the transposable elements *Minos* (Loukeris et al., 1995) and *piggyBac* (Handler et al., 1998).

Recently, the cloned *cinnabar* (*cn*) gene from *D. melanogaster* was shown to be able to rescue eye color in a strain of *Ae. aegypti* with a null mutation in the kynurenine hydroxylase gene, *kh^w* (Bhalla, 1968; Cornel et al., 1997). Larvae reared in water containing 3-hydroxykynurenine and embryos injected with the cloned *D. melanogaster cn* gene both showed transient rescue of eye color in both the larval and adult eyes. This confirmed that the white-eyed phenotype of the *Ae. aegypti kh^w* strain (previously called *we* or *white eye* by Bhalla, 1968) was due to a defect in the gene encoding kyureni*ne hydroxylase (Cornel et al., 1997). The *kh^w* strain was subsequently transformed with both the *Hermes* and *mariner* transposable elements carrying the *Drosophila* wild-type *cn* gene as a marker (Jasinskiene et al., 1998; Coates et al., 1998).

---

* Color Figure 1 follows p. 108.

4.6 CONCLUSIONS

Genes affecting the pigmentation of the insect compound eye offer a large number of advantages for use as markers for the selection of transformants. Strains carrying these mutations are available in many economically and medically important insect species. Also, eye color mutations are relatively easy to isolate in various insects because they are visible and are usually not associated with major reductions in fitness, especially in laboratory conditions. Many genes that affect the eye color have been cloned from various insects, including the *white* gene and the structural gene for tryptophan oxygenase (*vermilion*). Most of these genes are relatively small, in the order of 2 to 3 kb or less, and are thus suitable for use with the available transposable element vectors.

Another marker that has been proposed for use in insect transformation is the gene for the green fluorescent protein (GFP) from the jellyfish, *Aequorea victoria*. However, GFP requires the use of ultraviolet light to detect its expression. On the other hand, the expression of eye color genes such as *white* and the structural genes of the ommochrome synthesis pathway are easily observable to the naked eye and appear to have no deleterious effects on the growth and development of transformed insects.

Of the many eye color genes potentially useful as markers for insect transformation, the most promising ones appear to be the structural genes of the ommochrome biosynthesis pathway. The ommochromes appear to constitute the major group of shielding pigments in all insects studied. Two major genes in this pathway, those for tryptophan oxygenase (*vermilion*) and kynurenine hydrolase (*cinnabar*), offer many advantages as markers. They are relatively small, well characterized, and can be used in heterologous species. For example, the gene encoding tryptophan oxygenase in *An. gambiae* can rescue eye color in the *vermilion* mutant of *D. melanogaster* (Besansky et al., 1997), while the kynurenine hydroxylase gene (*cinnabar*) from *D. melanogaster* has been used as a marker in the kh^w strain of *Aedes aegypti* (Jasinskiene et al., 1998; Coates et al., 1998). These genes do not have major pleiotropic effects and are not deleterious even when present in multiple copies. Another major advantage is that they are not cell autonomous; i.e., their expression is not required in the pigment cells to see color. This is because their gene products or the products of the biochemical pathway on which they act can be translocated across cell membranes and even tissue boundaries. This property can be used to test the ability of these genes to rescue mutant phenotypes transiently in different species (Cornel et al., 1997). It also permits various potential transformation constructs with different marker promoters or of different sizes to be tested in transient assays rather than in laborious transformation assays.

Genes affecting the pteridine biosynthesis pathway are unlikely to be useful as markers in many insects for a number of reasons. Many of the known genes in this pathway affect purine biosynthesis and metabolism, and mutations in these genes have pleiotropic effects that appear to affect the fitness of the insect adversely. Also, because of the branched nature of the pteridine biosynthesis pathway, many mutations do not completely abolish the production of the pteridine pigments. Finally, the medically important dipteran family of Culicidae (mosquitoes), and possibly other insects, appears to lack visible pteridine pigments in the compound eye (Beard et al., 1995).

Transport group mutants like *white*, *brown*, and *scarlet* and granule group mutants like garnet of *D. melanogaster* are also potentially less useful than mutants in the ommochrome biosynthesis pathway. This is because they are cell autonomous, usually have pleiotropic effects, and are found as multimeric complexes; thus, they may have phylogenetic restraints.

We believe that genes involved in ommochrome synthesis coupled with the appropriate null mutations offer the most flexible and robust eye color markers for insect germline transformation.

APPENDIX

Further information on *D. melanogaster* eye color mutations and strains can be obtained from internet resources such as the Flybase (http://flybase.bio.indiana.edu) and the Berkeley *Drosophila*

Genome Project (http://www.fruitfly.org) Web pages. The Web page of Celera Genomics Corporation is http://www.celera.com. Information on mosquito strains and genes are available at the Mosquito Genomics WWW server (http://klab.agsci.colostate.edu/). Many interesting eye color mutations in *Anopheles* are described at the *Anopheles* Genetic Resource Information Project site (http://klab.AgSci.colostate.EDU/~mbenedic), which is maintained by Dr. Mark Q. Benedict of the Centers for Disease Control, Atlanta, GA.

ACKNOWLEDGMENTS

The authors wish to thank Rajyashree Ray and Ali Dana for help with the figures.

REFERENCES

Akhtar K, Sakai RK (1985) Genetic analysis of three new eye colour mutations in the mosquito, *Anopheles stephensi*. Ann Trop Med Parasitol 79(4):449–455

Baglioni C (1959) Genetic control of tryptophan peroxidase-oxidase in *Drosophila melanogaster*. Nature 184:1084–1085

Beard CB, Benedict MQ, Primus JP, Finerty V, Collins FH (1995) Eye pigments in wild-type and eye-color mutant strains of the African malaria vector *Anopheles gambiae*. J Hered 86:375–380

Benedict MQ, Chang H (1996) Rapid isolation of anopheline mosquito eye colour mutants based on larval colour change. Med Vet Entomol 10(1):93–96

Benedict MQ, Besansky NJ, Chang H, Mukabayire O, Collins FH (1996) Mutations in the *Anopheles gambiae Pink eye* and *White* genes define distinct tightly linked eye-color loci. J Hered 87:48–53

Besansky NJ, Fahey GT (1997) Utility of the white gene in estimating phylogenetic relationships among mosquitoes (Diptera: Culicidae). Mol Biol Evol 14(4):442–454

Besansky NJ, Bedell JA, Benedict MQ, Mukabayire O, Hilfiker D, Collins FH (1995) Cloning and characterization of the *white* gene from *Anopheles gambiae*. Insect Mol Biol 4(4):217–231

Besansky NJ, Mukabayire O, Benedict MQ, Rafferty CS, Hamm DM, McNitt L (1997) The *Anopheles gambiae* tryptophan oxygenase gene expressed from a baculovirus promoter complements *Drosophila melanogaster vermilion*. Insect Biochem Mol Biol 27(8–9):803–805

Bhalla SC (1966) Genetics of Some Mutants Affecting Tanning and Eye Color in *Aedes aegypti*. PhD dissertation. University of Notre Dame, Notre Dame, IN

Bhalla SC (1968) Genetic aspects of pteridines in mosquitoes. Genetics 58:249–258

Caspari EW, Gottlieb FJ (1975) The Mediterranean meal moth, *Ephestia kuhniella*. In King RC (ed) Handbook of Genetics Vol 3, Plenum, New York, pp 125–147

Coates CJ, Schaub TL, Besansky NJ, Collins FH, James AA (1997) The *white* gene from the yellow fever mosquito, *Aedes aegypti*. Insect Mol Biol 6(3):291–299

Coates CJ, Jasinskiene N, Miyashiro L, James AA (1998) *Mariner* transposition and transformation of the yellow fever mosquito, *Aedes aegypti*. Proc Natl Acad Sci USA 95:3742–3751

Colln K, Klett G (1978) Über die Funktion des *alb*-Locus in der Ommochrom-Endsynthese bei *Ephestia kühniella* Z. Wilhelm Roux Arch 185:127–136

Cornel AJ, Benedict MQ, Rafferty CS, Howells AJ, Collins FH (1997) Transient expression of the *Drosophila melanogaster cinnabar* gene rescues eye color in the *white eye*(WE) strain of *Aedes aegypti*. Insect Biochem Mol Biol 27(12):993–997

Dubash CJ, Sakai RK, Baker RH (1980) Genetic and linkage analyses of maroon eye and acid phosphatase I in *Culex tritaeniorhynchus*. Can J Genet Cytol 22(3):369–374

Dustman JH (1968) Pigment studies on several eye color mutants of the honeybee, *Apis mellifera*. Nature 219:950–952

Dustman JH (1975) Die Pigmentgranula in Komplexauge der Honigbiene *Apis mellifera* bei Wildtyp und verschiedenen Augenfarbmutanten. Cytobiologie 11:133–152

Elizur A, Vacek AT, Howells AJ (1990) Cloning and characterization of the *white* and *topaz* eye color genes from the sheep blowfly *Lucilia cuprina*. J Mol Evol 30(4):347–358

Ewart GD, Cannell D, Cox GB, Howells AJ (1994) Mutational analysis of the traffic ATPase (ABC) transporters involved in the uptake of eye pigment precursors in *Drosophila melanogaster*. Implications for structure–function relationships. J Biol Chem 269(14):10370–10377

Ferre J, Silva FJ, Real MD, Mensua JL (1986) Pigment patterns in mutants affecting the biosynthesis of pteridines and xanthommatins in *Drosophila melanogaster*. Biochem Genet 24(7–8):545–569

Gomez SP, Handler AM (1997) A *Drosophila melanogaster hobo-white+* vector mediates low frequency gene transfer in *D. virilis* with full interspecific *white+* complementation. Insect Mol Biol 6:1–8

Green MM (1949) A study of tryptophan in eye color mutants of *Drosophila*. Genetics 34:564–573

Grigolo A (1969) Kynurenine formamidase (formylase) activity in normal strains of adult *Musca domestica* and in strains mutant for eye color. Redia 51:169–178

Handler AM, Harrell RA (1999) Germline transformation of *Drosophila melanogaster* with the *piggyBac* transposon vector. Insect Mol Biol 8:449–457

Handler AM, McCombs SD, Fraser MJ, Saul SH (1998) The lepidopteran transposon vector, *piggyBac*, mediates germline transformation in the Mediterranean fruitfly. Proc Natl Acad Sci USA 95:7520–7525

Hartberg WK, Craig GB (1974) Three new mutants in *Aedes mascarensis*: *Currant-eye*, *small-antenna* and *yellow*. J Med Vet Entomol 11(4):447–454

Hazelrigg T (1987) The *Drosophila white* gene: a molecular update. Trends Genet 3:43–47

Howells AJ, Summers KM, Ryall RL (1977) Developmental patterns of 3-hydroxykynurenine accumulation in *white* and various other eye color mutant of *Drosophila melanogaster*. Biochem Genet 13:273–282

Jasinskiene N, Coates CJ, Benedict MQ, Cornel AJ, Rafferty CS, James AA, Collins FH (1998) Stable, transposon mediated tranformation of the yellow fever mosquito, *Aedes aegypti*, using the *Hermes* element from the housefly. Proc Natl Acad Sci USA 95:3753–3757

Klemenz R, Weber U, Gehring WJ (1987) The *white* gene as a marker in a new *P*-element vector for gene transfer in *Drosophila*. Nucl Acids Res 15(10):3947–3959

Langer H (1975) Properties and functions of screening pigments. In Snyder AW, Mengel R (eds) Photoreceptor optics. Springer Verlag, Berlin, pp 429–455

Lidholm DA, Lohe AR, Hartl DL (1993) The transposable element *mariner* mediates germline transformation in *Drosophila melanogaster*. Genetics 134:859–868

Lloyd V, Ramaswami M, Kramer H (1998) Not just pretty eyes: *Drosophila* eye-colour mutations and lysosomal delivery. Trends Cell Biol 8(7):257–259

Lohe AR, Hartl DL (1996) Germline transformation of *Drosophila virilis* with the transposable element *mariner*. Genetics 143:365–374

Loukeris TG, Livadaras I, Arca B, Zabalou S, Savakis C (1995) Gene transfer into the medfly, *Ceratitis capitata*, with a *Drosophila hydei* transposable element. Science 270:2002–2005

Lozovskaya ER, Nurminsky DI, Hartl DL, Sullivan DT (1996) Germline transformation of *Drosophila virilis* mediated by the transposable element *hobo*. Genetics 142:173–177

Mackay WJ, O'Donnell JM (1983) A genetic analysis of the pteridine biosynthetic enzyme guanosine triphosphate cyclohydrolase, in *Drosophila melanogaster*. Genetics 105:35–53

Mason GF (1967) Genetic studies on mutations in species A and B of the *Anopheles gambiae* complex. Genet Res 10:205–221

Milani R (1975) The house fly, *Musca domestica*. In King RC (ed) Handbook of Genetics, Vol 3. Plenum, New York, pp 337–399

Montell I, Rasmuson A, Rasmuson B, Holmgren P (1992) Uptake and incorporation in pteridines of externally supplied GTP in normal and pigment deficient eyes of *Drosophila melanogaster*. Biochem Genet 30(1–2):61–75

Mukabayire O, Cornel AJ, Dotson EM, Collins FH, Besansky NJ (1996) The tryptophan oxygenase gene of *Anopheles gambiae*. Insect Biochem Mol Biol 26(6):525–528

Narang S, Seawright JA (1982) Linkage relationships and genetic mapping in *Culex* and *Anopheles*. In Steiner WWM, Tabachnick WJ, Rai KS, Narang S (eds) Recent developments in the genetics of insect disease vectors. Stipes, Champaign, IL, pp 231–289

O'Brochta DA, Warren WD, Saville KJ, Atkinson PW (1996) *Hermes*, a functional non-drosophilid vector from *Musca domestica*. Genetics 142:907–914

O'Donnell JM, Mclean JR, ReynoldsER (1989) Molecular and developmental genetics of the *Punch* locus, a pterin biosynthesis gene in *Drosophila melanogaster*. Dev Genet 10(3):273–286

Ooi CE, Moreira JE, Dell'Angelica EC, Poy G, Wasssarman DA, Bonifacino JS (1997) Altered expression of a novel adaptin leads to defective pigment granule biogenesis in the *Drosophila* eye color mutant *garnet*. EMBO J 16(15):4508–4518

Reaume AG, Knecht DA, Chovnick A (1991) The rosy locus of *Drosophila melanogaster*: xanthine dehydrogenase and eye pigments. Genetics 129(4):1099–1109

Rong YS, Golic KG (1998) Dominant defects in *Drosophila* eye pigmentation resulting from a euchromatin-heterochromatin fusion gene. Genetics 150(4):1551–1556

Rubin GM, Spradling AC (1982) Genetic transformation of *Drosophila* with transposable element vectors. Science 218:348–353

Sakai RK, Chaudhry M, Baker RH (1980) An EMS induced mutation, rose eye, in *Culex quinquefasciatus*. J Hered 71(2):136–139

Searles LL, Voelker RA (1986) Molecular characterization of the *Drosophila vermilion* locus and its suppressible alleles. Proc Natl Acad Sci USA 83:404–408

Searles LL, Ruth RS, Pret A, Fridell RA, Ali AJ (1990) Structure and transcription of the *Drosophila melanogaster vermilion* gene and several mutant alleles. Mol Cell Biol 10:1423:1431

Shoup JR (1966) The development of pigment granules in the eyes of the wild type and mutant *Drosophila melanogaster*. J Cell Biol 29:223–249

Silva FJ, Escriche B, Ordono E, Ferre J (1991) Genetic and biochemical characterization of the little isoxanthopterin (lix), a gene controlling dihydropterin oxidase activity in *Drosophila melanogaster*. Mol Gen Genet 230(1–2):97–103

Summers KM, Howells AJ (1978) Xanthommatin biosynthesis in wild type and mutant strains of the australian sheep blowfly, *Lucilia cuprina*. BiochemGenet 10:1151–1163

Summers KM, Howells AJ, Pyliotis NA (1982) Biology of eye pigmentation in insects. In Berridge MJ, Treherue, JE, Wigglesworth VB (eds) Advances in Insect Physiology, vol 16. Academic Press, Orlando, FL, pp 119–166

Tearle RG, Belote JM, McKeown M, Baker BS, Howells AJ (1989) Cloning and characterization of the *scarlet* gene of *Drosophila melanogaster*. Genetics 122(3):595–606

Trepte HH (1978) *Ivory*: a recessive white eyed tryptophan metabolism mutant with intermediate F_2 and R_1 progenies in the flesh fly, *Sacrophaga barbata*. Theor Appl Genet 51:185–191

Walker JE, Saraste M, Runswick MJ, Gay NJ (1982) Distantly related sequences in the alpha- and beta-subunits of ATP synthase, myosin, kinases and other ATP-requiring enzymes and a common nucleotide binding fold. EMBO J 1(8):945–951

Warren WD, Palmer S, Howells AJ (1996) Molecular characterization of the *cinnabar* regionof *Drosophila melanogaster*: identification of the *cinnabar* transcription unit. Genetica 98(3):249–262

White LD, Coates CJ, Atkinson PW, O'Brochta DA (1996) An eye color gene for the detection of transgenic non-drosophilid insects. Insect Biochem Mol Biol 26(7):641–644

Wiederrecht GJ, Brown GM (1984) Purification and properties of the enzymes from *Drosophila melanogaster* that catalyze the conversion of dihydroneopterin triphosphate to the pyrimidodiazepine precursor of the drosopterins. J Biol Chem 259(2):14121–14127

Wilson TG, Jacobson KB (1977) Mechanism of supression in *Drosophila*. V. Localization of the purple mutant of *Drosophila melanogaster* in the pteridine biosynthetic pathway. Biochem Genet 15:321–332

Zwiebel LJ, Saccone G, Zacharopouloua A, Besansky NJ, Favia G, Collins FH, Louis C, Kafatos FC (1995) The *white* gene of *Ceratitis capitata*: a phenotypic marker for germline transformation. Science 270:2005–2008

5 Green Fluorescent Protein (GFP) as a Marker for Transgenic Insects

Stephen Higgs and David L. Lewis

CONTENTS

5.1 Introduction 94
5.2 Markers/Reporters for Selection of Transformed Insects 94
5.3 The Ideal Reporter 95
5.4 Green Fluorescent Protein 96
    5.4.1 Characteristics of Wild-Type GFP 96
    5.4.2 GFP Mutants and Related Proteins 97
    5.4.3 Why Use GFP as a Reporter for Insect Transgenesis? 99
        5.4.3.1 Positive Characteristics of GFP for Use in Insect Transgenesis 99
        5.4.3.2 General Problems with GFP 100
        5.4.3.3 Specific Problems with GFP When Expressed in Insects 100
5.5 Methods and Applications 100
    5.5.1 Expression of GFP from Viral Vectors 100
        5.5.1.1 Double Subgenomic Sindbis Virus 101
        5.5.1.2 Infection of Mosquitoes with dsSIN Virus 101
        5.5.1.3 Infection of Non-Dipteran Species of Arthropod with dsSIN/GFP Virus 101
        5.5.1.4 Infecting Larvae with Densoviruses 102
    5.5.2 Transformation of Insects with GFP as the Reporter of Transgenesis 102
    5.5.3 Visualization of GFP *in Vitro* 103
    5.5.4 Visualization of GFP *in Vivo* 103
    5.5.5 Potential Problems and Solutions 104
        5.5.5.1 *In Vitro* 104
        5.5.5.2 Histology 104
        5.5.5.3 *In Vivo* 105
Acknowledgments 106
Resources 106
    Products and Protocols 106
    Literature 107
    GFP Newsgroup 107
References 107

5.1 INTRODUCTION

Over the last few years, interest in genetically manipulating invertebrates has increased, and recently developed technologies have revolutionized the ability to engineer insects genetically. One of the critical issues for insect transformation is the ability to readily differentiate between transformed and untransformed individuals. At present, genetic manipulation lacks specificity in terms of being able to target genome sites for integration of heterologous DNA. As a result, a given transformation experiment may produce individuals that have different numbers of insertions and different insertional sites. Due to positional effect, the phenotypes of these different individuals may vary. Consequently, it is important to identify transformed individuals quickly so that separate, pure transgenic lines may be reared and tested for the desired phenotype.

5.2 MARKERS/REPORTERS FOR SELECTION OF TRANSFORMED INSECTS

When efforts to produce transgenic mosquitoes were begun, the available number of reporter genes was small. For *Anopheles gambiae*, *Aedes triseriatus*, and *Ae. aegypti*, the bacterial neomycin phosphotransferase gene (*neo*), which confers resistance to the aminoglycoside antibiotic Geneticin® (also known as G418 and neomycin), was chosen (McCrane et al., 1988; Miller et al., 1987; Morris et al., 1989). Potentially transformed mosquitoes were exposed in the larval stage to G418 added to the rearing water at doses that would kill most wild-type larvae, and so enrich for transformants containing the resistance gene. In similar experiments, the bacterial *hyg* gene encoding for resistance to hygromycin was also used (V. McGrane, personal communication). The two reagents are toxic to both prokaryotes and eukaryotes. As a selectable marker for mosquito transformation, it was necessary to determine dose–response curves by exposing larvae to different concentrations of antibiotic added to the water in which they were reared. Once this was established, the lethal concentration required to kill 50% of exposed wild-type larvae (LC_{50}) was used to select potential transformants. Using this procedure, researchers were able to identify transformed insects. The procedure has inherent problems; for example, by using the LC_{50}, 50% of wild-type individuals may be expected to survive. Although the surviving population should be "enriched" for transgenics, further time and effort are required to isolate the transformed individuals. In addition to resistance-based selection, nutritional selection may also be operating since aquatic bacteria populations upon which mosquito larvae feed may be killed. Furthermore, both G418 and hygromycin are hazardous to researchers and therefore require careful handling and disposal. When rearing 100 larvae per 500 ml of water, volumes of toxic waste can quickly escalate to problematic levels. However, the idea of selection based upon killing wild-type individuals remains attractive. Chemicals considered for selection have included, for example, paraoxan (OPD) and dieldrin (see Chapter 6 by ffrench-Constant and Benedict). Although these can be used to select at the adult stage, both are hazardous to researchers and have handling problems. For example, when using OPD, the chemical stock in methanol is solubilized in acetone. Vessels are surface-coated with the suspended OPD by swirling and allowing the acetone to evaporate. Insects are placed in the container and die as a result of contact exposure and adsorption of OPD through the tarsi. To inactivate OPD, vessels are immersed in a solution of NaOH (10 *M*). One problem is that toxicity is related to mosquito size (A. James, personal communication). This necessitates separate exposure parameters for males and females. A more serious and unexpected problem with OPD-based selection was that some mosquitoes may harbor bacteria with an OPD-resistance gene. Thus, when screened by PCR and Southern hybridization, some wild-type mosquitoes may be identified as transformants. Similar problems associated with bacteria have been experienced when β-galactosidase was used as a marker. The utility of other chemical-based selection systems (*Rdl* and NPT) are discussed in Chapter 6.

As discussed elsewhere (see Chapter 4 by Sarkar and Collins), phenotypic markers based on genes that determine the eye color in *Drosophila melanogaster* have been successfully used as markers of transgenesis in several species of Diptera. Since these genes are insect derived, it is known that they can function nonlethally in insects. However, for these to be useful as markers, the insect species of interest must have suitable eye color mutants in which the marker gene can cause a phenotypic change. Unfortunately, such mutants are unavailable for many, if not most, insects. Phenotypic changes in eye color may result from mutations at more than one allele (see Coates et al., 1997), and so some apparently ideal phenotypes may therefore be genetically unsuitable recipients for this type of eye color rescue. For example, in *Ae. aegypti*, at least one white-eyed strain (Higgs) exists that cannot be rescued by the *cinnabar* (*cn*) gene (M. Nguyen and A. James, unpublished). Fortunately, two species of insect in which researchers are interested, *Ae. aegypti* and the medfly *Ceratitis capitata*, pure breeding eye color mutants are available that are suitable recipients.

The genes *white* (*w*) and *cinnabar* (*cn*) from *D. melanogaster* have proved to be particularly useful as phenotypic markers. The *white* gene homologue has enabled identification of *C. capitata* transformed using the *Minos* element from *D. hydei* (Loukeris et al., 1995). The use of the *cinnabar* gene has been key in demonstrating the utility of the transposable elements *mariner* and *Hermes* for transformation of Diptera (Coates et al., 1998). The expression of this gene in *Drosophila* produces the cinnabar eye phenotype. The white-eyed *kynurenine hydroxylase-white* (kh^w) Rockefeller strain of *Ae. aegypti* (Bhalla, 1968; Jasinskiene et al., 1998) has a mutation that blocks the catalysis of kynurenine to 3-hydroxykynurenine and hence ommochrome pigment formation. This mutation is homologous to the *cinnabar* gene and so "rescue" through expression of the *cinnabar* gene is possible. Introduction of a single copy of the gene into this strain is sufficient to produce coloration, although positional effects influence coloration pattern and intensity (Coates et al., 1998; Jasinskiene et al., 1998). The *cinnabar* gene has proved to be inheritable through several generations and can facilitate selection against heterozygotes and revertants, the eyes of which are less intensely colored than those of homozygotes. Although eye color genes have a number of desirable characteristics as markers of transgenesis, their genetic specificity and the lack of suitable recipient strains in some insects limits their usefulness.

5.3 THE IDEAL REPORTER

For reasons outlined above, a marker system based upon either chemical selection or eye color phenotype is not ideal. Undoubtedly, as knowledge of molecular genetics and understanding of specific insects of interest increase, other insect genes will be identified that can be used as promoters or reporters for insect transgenesis. The ideal reporter would have a number of characteristics:

1. Easily detected with simple equipment
2. Not present in wild-type organisms or their associated symbionts
3. Detectable at an early developmental stage (preferably first instar larvae)
4. Nontoxic (presence does not reduce longevity or fecundity)
5. Stable (i.e., not prone to deletion or mutation to reduce functionality)
6. Heritable
7. Detection related to copy number
8. Expression from somatic tissues and germline integration distinguishable
9. Homozygotes and heterozygotes distinguishable
10. Requiring minimal handling of recipient
11. Not species-specific
12. Expression minimally influenced by insertion site into the organisms' genome

From this list, one might conclude that a phenotypic marker that can be seen with the naked eye is the marker of choice. For mosquito research, one gene and its product — green fluorescent protein (GFP) — has recently proved to be especially useful. This gene has been used as a marker for transgenesis in a variety of organisms, including species of plants, bacteria, yeast, nematodes, insects, fish, amphibians, and mammals (see Chalfie and Kain, 1998). Not only can it be used to identify transgenics, but it can be fused with a gene of interest and, through coexpression, be used to track the tissue-specific expression of an associated gene.

5.4 GREEN FLUORESCENT PROTEIN

GFP was first extracted from the jellyfish species *Aequorea aequorea* (Shimomura et al., 1962). Some reports (for example, Prasher et al., 1992) use the name *A. victoria*; however, it seems that there is insufficient evidence to consider this a separate species (Shimomura, 1998). A gene (*gfp 10*) encoding GFP was cloned by Prasher et al. (1992). At least five variants of GFP have been identified from *Aequorea* (Prasher et al., 1992; Inouye and Tsuji 1994; J. N. Watkins and A. K. Cambell, unpublished). A major accomplishment occurred in 1994 when experiments were reported that demonstrated the functional expression of a *gfp* gene in *Escherichia coli* (Inouye and Tsuji, 1994) and in both *E. coli* and *Caenorhabditis elegans* (Chalfie et al., 1994). The demonstration that GFP could be produced in the absence of jellyfish-specific cofactors was the critical event that made GFP a reporter of broad applicability.

5.4.1 CHARACTERISTICS OF WILD-TYPE GFP

There are now several varieties of GFP available for research. The following discussion refers to wild-type GFP (wt GFP), with characteristics of mutant GFPs being described in later sections. Wt GFP is a 27-kDa monomer of 238 amino acids, and emits green light ($\lambda_{max} = 509$) when excited with ultraviolet light ($\lambda_{max} = 395$). The wt GFP obtained from *Aequorea* has a relatively complex emission spectra, since two chemically distinct proteins are present. In addition to the major peaks described above, a second emission at 503 nm results from excitation at 475 nm. Light emission is due to a *p*-hydroxybenzylideneimidazolinone chromophore consisting of a cyclic tripeptide derived from a Ser-Tyr-Gly motif at positions 65 to 67 in the protein sequence. The nascent wt GFP does not fluoresce. For fluorescence to occur, a post-translational cyclization reaction together with oxidation of the Tyr-66 is required to generate the functional chromophore. Cofactors, substrates, or coexpressed genes are not required. Provided that molecular oxygen is present, the functional chromophore can, therefore, be formed in a wide variety of cell types and organisms so that wt GFP may be visualized *in situ*. The primary, secondary, tertiary, and quaternary structure of GFP have been elucidated (see reviews by Phillips, 1998; Tsien, 1998). Crystals may form as dimers or monomers, and are formed as an α-helix containing the chromophore, enclosed within a tightly packed β-can (Ormo et al., 1996; Yang et al., 1996).

Biochemical and physical properties of GFP have been reviewed by Ward (1998). Wt GFP is relatively stable within a wide range of environmental conditions. Fluorescence occurs within a pH range of 5.5 to 12, and in the presence of mild reducing agents. Chromophore formation is temperature sensitive, with fluorescence being more intense at relatively low temperatures (for example, 15°C vs. 37°C). This temperature sensitivity is due to the efficiency of protein folding, which declines as temperature increases. However, if the protein is allowed to mature at low temperature, fluorescence is retained at least to 65°C (Tsien, 1998). Sensitivity to some organic solvents means that certain histological procedures may influence or ablate fluorescence. Fully denatured GFP does not fluoresce, and so, if it is impractical to examine wt GFP in living tissues, fixatives, reducing agents, and other histological reagents should be tested carefully to establish their effects. Although photobleaching upon exposure to high-intensity illumination occurs, wt GFP is less susceptible than fluorescein. The photobleaching rate may be somewhat species specific.

Although wt GFP may be excited and visualized using commonly available fluorescence microscopes that are equipped with FITC filter sets, these do not provide optimal conditions.

Excitation filters for fluorescein produce light of approximately 450 nm, but for wt GFP the λ_{max} is 395 nm. Due to rapid photoisomerization and loss of the fluorescent signal, excitation at 395 nm is not recommended. Use of FITC filter sets therefore result in fluorescence from the 470-nm absorbance peak and consequently signal from wt GFP using FITC filter sets is suboptimal. Other characteristics of wt GFP also impose limitation of its usefulness, for example, the suboptimal folding at 37°C.

5.4.2 GFP Mutants and Related Proteins

For reasons outlined above, wt GFP has characteristics that compromise its usefulness in certain situations. Therefore, using techniques such as chemical mutagenesis and error-prone PCR, the GFP gene has been mutated to produce engineered GFPs with characteristics that are optimal for expression in a range of conditions when examined with FITC filter sets. Some mutants have been produced that fluoresce optimally with alternative filter sets. For truly optimal visualization, filter sets that precisely match the excitation and emission characteristics of each GFP variant are available (for example, from Chroma Technology Corp., 72 Cotton Mill Hill, Unit A9, Brattleboro, VT). Tsien (1998) classifies variants of GFP into two categories with seven classes:

A. Derived from polypeptides with Tyr at position 66
  Class 1. Wild-type mixture of neutral phenol and anionic phenolate
  Class 2. Phenolate anion
  Class 3. Neutral phenol
  Class 4. Phenolate anion with stacked π-electron system
B. Derived from a polypeptide with Trp, His, and Phe at position 66
  Class 5. Indole
  Class 6. Imidazole
  Class 7. Phenyl

The so-called red-shifted variants of wt GFP have excitation peaks between 488 and 490 nm (Table 5.1). This change is produced by substituting amino acids within or adjacent to those of the chromophore (for example, Phe-64 to Leu and Ser-65 to Thr substitutions). To optimize GFP expression for specific cell types and conditions, additional mutations have been generated. For expression in insects and insect cell lines, both wt GFP and variants have been used (Table 5.2). For general use in transgenic insects, the class 2 enhanced variant "EGFP" is probably the GFP of choice. All variants with the GFP notation emit green light; however, mutagenesis has been used to move the emission spectra to produce blue, yellow-green, and cyan emission (variants BFP, YFP, and CFP, respectively). Used in conjunction with GFP, these proteins can be used for simultaneously visualizing multiple cellular proteins to which each has been fused or in different tissues in conjuction with different promoters. Expression of BFP and YFP is being assessed for use in insects (A. Handler, personal communication).

A new red fluorescent protein (DsRed) that was isolated from an Indo-Pacific sea anemone-relative *Discosoma striata* has very recently become available (Matz et al., 1999). It has an emission maximum at 583 nm and excitation maximum at 558 nm, making it most easily distinguishable from GFP and the GFP variants. Specific filter sets are not as yet available, but it is easily detectable using rhodamine or propidium iodide filters (see Clontech catalog for additional information). Its use has not yet been reported for insects, and will not be further discussed here, although it clearly has great potential for use as a marker in insects and especially for double-labeling purposes. New fluorescent protein markers may be expected as well in the coming years.

TABLE 5.1
A Simplified Description of the Characteristics of GFP, GFP Variants, and DsRed

Plasmid	Major Emission (nm)	Excitation Maxima (nm)	Relative Brightness	Mutations	Accession No.
pGFP (Clontech)	395 (470 minor)	509	1	None (wild-type)	U17997
pGFPuv (Clontech)	395	509	18x	F99S, M153T Val-163 to Ala 5 Arg codons replaced to optimize expression in *E. coli*	U62636
pEGFP (Clontech)	488	507	35x	Phe-64 to Leu and Ser-65 to Thr >190 "humanized" mutations to maximize mammalian codon usage	U76561
pGreen Lantern-1 (Gibco-BRL)	490		40x	Ser-65 to Thr >80 "humanized" mutations to maximize mammalian codon usage	
pEBFP (Clontech)	380	440	n/a	4 AA substitutions >190 "humanized" mutations to maximize mammalian codon usage	n/a
pEYFP (Clontech)	513	527	n/a	4 AA substitutions >190 "humanized" mutations to maximize mammalian codon usage	n/a
pDsRed (Clontech)	583	558	n/a	None (wild-type)	AF168420
pDsRed1 (Clontech)	583	558	n/a	144 "humanized" mutations to maximize mammalian codon usage	n/a

Note: For a more detailed description see Tsien (1998) and Matz et al. (1999).

TABLE 5.2
Species of Insects and Insect Cell Lines in Which GFP Variants Have Been Expressed

GFP Variant	Insect Species	Delivery/Expression System	Ref.
GFP, S65T GFP	*Aedes aegypti*	dsSIN	Higgs et al., 1996
S65T GFP	*Ae. aegypti*	AeDNV	Afanasiev et al., 1999
EGFP	*Ae. aegypti*	Hermes	Pinkerton et al., 1999
GFP, S65T GFP	*Ae. triseriatus*	deSIN	S. Higgs (unpublished)
EGFP	*Anastrepha suspensa*	piggyBac	A. Handler (personal communication)
GFP	*Anopheles gambiae*	dsSIN	Higgs et al., 1996; Olson et al., 1998
S65T GFP	*An. stephensi*	dsSIN	S. Higgs and C. Barillas-Mury (unpublished)
EGFP	*Boactrocera dosalis*	piggyBac	A. Handler (personal communication)
EGFP	*Bobyx mori*	AcNPV	Yamao et al., 1999
EGFP	*Ceratitis capitata*	piggyBac	A. Handler (personal communication)
EGFP	*Culex quinquefasciatus*	Hermes	M. Allen, C. Levesque, D. O'Brochta, P. Atkinson (personal communication)
S65T GFP	*Cx. pipiens*	dsSIN	Higgs et al., 1996
EGFP	*Drosophila melanogaster*	dsSIN	A. Rayms-Keller (personal communication)
GFP	*D. melanogaster*	Pseudotyped retrovirus	J. Burns (personal communication)
EGFP	*D. melanogaster*	piggyBac	A. Handler (personal communication)
EGFP	*Oncopeltus fasciatus*	dsSIN	D. Lewis et al., 1999
EGFP	*Precis coenia*	dsSIN and SinRep5	D. Lewis et al., 1999
EGFP	*Papilio glaucus*	dsSIN	D. Lewis et al., 1999
EGFP	*Tribolium castaneum*	dsSIN and SinRep5	D. Lewis et al., 1999
GFP, S65T GFP	C6/36 (*Ae. albopictus*)	dsSIN	Higgs et al., 1996
S65T GFP	C6/36 (*Ae. albopictus*)	AeDNV	Afanasiev et al., 1999
S65T GFP	ATC-15 (*Ae. aegypti*)	dsSIN (MRE-16 version)	S. Higgs (unpublished)
S65T GFP	AP-61 (*Ae. pseudoscutellaris*)	dsSIN (MRE-16 version)	S. Higgs (unpublished)

5.4.3 WHY USE GFP AS A REPORTER FOR INSECT TRANSGENESIS?

From the characteristics described above, it is clear that GFP, in particular, EGFP, has a number of advantages over previously used selectable markers. There are, however, some undesirable characteristics and problems with the use of GFP. Both useful and problematic attributes of GFPs are described below, with solutions to some of the problems discussed in Section 5.5.5.

5.4.3.1 Positive Characteristics of GFP for Use in Insect Transgenesis

1. Expression and visualization is possible in a broad range of cell types and organisms including bacteria, protozoa, arthropods, nematodes, fish, amphibians, mammals.
2. No cofactors are required.
3. It is nontoxic and does not result in production of toxic waste.

4. Fluorescence may be visible from single-copy insertions.
5. Insertion site has relatively little influence on GFP synthesis.
6. High visibility allows for rapid screening of cells and organisms soon after transformation. Insects can therefore be screened at early stages to eliminate nontransformants and efforts can be focused on transformants.
7. Cells can be sorted by fluorescent-activated cell sorter (FACS) analysis allowing enrichment of cultures. This may be useful when testing GFP delivery based upon infectious agents (e.g., dsSIN and AeDNV).
8. Correct cloning into plasmids can be confirmed in bacterial vectors.
9. Stability may facilitate population mark–release–recapture studies — even if insects are dead when recovered (>14 days for EGFP; A. Handler, personal communication). Stability can permit precise localization in specimens prepared for histology. Antibodies are also available.
10. The relatively small size and activity from most positions within the reading frame allow GFP to be cloned to function as a fusion protein (N- and C-terminal).
11. The small size of the GFP coding region (approximately 714 nt) allows for testing in other systems (e.g., dsSIN), prior to tedious transformation procedures.
12. Primers are commercially available for PCR confirmation of insertion.

5.4.3.2 General Problems with GFP

1. Visualization requires a fluorescence microscope with suitable filters.
2. Requirement for oxygen precludes use in obligate anaerobes.
3. The relative stability of GFP may be problematic for protein trafficking research, but destabilized versions of GFP are available.
4. Insolubility may be undesirable, but EGFP is more soluble that wt GFP.

5.4.3.3 Specific Problems with GFP When Expressed in Insects

1. GFP fluorescence in tissues and internal organs may be obscured by the color of the insect exoskeleton. Fortunately, many immature stages are transparent or translucent.
2. GFP fluorescence may be confused by autofluorescence (for example, as associated with food, Malpighian tubules, cuticle and necrosis).
3. The intense illumination required for visualization may adversely affect insects, especially if long exposure for photography is required.

5.5 METHODS AND APPLICATIONS

This section deals with GFP expression in systems that are specifically appropriate for insect transgenesis studies. *In vitro* expression is included since it is often important to confirm successful cloning and expression prior to *in vivo* work. Some techniques, such as FACS sorting and GFP expression in plants, are not discussed. If required, these can be obtained from sources and references listed at the end of this chapter.

5.5.1 Expression of GFP from Viral Vectors

Prior to transforming mosquitoes, we tested expression of GFP in mosquitoes using infectious expression systems. We assumed that if GFP could not be visualized using these systems, then the low copy number expression anticipated from germline transformants would be unlikely. Although Sindbis (SIN) viruses have primarily evolved to infect mosquitoes, the double subgenomic SIN (dsSIN) viruses have been successfully used to express genes including GFP in other organisms (Table 5.2; Lewis, et al. 1999). Detailed protocols describing all aspects of using dsSIN viruses for gene expression in mosquitoes are available (Higgs et al., 1997; see Chapter 8 by Carlson et al.), and are only briefly given in this chapter. The dsSIN viruses used have been based upon the TE/3'2J vectors that have a limited capacity to infect mosquitoes when presented orally. Efforts to engineer the current dsSIN vectors to extend the host range, for example, by substituting the structural genes of TE/3'2J with those from the more orally infectious MRE strain are ongoing (see Chapter 8 by Carlson et al.). Densoviruses (see Chapter 8 by Carlson et al.), that have the potential to integrate into the mosquito genome have also employed EGFP as a marker.

5.5.1.1 Double Subgenomic Sindbis Virus

To determine if GFP could be expressed and visualized in mosquitoes, the dsSIN system was utilized to test both wt GFP and an enhanced red-shifted mutant GFP (S65T GFP). The bacteria used to host the plasmids for cloning purposes were screened by examining for GFP using ultraviolet illumination. Both pGFP and pEGFP plasmids contain an ampicillin-resistance gene for antibiotic selection. Once plated on agar, colonies have been allowed to grow and then are screened by ultraviolet transillumination with green colonies selected and transferred to liquid culture containing ampicillin. From these cultures, DNA was extracted and cloned into dsSIN vectors and virus was produced as previously described (Higgs et al., 1997; see Chapter 9 by Olson).

5.5.1.2 Infection of Mosquitoes with dsSIN Virus

Sindbis virus expression systems have been used to express GFP in several species of mosquito (see Table 5.2; Color Figure 2*), including *Ae. aegypti* (Higgs et al., 1996); *Ae. triseriatus* (S. Higgs, unpublished); *Anopheles gambiae* (Higgs et al., 1996; Olson et al., 1998); *An. stephensi* (S. Higgs and C. V. Barillas-Muray, unpublished), and *Culex pipiens* (Higgs et al., 1996). Initial studies were with mosquitoes that were intrathoracically inoculated with the virus, TE/3'2J/GFP (Higgs et al., 1996). Larvae were transferred to wet gauze pads and viewed under an Olympus SZH stereomicroscope. A simple inoculation apparatus (Rosen and Gubler, 1974) with glass capillary needles with a tip approximately 20 μm in diameter was used to intrathoracically inoculate 0.5 μl of virus. Larvae were immediately returned to rearing pans. At intervals, larvae, pupae, and adults that eclosed from the inoculated larvae were examined for GFP expression as described below.

5.5.1.3 Infection of Non-Dipteran Species of Arthropod with dsSIN/GFP Virus

In addition to its ability to infect mosquitoes productively, TE/3'2J/GFP has been found to infect a wide variety of other insect species, namely, the Lepidoptera *Precis coenia* (buckeye butterfly) and *Papilio glaucus* (tiger swallowtail butterfly), the coleopteran *Tribolium castaneum* (red flour beetle), and the hemipteran *Oncopeltus fasciatus* (milkweed bug). Just as in mosquitoes, infectability appears to be independent of the developmental stage of the animal. Injections of TE/3'2J/GFP into *P. coenia* larval and pupal stage animals using a Hamilton syringe or a glass needle have resulted in the expression of both GFP and viral E1 antigen in infected tissues (see Section 5.5.5.2). In addition, the embryos of beetles and milkweed bugs have been found to be susceptible to infection with TE/3'2J/GFP. Injection of embryos is a bit more problematic than injection of animals at later developmental stages because of their small size and the presence of a chorion that may be difficult to puncture or remove in some species. However, successful embryonic injections with good survival rates of these two species have been achieved, and GFP expression can be detected in both living or fixed embryos.

---

* Color Figure 2 follows p. 108.

As an alternative to inoculation, a new technique has been developed (Higgs et al., 1999; Lewis et al., 1999). By using the dsSIN vectors based upon the Malaysian MRE-16 strain (Seabaugh et al., 1998), C6/36 (*Ae. albopictus*) cells grown in T-25 flasks were infected at a multiplicity of infection of approximately 0.01. At 24 to 72 h postinfection, cells were resuspended by scraping, and newly hatched larvae were introduced into the flask and allowed to feed. At time intervals, larvae, pupae, and resulting adults were examined as described above. A similar technique to infect other aquatic arthropods orally, including the crustacean, *Artemia franciscana*, has been used. In this case, 2×10^6 C6/36 cells are infected with TE/3'2J/GFP at a multiplicity of infection of 1 in a 1.5-ml tube. At 24 to 26 h postinfection, freshly hatched *Artemia nauplii* are added and allowed to feed on the infected cells for 24 h. The *Artemia* are then removed from the tube and allowed to develop for an additional 24 h in 1% artificial seawater. By using this method, up to 10% of the animals showed GFP expression. This technique has proved to be effective and might be adapted to nonaquatic insects feeding upon moist diet if infected cells were placed on the surface with feeding larvae. Drying will almost certainly result in loss of viral infectivity. To infect the adult stage, mosquitoes are placed in a refrigerator for 10 to 20 min and, once cold-anesthetized, transferred to a chill table (BioQuip Products, Gardena, CA) and inoculated using an apparatus identical to that as used for larvae.

5.5.1.4 Infecting Larvae with Densoviruses

The parvovirus expression system based upon AeDNV has been engineered to express GFP in *Ae. aegypti* mosquitoes (Afanasiev et al., 1999). This expression system is described in Chapter 8 by Carlson and so it is not discussed in detail here. Briefly, C6/36 cell cultures were cotransfected with pUCA and pNS1-GFP, using lipofectin. At 48 h post-transfection, medium and cells were centrifuged at 32,000 rpm in a Beckman SW48 rotor to concentrate the viral transducing particles. The concentrated pellet was resuspended in 2 ml of water and sonicated for 30 s. Day-old mosquito larvae were added to the sonicated supernatant, maintained at 28°C for 24 h, and then transferred to approximately 1 liter of water and reared under standard conditions. Individuals were examined periodically and GFP was observed in larvae, pupae, and adults (Afanasiev et al., 1999; T. Ward, personal communication; see Color Figure 2D and E*).

5.5.2 TRANSFORMATION OF INSECTS WITH GFP AS THE REPORTER OF TRANSGENESIS

Techniques for insect transgenesis are thoroughly covered in other chapters. Several of these systems have successfully incorporated GFP as a reporter to identify transformed individuals. Three examples are the *Hermes* transposable element (see Chapter 12 by Atkinson and O'Brochta), the pseudotyped retroviruses (see Chapter 7 by Burns), and the *piggyBac* gene-transfer system (see Chapter 14 by Fraser).

Both *Ae. aegypti* and *C. quinquefasciatus* have been transformed using the *Hermes* element that has EGFP expressed from the Actin 5C promoter (Allen et al., 1999; Pinkerton et al., 2000). Color Figure 2* shows GFP expression in various species of insects.

In a collaboration among J. C. Burns (University of California, San Diego) and W. Brown and N. H. Patel (University of Chicago), germline transformation of *D. melanogaster* has been achieved using pseudotyped retroviruses (see Chapter 7 by Burns). Approximately 10 colony-forming units of pantropic retroviruses was inoculated into the perivitelline space of embryos at the 64-cell stage. EGFP was expressed from the MoMLV LTR. GFP was observed in the gut and neuroectoderm of some G_0 flies. G_0 flies were interbred and, of 600 G_1 embryos visually screened, GFP was observed in five individuals. PCR analysis using LTR-specific primers confirmed the integration in four animals.

---

* Color Figure 2 follows p. 108.

Germline transformation of *D. melanogaster* has been achieved with the *piggyBac* gene-transfer system (Handler and Harrell, 1999) using a dual-reporter system. The vector was marked with the *D. melanogaster* mini-*white* gene, together with EGFP regulated by the polyubiquitin promoter linked to the SV40 nuclear localizing sequence. Most of the 70 G_1 transformants (from seven G_0 lines) visibly expressed only GFP, with little or no eye pigmentation observable. It was therefore concluded that position effect suppression appears to have relatively little influence on EGFP expression, compared with the phenotypic expression of white. A *piggyBac*/GFP vector, pB[PUbnlsEGFP], has also been used to transform Carribean (*Anastrepha suspensa*), Mediterranean (*Ceratitis capitata*), and oriental (*Bactrocera dorsalis*) fruit flies (A. Handler, personal communication).

5.5.3 Visualization of GFP *in Vitro*

In our hands, visualization of GFP expressed from dsSIN and AeDNV viruses has been relatively straightforward. Prior to working with mosquitoes, our policy was to test for GFP expression *in vitro*. With the dsSIN vectors this is done by infecting cell cultures (mammalian and mosquito derived) and then at 12 to 48 h postinfection, examining these using an Olympus IMT-2 inverted microscope equipped for epifluorescence with FITC filter sets. Viewing cells through plastic tissue culture flasks is subject to some distortion of the light and sometimes autofluorescence problems (see below). Although GFP is visible in cells growing in tissue culture flasks, we prefer to infect cells and grow these on sterile glass coverslips (Gould et al., 1985). Uninfected control cells are prepared in parallel. Coverslips are removed from medium, washed three times in PBS, and placed cell-side down onto 1 drop of PBS:glycerol (1:9) placed on a microslide.

For our purposes, we have not used fixed cultures, but this is feasible. At an appropriate time postinfection, cells grown on coverslips should be washed three times in PBS and then fixed in 4% paraformaldehyde in PBS (pH 7.4 to 7.6) at room temperature for 30 min. These coverslips are examined as described above. If necessary, coverslips can be sealed using molten agarose or rubber cement. Nail polish is to be avoided since this inhibits GFP fluorescence. The GFP fluorescence will be retained for at least 2 weeks if the slides are stored at 4°C.

If working with noninfectious delivery systems, for example, with certain replicons or transposable elements, standard transfection procedures (for example, liposomes or electroporation) may be used to test for GFP expression in prokaryotic or eukaryotic cells.

5.5.4 Visualization of GFP *in Vivo*

To screen mosquitoes for GFP, the stage of interest is placed on a microscope slide and entrapped beneath a glass coverslip (in water for aquatic stages). We have found it useful to use contrasting transmitted back illumination. The most suitable illumination has been achieved using a standard FITC excitation filter, which gives a red background. Compound (Olympus BH-2), inverted (Olympus IMT-2), and stereo (SZX) microscopes have been employed. BH2 and IMT-2 microscopes were equipped for epifluorescence using standard FITC filters, whereas GFP-specific filters were fitted to the SZX. The SZX microscope has the advantage of zoom capability. For still photography, two types of camera have been used. An Olympus OM-4 Ti 35mm camera with 100ASA speed film (Kodak Elite 100) has been used. An Olympus DP10 digital camera has also been successfully employed, providing for easy transfer of images to a computer. This particular model has a built-in, real-time viewing screen and so can be easily transferred between microscopes since no computer linkup is required to review the captured image. For video recording, a model VPC-920 CCD camera was found to be suitable. Other manufacturers, for example, Leica, Nikon, and Zeiss, all produce microscopes suitable for GFP detection. Companies such as Kramer Scientific Corporation (5 Westchester Plaza, Elmsford, NY 10523) offer setups to convert microscopes for GFP visualization.

To aid rapid screening, we transfer larvae to 96-well tissue culture plates, with one larva per well in 50 to 100 µl of water. The plate is placed on the stage of the IMT-2 inverted microscope and a mechanical stage is used to move across each row systematically. In this way, 96 larvae can be screened in just a few minutes. A stereomicroscope would likely further reduce the time required.

5.5.5 POTENTIAL PROBLEMS AND SOLUTIONS

5.5.5.1 *In Vitro*

We have experienced few difficulties when screening for GFP expression *in vitro* (vertebrate and mosquito cells). Autofluorescence associated with cell types and certain tissues, for example, the cell line derived from *Ae. pseudoscutellaris* (AP-61), may be a distracting yellow when examined with FITC filters, but the *Ae. albopictus* line (C6/36) is not. Autofluorescence may be caused by various factors. In live mammalian cells, autofluorescence may be caused by flavin coenzymes and NADH bound to mitochondria and may even be influenced by cell type and culture history (Aubin, 1979). If this is a problem, since autofluorescence may be wavelength specific, it may be necessary to use either alternative filter sets or a different GFP variant.

5.5.5.2 *Histology*

The ability to visualize GFP expression from tissues and cells within a transformed animal is essential for evaluating the spatial and temporal controls regulating expression resulting from germline or somatic cell transformation events. The physical properties of GFP make it a convenient histological reagent. It is resistant to formaldehyde and gluteraldehyde fixation, and this enables visualization of fine cellular detail in fixed tissues. In addition, visualization of GFP does not require the addition of exogenous reagents and GFP fluorescence is relatively resistant to photobleaching. However, certain characteristics of GFP must be considered prior to processing. For example, absolute ethanol quenches all GFP fluorescence. It has also been reported that certain nail polishes that are used to seal coverslips onto microscope slides also quench GFP fluorescence (Chalfie et al., 1994; Wang and Hazelrigg, 1994), although we have not experienced this.

Although GFP can be visualized using conventional epifluorescence microscopy, confocal microscopy will yield much higher resolution because it eliminates interference from fluorescence emanating from tissue outside the region of interest. This enables the investigator to pinpoint precisely the cells that are expressing GFP. Various methods and applications involving confocal microscopy have been described in detail elsewhere and will only be described briefly here (Paddock, 1999).

Another great advantage of the confocal microscope is the ability to collect images simultaneously of macromolecules that fluoresce at different wavelengths. Most often, up to three different macromolecules are detected and the digital image of each is collected and viewed in the red, green, or blue channels of an RGB monitor. The image can be viewed as a single RGB color image, with any overlap of the probes seen as a different additive color. Since the images collected are stored as digital files, they can be modified using digital image-processing software such as Photoshop (Adobe Systems, Inc.). With this software, manipulations such as adding color to single-channel images, rearranging colors in multichannel images, as well as altering the size of the final image and adding of text can be performed. Detailed protocols for retrieving and presenting confocal images can be found elsewhere (Halder and Paddock, 1999).

We have used multiple labeling to follow infection of TE/3'2J/GFP in tissues of various insects. By using a Bio-Rad MRC1024 Laser Scanning Confocal Microscope, GFP is detected in the green (fluorescein) channel, viral E1 protein in the red (rhodamine/Cy 3) channel, and stained nuclei in the blue (Cy 5) channel. For immunostaining, the tissue of interest is dissected and fixed for 30 min on ice in fix buffer containing 0.1 M PIPES, pH 6.9/1 mM EGTA/2 mM MgSO$_4$/1.0% Triton X-100 and 3.7% formaldehyde. After three washes in PBS, the tissue is incubated for 1 h at 4°C in immunostaining buffer containing 50 mM Tris, pH 6.8/150 mM NaCl/0.5% Nonidet-P40/5 mg/ml BSA to reduce nonspecific binding by the antibody. The tissue is then incubated with a 1:200 dilution of mouse anti-E1 monoclonal antibody overnight at 4°C in the same buffer. Unbound antibody is removed from the tissue by rinsing with immunostaining buffer six times over 1 h on ice. The secondary antibody, in this case Cy 3 conjugated donkey antimouse Fab fragments (Jackson ImmunoResearch Laboratories, Inc.), is added to the rinsed tissue at a dilution of 1:400 in immunostaining buffer. We have found that using Fab fragments of the secondary antibody as opposed to whole IgG reduces nonspecific binding. After incubation at 4°C for 2 to 16 h, the unbound secondary antibody is removed by washing the tissue six times over 1 h on ice with immunostaining buffer. The tissue is then incubated at least 2 h at 4°C in mounting buffer containing 50 mM Tris pH 7.0/150 mM NaCl/30% glycerol. The nucleic acid stain TO-PRO-3 (Molecular Probes) is included at a dilution of 1:1000 to visualize nuclei. The TO-PRO-3 is rinsed from the tissue once with mounting buffer and the tissue is mounted on a slide under a glass coverslip. Nail polish is used to seal the coverslip to the slide in these preparations since we have not observed any quenching of the GFP signal. For staining embryos, the fixation is performed at room temperature in a two-phase mixture of heptane and fix buffer without Triton X-100. The heptane is used to permeabilize the vitelline membrane and this allows access to the embryonic tissue by the antibodies. After removing the heptane with three washes of PBS, all other steps are performed as described above.

In infected cells of embryonic, larval, and pupal stage animals, GFP is found in the cytoplasm and cytoplasmic extensions as well as in nuclei. The presence of GFP in the nucleus is most likely due to its small size (27 kDa), and not because of active transport since it lacks a recognizable nuclear localization signal. Infected cells also express the viral E1 protein, which has a subcellular location distinct from that of GFP, being present both in the endoplasmic reticulum and the cell membrane.

Although the high level of GFP expression achieved in cells infected with TE/3'2J/GFP results in easily detectable levels of GFP, other expression systems may not give this result. In these cases, using polyclonal anti-GFP antibodies (Clontech, Palo Alto, CA) and fluorescein-labeled secondary antibodies will increase the fluorescent signal observed at least fivefold. Anti-GFP antibodies may also be used for immunofluorescence analysis of fixed tissues or in immunoblotting tests in which GFP fluorescence has been destroyed.

5.5.5.3 *In Vivo*

The main problem experienced when examining living mosquitoes using standard FITC filters for GFP expression is autofluorescence. The easiest way to determine if this may be a problem is to examine large numbers of control animals (noninfected or nontransformed) reared identically to experimental insects. Familiarity with normal insects observed under appropriate conditions of illumination may save much time — and unwarranted excitement! With mosquitoes, we have observed three sources of autofluorescence. One is the food used for mosquito rearing. Certain fish foods and some laboratory rodent diets have components that may glow bright green under ultraviolet illumination and appear identical to GFP fluorescence. This is presumably caused by plant materials such as cellulose. As a result, the larval gut may appear green. The obvious localization of this fluorescence and restriction within the gut should immediately cause the researcher to suspect that the apparent GFP signal may be false. For mosquitoes, one remedy for this potential problem is to feed the larvae a diet without fluorescent components. One such suitable diet has been concocted by P. W. Atkinson (personal communication). This consists of 60 g of "Milkbone" original dog biscuits ground to a very fine powder, mixed with 30 g of "Wheast" (Red Star Specialty Nutrex 55). This mix is added to larval rearing pans as usual.

A second cause of autofluorescence seen in both larvae and adults results from crystals within the Malpighian tubules. We have not resolved this problem although, since dsSIN rarely infects these excretory organs, we usually consider that fluorescence associated with the tubules as autofluorescence and not GFP. It can, unfortunately, be indistinguishable from the fluorescence emission typical of GFP. Again, the specific and restricted distribution of the signal should make the researcher cautious in interpreting data.

Finally the chitinous exoskeleton may autofluoresce. Although distracting, this tends to be clearly more yellow than GFP green fluorescence. In more colorful insects, however, this may not be true. Similar problems of autofluorescence have been reported for other organisms by other researchers (Chalfie et al., 1994; Niswender et al., 1995). Long-pass emission filters or DAPI filter sets may allow GFP and autofluorescence to be distinguished (Brand, 1995; Pines, 1995). In addition, fluorescence due to GFP will only be seen using FITC filter sets, whereas autofluorescence will often be seen using both FITC and Rhodamine/Cy3 filter sets. The intense illumination required to visualize GFP has occasionally proved fatal to the adult mosquito during examination. A high-power objective that produces a highly concentrated beam of light proved most lethal, especially if adults were examined dry with a coverslip to protect them. For larvae examined immersed in water, no adverse effects have been noticed.

ACKNOWLEDGMENTS

We thank Steve Kain for useful discussions and Peter Atkinson, Al Handler, and Todd Ward for discussions and photographs. Our thanks to Dr. Marco DeCaMillis for work on embryo injections and processing.

RESOURCES

The development of GFP as a reporter gene has coincided with the advent of widespread electronic communication. Information on all aspects of GFP use, related products, and references is available in various World Wide Web sites, which have the advantage of being constantly updated. Mention of companies, trade names, proprietary products, and equipment does not constitute a guarantee or warranty by the authors and does not imply approval to the exclusion of other products that can be equally suitable. Web sites include:

PRODUCTS AND PROTOCOLS

http://www.aurorabio.com/
http://www.chroma.com/
http://www.clontech.com/
http://www.invitrogen.com/
http://www.jacksonimmuno.com/
http:/www.kairos-scientific.com/
http://www.lifetech.com/
http://www.molecularprobes.com/
http://www.probes.com/
http://www.packardinst.com/
http://www.pharminogen.com/
http://www.qbi.com/
http://www.turnerdesigns.com/

LITERATURE

http://www.clontech.com/clontech/GFPRefs.html

GFP NEWSGROUP

http://www.bio.net/hypermail/FLUORESCENT-PROTEINS/

REFERENCES

Afanasiev BN, Ward TW, Beaty BJ, Carlson JO (1999) Transduction of *Aedes aegypti* mosquitoes with vectors derived from *Aedes* densovirus. Virology 257:62–72

Aubin JE (1979) Autofluorescence of viable cultured mammalian cell. J Histochem Cytochem 27:360–443

Bhalla SC (1968) White eye, a new sex-linked mutant of *Aedes aegypti*. Mosquito News 28:380–385

Brand A (1995) GFP in *Drosophila*. Trends Geyet 11:324–325

Chalfie M, Kaine S (1998) GFP Green Fluorescent Protein Properties, Applications and Protocols. Wiley–Liss, New York, 385 pp

Chalfie M, Tu Y, Euskirchen G, Ward WW, Prasher DC (1994) Green fluorescent protein as a marker for gene expression. Science 263:802–805

Chanas AC, Ellis DS, Stamford S, Gould EA (1982). The interaction of monoclonal antibodies directed against the envelope glycoprotein E1 of Sindbis virus with virus-infected cells. Antiviral Res 2:191–201

Coates CJ, Schaub TL, Besansky NJ, Collins FH, James AA (1997) The white gene from the yellow fever mosquito, *Aedes aegypti*. Insect Mol Biol 6:291–299

Coates CJ, Najinskiene N, Miyashiro L, James AA (1998) *Mariner* transposition of the yellow fever mosquito, *Aedes aegypti*. Proc Natl Acad Sci USA 95:3748–3751

Gould EA, Buckley A, Cammack N (1985) Use of a biotin-streptavidin interaction to improve flavivirus detection by immunofluorescence and ELISA tests. J. Virol Methods 11:41–48

Halder G, Paddock SW (1999) Presentation of confocal images. In Paddock SW (ed) Confocal Microscopy, Methods and Protocols. Humana Press, Totawa, NJ, pp 373–384

Handler AM, Harrell RA (1999) Germline transformation of *Drosophila melanogaster* with the *piggyBac* transposon vector. Insect Mol Biol 8:449–457

Higgs S, Traul D, Davis BS, Kamrud KI, Wilcox C, Beaty BJ (1996) Green fluorescent protein expressed in living mosquitoes, without the requirement of transformation. BioTechniques 21:660–664

Higgs S, Olson KE, Kamrud KI, Powers AM, Beaty BJ (1997) Viral expression systems and viral infections in insects. In Crampton JM, Beard CB, Louis C (eds) The Molecular Biology of Disease Vectors: A Methods Manual. Chapman & Hall, London, pp 459–483

Higgs S, Oray CT, Myles K, Olson KE, Beaty BJ (1999) Infecting larval arthropods with a chimeric double subgenomic Sindbis virus vector to express genes of interest. BioTechniques 27:908–911

Inouye S, Tsuji FI (1994) *Aequorea* green fluorescent protein. Expression of the gene and fluorescence characteristics of the recombinant protein. FEBS Lett 341:277–280

Jasinskiene N, Coates CJ, Benedict, MQ, Cornel AJ, Rafferty CS, James AA, Collins FH (1998) Stable transformation of the yellow fever mosquito, *Aedes aegypti*, with the *Hermes* element from the housefly. Proc Natl Acad Sci USA 95:3743–3747

Lewis DL, De Camillis MA, Brunetti, CR, Halder G, Kassner VA, Selegue JE, Higgs S, Carroll SB (1999) Ectopic gene expression and homeotic transformations in arthropods using recombinant Sindbis viruses. Current Biology 9:1279–1287

Loukeris TG, Livardaras I, Arca B, Zabalou S, Savakis C (1995) Gene transfer into the medfly, *Ceratitis capitata*, with a *Drosophila hydei* transposable element. Science 270:2002–2005

Matz MV, Fradkov AF, Labas YA, Savitsky AP, Zaraisky AG, Markelov ML, Lukyanov SA (1999) Fluorescent proteins from nonbioluminescent Anthozoa species. Nat Biotechnol 17:969–973

McGrane V, Carlson JO, Miller BR, Beaty BJ (1988) Microinjection of DNA into *Aedes triseriatus* ova and detection of integration. Am. J Trop Med Hyg 39: 501–510

Miller LH, Sakai RK, Romans P, Gwadz RW, Kantoff P, Coon HG (1987) Stable integration and expression of a bacterial gene in the mosquito *Anopheles gambiae*. Science 237:779–781

Morris AC, Eggleston P, Crampton JM (1989) Genetic transformation of mosquito *Aedes aegypti* by microinjection of DNA. Med Vet Entomol 3:1–7

Niswender KD, Blackman SM, Rohde L, Magnuson MA, Piston DW (1995) Quantitative imaging of green fluorescent protein in cultured cells: comparison of microscopic techniques, use in fusion proteins and detection limits. J Microsc 180:109–116

Olson KE, Beaty BJ, Higgs S (1998) Sindbis virus expression systems for the manipulation of insect vectors. In Miller LK and Ball A (eds) The Insect Viruses. Plenum, New York, pp 371–404

Ormo M, Cubitt A, Kallio K, Gross L, Tsien R, Remington S (1996) Crystal structure of the *Aequorea victoria* green fluorescent protein. Science 273:1392–1395

Paddock SW (1999) An introduction to confocal imaging. In Paddock SW (ed) Confocal Microscopy, Methods and Protocols. Humana Press, Totawa, NJ, pp 1–34

Phillips GN Jr (1998) The three-dimensional structure of green fluorescent protein and its implications for function and design. In Chalfie M, Kaine S (eds) GFP Green Fluorescent Protein Properties, Applications and Protocols. Wiley–Liss, New York, pp 77–96

Pines J (1995) GFP in mammalian cells. Trends Genet 11:326–327

Pinkerton AC, Michel K, O'Brochta DA, Atkinson PW (2000) Green fluorescent protein as a genetic marker in transgenic *Aedes aegypti*. Insect Mol Biol 9:1–10

Prasher DC, Eckenrode VK, Ward WW, Prendergast FG, Cormier MJ (1992) Primary structure of *Aequorea victoria* green-fluorescent protein. Gene 111:229–233

Rosen L, Gubler D (1974) The use of mosquitoes to detect and propagate dengue viruses. Am J Trop Med Hyg 23:1153–1160

Seabaugh RC, Olson KE, Higgs S, Carlson JO, Beaty BJ (1998) Development of a chimeric Sindbis virus with enhanced *Per Os* infection of *Aedes aegypti*. Virology 243:99–112

Shimomura O (1998) The discovery of green fluorescent protein. In Chalfie M, Kaine S (eds) GFP Green Fluorescent Protein Properties, Applications and Protocols. Wiley–Liss, New York, pp 3–15

Shimomura O, Johnson FH, Saiga Y (1962) Extraction, purification and properties of aequorin, a bioluminescent protein from a luminous hydromedusan, *Aequorea*. J Cell Comp Physiol 59:223–240

Tsien RY (1998) The green fluorescent protein. Annu Rev Biochem 67:509–544

Tsien RY, Prasher D (1998) Molecular biology and mutation of green fluorescent protein. In Chalfie M, Kaine S (eds) GFP Green Fluorescent Protein Properties, Applications and Protocols. Wiley–Liss, New York, pp 97–117

Wang S, Hazelrigg T (1994) Implications for *bcd* mRNA localization from spatial distribution of *exu* protein in *Drosophila* oogenesis. Nature (London) 369:400–403

Ward WW (1998) Biochemical and physical properties of green fluorescent protein. In Chalfie M, Kaine S (eds) GFP Green Fluorescent Protein Properties, Applications and Protocols. Wiley–Liss, New York, pp 45–75

Yamao M, Katayama N, Nakazawa H, Yamakawa M, Hayashi Y, Hara S, Kamei K, Mori H (1999) Gene targeting in the silkworm by use of a baculovirus. Genes Dev 13:511–516

Yang F, Moss LG, Phillips GN Jr (1996) The molecular structure of green fluorescent protein. Nat Biotechnol 14:1246–1251

Color Figures

COLOR FIGURE 1 Eye color phenotypes of wild-type, mutant, and transgenic insects marked with the wild-type allele of the mutations. *Drosophila melanogaster* (A) wild-type; (B) *white* mutant host; and (C and D) *piggyBac/white+* transformants; (E) *Ceratitis capitata piggyBac/white+* transformant and (F) *white eye* host strain; (G) *Aedes aegypti* wild type; (H) kynurenine hdroxylase-white (kh^w) mutant host; and (I and J) *Hermes/cinnabar+* transformants; (K) *Anagasta kuehniella* wild-type; and (L) *yellow* mutant. The kh^w mutation is complemented by the *D. melanogaster cinnabar* wild-type allele and the *yellow* mutation, defective for tryptophan oxygenase, is complemented by the *D. melanogaster vermilion* wild-type allele. Photo credits: A–D (A. Handler), E and F (S. McCombs), G–J (A. James), and K and L (P. Shirk).

COLOR FIGURE 2 GFP expressed in different insects. (A) Larva of *Ae. aegypti* infected with TE/3'2J/GFP virus (uninfected control on right); (B) Pupa of *Anopheles gambiae* infected with TE/3'2J/recombinant GFP; (C) adult *Ae. aegypti* with TE/3'2J/GFP; (D) larva of *Ae. aegypti* infected with recombinant densovirus (AeDNV); (E) pupa of *Ae. aegypti* infected with recombinant densovirus (AeDNV); (F) section of embryo of *Tribolium castaneum* infected with TE/3'2J/GFP; (G) wing of adult *Papillio glaucus* infected with TE/3'2J/GFP; (H) eggs of *Anastrepha suspensa* transformed with *piggyBac* vector pB[PUb nls EGFP] (nontransformed control in upper left); (I) larva of *A. suspensa* transformed with *piggyBac* vector (control on right); (J) adult of *A. suspensa* transformed with *piggyBac* vector (control on right); (K) larva of *Ae. aegypti* transformed with *Hermes*; (L) adult *Ae. aegypti* transformed with *Hermes*. Photo credits: A and C (Higgs, S. et al., *Biotechniques* 21, 660–664, 1996. With permission). B (courtesy of S. Higgs). D and E (courtesy of T. Ward). F and G (courtesy of D. Lewis). H–J (courtesy of A. Handler). K and L (courtesy of A. Pinkerton and P. Atkinson).

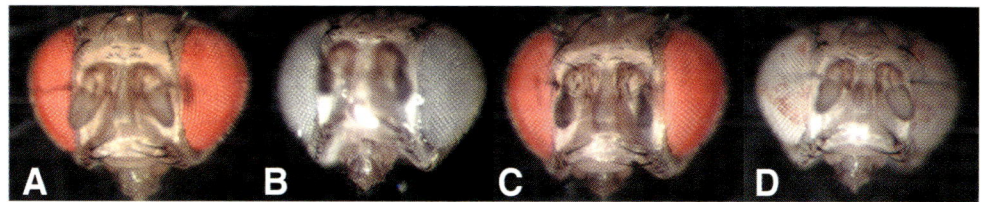

COLOR FIGURE 1

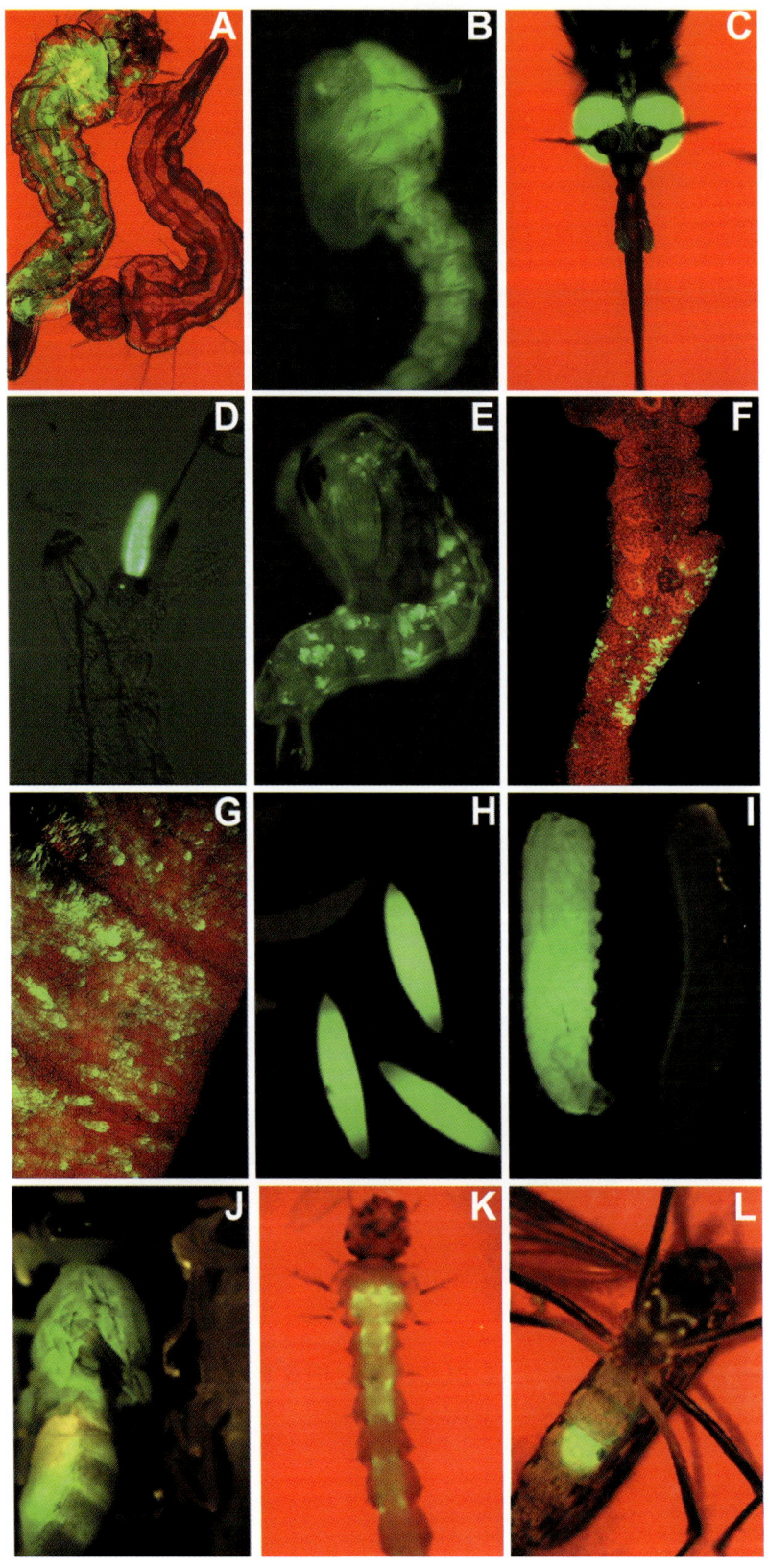

COLOR FIGURE 2

6 Resistance Genes as Candidates for Insect Transgenesis

Richard H. ffrench-Constant and Mark Q. Benedict

CONTENTS

6.1 Introduction 109
6.2 Candidate Genes 110
    6.2.1 Insect Genes 110
        6.2.1.1 Acetylcholinesterase 110
        6.2.1.2 The *para* Voltage-Gated Sodium Channel 111
        6.2.1.3 Metabolic Genes 111
    6.2.2 Noninsect Genes: *Neomycin* and *Opd* 111
        6.2.1.1 *neo* 112
        6.2.1.2 *opd* 113
6.3 *Rdl* as a Case Study 115
    6.3.1 Cloning and Expression of a Functional Mosquito Receptor 115
    6.3.2 *P*-Element Mediated Transformation of *Drosophila* — What Can Be Learned? ... 115
    6.3.3 Defining the *Rdl* Promoter in *Drosophila* 117
    6.3.4 Defining the *Rdl* Promoter in Mosquitoes 118
    6.3.5 What Will Be Needed to Transform Large Genes in Mosquitoes? 118
References 119

6.1 INTRODUCTION

Initial interest in insecticide resistance genes in relation to non-drosophilid insect transgenesis focused on the need for selectable markers to isolate putative germline transformation events rapidly (Thompson et al., 1993). Given the recent success of eye color genes and green fluorescent protein in providing phenotypes that can be visually screened for (see Chapter 4 by Sarkar and Collins and Chapter 5 by Higgs and Lewis), the need for drug-resistant selectable markers is reduced somewhat. It is therefore the purpose of this chapter not only to summarize attempts to use these genes as selectable markers themselves, but also to emphasize the extent to which we fully understand these genes, which still represent some of the only non-drosophilid marker genes in which the molecular basis of the phenotype we are trying to rescue is known. In other words, where the nature of both the gene product and the mutation causing the mutant phenotype is known. In this respect, resistance genes illustrate some of the critical practical problems we face, beyond those of vector integration, in actually transforming defined phenotypes such as disease refractoriness in mosquitoes.

Developing transformation methods for insects is a labor-intensive effort, so when possible all reagents and assay procedures should be tested sufficiently before dedicated transformation attempts are made. This is imperative because a number of unknown variables initially confront the researcher: (1) What is the mortality induced by introduction and expression of the candidate DNA? (2) What is the efficiency of integration of the exogenous DNA? (3) What is the rate of false positives and negatives from marker gene expression? (4) What, if any, are the effects of DNA introduction and integrations on fertility? Understanding as many of these factors as possible is essential to estimate the effort one must commit to set up a protocol capable of detecting a given frequency of transformants (i.e., survivors of a discriminating dose of antibiotic or insecticide).

As part of this approach, it is important to select marker genes whose biochemical activity and selectability can be demonstrated easily before extensive transformation efforts are made. In this context, marker genes that can be assayed by transient expression in embryos or cultured cells are often an advantage. While gene expression in transient assays may not be on levels similar to that of single-copy expression of a marker gene, assuring biological activity of the marker gene in a relevant system is crucial for success. In many situations, the only choice for testing markers for non-drosophilids is initial testing in *Drosophila*, in which efficient transformation systems and superior co-markers are available.

6.2 CANDIDATE GENES

A number of genes conferring resistance to xenobiotics have been examined as candidates for selectable markers. These include both insect and non-insect genes and also genes that have been proposed for either whole-animal transformation or for use in tissue culture systems.

6.2.1 INSECT GENES

Candidate insecticide resistance genes from insects include the three major targets of conventional insecticides as well as a range of metabolic enzymes capable of xenobiotic metabolism. The conventional targets of insecticides are (1) the *Resistance to dieldrin* (*Rdl*)–encoded aminobutyric acid receptor; (2) insect acetylcholinesterase (AChE) encoded by the gene *Ace*; and (3) the voltage-gated sodium channel gene encoded by the gene *para*. *Rdl* will form the main focus of this chapter; however, the applicability of the latter two examples will first be briefly discussed.

6.2.1.1 Acetylcholinesterase

AChE hydrolyzes acetylcholine at excitatory synapses in insects, and this important enzyme can be blocked by a range of organophosphorus or carbamate insecticides. The *Ace* gene encoding the enzyme was originally cloned from *Drosophila* (Hall and Spierer, 1986), but *Ace* homologues have now been isolated from a range of insects including houseflies (Williamson et al., 1992; Feyereisen, 1995) and mosquitoes (Hall and Malcolm, 1991; Anthony et al., 1995). Like all of the three major target site genes, the *Ace* gene is large and in *Drosophila* the coding and noncoding regions are spread over 34 kb (Hoffmann et al., 1992). However, the lethal phenotypes associated with this locus can be rescued in *Drosophila* using a minigene of only 5.8 kb, consisting of 1.5 kb of 5'-end flanking DNA fused to the intronless open reading frame. This construct not only rescues the lethal phenotypes, but also appears to give the correct spatial pattern of gene expression (Hoffmann et al., 1992).

The mutations that render AChE insensitive to insecticide inhibition have been well studied in *Drosophila* (Mutero et al., 1994), houseflies (Feyereisen, 1995), and also the mosquito, *Aedes aegypti*, which was used as a model system for site-directed mutagenesis of the enzyme (Vaughan et al., 1997). In all species examined, a number of different amino acid replacements confer relatively low levels of insensitivity on their own but can combine to create more resistant alleles. The most insensitive alleles therefore carry two or more mutations (Fournier et al., 1996). Although only one allele has been used to transform *Drosophila*, and gave low levels of resistance that would not be useful as a selectable marker, insensitive *Ace* alleles *have* been used in the genetic sexing of mosquitoes and may therefore still present some practical use in insect transformation. In one example, an insensitive *Ace* allele in *Anopheles albimanus* was linked to the Y chromosome via a radiation-induced translocation, and genetic recombination was subsequently suppressed by inversions (Seawright et al., 1978). Thus, resultant males always carry one resistant copy of the gene and are always heterozygous (*R/S*) for resistance. Appropriately, this strain was termed "Macho." In the case of *An. albimanus*, chemical selection for heterozygosity is possible allowing 99.7% of females to be killed with a discriminating dose of insecticide, with the surviving males being subsequently irradiated for sterile insect release programs. Subsequent biochemical analysis of insensitive AChE in mosquitoes has shown highly resistant enzymes in a range of different species (Ayad and Georghiou, 1975; ffrench-Constant and Bonning, 1989). Thus, if semidominant resistant heterozygotes can be selected in other species of insects, insensitive *Ace* still represents a potentially useful selectable marker.

6.2.1.2 The *para* Voltage-Gated Sodium Channel

The voltage-gated sodium channel encoded by homologues of the *Drosophila* gene *para* is the target site for DDT and pyrethroids. At first, this might seem an attractive candidate for a marker gene for genetic transformation as the resistance-associated mutations that cause *knockdown resistance* (*kdr*) to pyrethroids (Miyazaki et al., 1996; Williamson et al., 1996) and the allelic double mutant *super-kdr* (Williamson et al., 1996) appear conserved among a range of pest species, including the malaria vector *A. gambiae* (Martinez-Torres et al., 1998). However, rescue of mutant alleles of this locus has not been achieved by transformation in *Drosophila*, and the large size of the genomic transcription unit, over 60 kb (Loughney et al., 1989), puts it beyond the reach of even a cosmid transformation vector. This problem of the large size of some transcription units and the associated difficulty in identifying the necessary promoter and regulatory elements will be discussed below in the specific context of *Rdl*. In summary, although an important and potentially attractive marker gene for transformation, the use of *para* constructs awaits the development of superior vectors capable of carrying large (over 40-kb) inserts.

6.2.1.3 Metabolic Genes

Candidate marker genes drawn from examples of metabolic insecticide resistance, although again initially attractive, may pose substantial problems given the poor understanding of their transcriptional control (ffrench-Constant et al., 1998). Thus, although it might appear that one could take candidate cytochrome P-450, esterase, or glutathione-*S*-transferase structural genes that are capable of metabolizing insecticides and use them as selectable makers, phenotypic resistance is associated with upregulation of these enzymes. For example, although we are beginning to understand how particular *Drosophila* P-450s can be mutated to increase insecticide metabolism (Berge et al., 1998), expression of these genes is controlled by regulatory genes whose function and cloning remains elusive (ffrench-Constant et al., 1998). With the availability of such cloned regulatory loci, they will themselves become candidates for transformation. But again, these may need to be expressed in combination with mutated structural genes to confer high levels of resistance.

6.2.2 NONINSECT GENES: *NEOMYCIN* AND *OPD*

Only a few species of insects, most prominently drosophilids, have been studied sufficiently that numerous mutant strains and well-characterized complementing genes are available for use as genetic transformation markers. In contrast, such combinations are not available for the biologically diverse species of insects of interest in recent genetic transformation efforts. Therefore, the primary impetus for developing and applying noninsect (primarily bacterial) markers to insect transformation has been the lack of cloned insect genes and appropriate mutant strains of the test species in which complementation can be observed.

Bacterial genes have several characteristics that make them attractive candidates as transformation markers: the genetics, transcription units, and translation products are usually well characterized. They are typically compact, intronless, widely available, and numerous substrates against which they are active have been described. These characteristics contrast sharply with most insect genes whose full-length cDNAs must be isolated or whose overall structure must be determined before functional expression systems can be designed and assembled. The most desirable bacteria-derived markers would not require host-specific genotypes or phenotypes and could be considered for use in a variety of animal species and/or plants.

Bacterial transcription and translation units are relatively predictable, and, therefore, the probability of effective eukaryotic expression in a spatially and developmentally regulated manner can be controlled by selection of either an inducible or constitutive eukaryotic promoter. Examples of the former have been the heat-shock promoters of the conserved *D. melanogaster Hsp70* (Pelham, 1982) and *Hsp82* (Xiao and Lis, 1989) genes. The *Hsp70* promoter differs from the *Hsp82* in that it is more highly inducible and has lower levels of basal expression. Most researchers have chosen to use constitutive promoters with generally high levels of developmental and tissue expression, for example, actin (Chung and Keller, 1990), polyubiquitin (Lee et al., 1988), and the baculovirus hr5-IE1 (Pullen and Friesen, 1995) to express bacterial genes.

Several noninsect genes have been tested as selectable markers, and these are primarily bacterial genes capable of metabolizing xenobiotics. This section discusses two that have achieved some success, *neo* and *opd*.

6.2.1.1 *neo*

The *neo* gene encodes neomycin phosphotransferase (EC 2.1.7.95), which is often called simply NPT. Several analogues of neomycin are commonly used for selections, particularly G418 and gentamicin sulfate (Genticin™ and Gentimycin™). Since the first report of the use of *neo* as a selectable marker in *D. melanogaster* (Steller and Pirrotta, 1985), molecular entomologists have attempted to apply it to transformation of other insects. The *neo* gene has several advantages including its small size (approximately 1 kb), the wide range of plasmid constructs available, potential for transient expression (Okano et al., 1992; see below), and the ability to test and apply it to both whole animals and cultured cells (Maisonhaute and Echalier, 1986). On the negative side, the cost of using relatively high concentrations of neomycin can be prohibitive and the insects must be submerged in the antibiotic (typically applied during the immature stages when continuous contact in food medium or water is required).

Transient expression of *neo* in whole animals was demonstrated by McInnis et al., (1990) in the medfly *Ceratitis capitata*. Embryos were injected with pUC*hsneo*, a pUC plasmid carrying the *neo* gene driven by the *Hsp70* promoter (Steller and Pirrotta, 1985) and increased resistance to Genticin was observed among the hatching larvae, more so if larvae were heat-shocked. Continued selection on G418 and elevated levels of resistance led the investigators to hope that genetic transformation by the *P*-element vector had occurred. But, as has been commonly observed with the *neomycin* marker, neomycin resistance in subsequent generations can occur, but without genetic transformation by transposable elements carrying *neo*.

The first successful transformation of the African malaria vector mosquito, *An. gambiae*, was accomplished using *neo* as the selectable marker (Miller et al., 1987). The pUC*hsneo* plasmid was injected into preblastoderm embryos and a single transformed line was recovered after selection during the aquatic larval stage (in spite of an apparently suboptimal heat-shock regime). Subsequently, by using the same transformed line as an experimental tool, increased levels of resistance and discrimination between resistant and susceptible animals were observed when the heat-shock temperature was raised from 37 to 41°C (Sakai and Miller, 1992). Although initial success and this technical improvement were encouraging, follow-up successes were not forthcoming, due in part to the expense of the reagent, difficulties in providing high doses in the different types of culture media that various insects require, and finally the availability of superior markers.

Additional problems often plague *neo* selection. However, many of these have only been related anecdotally by members of the *Drosophila* community who report chronic false positives appearing in lines that do not carry the *neo* gene (Ashburner, 1989). We also have experienced this problem during the development of the *opd* marker discussed below (Benedict et al., 1994). In these experiments, a doubly selectable *P*-element plasmid carrying the *hsneo* and *hsopd* genes was injected into *D. melanogaster*, and G418 selection yielded 35 lines that were clearly positive by Southern hybridization and whose G418 resistance had been mapped to one of the chromosomes. Sublines of transformants were maintained under selection, and single-pair lines were established and also kept under G418 selection. Genetic and molecular analysis demonstrated that, of these, only 19 had *hsneo*-containing DNA. We concluded that even though *neo* expression may have provided the resistance required for initial isolation of the lines, *neo* was subsequently lost and another mechanism was apparently providing resistance.

In summary, *neo* has enjoyed success in several laboratories. However, the expense and complications (e.g., maintaining a suitable discriminating dose) of using *neo* as a selectable marker make it worth consideration only if use of a more readily selectable marker is not possible.

6.2.1.2 *opd*

The *opd* gene, isolated from *Pseudomonas diminuta* and *Flavobacterium* sp. (Mulbry et al., 1987; McDaniel et al., 1988; Serdar et al., 1989) encodes a phosphotriesterase, organophosphate hydrolase (OPH, EC 3.1.8.1), also widely called parathion hydrolase. This enzyme metabolizes (hydrolyzes) numerous organophosphorus-type insecticides (AChE inhibitors), including paraoxon, ethyl- and methyl-parathion (Dumas et al., 1989; 1990a), and the nerve gases sarin (Dumas et al., 1990a) and VX (Hoskin et al., 1995).

The potential advantages of using bacterial genes as eukaryotic markers are prominently exhibited by the *opd* gene. The coding region is only 1.0 kb in length, the activity rates against numerous substrates have been determined, and the mature native protein has been purified (Dumas et al., 1989). Moreover, a simple biochemical assay for OPH activity, based on the hydrolysis of paraoxon or parathion, can be performed by spectrophotometry to detect the yellow-colored product, paranitrophenol. Surprising persistence of elevated paraoxon resistance over several days after induction by heat shock (see below) was attributed to apparent stability of the OPH protein (Phillips et al., 1990; Benedict et al., 1994). Indeed, such long-term stability of the enzyme has since been demonstrated experimentally (Grimsley et al., 1997). This unexpected characteristic means that a given transcription level potentially yields a relatively high level of enzyme.

The potential of *opd* as a selectable marker for insect transformation was first shown when *D. melanogaster* transformed with the *opd* gene (driven by an H*sp70* promoter) had increased resistance to paraoxon as determined by adult bioassay (Phillips et al., 1990). However, only a few lines were examined, and low levels of resistance were observed in spite of high levels of enzyme activity. Perhaps contributing to the low level of resistance observed was the presence of the region encoding the native bacterial membrane anchor sequence. Benedict et al. (1994) subsequently improved expression of the *opd* gene by removing the sequences that encoded the bacterial membrane anchor or by replacing them with a secretory signal sequence. Although transcript levels were similar from both types of genes, the former consistently provided higher levels of resistance, whereas the latter conferred levels of resistance similar to that observed by Phillips et al. (1990). Subsequently, rates of detection of both false-positive and false-negative individuals were determined to be similar to those seen with the *Drosophila white* gene (Benedict et al., 1995) and a general-purpose transformation *P*-element vector was therefore designed.

Potential for *opd*-conferred paraoxon resistance in other insects was demonstrated in a caterpillar pest, the fall armyworm, *Spodoptera frugiperda,* by expressing *opd* in a baculovirus system (Dumas et al., 1990b). In their experiments, 280-fold resistance was observed; however, it is difficult to use their results to predict single-copy expression, such as would be used in transformed insect selection.

While routine selection of *opd*-transformed individuals has been performed by exposing adults to toxicant in paraoxon-coated glass scintillation vials, we have conducted preliminary experiments that indicate it is also possible to perform larval selection. *D. melanogaster* w^{1118} or heterozygous males from four strains that contain an autosomal *P*-element insertion marked with both *hsopd* and mini-*white* were crossed to w^{1118} females. This scheme predicts that in the absence of larval selection a 1:1 ratio of white-eyed (paraoxon susceptible) and wild-type (resistant) progeny will be observed. On the other hand, greater numbers of wild-type than white-eye progeny will be observed from heterozygous males in paraoxon treated groups. Matings were performed in vials containing 0, 0.1, 1.0, or 5.0 ppm paraoxon-dissolved in 1% ethanol used to formulate the Carolina Biological Supply Instant Blue medium. Larvae were heat-shocked every other day for 1 h at 37°C in a water bath, with G_1 progeny subsequently examined for eye color.

Fortunately, *D. melanogaster* parental adults were not killed at paraoxon concentrations in the media that prevented all development of progeny to adulthood. Therefore, as when neomycin is supplied in *Drosophila* media, parents survive long enough to mate and oviposit, even though their progeny may not survive. Similar mortality rates were observed among parental males regardless of whether they carried the *hsp70-opd* gene or not, so *hsp70-opd* expression does not seem to increase adult longevity significantly under these conditions. No progeny were obtained from the 1.0- or 5.0-ppm vials. Nontreated control animals contained equal numbers of wild-type and white-eye progeny regardless of whether they were heat-shocked or not. The 0.1-ppm treatment yielded numbers of progeny that indicated low-discrimination selection was occurring; 66 and 62 wild-type adults arose from the heat-shocked and non-heat-shocked groups, respectively, and only 10 and 30 white-eye adults. One stock yielded only wild-type progeny and from the heat-shocked group only. Clearly, the dose rate and/or heat-shock regimen was not adequate to eliminate all nontransformed flies completely, but it did provide enrichment. A higher dose between 0.1 and 1.0 ppm may provide more useful discrimination.

The features of the current *opd* selection and inheritance have both positive and negative aspects. Resistance due to *opd* is inherited as a semidominant allele, thus sufficiently high selection with paraoxon permits purification of homozygous insertion lines without the use of balancer chromosomes. In contrast to *neo*, toxicity is acute and it should be possible to apply toxicants topically, by dipping or as an aerosol. A significant improvement on the *opd* gene would be to use a promoter with a high level of constitutive activity since the heat-shock regimen necessary to induce the *hsp70* promoter are at near-lethal levels. This could also simplify and make selection of immature stages more effective.

While both false-negatives and false-positives arising during adult selection were not problematic, the marker gene has been received with reservations by many workers, primarily due to its novel nature. Some researchers express the view that it is more acceptable in the context of biosafety to use a native insecticide-resistance marker that originated in the insect being transformed (e.g., *Rdl* or *Ace*). However, introducing a truly novel gene into insects that may increase their resistance to insecticides in the event of accidental release is also a concern, as discussed elsewhere (Chapter 20 by Young et al.). Moreover, several laboratories that have inquired about using this marker had reservations about the personal safety of the AChE inhibitor, paraoxon, for researchers. With a high mammalian dermal toxicity, paraoxon requires caution when preparing stock dilutions from technical concentrations, although the working concentrations and amounts used for treating selection containers make accidental exposure to toxic levels of working solutions unlikely. Nevertheless, some institutions require routine monitoring of blood AChE levels in workers who use these compounds.

6.3 *Rdl* AS A CASE STUDY

Having briefly reviewed the past and likely future utility of resistance genes as selectable markers, the remainder of this chapter will focus on what we can hope to learn by using resistance genes as *candidate* genes *themselves*. The use of these genes as model systems in the laboratory neither implies nor infers their potential release in the field, merely that they represent some of the best-understood genes in pest insects and therefore provide a perfect focus for testing the necessary attributes of workable transformation systems.

In preparing a gene for a transformation attempt, or in trying to understand why one may have failed, we need to know the answer to several critical questions:

1. Does the open reading frame encode a functional copy of the gene?
2. Is the nature of the mutation(s) underlying the mutant phenotype known?
3. What is the size of the complete transcriptional unit and where are the necessary promoter or regulatory elements?
4. What is the role of the alternative splicing commonly seen in insect genes (e.g., *Rdl* and *para*)?

Here, these questions will be examined within the specific context of the *Rdl* gene and resistance to the cyclodiene insecticide, dieldrin. The section draws on data from *Drosophila* transformation attempts and also examines what preliminary experiments can be performed in non-drosophilids to begin to address questions about the location of the necessary gene promoter elements.

6.3.1 CLONING AND EXPRESSION OF A FUNCTIONAL MOSQUITO RECEPTOR

We isolated an *Aedes aegypti Rdl* cDNA clone from an adult cDNA library using the *D. melanogaster* clone as a probe in a low stringency screen (Thompson et al., 1993). Not only does the predicted amino acid sequence of the mosquito receptor subunit show a high degree of identity with the *Drosophila* clone, but the same amino acid (alanine302 in *Drosophila*) is replaced with a serine in dieldrin-resistant strains (Thompson et al., 1993). To answer the first question raised above and test if this cDNA encodes a functional γ-aminobutyric acid (GABA) receptor subunit, we expressed the *Ae. aegypti Rdl* cDNA in baculovirus-infected insect cells (Shotkoski et al., 1994). Patch clamp analysis of infected cells showed GABA-activated currents that increased linearly with voltage and reversed at or about 0 mV, as expected for chloride-selective channels in symmetrical chloride solutions. These channels could be blocked by application of 10 μ*M* picrotoxinin (PTX), a GABA receptor antagonist that acts at the same binding site as dieldrin. To prove the functionality of the putative resistance-associated mutation, we compared the level of block achieved in channels mutated to alanine302 > serine. These channels required 1 m*M* PTX to achieve the equivalent level of block, proving that the alanine302 > serine replacement is functionally associated with resistance (Shotkoski et al., 1994).

6.3.2 P-ELEMENT MEDIATED TRANSFORMATION OF *DROSOPHILA* — WHAT CAN BE LEARNED?

Having isolated functional *Rdl* cDNAs both from *D. melanogaster* and from *Ae. aegypti* and having shown that the same mutation confers resistance in both species, we performed germline transformation with the fruit fly *Rdl* to determine if selectable levels of resistance could be conferred. During the initial characterization of the *Rdl* gene in *Drosophila*, we had proved the cloning of *Rdl* by rescuing the susceptible phenotype of the locus using a cosmid transformant (ffrench-Constant et al., 1991). We used a 40-kb cosmid (cosmid 6), encompassing most of the susceptible *Rdl* locus (Figure 6.1), and inserted it into the fruit fly germline using *P*-element-mediated transformation (Spradling, 1986). Note that, even for *Drosophila* transformation, such

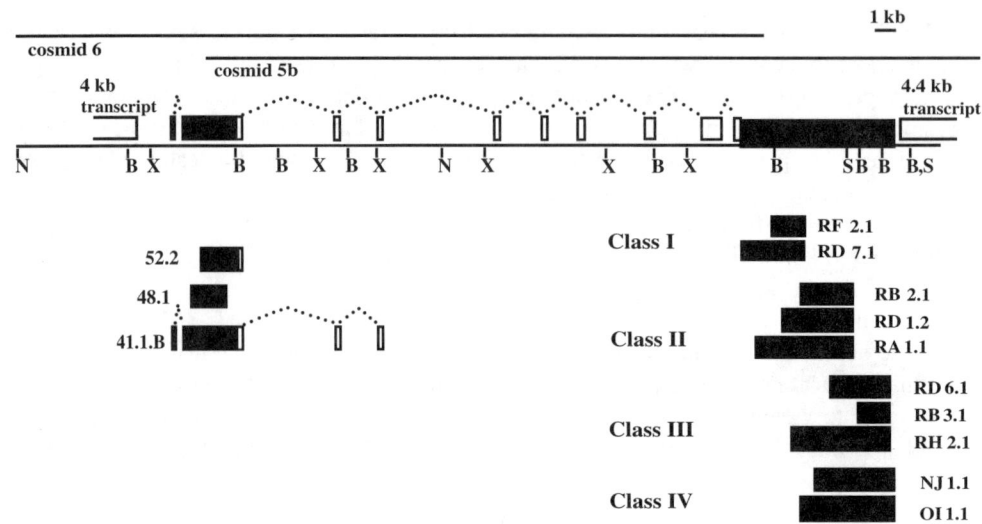

FIGURE 6.1 Diagram illustrating the size and complexity of the *Rdl* locus in *Drosophila*. Note that the entire locus does not fit within the 40-kb cosmid 6 and that much of the 3' untranslated region is missing. Note also the large number of cDNAs isolated corresponding to the 3' end of the gene. Exons 3 and 6 each have two alternative exons, termed A and B, and C and D, respectively (not shown in diagram). There are therefore four known versions of the *Rdl* mRNA containing the different alternative exons: 3A/6C, 3A/6D, 3B/6C, and 3B/6D.

an event occurs at a relatively low frequency (1%, compared with >10% often achieved with smaller constructs; see Spradling, 1986). When combined with a single copy of the resistance gene (paired with a deficiency deleting the other native copy), transgenic flies are as susceptible as heterozygotes, *R/S* (Figure 6.2), proving that the cosmid carries a functional copy of the susceptible gene (ffrench-Constant et al., 1991).

Having shown that cosmid 6 carries a genomic copy of the susceptible gene, we now needed to identify the 5' or 3' DNA flanking the open reading frame that would contain sufficient promoter and/or regulatory elements to drive cDNA expression. Interestingly, an overlapping cosmid (termed cosmid 5B), that lacks 5.7 kb of the total 9.2 kb of flanking 5' sequence in cosmid 6 (see Figure 6.1), failed to rescue the susceptible phenotype and thus appears to lack sufficient 5' DNA to restore proper gene function (ffrench-Constant et al., 1991). To test this hypothesis, we fused the 5' flanking DNA from cosmid 6 onto an intronless cDNA and repeated the rescue experiment (Stilwell et al., 1995). The 9.2 kb of 5' flanking DNA was capable of partial rescue of susceptibility and, in fact, gave equivalent levels of rescue relative to the complete cosmid (Figure 6.2). This 5' flanking DNA therefore contains sufficient promoter and/or regulatory elements to give partial rescue. This construct also rescues the lethal phenotype associated with the locus whereby *Rdl* null embryos fail to hatch and die during late development (Stilwell et al., 1995). It is important to note that equivalent levels of rescue were achieved with either of two of the four possible different *Rdl* alternative splice forms (ffrench-Constant and Rocheleau, 1993) and that combining two different constructs in the same fly also did not improve rescue (Stilwell et al., 1995).

We also tested this same cDNA construct with the resistance-associated mutation and again achieved partial rescue of the resistant phenotype. Combining two copies of the resistant transgene (*R'*) in susceptible flies (*S/S*), i.e., a genotype of *R'/R'*; *S/S*, allowed survival for longer times on insecticide than what was observed for normal susceptible flies (Stilwell et al., 1995). These experiments, although encouraging, leave many questions unanswered. Namely,

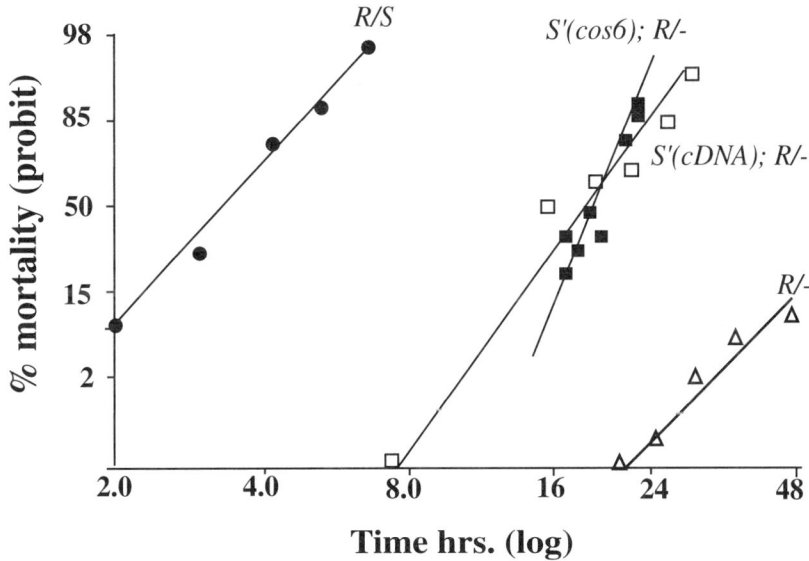

FIGURE 6.2 Dose–response curves comparing the level of rescue of susceptibility achieved using either all of cosmid 6, $S'(cos6)$ or a construct containing 5' flanking DNA and the intronless cDNA, $S'(cDNA)$. Note that the level of rescue achieved by the cDNA construct is equivalent to the complete 40-kb cosmid but that *both* are partial levels of rescue, i.e., intermediate between the fully resistant $Rdl^R/Df29A6$ labeled R/-, and the heterozygous Rdl^R/Rdl^S labeled R/S.

1. Why is rescue incomplete and what elements are missing from the genomic and cDNA constructs?
2. What is the role of the different alternative splice forms found at the locus and is transformation of one splice form alone responsible for the low levels of rescue?

6.3.3 DEFINING THE *RDL* PROMOTER IN *DROSOPHILA*

Given the low levels of rescue achieved in *D. melanogaster* with either genomic cosmid constructs or cDNAs fused to large lengths of 5' flanking DNA, we needed to undertake a more detailed analysis of the fruit fly transcription unit and also to begin to develop techniques for analysis of the mosquito promoter.

We have recently completed a detailed analysis of the *D. melanogaster Rdl* transcription unit that begins to reveal the potential reasons for the difficulty in rescuing phenotypes associated with such a large and complex locus (Stilwell and ffrench-Constant, 1998). The full extent of the transcription unit was estimated by identifying the position of adjacent transcriptional units at the 5' and 3' end of the locus. This showed that the full *Rdl* transcription unit was over 50 kb in length and that the 40 kb of cosmid 6 was therefore unlikely to contain all of the locus. Isolation of several cDNAs from the 5' and 3' end of the locus revealed that cosmid 6 lacks most of the 5-kb-long 3' untranslated DNA from the mature 8.8-kb message (Stilwell and ffrench-Constant, 1998). This potentially explains why previous constructs have failed to give full rescue and means that a single splice form of *Rdl* would require an 8.8-kb cDNA construct alone (1.8 kb of 5' flanking DNA, 2 kb of open reading frame, and 5 kb of 3' untranslated region).

Although this analysis revealed the large extent of 3' DNA missing from previous transformation constructs, deletion analysis of the 5' flanking DNA and subsequent re-transformation and examination of *in situ* hybridization patterns revealed that only 3.5 kb of upstream DNA was necessary to drive the proper pattern of *Rdl* transcription in embryos and adult brains. This region was shown to contain the transcriptional start point and also other regulatory motifs common to other ion channel genes (Stilwell and ffrench-Constant, 1998). This shows that some truncation of the 5′ DNA may be possible, but also suggests that the long 3′ untranslated region may play a critical role in message stability or trafficking and that its inclusion in future constructs may be essential for full rescue.

6.3.4 Defining the *Rdl* Promoter in Mosquitoes

Following a description of the *Rdl* transcription unit in *D. melanogaster* and identification of putative promoter and regulatory elements via deletion analysis and subsequent P-element transformation, we asked, "How can we undertake a similar analysis in non-drosophilids?" In the initial absence of workable transformation systems in mosquitoes, we used *in vivo* analysis of reporter gene constructs to assay *Rdl* promoter activity in *Ae. aegypti*.

In this analysis we constructed a series of deletions across the 5′ flanking DNA of the *Ae. aegypti Rdl* locus and fused each of these to a luciferase reporter gene (Shotkoski et al., 1996). Each of these constructs was tested for luciferase activity following direct injection of the construct into developing embryos. The latter was performed using the same techniques routinely used for transformation (see Chapter 12 by Atkinson and O'Brochta and Chapter 1 by Handler). By assaying activity in this range of deletions, we were able to determine that most of the promoter activity was contained in a 2.5-kb fragment immediately to the 5′ end of the *Rdl* start codon. Interestingly, primer extension and sequence analysis subsequently revealed three closely linked transcription initiation sites within this region. This region also contained several consensus regulatory elements shared by other genes expressed in the nervous system (Shotkoski et al., 1996). Thus, although higher levels of activity were expressed by 7.2 kb of 5′ flanking DNA, and the fact that no spatial information is given by this analysis (i.e., if the gene is expressed in the same tissues as is observed *in vivo*), much of the reporter gene activity was attributable to only this 2.5-kb promoter fragment. Following the identification of this fragment, we used this 5′ flanking DNA fused to the resistant *Rdl* cDNA in transformation experiments with the *Hermes* vector (S. Doll, M. deCamillis, D. O'Brochta, B. Christensen, A. James, and R. ffrench-Constant, unpublished). Although inserted copies of *Rdl* were detected by both Southern analysis and inverse PCR in insects surviving insecticide exposure, no stable transformants were isolated at that time and these experiments were discontinued.

These experiments, therefore, demonstrate some of the difficulties of transforming large mosquito genes. In summary, these are (1) the presence of large transcription units with numerous introns; (2) the associated long stretches of 5′ and 3′ flanking DNA whose function is often uncertain, e.g., the long 3′ tail of *Rdl*; and (3) the difficulty of identifying minimal 5′ and 3′ pieces of DNA without routine analysis of deletions by transformation.

6.3.5 What Will Be Needed to Transform Large Genes in Mosquitoes?

In conclusion, the routine transformation and analysis of large mosquito genes will clearly await the development of transformation vectors more equivalent to those used in *D. melanogaster*. First, vectors will need to incorporate at least 40 kb of DNA to give transformation of genomic gene copies contained in cosmids (also a critical step in identifying genes via chromosomal walking). Second, transformation itself and the associated animal husbandry will need to be sufficiently routine to allow for the repeated analysis of different deletion constructs in identifying promoter and regulatory elements. Finally, we will need to focus more attention on the putative roles of the alternative splice forms commonly seen in these genes and to assess their possible role in restoring full rescue of transformed phenotypes.

REFERENCES

Anthony N, Rocheleau T, Mocelin G, Lee H-J, ffrench-Constant RH (1995) Cloning, sequencing and functional expression of an acetylcholinesterase gene from the yellow fever mosquito *Aedes aegypti*. FEBS Letts 368:461–465

Ashburner M (1989) Drosophila: A Laboratory Handbook. Cold Spring Harbor Laboratory Press, Cold Spring Harbor, NY

Ayad H, Georghiou GP (1975) Resistance to organophosphates and carbamates in *Anopheles albimanus* based on reduced sensitivity of acetylcholinesterase. J Econ Entomol 68:295–297

Benedict MQ, Scott JA, Cockburn AF (1994) High-level expression of the bacterial *opd* gene in *Drosophila melanogaster*: improved inducible insecticide resistance. Insect Mol Biol 3:247–252

Benedict MQ, Salazar CE, Collins FH (1995) A new dominant selectable marker for genetic transformation; *Hsp70-opd*. Insect Biochem Mol Biol 25:1061–1065

Berge J-B, Feyereisen R, Amichot M (1998) Cytochrome P450 monooxygenases and insecticide resistance in insects. Philos Trans R Soc London, Ser B 353:1701–1705

Chung YT, Keller EB (1990) Regulatory elements mediating transcription from the *Drosophila melanogaster* actin 5C proximal promoter. Mol Cell Biol 10:206–216

Dumas DP, Caldwell SR, Wild JR, Raushel FM (1989) Purification and properties of the phosphotriesterase from *Pseudomonas diminuta*. J Biol Chem 264:19659–19665

Dumas DP, Durst DH, Landis WG, Raushel FM, Wild JR (1990a) Inactivation of organophosphorus nerve agents by the phosphotriesterase from *Pseudomonas diminuta*. Arch Biochem Biophys 277:155–159

Dumas DP, Wild JR, Raushel FM (1990b) Expression of *Pseudomonas* phosphotriesterases activity in the fall armyworm confers resistance to insecticides. Experientia 46:729–731

Feyereisen R (1995) Molecular biology of insecticide resistance. Toxicol Lett 82/83:83–90

Fournier D, Berrada S, Bongibault V (1996) Molecular genetics of acetylcholinesterase in insecticide-resistant *Drosophila melanogaster*. Molecular Genetics and Evolution of Pesticide Resistance. American Chemical Society, Washington, D.C., pp 17–27

ffrench-Constant RH, Bonning BC (1989) Rapid microtitre plate test distinguishes insecticide resistant acetylcholinesterase genotypes in the mosquitoes *Anopheles albimanus*, *An. nigerrimus* and *Culex pipiens*. Med Vet Entomol 3:9–16

ffrench-Constant RH, Mortlock DP, Shaffer CD, MacIntyre RJ, Roush RT (1991) Molecular cloning and transformation of cyclodiene resistance in *Drosophila*: An invertebrate $GABA_A$ receptor locus. Proc Natl Acad Sci USA 88:7209–7213

ffrench-Constant RH, Park Y, Feyereisen R (1998) Molecular biology of insecticide resistance. In Puga A, Wallace KB (eds) Molecular Biology of the Toxic Response. Taylor & Francis, Philadelphia, pp 533–551

ffrench-Constant RH, Rocheleau T (1993) *Drosophila* γ-aminobutyric acid receptor gene *Rdl* shows extensive alternative splicing. J Neurochem 60:2323–2326

Grimsley JK, Scholtz JM, Pace CN, Wild JR (1997) Organophosphorus hydrolase is a remarkably stable enzyme that unfolds through a homodimeric intermediate. Biochemistry 36:14366–14374

Hall LM, Malcolm CA (1991) The acetylcholinesterase gene of *Anopheles stephensi*. Cell Mol Neurobiol 11:131–41

Hall LM, Spierer P (1986) The *Ace* locus of *Drosophila melanogaster*: structural gene for acetylcholinesterase with an unusual 5' leader. EMBO J 5:2949–54

Hoffmann F, Fournier D, Spierer P (1992) Minigene rescues acetylcholinesterase lethal mutations in *Drosophila melanogaster*. J Mol Biol 223:17–22

Hoskin FC, Walker JE, Dettbarn WD, Wild JR (1995) Hydrolysis of tetriso by an enzyme derived from *Pseudomonas diminuta* as a model for the detoxication of O-ethyl S-(2-diisopropylaminoethyl) methylphosphonothiolate (VX). Biochem Pharmacol 49:711–715

Lee HS, Simon JA, Lis JT (1988) Structure and expression of ubiquitin genes of *Drosophila melanogaster*. Mol Cell Biol 8:4727–4735

Loughney K, Kreber R, Ganetzky B (1989) Molecular analysis of the *para* locus, a sodium channel gene in *Drosophila*. Cell 58:1143–1154

Maisonhaute C, Echalier G (1986) Stable transformation of *Drosophila* Kc cells to antibiotic resistance with the bacterial neomycin resistance gene. FEBS Letts 197:45–49

Martinez-Torres D, Chandre F, Williamson MS, Darriet F, Berge JB, Devonshire AL, Guillet P, Pasteur N, Pauron D (1998) Molecular characterization of pyrethroid knockdown resistance *(kdr)* in the major malaria vector *Anopheles gambiae* ss. Insect Mol Biol 7:179–184

McDaniel CS, Harper LL, Wild JR (1988) Cloning and sequencing of a plasmid-borne gene *(opd)* encoding a phosphotriesterase. J Bactiol 170:2306–2311

McInnis DO, Haymer DS, Tam SYT, Thanaphum S (1990) *Ceratitis capitata* (Diptera, Tephritidae) — transient expression of a heterologous gene for resistance to the antibiotic geneticin. Ann Entomol Soc Am 83:982–986

Miller LH, Sakai RK, Romans P, Gwadz RW, Kantoff P, Coon HG (1987) Stable integration and expression of a bacterial gene in the mosquito *Anopheles gambiae*. Science 237:779–781

Miyazaki M, Ohyama K, Dunlap DY, Matsumura F (1996) Cloning and sequencing of the para-type sodium channel gene from susceptible and *kdr*-resistant German cockroaches (*Blatella germanica*) and the house fly (*Musca domestica*). Mol Gen Genet 252:61–68

Mulbry WW, Kearny PC, Nelson JO, Karns JS (1987) Physical comparison of parathion hydrolase plasmids from *Pseudomonas diminuta* and *Flavobacterium* sp. Plasmid 18:173–177

Mutero A, Pralavorio M, Bride J-M, Fournier D (1994) Resistance-associated point mutations in insecticide-insensitive acetylcholinesterase. Proc Natl Acad Sci USA 91:5922–5926

Okano K, Miyajima N, Takada N, Kobayashi M, Maekawa H (1992) Basic conditions for the drug selection and transient gene expression in the cultured cell line of Bombyx mori. In Vitro Cell Dev Biol 28A:779–781

Pelham HRB (1982) A regulatory upstream promoter element in the *Drosophila* HSP70 heat-shock gene. Cell 30:517–528

Phillips JP, Xin JH, Kirby K, Milne CP, Krell P, Wild JR (1990) Transfer and expression of an organophosphate insecticide-degrading gene from *Pseudomonas* in *Drosophila melanogaster*. Proc Natl Acad Sci USA 87:8155–8159

Pullen SS, Friesen PD (1995) Early transcription of the *ie-1* transregulator gene of *Autographa californica* nuclear polyhedrosis virus is regulated by DNA sequences within its 5' noncoding leader region. J Virol 69:156–165

Sakai RK, Miller LH (1992) Effects of heat shock on the survival of transgenic *Anopheles gambiae* (Diptera: Culicidae) under antibiotic selection. J Med Entomol 29:374–375

Seawright JA, Kaiser PE, Dame DA, Lofgren CA (1978) Genetic method for the preferential elimination of females of *Anopheles albimanus*. Science 200:1303–1304

Serdar CD, Murdock DC, Rohde MF (1989) Parathion hydrolase gene from *Pseudomonas diminuta* MG: subcloning, complete nucleotide sequence, and expression of the mature portion of the enzyme in *Escherichia coli*. Bio/Technol 7:1151–1155

Shotkoski F, Lee H-J, Zhang H-G, Jackson MB, ffrench-Constant RH (1994) Functional expression of insecticide-resistant GABA receptors from the mosquito *Aedes aegypti*. Insect Mol Biol 3:283–287

Shotkoski F, Morris AC, James AA, ffrench-Constant RH (1996) Functional analysis of a mosquito γ-aminobutyric acid receptor gene promoter. Gene 168:127–133

Spradling AC (1986) *P*-element-mediated transformation. In Roberts DB (eds) *Drosophila*: A Practical Approach. IRL Press, Washington, D.C., Chapt 8

Steller H, Pirrotta V (1985) A transposable P vector that confers selectable g418 resistance to *Drosophila* larvae. EMBO J 4:167–171

Stilwell GE, ffrench-Constant RH (1998) Transcriptional analysis of the *Drosophila* GABA receptor gene *Resistance to dieldrin*. J Neurobiol 36:468–484

Stilwell GE, Rocheleau T, ffrench-Constant RH (1995) GABA receptor minigene rescues insecticide resistance phenotypes in *Drosophila*. J Mol Biol 253:223–227

Thompson M, Shotkoski F, ffrench-Constant R (1993) Cloning and sequencing of the cyclodiene insecticide resistance gene from the yellow fever mosquito *Aedes aegypti*. FEBS Lett 325:187–190

Vaughan A, Rocheleau T, ffrench-Constant R (1997) Site-directed mutagenesis of an acetylcholinesterase gene from the yellow fever mosquito *Aedes aegypti* confers insecticide insensitivity. Exp Parasitol 87:237–244

Williamson MS, Moores GD, Devonshire AL (1992) Altered forms of acetylcholinesterase in insecticide-resistant houseflies (*Musca domestica*). In Shafferman A, Velan B (eds) Multidisciplinary Approaches to Cholinesterase Functions. Plenum Press, New York, pp 83–86

Williamson MS, Martinez-Torres D, Hick CA, Devonshire AL (1996) Identification of mutations in the housefly *para*-type sodium channel gene associated with knockdown resistance (*kdr*) to pyrethroid insecticides. Mol Gen Genet 252:51–60

Xiao H, Lis JT (1989) Heat shock and developmental regulation of the *Drosophila melanogaster hsp83* gene. Mol Cell Biol 9:1746–1753

Section IV

Viral Vectors

7 Pantropic Retroviral Vectors for Insect Gene Transfer

Jane C. Burns

CONTENTS

7.1	Introduction	126
7.2	Overview of Retroviral Vectors	126
	7.2.1 Pseudotyping to Create Pantropic Retroviral Vectors	126
7.3	Producing Pantropic Retroviral Vectors	127
7.4	Genetic Organization of Vectors	127
	7.4.1 Monocistronic Vectors	127
	7.4.2 Dicistronic Vectors	128
7.5	Infection of Insect Cell Lines	128
7.6	Infection of Embryos	128
7.7	Infection of Larvae	128
7.8	Conclusion	130

Appendix 131
A Generation of Pantropic Vectors by Transient Transfection Using Calcium Phosphate Coprecipitation 131
B Generation of Pantropic Vectors by Creating a Stable Producer Cell Line 132
    B.1 Selection of 293-gag-pol Producer Cell Clones 132
    B.2 Production of Vector 133
    B.3 Harvesting Virus 133
C Ultracentrifugation of Vector 133
    Day 1 133
    Day 2 134
D PCR Amplification of Provirus 134
    D.1 Retroviral LTR PRIMERS 134
    D.2 Actin Primers 134
    D.3 PCR Assay for Actin and LTR 134
E Protocol for Luciferase Assays in Mosquito Cell Lines and Mosquito Adults 135
    E.1 Cell Lines 135
    E.2 Whole Mosquitoes 135
F Reagents and Solutions 135
    F.1 Reagents for Calcium Phosphate Coprecipitation 135
        2 × HBS 135
        2 M CaCl2 135
        1:10 TE (1 mM Tris/0.1mM EDTA) 136
    F.2 Solutions for Tissue Culture 136
        Polybrene (hexadimethrine bromide) 136

G418 solution 136
Acknowledgments 136
References 136

7.1 INTRODUCTION

To provide a new tool for genetic manipulation in insects, we developed a novel class of retroviral vectors based on the Moloney murine leukemia virus (MoMLV) that contain the envelope glycoprotein of vesicular stomatitis virus (VSV-G). These replication-incompetent, pantropic retroviral vectors have a broadened host cell range that includes insect cells and can be concentrated to the high titers ($>10^9$ infectious units/ml) needed for microinjection studies. Such vectors have been demonstrated to attach, uncoat, reverse-transcribe, and integrate as a DNA provirus into the chromosomes of dipteran and lepidopteran cells.

7.2 OVERVIEW OF RETROVIRAL VECTORS

Retroviral vectors based on the murine and avian leukemia viruses have become standard tools for the introduction and expression of foreign genes in mammalian cells, both *in vitro* and *in vivo* (Anderson, 1998). To transform a *retrovirus* into a *retroviral vector*, the coding sequences for the structural gene (*gag*), the reverse transcriptase gene (*pol*), and the envelope protein gene (*env*) are removed and replaced with coding sequences of interest under the control of the retroviral promoter (long terminal repeat, LTR) or an internal promoter. These replication-incompetent vectors are produced in packaging cell lines that express the retroviral proteins needed for the assembly of an infectious particle. The vector particles can be recovered from the culture supernatant and used to infect target cells.

A decade of experience with these replication-incompetent vectors has demonstrated that they can be produced safely in the laboratory. Because they are incapable of self-propagation, there are no containment issues for infected cells harboring the integrated provirus, other than matters relating to recombinant DNA. These vectors are now widely used for human gene therapy protocols and have been intravenously administered to humans with no untoward effects.

7.2.1 Pseudotyping to Create Pantropic Retroviral Vectors

To broaden the host range of retroviral vectors, we created Moloney murine leukemia virus (MoMLV)-based pseudotyped vectors that contain the envelope glycoprotein of vesicular stomatitis virus (VSV-G) substituted for the retroviral envelope protein (Burns et al., 1993). The VSV-G protein binds to phospholipid moieties in the cell membrane, thus circumventing the need for a specific protein receptor molecule (Mastromarino et al., 1987). These modified retroviral vectors, therefore, have an expanded host cell range (pantropic) and, unlike their amphotropic counterparts containing the native retroviral envelope protein, can be concentrated to titers $>10^9$ cfu/ml by ultracentrifugation. Biological containment issues are the same as for the amphotropic vectors since the pantropic vectors are also replication incompetent. We have demonstrated that these vectors can attach, uncoat, reverse-transcribe, and stably integrate into the genome of dipteran and lepidopteran cell lines (Matsubara et al., 1996; Franco et al., 1998; Teysset et al., 1998) and embryos (Jordan et al., 1998; Franco et al., 1998). In other animals, such as the zebrafish, cows, and the surfclam, *Mulinia lateralis*, we have exposed early embryos to high-titer vector stocks to create transgenic organisms (Lin et al., 1994; Lu et al., 1996; Chan et al., 1998). Proceeding even farther down the phylogenetic tree, pantropic vectors have been used to create transgenic amoeba (Que et al., 1999).

Two general limitations to the use of the vector to create transgenic organisms are (1) the need for the vector to traverse a cell membrane to uncoat and (2) the requirement that the target cell divide within several hours after infection. In the first case, the virus particle must uncoat in an endocytic vesicle after being endocytosed by the target cell. The vector particle is not capable of crossing other barrier membranes, such as the chorion, nor can the particles infect across barrier cells, such as nurse cells, to reach the final target. This means, therefore, that the particle must be delivered to the surface of the target cell in sufficient quantity to make an infection event likely. In the case of a nondividing cell, successful entry of the retroviral vector into the cell may still not result in stable transduction because the MoMLV-based vectors do not contain nuclear targeting signals and thus depend on the breakdown of the nuclear envelope just prior to cell division to gain access to the nucleus. Lentiviral vectors, which do contain nuclear targeting sequences, can also be pseudotyped with VSV-G (Poeschla et al., 1997), and preliminary data suggest that these vectors also readily transduce mosquito cells (J. Burns, unpublished). Further studies are in progress to determine the *in vitro* and *in vivo* infection efficiency of these vectors.

7.3 PRODUCING PANTROPIC RETROVIRAL VECTORS

Detailed protocols for producing pantropic vectors are provided as an Appendix to this chapter. All the reagents, cells lines, and plasmids necessary to create the pseudotyped particles are available from Clontech Laboratories, Palo Alto, CA. Vector stocks may be produced by creating a stable producer cell line in the 293-gag-pol cells that express high levels of the retroviral proteins needed for packaging of particles (Yee et al., 1994a). To produce virus, 293-gag-pol cells containing an integrated provirus are transfected with a plasmid encoding the VSV-G protein and the vector is harvested from the supernatant until the cells die from syncytium formation caused by the VSV-G (Yee et al., 1994b). Alternatively, vector stocks can be transiently produced by cotransfection of the plasmid carrying the retroviral construct and the plasmid encoding the VSV-G. Once vector stocks are produced, they may be used directly, stored at –70°C for several years, or concentrated by ultracentrifugation to obtain higher titers and to remove factors in the tissue culture supernatant that may be toxic to the target cells.

7.4 GENETIC ORGANIZATION OF VECTORS

7.4.1 MONOCISTRONIC VECTORS

The retroviral particle can contain between 10 and 13 kb of sequence and still be efficiently packaged. The coding sequence of interest can be expressed in insect cells from the retroviral LTR or from an internal promoter such as the *Drosophila hsp70* promoter (Franco et al., 1998; Jordan et al., 1998) (Figure 7.1). Vectors expressing a selectable marker (e.g., neomycin or hygromycin phosphotransferase) are useful to select transduced cells and to make a stable packaging cell line. Alternatively, vector can be produced by the transient transfection method (see Appendix), and both cistrons in the vector can be coding sequences of interest.

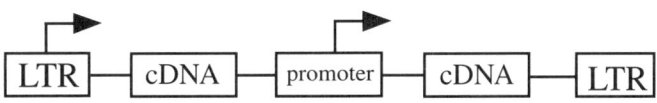

FIGURE 7.1 Genetic organization of a retroviral vector. Genes of interest (cDNA) can be expressed from the MoMLV LTR as well as from an internal promoter of choice. The polyadenylation signal for both of these mRNAs resides in the 3′ LTR.

7.4.2 DICISTRONIC VECTORS

Cap-independent translation occurs in many invertebrate cells and allows two cistrons separated by an internal ribosomal entry site (IRES) to be translated from the same mRNA. We are currently evaluating whether the polio virus 5′ untranslated region (Adam et al., 1991) and the *Drosophila antennapedia* exon DE (Oh et al., 1992) function as an IRES in mosquito cell lines. The ability to construct vectors with dicistronic genetic organization would give increased flexibility to the design of vectors for gene expression in mosquitoes (Figure 7.2).

7.5 INFECTION OF INSECT CELL LINES

Cell lines from several different dipteran and lepidopteran species have been stably transduced with the pantropic retroviral vectors. High levels of reporter gene expression (e.g., firefly and *Renilla* luciferase) have been obtained (Franco et al., 1998), and transduced cells have been successfully selected in hygromycin (Matsubara et al., 1996). The amount of luciferase expression is proportional to the amount of vector used to infect the cells (Figure 7.3). Many different reporter gene vectors are currently available for testing expression in insect cells (Figure 7.4).

To increase the infection efficiency of the pantropic vectors, we examined a number of different parameters. The use of polycations, polybrene, protamine sulfate, and poly-L-lysine, to overcome the negative charge on the surface of the virus particles and on the target cells increased the infection efficiency only slightly. For infection of mosquito cell lines, a 1.5- to 2.0-fold increase in infection efficiency was observed. More significant increases in efficiency were noted after adopting a centrifugation or flow-through protocol for infecting the cells. In the flow-through method, target cells or tissues are placed in tissue culture well inserts (Transwell inserts, Costar) and virus, diluted in medium plus polybrene (2 μg/ml), is allowed to flow past the cells. For a single well of a 6-well plate with 1 to 5×10^5 cells, we typically use 10^4 cfu of vector in 1 ml medium containing 2 μg/ml of polybrene. The directed flow of virus particles increases the likelihood of contact with the target cell surface (Palsson and Andreadis, 1997). Using this type of protocol, we demonstrated a five- to ten-fold increase in infection efficiency of MOS-55 (*Anopheles gambiae*) cells. A similar effect can also be achieved with 1 to 2 h of low-speed centrifugation ($1000 \times g$) of cells plus virus and polybrene (Figure 7.5).

7.6 INFECTION OF EMBRYOS

In different mosquito species, microinjection of concentrated vector stocks into the early embryo has resulted in at least somatic infection in 5 to 20% of G_0 animals depending on the titer of the vector, the mosquito species, and the experience of the individual doing the microinjection. Limited gene amplification studies have suggested that the transgene can be passed on to subsequent generations. Large-scale screens for transgenic G_1 and G_2 mosquitoes must await improved phenotypic markers for transgenic embryo detection. Studies with the green fluorescent protein as a dominant phenotypic marker suggest that levels of stable expression are below detection thresholds.

7.7 INFECTION OF LARVAE

Somatic infection of various tissues in developing larvae has been demonstrated in *D. melanogaster*, *Aedes triseriatus*, *Culex tarsalis*, *An. gambiae*, and *Manduca sexta* (Jordan et al., 1998; Franco et al., 1998). Expression of β-galactosidase and firefly luciferase has been documented in adults following microinjection during larval stages. Larvae also have been infected via the oral route with pantropic retroviral vector administered in the swim water. Applications of these vectors to *in vivo* studies of promoter function and regulation appear promising. In

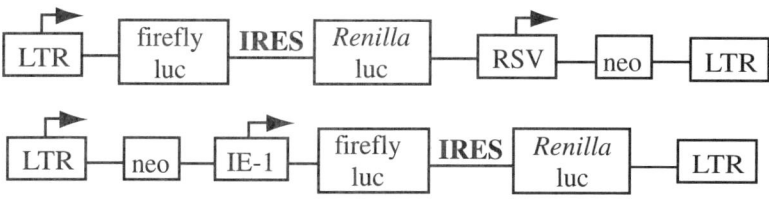

Key to Nomenclature

LTR	MoMLV long terminal repeat
RSV	Rous sarcoma virus LTR
neo	neomycin phosphotransferase
IE-1	baculovirus immediate early-1 promoter
luc	luciferase
IRES	internal ribosomal entry site

FIGURE 7.2 Genetic organization of a dicistronic vector. Cap-independent translation in insect cells can be mediated by the poliovirus IRES. Vectors carrying this sequence between the firefly and *Renilla* luciferase cDNAs demonstrate expression of both reporter genes. Use of an IRES provides greater flexibility in the design of vectors for expression in insect cells.

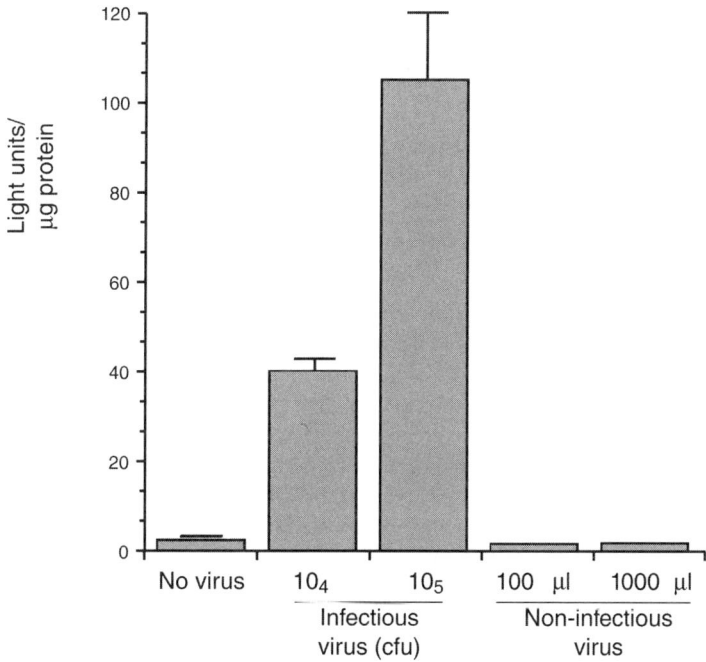

FIGURE 7.3 Infection of MOS-55 cells (*An. gambiae*) with vector LNhsp70lucL (firefly luciferase expressed from *Drosophila hsp70* promoter). Triplicate wells of MOS-55 cells (80% confluent) in a six-well plate were exposed to the different solutions as shown. Polybrene (2 µg/ml) was added to all wells. At 24-h postinfection, medium was replaced with normal growth medium. At 72-h postinfection, cells were lysed and assayed for total protein and luciferase activity (see Appendix for detailed protocol). No virus: polybrene and medium alone. Infectious virus: pseudotyped LNhsp70lucL vector at either 10^4 or 10^5 cfu/well. Noninfectious virus: nonpseudotyped (bald) LNhsp70lucL vector. The bald virus negative control is produced from packaging cells that do not contain an envelope glycoprotein. The particles contain a functional nucleocapsid but are noninfectious as a result of the absence of an envelope protein. Bars represent mean constitutive luciferase activity/µg protein of triplicate wells ±1 standard deviation.

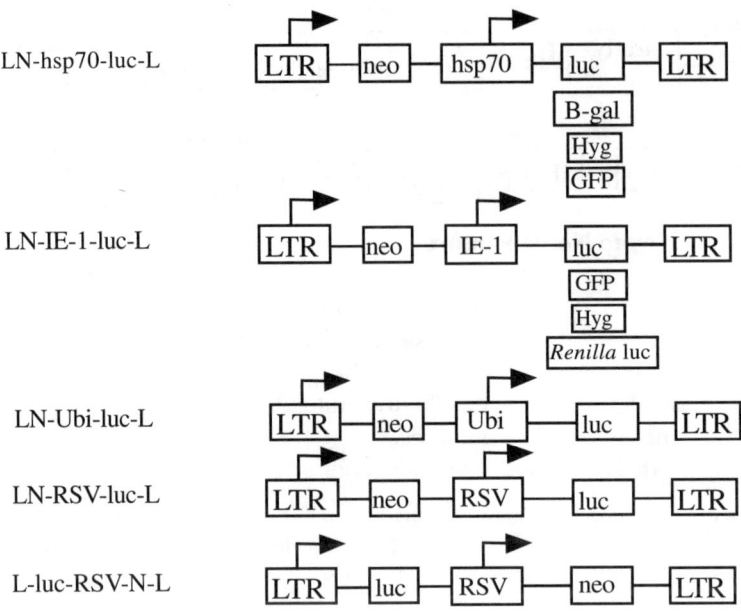

FIGURE 7.4 Genetic organization of pantropic retroviral vectors available for testing reporter gene expression in insect cells.

addition, use of the vectors for somatic infection of specific tissues should allow rapid evaluation of foreign gene expression as a first step to designing other types of vectors for creating transgenic organisms.

7.8 CONCLUSION

In summary, these pantropic vectors represent an important innovation that permits the use of retroviral vectors to introduce foreign genes stably into nonmammalian cells. Vectors can be constructed easily to compare gene expression mediated by different promoters. This provides a way to assess promoter function from a single, stably integrated copy of the provirus. In addition, the vectors may provide insight into regulation of gene expression *in vivo* in different tissues such as the midgut, salivary gland, and fat body. These vectors should, in principle, be useful vehicles for germline transformation in a number of different insect species in which vector delivery to the primordial germline cells can be achieved.

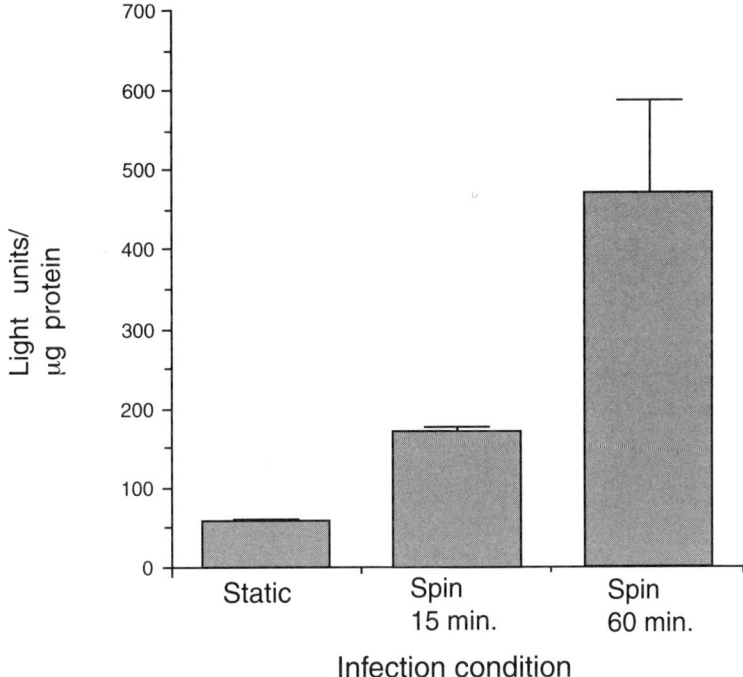

FIGURE 7.5 Pantropic retroviral vector infection of MOS-55 cells by different methods. Cells were grown to 80% confluence in a six-well plate were exposed in quadruplicate to 10^5 cfu of the vector LNhsp70lucL (MoMLV LTR-neo-*Drosophila hsp70* promoter-firefly luciferase-LTR; Jordan et al., 1998) and polybrene (2 µg/ml) under the following conditions: (1) Static infection: Cells were incubated at room temperature with vector plus polybrene for 24 h, at which time medium was replaced; (2) Centrifugation: Cells were centrifuged at room temperature at $1000 \times g$ for either 15 min or 1 h in the presence of virus plus polybrene. Medium was replaced at 24 h. For both infection conditions, cells were harvested at 48-h postinfection, lysed in cell lysis buffer (Analytical Luminescence Laboratory), and assayed for luciferase activity and total cellular protein (Biorad) according to the manufacturers' instructions.

APPENDIX

A   GENERATION OF PANTROPIC VECTORS BY TRANSIENT TRANSFECTION USING CALCIUM PHOSPHATE COPRECIPITATION

Note: The advantages of this transient transfection method are that (1) small quantities of vector can be quickly produced and (2) the impact of reverse transcription errors resulting in a defective viral genome (<10%) is minimized. The disadvantage is that the titers are generally 0.5 to 1.0 log lower by this method than by selecting a stable producer clone (see Section B below).

1. Grow 293-gag-pol cells in DMEM high glucose/10% FCS until monolayer is 80 to 90% confluent. Allow cells to grow on plate for at least 48 h before transfection. Replace medium with fresh DMEM/10% FCS before transfection procedure.
2. Transfection: For 1 ml of precipitate (1 ml/10-cm plate, 5 ml/T225 flask), add 1:10 sterile TE pH 8.0 + plasmid to equal 437 µl.
3. Add a total of 20 µg of plasmid DNA so that the molar ratio of pCMV-VSV-G and retroviral plasmid is 1:1. For example, to produce LLRNL vector, use 11.6 µg of pLLRNL (8.7 kb) and 8.4 µg of pCMV-VSV-G (6.3 kb).

4. Add 63 μl of 2 M CaCl$_2$ by single drops while swirling the DNA mixture gently.
5. Add 500 μl of 2 × HBS to DNA mixture while gently swirling.
6. Leave at room temperature for 30 min to allow precipitate to form.
7. Gently mix by inverting five times and add precipitate to plate by single drops, swirling after each addition.
8. Allow precipitate to stay on cells for 6 to 8 h while incubating at 37°C/ 10% CO$_2$. Do not leave 293 cells with precipitate longer, or cells will come off the plate and die.
9. After 6 to 8 h, gently aspirate medium and replace with 8 ml of DMEM/5% FCS.

B GENERATION OF PANTROPIC VECTORS BY CREATING A STABLE PRODUCER CELL LINE

B.1 Selection of 293-gag-pol Producer Cell Clones

1. Cotransfect 293-gag-pol cells with 20 μg of the retroviral plasmid and pCMV-VSV-G as described above (Appendix A, steps 1 through 9).
2. At 48-h post-transfection, harvest the supernatant, filter through a 0.45-μm filter, and infect two fresh plates of 293-gag-pol cells with 0.1 and 1.0 ml of supernatant and polybrene (8 μg/ml), respectively.
3. At 12-h postinfection, replace medium with DMEM high glucose with 10% FCS and 400 μg/ml G418 (neomycin analogue). Nontransduced cells will begin to die in selection 24 to 48 h after adding the G418. Individual colonies, each representing infection of a single cell, will become visible at 7 to 10 days after selection.
4. To pick clones, choose a plate that has an appropriate number of colonies (e.g., not too crowded) and aspirate all medium from the plate. Draw up 3 μl of medium in an Eppendorf pipette and vigorously pipette up and down on an individual colony to remove the cells. Transfer the 3 μl to a well in a 24-well plate until all the colonies are harvested.

Note: Avoid picking colonies that are too close to their neighbor. Colonies growing in the center of the plate are more likely to be unique than colonies growing around the edge of the plate. Working quickly is important to avoid desiccation of the colonies. Turning off the blower on the tissue culture hood may slow the rate of desiccation.

5. Grow colonies in 24-well plate in presence of G418 (400 μg/ml) and passage into a duplicate 24-well plate. Different colonies will have different growth rates, and not all clones will be ready to transfect at the same time.
6. When clones in one of the duplicate 24-well plates are 80 to 90% confluent, replace medium with DMEM high glucose with 10% FCS but without G418.
7. For 16 clones, make 1 ml of transfection mixture with 20 μg of pCMV-VSV-G plasmid (see Appendix A, steps 2 through 5). Each well needs 60 μl of mixture, so if there are more than 16 clones (60 μl × 16 well = 960 μl), calculate and make more mixture.
8. Slowly add 60 μl of transfection mixture to each well while swirling.
9. Incubate the plate for 6 to 8 h at 37°C in 10% CO$_2$. Remove precipitate and replace with normal growth medium (500 μl/well) without G418.
10. Incubate for 48 h. To titer the supernatants, grow 208F cells (or NIH 3T3 cells) in a 24-well plate to 25% confluency.
11. Harvest 2 μl from each 293 clone, taking care not to touch the monolayer. Infect each well of 208F cells with supernatant from a different 293 clone and label the wells. Add polybrene (8 μg/ml) to each well of infected cells.
12. Incubate titer plate for 12 to 24 h. Replace medium with G418-containing DMEM.
13. Incubate for 6 to 8 days with media changes to remove dead cells.

14. Fix the 208F plate with methanol × 10 min at room temperature. Rinse plate with water and add Giemsa stain (2 ml commercial stain + 50 ml PBS). Stain for 5 to 15 min. Rinse and photocopy plate to keep record of different clones. Wells infected with the most vector will have the largest patches of blue cells. Do not bother counting colonies. Simply choose the best two to three producer clones based on this titer estimate.
15. To choose the best 293-gag-pol producer clone, determine a more precise titer by growing the two to three winning clones in a 10-cm plate, transfecting with pCMV-VSV-G, infecting 208F cells with filtered supernatant, and selecting colonies in G418-containing medium.

Note: Deletions can occur due to reverse transcriptase errors and all clones should be checked by PCR for full-length, integrated provirus as well as for levels of transgene expression. The protocol offered here avoids the use of the PA317 packaging cell line, which has the theoretical problem of containing endogenous murine retroelements that could lead to formation of recombinant, replication-competent virus.

B.2   Production of Vector

1. Grow the best 293-gag-pol producer clone in DMEM high glucose with 10% FCS and G418 (400 µg/ml) until monolayer is about 80 to 90% confluent.
2. Transfect cells with 20 µg pCMV-VSV-G (for a 10-cm plate) by calcium phosphate coprecipitation as described in Appendix A, steps 2 through 5.

B.3   Harvesting Virus

1. Beginning 48 h after transfection precipitate was initially placed on cells, collect supernatant from each plate with a 20-cc syringe and pass through a 0.45-µm filter to remove cellular debris. Store filtered supernatants at –70°C until ready to use. Supernatants can be frozen and thawed up to three times without significant loss in titer.
2. Replace 6 to 8 ml of medium (without G418) onto 293-gag-pol producer cells and harvest virus again 24 h later. Harvesting virus at more frequent intervals does not seem to increase the overall yield.
3. Harvest virus supernatant at 48, 72, and 96 h. Titers of virus drop precipitously more than 4 days post-transfection.

C   ULTRACENTRIFUGATION OF VECTOR

Note: The stability of the VSV-G envelope protein permits physical concentration of the vectors by a variety of methods. Ultracentrifugation can be expected to yield a 2 log increase in titer after one round of centrifugation and an additional 1 log increase in titer after a second round of centrifugation.

Day 1

1. Place the ultracentrifuge tubes overnight under ultraviolet light in the tissue culture hood to sterilize.
2. Add the virus (13.5 ml/tube for SW41 rotor, Beckman Ultraclear Cat 344059, 14 × 89 mm) and centrifuge for 90 to 120 min at 25,000 rpm at 4°C with the brake on (= 100,000 × g). Any clear ultracentrifuge tube/rotor can be used. A swinging bucket rotor is preferred since after one round of ultracentrifugation, the pellet is frequently invisible.

Speeds and Conditions for other Rotors
Beckman SW28: 37.5 ml, swinging bucket, 24,000 rpm × 2 h at 4°C, brake on.
Resuspend second pellet in 30 µl unless pellet is too big, then increase volume slightly.
Beckman Ti45: 65 ml/tube, spin at 35,000 rpm × 2 h at 4°C, brake on.

3. Decant the supernatant, dry the tubes with Kimwipes wrapped around sterile forceps, taking care not to disturb the pellet. Place tubes immediately on ice.
4. Add 30 µl of 0.1 × Hank's balanced salt solution (diluted with ddH$_2$O or PBS; this has not been well studied and both ways seem to work), cover the tubes with Parafilm, and hold them at 4°C overnight. This allows the pellet to resuspend slowly.

Day 2

1. Resuspend the pellet several times with a P200 pipetman. Try to avoid creating bubbles.
2. For a second round of concentration (necessary to achieve titers of 10^9 cfu/ml), pool pellets from six SW41 tubes into a single tube, bring up to volume with PBS, and repeat the centrifugation. Resuspend pellet next day in 30 µl of 0.1 × Hank's and store at –70°C.
3. To titer virus, make a 10^{-5} dilution of concentrated virus in media and infect 208F cells with 50 µl. Remember to add polybrene (final concentration, 8 µg/ml) when adding virus.

D PCR AMPLIFICATION OF PROVIRUS

D.1 Retroviral LTR PRIMERS

Upstream primer
 MoMLV U3 region, BP 8148–8167
 5' AG GAC CTG AAA TGA CCC TGT 3'
Downstream primer
 Complementary to MMSV U5 region, BP 108-125
 5' G GGT AGT CAA TCA CTC AG 3'
 Amplicon size 244 bp

D.2 Actin Primers

Upstream primer
 5' CCC AGA GCA AGA GAG GTA 3'
Downstream primer
 5' TCC AGA CGG AGG ATG GCG GT 3'

D.3 PCR Assay for Actin and LTR

To detect the presence of amplifiable DNA from mosquito extracts, positive control primers were selected from a consensus alignment of the muscle actin gene from several different invertebrates (Jordan et al., 1998). The predicted amplicon size based on these sequences is 336 bp. The primers (0.1 µM each) were added to a 25-µl reaction mixture containing 10 mM Tris (pH 8.3), 1.5 mM MgCl$_2$, 50 mM KCl, 200 µm each dNTP, and 0.2 µl *Taq* DNA polymerase (AmpliTaq, 5U/µl, Perkin Elmer, Foster City, CA). These optimal buffer conditions were determined after testing a range of concentrations for the primers (0.1, 0.25, and 0.5 µM) and magnesium (1.5 and 2.5 mM). Optimal thermal cycling conditions were 94°C × 30 s, 55°C × 30 s, and 72°C × 60 s for 40 cycles.

For the LTR PCR assay, the optimized buffer and thermal cycling conditions were the same as for actin except for the primer concentration (0.5 µM) and the annealing temperature (60°C).

E PROTOCOL FOR LUCIFERASE ASSAYS IN MOSQUITO CELL LINES AND MOSQUITO ADULTS

E.1 Cell Lines

1. Harvest cell lysate at least 72 h postinfection (from start of infection) according to manufacturer's protocol (Analytical Luminescence Laboratory, Enhanced Luciferase Assay Kit, Ann Arbor, MI; Promega reagents are also fine).
2. Aspirate medium and wash monolayer twice with PBS.
3. Add 150 µl of cell lysis buffer/well for six-well plate (60 µl/well for 24-well plate) prepared according to manufacturer's instructions. Incubate at 4°C for 15 to 20 min.
4. Spin out cell pellet and debris in Eppendorf centrifuge for 5 min at 4°C or 3 min at room temperature at 13,000 rpm (maximum speed). Remove 120 µl supernatant and freeze at −70°C for later assay or continue. Save pellet for DNA extraction /PCR, if needed. Allow cell lysate to warm to room temperature before continuing.
5. Use 100 µl (40 µl from 24-well plate) of clarified cell lysate for luciferase assay and 5 µl for protein assay (Bio-Rad dye reagent). Calculate light units/µg protein.

E.2 Whole Mosquitoes

1. Freeze single mosquitos in Eppendorf tubes on dry ice.
2. Grind frozen, dry mosquito with disposable pestle (separate pestle for each tube) without allowing to thaw.
3. Add 150 µl of cell lysis buffer at room temperature. Grind again in lysis buffer. Incubate at 4°C × 10 to 15 min.
4. Spin out debris as above and transfer 100 µl supernatant to assay tubes. Mosquito pellet can be used for DNA extraction and PCR analysis.

F REAGENTS AND SOLUTIONS

F.1 Reagents for Calcium Phosphate Coprecipitation

Note: Store all reagents at 4°C in sterile tubes.

2 X HBS

For 200 ml, weigh out the following:

2.382 g HEPES (50 mM)
3.280 g NaCl (280 mM)
42.400 mg Na_2HPO_4 (1.5 mM)

Dissolve in 180 ml ddH_2O. Adjust pH to 7.12 with 5 N NaOH. pH of final solution is *critical*! Add ddH_2O to equal 200 ml. Filter-sterilize with 0.2 µm filter.

2 M $CaCl_2$

For 100 ml of solution:

29.41 g $CaCl_2 \cdot 2H_2O$

Dissolve in ddH_2O to make 100 ml. Filter-sterilize with 0.2 µm filter.

1:10 TE (1 mM Tris/0.1mM EDTA)

For 50 ml: Add 50 μl of 1 M Tris, pH 8.0 and 10 μl 500 mM EDTA, pH 8.0, to 50 ml ddH_2O. Autoclave to sterilize.

F.2   Solutions for Tissue Culture

Polybrene (hexadimethrine bromide)

Available from Sigma. Powder (stored at 0 to 5°C) is soluble in ddH_2O up to 10%. Make a 4 mg/ml stock solution. Filter-sterilize (0.2 μm filter). Recommended concentrations for cell line infection: 2 μg/ml for mosquito cell lines, 6 to 8 μg/ml for 293-gag-pol and 208F cell lines.

G418 solution

Need 0.1 M HEPES to dilute G418 (Gibco BRL). For 400 ml 0.1 M HEPES (free acid, Sigma, MW 238.3):

1. Dissolve 9.52 g in 350 ml ddH_2O.
2. Adjust pH with 5 N NaOH to pH 7.45.
3. Adjust final pH with 0.1 N NaOH to 7.55.
4. Q.S. to 400 ml.

To make G418 stock:

1. Calculate volume of 0.1 M HEPES to add to G418 powder to yield 40 mg/ml of active compound. Purity of G418 is typically between 500 and 700 μg/mg. For example, if potency of G418 is 661 μg/mg, then 40 mg/ml = 661 μg/x, where x = 16.52 ml 0.1 M HEPES.
2. Filter-sterilize (0.2 μm filter) in hood and aliquot into 15-ml polypropylene centrifuge tubes. Store frozen at –20°C until ready to use. Keep at 4°C × 2 to 3 months.

ACKNOWLEDGMENTS

The work reviewed here was supported in part by the following grants to JCB: the UNDP/World Bank/WHO Special Programme for Research in Tropical Diseases (TDR) Project 950469, NIH–NIAID R01AI37671, and the University of California University-wide Mosquito Research Program 98-003-2.

REFERENCES

Adam MA, Ramesh N, Miller AD, Osborne WRA (1991) Internal initiation of translation in retroviral vectors carrying picornavirus 5′ nontranslated regions. J Virol 65:4985–4990

Anderson WF (1998) Human gene therapy. Nature 392:25–30

Burns JC, Friedman T, Driever W, Burrascano M, Yee JK (1993) VSV-G pseudotyped retroviral vectors: concentration to very high titer and efficient gene transfer into mammalian and nonmammalian cells. Proc Natl Acad Sci 90:8033–8037

Chan AWS, Homan JE, Ballou LU, Burns JC, Bremel RD (1998) Transgenic cattle produced by reverse-transcribed gene transfer in oocytes. Proc Natl Acad Sci USA 95:14028–14033

Franco MD, Rogers M, Shimizu C, Shike H, Vogt RG, Burns JC (1998) Infection of Lepidoptera with a pseudotyped retroviral vector. Insect Biochem Mol Biol 28:819–825

Jordan TV, Shike H, Boulo V, Cedeno V, Fang Q, Davis BS, Jacobs-Lorena M, Higgs S, Fryxell KJ, Burns JC (1998) Pantropic retroviral vectors mediate somatic cell transformation and expression of foreign genes in dipteran insects. Insect Mol Biol 7:215–222

Lin S, Gaiano N, Yee J-K, Culp P Burns JC, Friedmann T, Hopkins N (1994) Integration and germline transmission of a pseudotyped retroviral vector in zebrafish. Science 265:666–669

Lu JK, Chen TT, Allen SK, Matsubara T, Burns JC (1996) Production of transgenic dwarf surfclams, *Mulinia lateralis* with pantropic retroviral vectors. Proc Natl Acad Sci USA 93:3482–3486

Mastromarino P, Conti C, Goldoni P, Hauttecoeur B, Orsi N (1987) Characterization of membrane components of the erythrocyte involved in vesicular stomatitis virus attachment and fusion at acidic pH. J Gen Virol 68:2359–2369

Matsubara T, Beeman RW, Shike H, Besansky NJ, Mukabayire O, Higgs S, James AA, Burns JC (1996) Pantropic retroviral vectors integrate and express in cells of the malaria mosquito, *Anopheles gambiae*. Proc Natl Acad Sci USA 93:6181–6185

Oh SK, Scott MP, Sarnow, O (1992) Homeotic gene *Antennapedia* mRNA contains 5'-noncoding sequences that confer translational initiation by internal ribosome binding. Genes Dev 6:1643–1653

Palsson B, Andreadis S (1997) The physico-chemical factors that govern retrovirus-mediated gene transfer. Exp Hematol 25:94–102

Poeschla EM, Wong-Staal F, Looney DJ (1998) Efficient transduction of nondividing human cells by feline immunodeficiency virus lentiviral vectors. Nat Med 4:354–357

Que X, Kim D, Alagon A, Hirata K, Shike H, Shimizu C, Burns JC, Reed SL (1999) Pantropic retroviral vectors mediate stable gene transfer and expression in *Entamoeba histolytica*. Mol Biochem Parasitol 99:237–245

Teysset L, Burns JC, Shike H, Sullivan BL, Bucheton A, Terzian C (1998) A Moloney murine leukemia virus-based retroviral vector pseudotyped by the insect retroviral *gypsy* envelope can infect *Drosophila* cells. J Virol 72:853–856

Yee J-K, Friedmann T, Burns JC (1994a) Generation of high-titer pseudotyped retroviral vectors with very broad host range. Methods Cell Biol 43:99–112

Yee J-K, Miyanohara A, LaPorte P, Bouic K, Burns JC, Friedmann T (1994b) Generation of high titer, pantropic retroviral vectors: efficient gene transfer into hepatocytes. Proc Natl Acad Sci USA 91:9564–9568

8 Densonucleosis Viruses as Transducing Vectors for Insects

Jonathan Carlson, Boris Afanasiev, and Erica Suchman

CONTENTS

8.1 Introduction 139
8.2 Biology of Densonucleosis Viruses 140
8.3 Molecular Biology of Densonucleosis Viruses 141
    8.3.1 Densonucleosis Virus Replication 141
    8.3.2 Densonucleosis Virus Genomic Organization 142
    8.3.3 Densonucleosis Virus Gene Expression 143
        8.3.3.1 AeDNV Gene Expression 143
            8.3.3.1.1 Nonstructural Gene Promoter 144
            8.3.3.1.2 Structural Gene Promoter 145
            8.3.3.1.3 Transactivation of AeDNV Promoters 145
        8.3.3.2 JcDNV Gene Expression 145
8.4 Development of Densonucleosis Virus Transducing Systems 146
    8.4.1 Construction of Densonucleosis Virus Packaging Cell Lines 148
    8.4.2 Packaging with a Sindbis Virus Helper 148
    8.4.3 Transduction of Mosquito Larvae 149
8.5 Densonucleosis Viruses as Potential Transformation Vectors for Insects 150
    8.5.1 Potential for Integrative Transformation by AeDNV 152
    8.5.2 Potential for Episomal Transformation 156
8.6 Summary 157
Acknowledgments 157
References 157

8.1 INTRODUCTION

Transduction is the use of viruses to package and deliver genes of interest to cells by the infectious process. Transduction was discovered in bacteria, and has greatly facilitated genetic analysis and manipulation in several bacterial species. More recently, transduction has been used in mammalian systems, particularly in gene therapy applications. Retrovirus, adenovirus, poxvirus, and parvovirus vectors have all been developed and have proved useful for specific applications. The use of viruses to transduce genes of interest in insects is less well explored, although the potential for laboratory studies and control of agricultural pests and vector-borne disease seems very attractive.

    The virus families available for use in insects include both RNA and DNA viruses. Many of the well-characterized RNA viruses known to infect insects are arboviruses, which infect both the vector insect and a vertebrate host. Indeed, virus vectors based on the alphaviruses (Sindbis virus and Semliki Forest Virus) are effective expression vectors in both mosquito systems and in mammalian cells. While these RNA virus vector systems have proved to be valuable laboratory tools, the potential for applications outside the laboratory seems limited because of their ability to infect vertebrates. The DNA viruses are more restricted in their host range, probably because of greater dependence on host functions. The DNA viruses that seem to have potential for insect transduction are the baculoviruses, polydnaviruses, parvoviruses, iridoviruses, and the entomopoxviruses. The baculoviruses of lepidopterans have been very successfully exploited as laboratory expression systems and in biocontrol applications. However, aside from these well-studied viruses, little has been done to develop transducing viruses for other insect systems.

Potential transducing viruses for vector insects such as mosquitoes are more limited. Although baculoviruses, iridoviruses, and entomopoxviruses of mosquitoes have been reported, they are as yet poorly characterized and not yet amenable to molecular biological manipulation. Polydnaviruses have so far only been shown to replicate in the ovaries of parasitic wasps. In contrast, the parvoviruses of insects (densoviruses) are relatively well characterized. Since they have a nonenveloped icosahedral structure, they are among the most stable viruses in the environment. Densoviruses appear to be relatively widespread in nature, but at the same time they are relatively restricted in their host specificity. This makes them attractive candidates for use in biological control strategies.

Our laboratory has been able to develop the densovirus of the mosquito *Aedes aegypti* (AeDNV) as a gene transfer vector that is able to transduce genes into mosquito larvae by natural routes of infection in their aquatic habitat. This opens the potential for introducing genes that increase the virulence of the virus or perhaps modify the life cycle or behavior of the mosquito into natural populations. Some of the mosquito densoviruses have been reported to be transmitted transovarially, which could potentially be exploited in the implementation of control strategies. The densovirus of the lepidopteran *Junonia coenia* (JcDNV) has also been developed as a gene transfer vector. This chapter will concentrate on these two viruses.

8.2 BIOLOGY OF DENSONUCLEOSIS VIRUSES

Densonucleosis viruses or densoviruses are so named because of the enlarged hypertrophied appearance of the nuclei in infected cells. This is presumably due to the fact that these viruses are replicated and assembled in the nucleus. The densoviruses are members of the family Parvoviridae. The Parvoviridae consist of two subfamilies, the Parvovirinae and the Densovirinae. The Parvovirinae are the parvoviruses of vertebrates. These include the minute virus of mice (MVM) and the adeno-associated viruses (AAV), which are the most thoroughly studied members of the family. Understanding of the molecular biology of parvoviruses is mainly based on the studies of these two viruses and their close relatives. The Densovirinae are the parvoviruses of arthropods, mainly insects. There are three genera in the Densovirinae: *Densovirus, Iteravirus,* and *Brevidensovirus*. The genus *Densovirus* contains the *Junonia coenia* densovirus (JcDNV) and the *Galleria mellonella* densovirus (GmDNV). The genus *Iteravirus* contains the *Bombyx mori* densovirus (BmDNV), and the *Brevidensovirus* genus contains the *Ae. aegypti* densovirus (AeDNV) and the *Ae. albopictus* densovirus (AaPV).

The AeDNV was isolated from a laboratory colony of *Ae. aegypti,* and another *Aedes* densovirus (AthDNV) has been isolated from *Ae. aegypti* and *Ae. albopictus* colonies in Thailand. This virus has also been detected in numerous natural populations in Thailand (P. Kittayapong, personal communication). As with the mammalian parvoviruses, many of the mosquito densoviruses have been isolated as contaminants in cell culture. A presumed densovirus was detected by electron microscopy in *Ae. pseudoscutellaris* (Ap-61) cells (Gorziglia et al., 1980). The *Ae. albopictus* virus (AaPV) was isolated from *Ae. albopictus* C6/36 cells (Jousset et al., 1993). Other viruses were isolated from *Haemogogus equinus, Toxorhynchites amboinensis,* and *Culex theileri* mosquito cell lines (O'Neill et al., 1995).

The AeDNV has been thoroughly characterized at the organismal level by Buchatsky (1989). The virus infects mosquitoes of the genera *Aedes, Culex,* and *Culiseta*. Multiple tissues including fat body, Malpighian tubules, imaginal disks, muscles, and others become infected in larvae, pupae, and adults. Infected animals are killed in a dose-dependent manner. In general, the higher the titer of the infecting virus and the earlier in the life cycle the infection occurs, the greater the morbidity and mortality caused by the virus. There are several reports of transovarial transmission of mosquito densoviruses (Buchatsky, 1989; O'Neill et al., 1995; Barreau et al., 1997). Thus, it is likely that the viruses are transmitted horizontally in pools where larvae undergo development and with the virus spread from pool to pool by transovarial transmission.

8.3 MOLECULAR BIOLOGY OF DENSONUCLEOSIS VIRUSES

Parvoviruses are icosahedral particles approximately 20 nm in diameter (Figure 8.1). The particles consist of DNA and protein, lack a lipid envelope, and are not glycosylated. They are relatively resistant to extremes in pH, temperature, etc. and are relatively stable in the environment (Buchatsky, 1989). The early events in infection involving binding to the cell, entry into the cell, and uncoating of the viral genome are poorly understood. The cellular receptors for AeDNV and JcDNV are unknown.

8.3.1 DENSONUCLEOSIS VIRUS REPLICATION

The genomes of parvoviruses are linear single-stranded DNA molecules between 4000 and 6000 nucleotides in length. Depending on the virus, either the negative-sense strand (complementary to mRNA) is predominantly packaged into virus particles (e.g., AeDNV) or both strands are packaged with about equal efficiency into virus particles (e.g., JcDNV). The ends of parvoviral genomes contain terminal inverted repeat (TIR) sequences predicted to fold into T-shaped or Y-shaped secondary structures. In some viruses, such as JcDNV, the TIRs are the same on both ends of the genome. In others, such as AeDNV, the two TIRs differ in sequence. During infection, after the

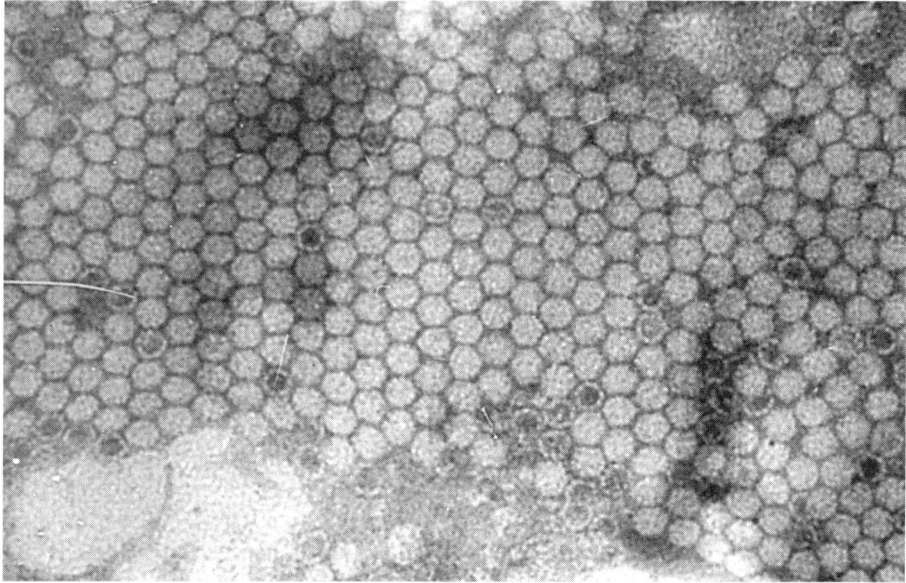

FIGURE 8.1 Electron micrograph of the AeDNV. (Courtesy of Prof. L. Buchatsky, Kiev, Ukraine.)

DNA has been uncoated, the 3' end acts as a primer for cellular DNA polymerase to synthesize a positive-sense DNA strand, complementary to the genome, thereby forming a double-stranded monomer replicative form (RF) of the parvovirus DNA (Berns, 1996). This molecule has a hairpin-like structure, and the 3' end can fold back on itself because of the TIR. Extension of the 3' end by DNA polymerase results in displacement of the original genomic strand, opening of the hairpin, and synthesis of a double-stranded head-to-head dimer RF DNA. Higher-order RF multimers can be formed by the same mechanism. The single-strand progeny genomes are presumably cleaved from the RF DNA and packaged into virus particles (Berns, 1996).

8.3.2 Densonucleosis Virus Genomic Organization

The genomes of two members of the *Brevidensovirus* genus, AeDNV and AaPV (Afanasiev et al., 1991; Boublik et al., 1994) and two members of the *Densovirus* genus, JcDNV and GmDNV (Dumas et al., 1992; Genbank accession number L32896 for GmDNV), have been completely sequenced. The genomic organizations of the members of these two genera are quite different. The two *Brevidensovirus* genomes are about 4000 bases in length, and are similar in organization to the mammalian parvoviruses in which the genes are all on one strand (Figure 8.2). Two overlapping open reading frames occupy the leftward two thirds of their genomes. These are the genes for nonstructural proteins designated NS1 and NS2. The longer reading frame encodes a protein predicted to be approximately 790 amino acids in length. The protein contains a potential nucleotide binding motif, and is the presumed equivalent of the NS1 protein of the mammalian parvoviruses. By analogy with the mammalian parvoviruses, NS1 is predicted to have site-specific nicking activity and helicase activity and to be necessary for replication and packaging of viral genomes. The reading frame predicted to encode NS2 lies completely within the NS1 reading frame in a +1 frameshift with respect to the NS1 frame. The predicted NS2 protein is approximately 360 amino acids in length. The function(s) of the NS2 proteins of mammalian parvoviruses are poorly under-

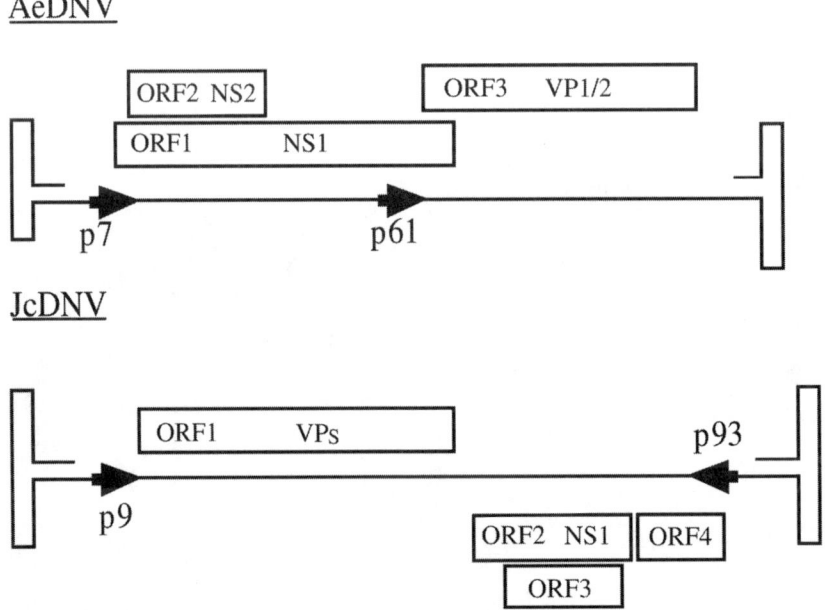

FIGURE 8.2 Genomic organization of AeDNV and JcDNV. The open reading frames indicated by open boxes are described in the text. Promoters are indicated by arrows. Predicted TIR secondary structures are not drawn to scale.

stood, and the function of the predicted AeDNV NS2 protein is completely unknown. The rightward one third of the AeDNV and AaPV genomes is occupied by an approximately 360-codon open reading frame that is predicted to be the gene for the structural proteins of the virion (VP1 and VP2). VP1 and VP2 are approximately 40,000 and 38,000 Da in molecular weight, respectively, and differ only in their amino termini. VP2 may be generated from VP1 by proteolytic cleavage, or, alternatively, the two proteins may result from initiation of translation at different initiation codons. An infectious clone of the AeDNV genome (pUCA) has been constructed by assembling a complete genome in the multiple cloning site of the plasmid vector pUC19 (Afanasiev et al., 1994). When pUCA is transfected into C6/36 cells, the genome is excised from the plasmid, presumably mediated by the NS1 protein, and an infection begins, ultimately resulting in the production of virus.

In contrast, the members of the *Densovirus* genus, JcDNV and GmDNV, have their structural and nonstructural protein genes coded on different strands of their 6000 base genomes (see Figure 8.2). A reading frame of 810 codons, predicted to encode the four structural proteins of the virion (VP1-4), occupies the 5' half of one strand of the genome. These four proteins range in size from 101,000 Da for VP1 to 49,000 Da for VP4. As in other parvoviruses, the structural proteins differ only in their amino termini. This is probably due to initiation of translation at different AUG codons or perhaps post-translational cleavage of VP1. The other half of the genome contains three open reading frames on the opposite strand from the VP reading frame, and these are predicted to encode nonstructural proteins of 545, 275, and 186 amino acids. The longest of these proteins has a predicted nucleotide binding motif and presumably functions as the NS1 protein for the virus. The function of the other two nonstructural proteins is unknown. An infectious clone (pBRJ) of the JcDNV genome has been constructed in the plasmid pBR322 (Dumas et al., 1982). When this plasmid is transfected into appropriate lepidopteran cell lines or injected into lepidopteran larvae, infectious virus is produced.

8.3.3 DENSONUCLEOSIS VIRUS GENE EXPRESSION

8.3.3.1 AeDNV Gene Expression

Analysis and comparison of the sequences of AeDNV and AaPV led to the prediction that promoters would be located at approximately 7 and 61 map units (% of nucleotide sequence) from the left end of the virus because of potential TATA sequences at these positions (Afanasiev et al., 1991; Boublik et al., 1994). Northern blot analysis of RNA from AeDNV-infected C6/36 *Ae. albopictus* cells indicated that there are two major RNA molecules of approximately 3500 and 1200 nucleotides from the AeDNV genome (Kimmick, 1998). These two transcripts are the sizes expected from the P7 and P61 promoters, respectively. The P7 transcript is the presumed mRNA for NS1 and NS2, and the P61 transcript is the presumed mRNA for the virion structural proteins VP1 and VP2. Potential polyadenylation signals are found in several places in the genome, including one just downstream of the *VP* gene near the right end. However, the sizes of the AeDNV RNA molecules suggest that the one near the right end of the genome is the primary, if not the only, one that is used.

Expression of the viral genes was studied by fusing the various open reading frames with the *Escherichia coli* β-galactosidase gene (*lacZ*) or the *E. coli* β-glucuronidase (*GUS*) gene. After transfection of the *lacZ* constructs into C6/36 cells, β-galalactosidase expression could be detected from the *NS1, NS2,* and *VP* gene fusions (Afanasiev et al., 1994; Kimmick et al., 1998). Expression could be detected either histochemically by staining with the chromogenic substrate X-gal (5-bromo-4-chloro-3-indolyl-β-D-galactoside), by quantitative spectrophotometric assay of cell lysates using the substrate ONPG (*o*-nitrophenyl-β-D-galactoside), or by a luminometric assay using the Galactolight reagent kit (Tropix). Similarly, expression of GUS fusion proteins could be detected histochemically by staining with X-gluc (5-bromo-4-chloro-3-indolyl-β-D-glucuronic acid) or by a quantitative luminometry assay using the GUSlight reagent kit (Tropix).

8.3.3.1.1 Nonstructural gene promoter

The nonstructural proteins NS1 and NS2 are both expressed from the promoter P7. The translational reading frame for NS2 is contained in the NS1 reading frame, but it is shifted by one nucleotide. Quantitation of the levels of expression of NS1 and NS2 *lacZ* fusion proteins indicated that both proteins were expressed in comparable amounts (Kimmick et al., 1998). The presumed NS1 AUG initiation codon is just downstream of the predicted 5' end of the P7 transcript. The presumed NS2 AUG initiation codon is 73 nucleotides downstream of the NS1 AUG (Figure 8.3a). There are no consensus RNA splice sequences in this portion of the AeDNV genome, so it appears that translation must initiate from the P7 mRNA at both AUG initiation codons with approximately equal efficiency. This is relatively unusual for eukaryotic mRNAs, since they are usually monocistronic and the first AUG in the mRNA is usually where translation begins. Interestingly, reporter genes could only be expressed from P7 as gene fusions with *NS1* or *NS2*. When a *GUS* gene was inserted downstream of the P7 promoter in such a way that it had to be translated from its own AUG initiation codon, no detectable GUS protein expression was detected after transfection into C6/36 cells. Although GUS protein production was reduced more than 100-fold relative to an *NS2–GUS* gene fusion

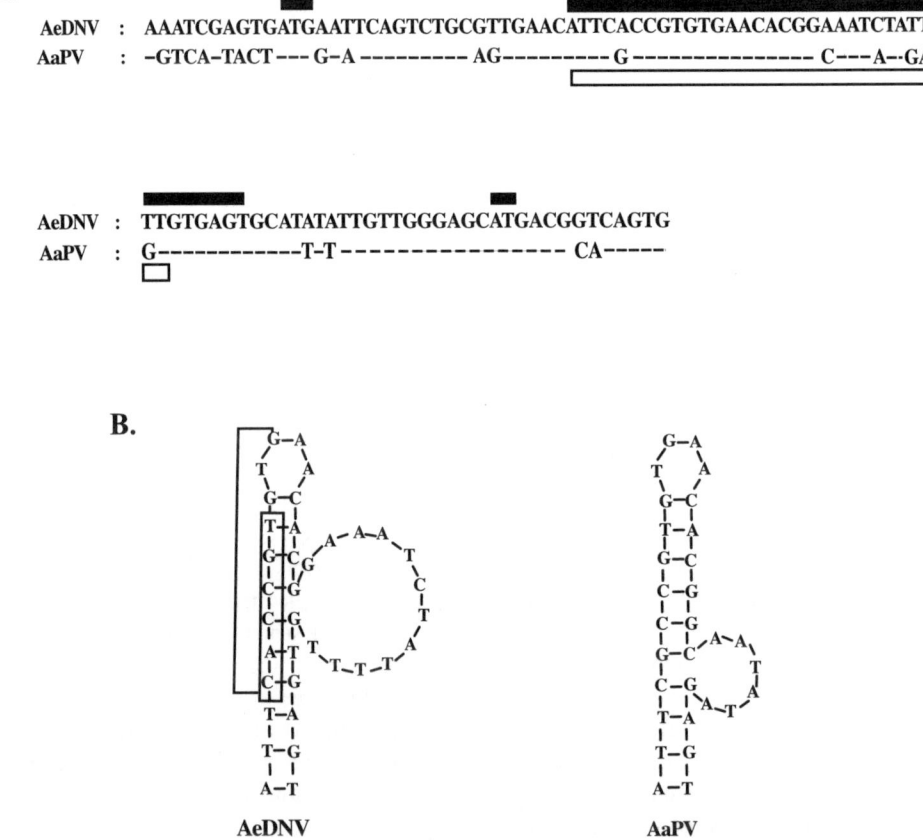

FIGURE 8.3 Structure of the P7 mRNA. (a) The sequences of the AeDNV and AaPv genomes in the regions of the predicted 5' ends of the P7 transcripts are aligned with one another. The conserved putative translational initiation codons for NS1 and NS2 are overlined by short filled boxes. The sequences predicted to fold into secondary structures are indicated by long filled and open boxes for AeDNV and AaPV, respectively. (b) The predicted secondary structures are indicated. In the AeDNV secondary structure, the *Dra*III restriction site used for mutagenesis to disrupt the secondary structure is indicated by a bracket. The six bases deleted by mutagenesis are boxed.

construct, P7 transcript levels, as measured by RNase protection, differed by less than a factor of two (Kimmick et al., 1998). Since this construct differed from the *NS2–GUS* construct only in the sequences between the 5'end of the mRNA and the GUS coding sequence, a sequence just downstream of the 5' end of the P7 transcript was apparently necessary for translation of the P7 mRNA. Comparison of the AeDNV and AaPV sequences showed that sequences predicted to fold into a stem–loop secondary structure with a bulge in the stem are present in both viruses in the 73 nucleotides between the NS1 and NS2 AUG codons (Figure 8.3b). Site-directed mutagenesis of this sequence produced a 6-base deletion that is predicted to disrupt the secondary structure. This mutation drastically reduced expression of both NS1 and NS2 fusion proteins. Although protein expression is dramatically reduced by the deletion, the amount of RNA transcribed from P7 is not affected. Thus, there appears to be a novel translational regulation element requiring an RNA secondary structure in the 5' portion of the RNA transcript that controls expression from the P7 promoter (Kimmick et al., 1998).

8.3.3.1.2 Structural gene promoter
Similar analysis using *VP–lacZ* fusion gene constructs is under way to characterize the structural gene promoter. Initial predictions based on nucleotide sequence analysis suggested that the structural gene promoter was likely to be defined by a TATA-box sequence at 61 map units. However, recent experiments using oligonucleotide primer extension by reverse transcriptase indicate that the 5' end of the structural protein mRNA is actually about 80 nucleotides upstream of the predicted site at map unit 59 (T. Ward, unpublished results). This site is just downstream of another potential TATA-box sequence and also has a CAGT sequence characteristic of transcription initiation sites of a large number of arthropod transcription units (Cherbas and Cherbas, 1993). Deletion analysis of the promoter is consistent with this interpretation since removal of a restriction fragment containing the initiation site destroys the activity of the structural gene promoter. Unlike P7, both fusion and nonfusion proteins can be efficiently expressed from the structural gene promoter.

8.3.3.1.3 Transactivation of AeDNV promoters
Constitutive basal expression was easily detectable from both the P7 and P61 promoters using the reporter gene fusion constructs described above. In mammalian parvoviruses the structural gene promoter is strongly induced by the NS1 protein. Cotransfection of the *NS1, NS2,* and *VP* reporter gene fusion constructs with complementing plasmids carrying the AeDNV nonstructural protein genes *NS1* and *NS2* resulted in transactivation of expression of all the fusion protein genes (Afanasiev et al., 1994; 1999). This transactivation can be mediated by the infectious clone pUCA or by nonreplicating constructs capable of expressing only the nonstructural proteins. A plasmid designed to express only NS2 does not mediate transactivation. Thus, as in mammalian parvoviruses, the NS1 protein is involved in both replication and gene expression. The amount of transactivation depended on the details of the experiment, but generally ranged between 7- and 20-fold. Although both the P7 and P61 promoters are inducible in the appropriate context, the minimal constitutive P61 promoter is not transactivatable by NS1. However, transactivatability can be restored by adding the left, right, or both TIR sequences of the virus back to the construct (T. Ward, unpublished results). These sequences are likely to contain NS1 binding sites since they are involved in NS1-mediated replication and packaging functions. The molecular mechanism of NS1-mediated transactivation of the promoters is as yet unknown.

8.3.3.2 JcDNV Gene Expression

The structural and nonstructural protein genes of JcDNV are expressed from promoters defined by TATA-box sequences located just inside the TIR sequences on either end of the genome. According to a preliminary report on the closely related GmDNV, two major transcripts of approximately 1.8 and 2.4 kb encode the nonstructural and structural proteins, respectively (Tal and Attathom, 1993).

This suggests that the transcription units on either strand extend from the promoters near the genomic ends to predicted polyadenylation signals in both strands near the center of the genome.

Expression from the structural gene promoter of JcDNV has been studied by construction of a *lacZ* gene fusion construct with the *VP* gene (Giraud et al., 1992). The site of the fusion of the two genes was downstream of the VP4 AUG initiation codon. When this fusion construct was transfected into appropriate lepidopteran cells, β-galactosidase fusion proteins were produced that had amino termini corresponding to VP2, VP3, and VP4. Expression of β-galactosidase from the structural gene promoter did not require nonstructural protein expression but it was higher in the presence of them. Thus the expression of the structural proteins seems similar to AeDNV and other parvoviruses. Expression from the JcDNV nonstructural gene promoter has not been studied in detail, but a construct designed to express chloramphenicol acetyl transferase (CAT) from the GmDNV nonstructural promoter has been made. When this construct was injected into lepidopteran larvae, CAT expression was observed (Tal and Attathom, 1993).

8.4 DEVELOPMENT OF DENSONUCLEOSIS VIRUS TRANSDUCING SYSTEMS

Several parvoviruses have been used to develop transducing systems including AAV, MVM, and LuIII, as well as the densoviruses reviewed here (Corsini et al., 1996a). The same general strategy has been followed for all of these systems. A transducing genome is constructed based on the infectious clone by replacing viral genes with the gene of interest (*GOI*) while retaining the TIRs that are required for replication and packaging of the genome. The products of the genes that are replaced by the *GOI* must be supplied by a helper construct. When both the transducing genome and the helper construct are introduced into the same cell, the two constructs complement one another and produce transducing particles that contain the transducing genome packaged in the virus coat.

AeDNV and JcDNV transducing genomes have been constructed in which the nonstructural protein genes or the structural protein genes or both have been replaced by reporter genes (Afanasiev et al., 1994; 1999; Giraud et al., 1992). Since the virus particle is icosahedral, there is a finite limit to the length of DNA that can be packaged in a particle. In the case of AeDNV, *lacZ* fusion constructs such as p61*lacZ* (Figure 8.4), which retained the left and right ends could be packaged into virus particles if their length did not exceed the wild-type length by more than about 8% (Afanasiev et al., 1994). These transducing particles could then deliver the *lacZ* gene to fresh cells, and β-galactosidase expression could be detected by histochemical staining with X-gal. This process could be blocked by antiserum against the virus, thus demonstrating that it was actual particle-mediated transduction (Afanasiev et al., 1994). In the case of JcDNV, a *lacZ*-containing transducing genome of approximately 9 kb (~50% larger than the wild-type JcDNV genome) was apparently packaged and delivered to fresh cells (Giraud et al., 1992). Whether this is due to a significantly larger volume of the virus particle, or an alternative particle structure for JcDNV is not known.

One particularly useful transducing genome (p7NS1–GFP) was constructed by fusing the gene for the green fluorescent protein (GFP) from the jellyfish *Aequorea victoria* in frame to the last codon of the AeDNV *NS1* gene, thereby replacing the AeDNV *VP* gene (see Figure 8.4). The NS1–GFP fusion protein produced by this construct appears to have all of the properties of the normal NS1 protein (Afanasiev et al., 1999). It is capable of mediating the excision of the transducing genome from its plasmid vector when transfected into mosquito cells, and it transactivates both of the AeDNV promoters. When provided with a complementing helper providing VP, the NS1–GFP fusion protein mediates packaging of the transducing genome into particles. The fusion protein is localized to the nuclei in transduced cells, as is the case with other parvoviral NS1 proteins, and transduced cells are readily detected while still living by fluorescence microscopy.

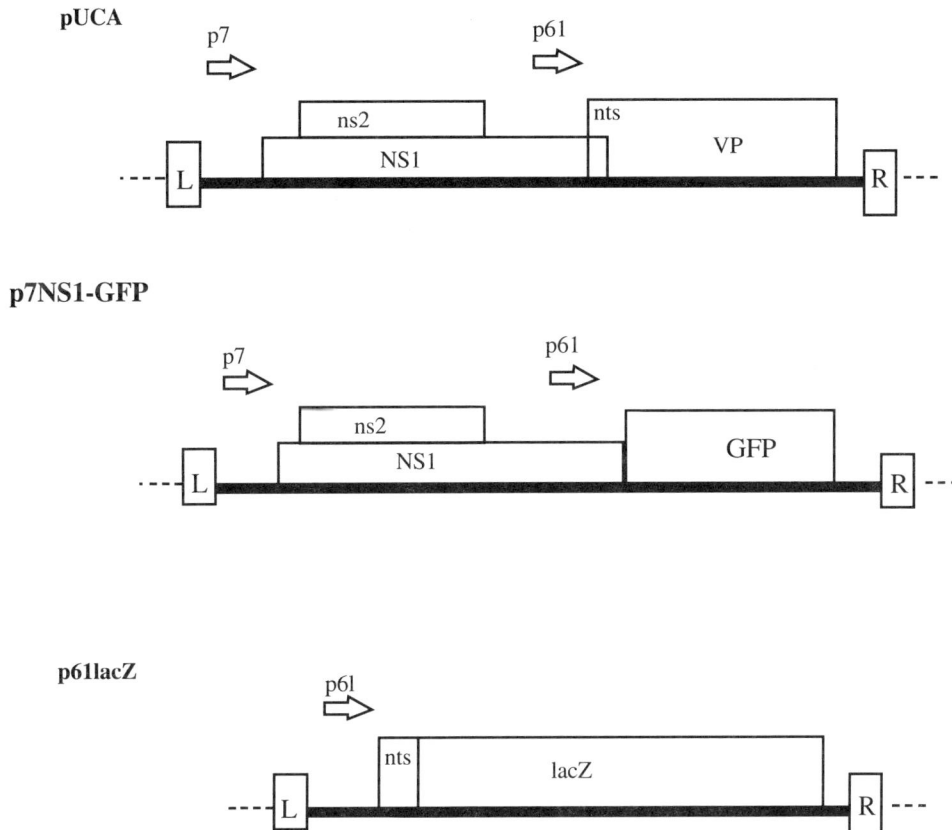

FIGURE 8.4 Structures of AeDNV transducing genomes and the pUCA helper. The left and right TIR sequences are depicted as boxes containing L or R, respectively. Plasmid vector sequences are shown as broken lines. The nuclear targeting signal of the VP protein is designated as nts.

The nature of the helper used in transducing experiments depends on the design of the experiment, and the design of the transducing genome. The full infectious clone pUCA (see Figure 8.4) can supply the nonstructural proteins (NS1 and NS2), and the virion structural proteins (VP), and can therefore act as a helper for any transducing genome. All of these proteins are necessary for excision, replication, and packaging of a genome in which all of the viral genes are replaced by the *GOI*. A potential drawback of pUCA as a helper for some experimental applications is that the transducing particle preparation will be contaminated with wild-type AeDNV particles because these are likely to be produced at least as efficiently as transducing particles. This wild-type viral contamination can be significantly reduced by removing the TIR sequences from the infectious clone. This should prevent excision of the helper from the plasmid vector and subsequent replication and packaging. If the transducing genome still retains intact *NS1/NS2* genes or intact *VP* genes, a helper construct that provides only the missing function is sufficient for generation of transducing particles. Recombination between transfecting plasmids resulting in generation of viable "wild-type" virus is easily demonstrated when cotransfection of mutually complementing plasmids is performed. Thus, one must always assume that preparations of transducing particles produced by cotransfection will contain some wild-type virus.

The conventional method of producing AeDNV transducing particles is by cotransfection of C6/36 (*Aedes albopictus*) cells with the plasmid carrying the transducing genome, and a helper plasmid which supplies the missing viral genes. While this results in the production of an easily detectable number of transducing particles (10^3 to 10^4), it is limited by the frequency of cotransfection, and more efficient methods are continually being sought. Our laboratory has investigated several different mosquito cell lines to see if any of them were more permissive for growth of the virus and therefore potentially a better producer of transducing particles. These included AP61 (*Ae. pseudoscutellaris*), ATC10 (*Ae. aegypti*), and MOS55 (*Anopheles gambiae*). Unfortunately, none was superior to C6/36 in the production of virus.

8.4.1 CONSTRUCTION OF DENSONUCLEOSIS VIRUS PACKAGING CELL LINES

Production of recombinant retroviruses is facilitated by the availability of packaging cell lines that constitutively produce components of the virus that are missing from the retroviral vector construct. These packaging lines are constructed by stable transformation of appropriate cell lines with recombinant constructs designed to express the desired viral component. We have also constructed packaging cell lines for AeDNV (unpublished). These are C6/36 cells stably transformed with helper DNA plasmids designed to express either the virus capsid proteins (VP) or both the VP and the nonstructural proteins NS1 and NS2. Stably transformed lines were derived by including a selectable marker such as a gene conferring resistance to an antibiotic such as hygromycin or zeocin on the helper plasmid. Cells that are capable of continuous growth in medium containing the appropriate antibiotic carry the plasmid as part of their genome. These cell lines produce viral capsid protein since they are positive for viral antigen by immunofluorescence assay using antibody to AeDNV particles. These lines would be predicted to rescue, replicate, and encapsidate transducing genomes after they are transfected into the cells. Although transducing particles were produced from some of these lines, more were produced by the cotransfection procedure described above. Thus, at the present time no efficient packaging cell line for densoviruses is available.

8.4.2 PACKAGING WITH A SINDBIS VIRUS HELPER

Production of transducing particles by the cotransfection method results in generation of wild-type virus as well as transducing particles because of recombination between transducing genomes and helper constructs. For some applications, pure preparations of transducing particles without virus contamination are desirable. One method of eliminating recombination resulting in wild-type virus would be to supply one of the genes necessary for helper function as RNA rather than DNA. Sindbis virus expression vectors are well suited for these applications since they readily infect mosquito cells, and establish persistent rather than lytic infections in mosquito cells (Xiong et al., 1989; Higgs et al., 1993; Bredenbeek et al., 1993). Furthermore, it is possible to infect essentially 100% of all cells in a culture if the multiplicity of infection is high enough. The level of expression of the gene of interest by Sindbis expression vectors compares favorably with DNA-based expression vectors. Plasmid clones of the Sindbis genome are available for production of recombinant Sindbis genomes. Full-length Sindbis RNA transcribed from these plasmids can be transfected into mammalian cells (BHK-21) to produce infectious Sindbis virus. Two types of Sindbis vectors are available, replicon vectors and double subgenomic vectors (Xiong et al., 1989; Higgs et al., 1993; Bredenbeek et al., 1993). In the case of the replicon vectors the gene of interest replaces the genes for the Sindbis structural proteins, and is transcribed from the promoter for the subgenomic RNA. Packaging of the replicon into Sindbis virus particles requires that the Sindbis structural proteins be supplied by a complementing construct. The double subgenomic (dsSIN) vectors contain all of the Sindbis genome plus a second subgenomic promoter that transcribes the gene of interest. The dsSIN vectors are therefore capable of producing viable recombinant virus particles. Due to the packaging limits of the Sindbis particle, the dsSIN vectors are restricted to *GOI* inserts of less than ~1.5 kb, while the replicon vectors can express *GOI* of up to approximately 5 to 6 kb.

The use of Sindbis vectors to supply helper functions for production of parvovirus transducing particles was initially tried with the mammalian parvovirus LuIII (Corsini et al., 1996b). The LuIII

NS1 protein was delivered with a SIN replicon vector to mammalian cells along with a plasmid carrying a LuIII transducing genome and a second plasmid providing LuIII VPs. Production of low amounts of LuIII transducing particles was detected. Optimization of this three-component system was complicated by the fact that Sindbis and LuIII NS1 are toxic to mammalian cells.

For use of Sindbis as a helper in the AeDNV system, another strategy was tried. A double subgenomic Sindbis virus vector that expresses the AeDNV VPs (TE/3'2J/VP; Figure 8.5) was constructed (Allen-Miura et al., 1999). This vector expressed AeDNV capsid antigen in both BHK-21 mammalian cells and C6/36 mosquito cells as assessed by indirect immunofluorescence assay using rabbit antiserum against AeDNV. Interestingly, in the C6/36 cells, the antigen accumulated in the nuclei of the cells as expected for densovirus antigen, while in the BHK-21 cells the antigen remained in the cytoplasm. This suggests that there is a significant difference between the nuclear localization signals of mammals and mosquitoes. This would also represent a significant block to productive replication of AeDNV in mammalian cells.

The TE/3'2J/VP helper should be effective in packaging any transducing genome with an intact *NS1* gene and the *GOI* replacing the *VP* gene. One such transducing genome is p7NS1–GFP, which has the GFP gene fused to the carboxy-terminus of the AeDNV NS1 protein (see Figure 8.4). When cells are transfected with the plasmid carrying the p7NS1–GFP transducing genome and then infected with the TE/3'2J/VP virus, transducing particles are produced in numbers that are comparable with the two-plasmid system (Allen-Miura et al., 1999). Optimization of this system is under way to determine its capabilities for production of transducing particles.

TE/3'2J/VP

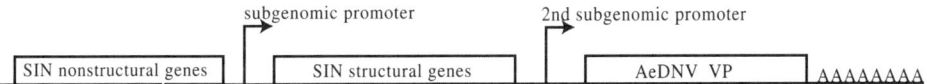

FIGURE 8.5 Structure of the TE/3'2J/VP Sindbis helper viral genome. The locations of the two subgenomic promoters are shown by arrows. The subgenomic mRNAs extend from the respective promoters to the 3' poly A.

8.4.3 TRANSDUCTION OF MOSQUITO LARVAE

Mosquito larvae can be readily infected by densovirus grown in cell culture by introducing larvae into the medium from the infected cells (Barreau et al., 1994). This protocol can also be used for exposing larvae to transducing particles. Alternatively, infected or transfected cells can be lysed by freezing and thawing or by sonication, and virus or transducing particles can be concentrated by centrifugation. Concentrated virus or transducing particles can then be introduced into the water in which the larvae are reared. Infection by these procedures in the laboratory probably mimics a natural infection since larvae are thought to be infected in their rearing sites in nature. The demonstration of transduction of larvae was initially attempted by packaging transducing genomes carrying the *lacZ* gene or the *GUS* gene. However, convincing demonstration of transduction in whole larvae by *lacZ* or *GUS* transducing particles was unsuccessful because endogenous background β-galactosidase and β-glucuronidase activities obscured the detection of the transduced enzymes. Successful transduction of living larvae was demonstrated with particles carrying the p7NS1–GFP transducing genome. Fortunately, mosquito larvae are relatively transparent, and transduction was readily observed in living larvae by visualization of GFP by fluorescence microscopy (Afanasiev et al., 1999; see Color Figure 2*). The natural route of infection had always been assumed to be oral through the digestive tract. Thus, the first cells to show evidence of transduction would be expected to be the cells of the gut. Although transduction of midgut cells has occasionally been detected, the

---

* Color Figure 2 follows p. 108.

first signs of transduction are most frequently noted in the anal papillae (Figure 8.6a). In some experiments the cells at the base of bristles are also frequently transduced. This was unexpected and may indicate new routes of infection for this virus. Most of the preparations of transducing particles used in these experiments were made using pUCA as the helper and, therefore, contain large amounts of wild-type virus as well as the transducing particles. Coinfection of a cell with a transducing particle and a wild-type virus would allow packaging of newly replicated transducing genomes, and subsequent spread of transducing particles to secondary target tissues. Indeed, many of the transduced larvae go on to show more-disseminated infections with GFP fluorescence evident in many tissues such as fat body, Malpighian tubules, muscles, and others (Figure 8.6b). Preparations of particles made with the Sindbis helper or the packaging cell line have, so far, only shown evidence of infection in anal papillae, in support of our hypothesis that they are primary routes of infection.

The structure of a papilla is a single layer of cells with the chitinous cuticle on one side and the hemocoel on the other side. In mosquito larvae the layer of cells is a syncytium with all the cells connected by cytoplasmic bridges (Garrett et al., 1984). As a consequence of the syncytial structure, the NS1–GFP fusion protein and probably the p7NS1–GFP transducing genome can be spread to all of the nuclei in the syncytium without packaging into virus particles and reinfection. This probably accounts for the fact that usually all the nuclei of a transduced papilla fluoresce brightly. One of the functions of the anal papillae is to regulate ion concentrations in the hemolymph. The syncytium of cells forming a papilla has a high surface area in contact with the hemolymph on the inside and the cuticle and aqueous environment on the other side. It is therefore ideally situated for mediating the trafficking of ions between the environment and the hemolymph. The cuticle covering the anal papillae may be comparatively thin and perhaps can allow passage of virus particles as well as ions. This route of infection may be exploitable for transducing genes whose products could be delivered to many parts of the insect by the circulatory system. Similarly, the sites where bristles come through the cuticle may have gaps in the cuticle large enough to allow passage of the virus particle and subsequent infection of the underlying cells. Larvae may also be more susceptible to infection during molting between instars.

Larvae that show evidence of a disseminated transduction/infection often exhibit symptoms of pathogenesis, and many die. However, some pupate and a few eclose as adults. Generally, many tissues exhibit green fluorescence in these individuals. To date, transduction has been detected in larvae of *Ae. aegypti* and but not in *Ae. triseriatus*. This is somewhat unexpected since the reported host range of AeDNV also includes species from the genera *Culex* and *Culiseta* (Buchatsky, 1989). Unexpectedly, when *Anopheles gambiae* larvae were exposed to AeDNV NS1-GFP transducing particles, transduction was observed (M. Edwards and M. Jacobs-Lorena, personal communication). In the case of the *Anopheles* larvae, transduction was only observed in the anal papillae and bristle cells. No evidence of disseminated infection was seen. Clearly, there are large gaps in knowledge regarding the host range, the tissue tropisms, the nature of densovirus receptor(s), and the distribution of virus receptors within and among mosquitoes. Nevertheless, we have demonstrated that densoviruses can be used to transduce living mosquitoes and that this can be accomplished without resorting to artificial means of introduction such as microinjection.

8.5 DENSONUCLEOSIS VIRUSES AS POTENTIAL TRANSFORMATION VECTORS FOR INSECTS

Adeno-associated virus (AAV) is not capable of a productive infection of most cell types without coinfection with an adenovirus or herpesvirus as a helper virus. The functions that are missing in AAV and are provided by the helper are poorly understood (Berns, 1996). When human cells are infected with AAV in the absence of helper, the AAV genome has been shown to integrate into a specific site on chromosome 19 of the human genome. This ability is highly attractive for human gene therapy applications, and the process has been investigated in some detail during the last few

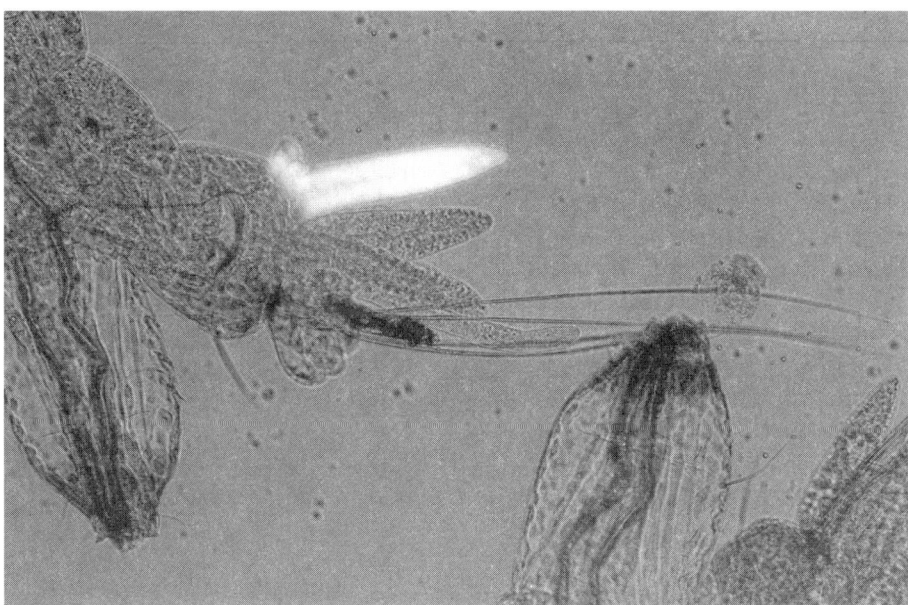
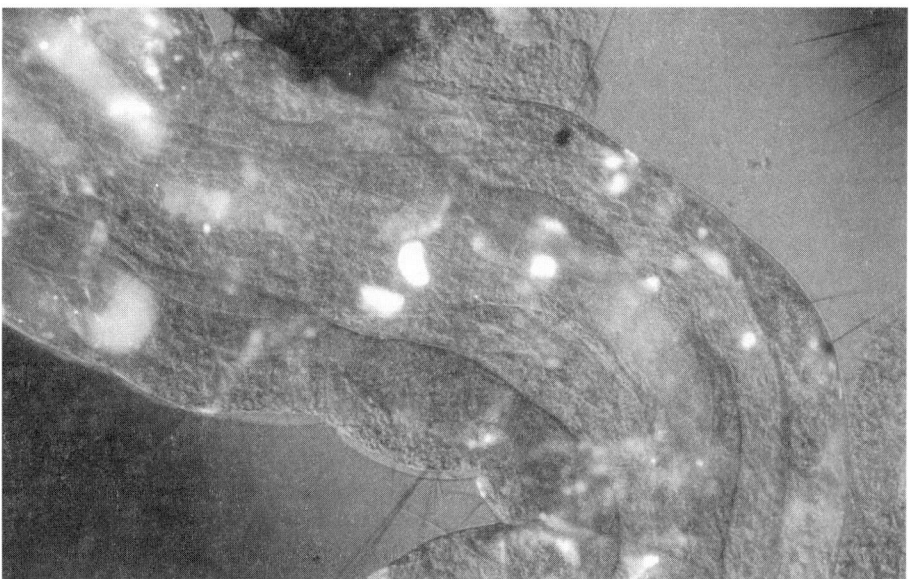

FIGURE 8.6 Transduction of larvae by transducing particles containing the p7NS1–GFP genome. (a) The bright object in the upper center part of the figure is an anal papilla showing strong green fluorescence. (b) Several fluorescent (some with brightly fluorescent nuclei) cells are evident in the center of the figure. Weakly fluorescent muscle fibers are seen in the upper left corner. See also Color Figure 2.*

years (Kotin, 1994; Linden et al., 1996a; 1996b). Site-specific integration seems to require the AAV rep protein (equivalent to NS1), and a sequence in the cellular genome that resembles an AAV replication origin. The rep protein presumably targets the integration to the chromosomal "origin," and may participate in the integrative recombination event in some way. Integration of recombinant

---

* Color Figure 2 follows p. 108.

AAV vectors lacking the *rep* gene also occurs; however, the site specificity is lost and the insertions occur at random sites in the genome (Yang et al., 1997). This seems to involve interactions of the hairpin end structures of the viral genome with unknown cellular recombination machinery, and does not involve NS1.

Other parvoviruses are not thought to integrate into their host cell genomes. However, recently, it was shown that the minute virus of mice (MVM), an autonomous parvovirus, is capable of integrating into a cloned MVM origin maintained on an episomal plasmid in cultured cells (Corsini et al., 1997). The targeted integration required the MVM NS1 protein. Although the system used to demonstrate MVM integration is highly artificial, it suggests that, under the appropriate conditions, integration events involving parvovirus replication origins can be mediated by parvoviral NS1 proteins. Can similar recombination/integration events be demonstrated with densoviruses? Studies with JcDNV transducing genomes carrying the *lacZ* gene indicated that they can persist in cell cultures for at least 25 subcultures (Giraud et al., 1992). This suggests that there is some mechanism for stable maintenance of the transducing genome, although the state of the DNA, integrated or episomal, was not determined.

8.5.1 Potential for Integrative Transformation by AeDNV

Are AeDNV transducing genomes capable of integrating into the mosquito genome? To address this question, a transducing genome containing a dominant selectable marker was constructed in the plasmid pUC19. This construct, p7hyg (Figure 8.7a), contains the viral left and right ends, but all of the viral genes were deleted and replaced with the hygromycin phosphotransferase gene (*hyg*) fused to the *NS2* gene and transcribed from p7 promoter. Thus far, it has not been possible to isolate hygromycin-resistant cell lines by infection with transducing particles carrying the p7hyg genome, generated by cotransfection of the plasmid carrying the p7hyg genome with helper constructs. The efficiency with which this genome is packaged into particles is unknown, so the reason for the lack of hygromycin-resistant transductants could be due to low titers of transducing particles. In an attempt to eliminate putative packaging efficiency problems and answer the question of whether the p7hyg genome can integrate, C6/36 *Ae. albopictus* cells were transfected with p7hyg. Transfections were performed both with and without a helper vector that provided the NS1 protein required for excision of the *hyg* transducing genome from the plasmid, replication, and perhaps integration into the mosquito genome. Hygromycin-resistant colonies were isolated and expanded in medium containing hygromycin, and several cell lines were developed. The generation of hygromycin-resistant cell lines did not depend on the presence of the helper plasmid supplying NS1.

The copy numbers of plasmid in these lines, estimated by Southern blot hybridization, ranged widely from several thousand to one. Some of the cell lines contained extremely high copy numbers of plasmid arranged in tandem linear arrays that formed double minute chromosomes similar to those previously described (Monroe et al., 1992). To characterize integration events, one line was chosen, which showed five apparent integration events on Southern blot. A genomic

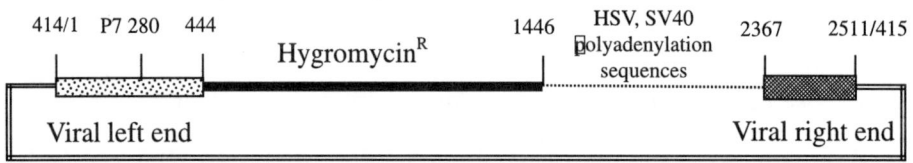

FIGURE 8.7a The plasmid containing the p7hyg transducing genome. The transducing genome is shown as a linear structure from nucleotide 1 to 2511 with the various components of the clone designated by different shading. The pUC19 plasmid vector is shown as a double line.

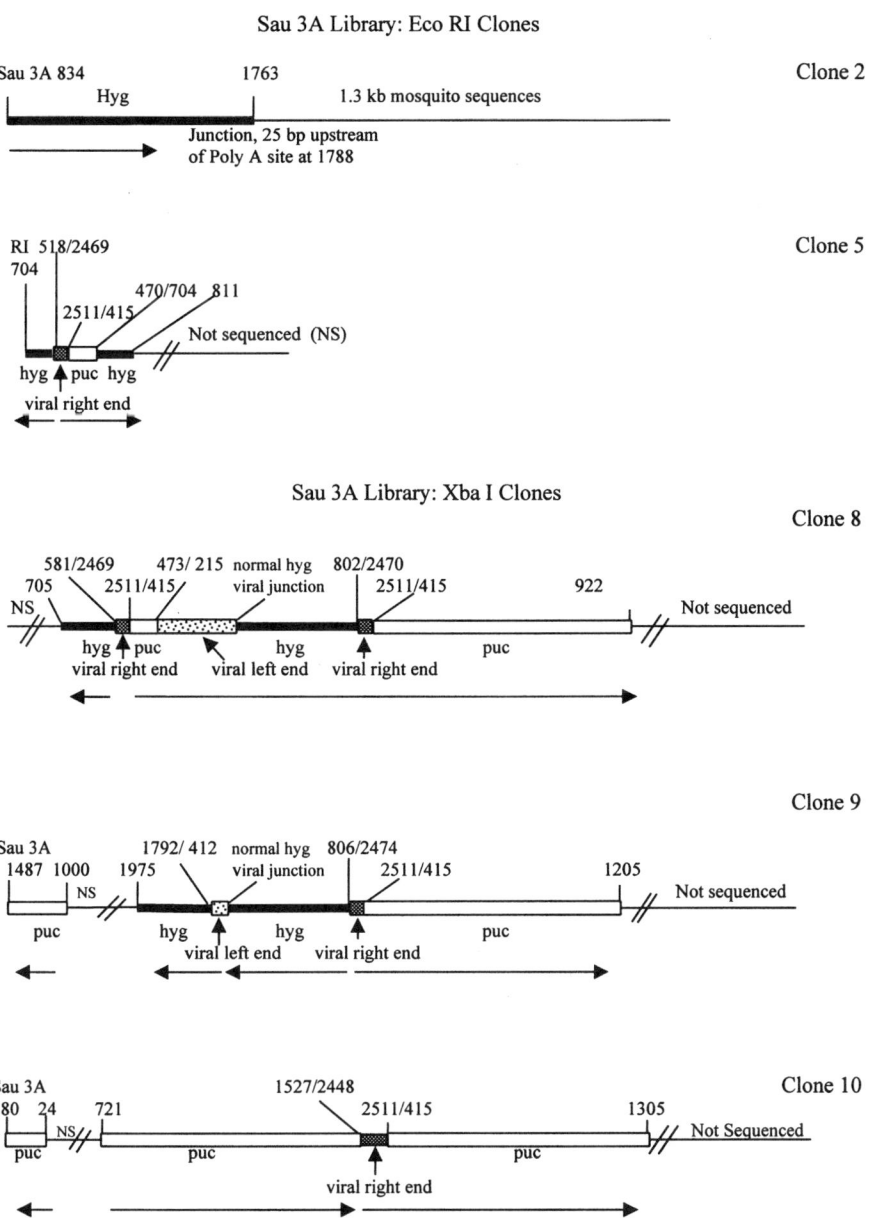

FIGURE 8.7b Summary of partial sequence analysis of five subclones described in the text. The shading of the sequenced portions of the subclones is the same as in Figure 8.7a. Clones 2 and 5 were created by using *Eco*RI; clones 8, 9, and 10 were created by using *Xba*I. Clones 5, 8, 9, and 10 contained viral sequences in which the right end is truncated between bp 2448 and 2470. Clones 8 and 5 contained the same rearrangement of the right end, although the other parts of these two clones look substantially different.

library was constructed from the DNA of this line in a lambda phage vector. This library was screened by plaque hybridization with a p7hyg-specific radiolabeled probe; 1.6×10^6 plaques were screened and six positive clones were isolated. These plaques were then hybridized to a densovirus-specific probe to determine whether they contained any viral sequences and five of the six hybridized to the densovirus probe. Fragments from these lambda clones were then subcloned into pUC19 for sequencing.

Figure 8.7b summarizes partial sequence analysis of five of the subclones. Clone 2 appears to be a junction fragment from an integration event involving illegitimate recombination. It contains about 930 bp of the *hyg* gene and 1.3 kb of sequence that was not in the p7hyg construct, and is therefore probably mosquito DNA. The right end of the virus is missing and the integrative recombination event occurred after the end of the coding region of the *hyg* gene, but 25 bp before the polyadenylation signal. Clone 5 contains a rearranged virus in which pUC19 sequences at the *Eco*RI site at 470 are joined to the *Eco*RI site at 704 in the virus. This led to worry that this was an artifact of subcloning. However, this same rearrangement has also been obtained by inverse PCR cloning, so we believe it is correct. All subsequent subcloning was done using *Xba*I, which does not cut within the viral construct, thereby eliminating this potential problem. Clones 8, 9, and 10 were found to contain greatly rearranged versions of the original p7hyg construct, and they all contained pieces of the pUC19 cloning vector. As of yet no junction fragments with mosquito DNA have been found in these clones. Most of the rearrangements appear to be illegitimate recombination events involving the right end of the virus. Figure 8.7c shows the sites of these recombination events as they would appear on the hairpin loops predicted to form in the TIR of the right end. Since these recombination events occurred within the terminal hairpins of the virus, NS1 may have been involved, but this remains to be proved. The transducing genomes in these four clones appear not to have been properly excised from the pUC19 cloning vector during the initial transfection. It is possible that these clones are the products of aberrant excision events. Little is known about the mechanism of excision of parvovirus genomes from plasmids aside from the fact that NS1 is somehow involved. The fraction of transfecting plasmids that undergo an excision event resulting in production of virus is also unknown. It would be interesting to determine whether liberating the virus genome from the plasmid vector prior to transfection affects rearrangement and/or integration of this transducing genome.

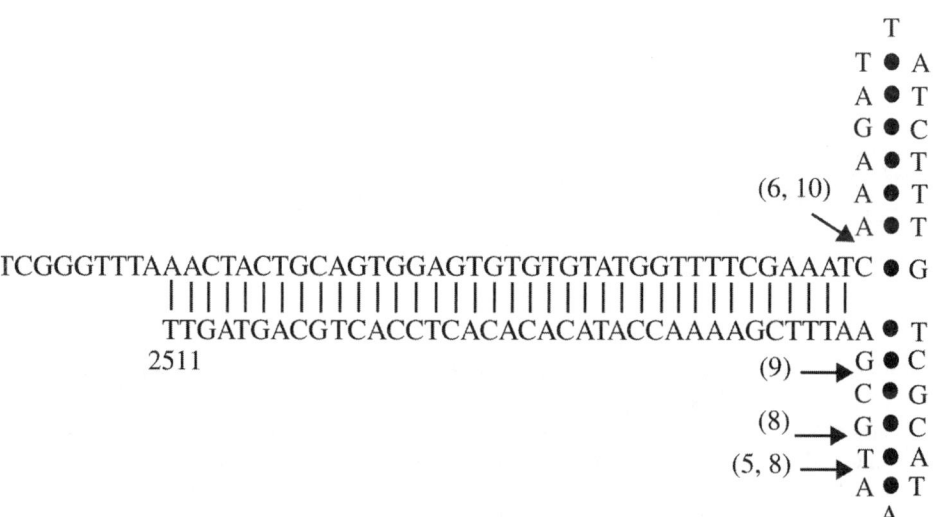

FIGURE 8.7c The sites where illegitimate recombination events occurred between the right end of the transducing genome and other sequences in clones 5, 8, 9, and 10. The sites are shown on the hairpin loops predicted from the sequence of the right TIR.

The rearrangements of the transducing genome may not be surprising in light of experiments in which a number of integration events of AAV were examined. Although AAV integrates into the human genome in a site-specific manner, the majority (34 of the 43, or 79%) of AAV integration events were shown to have deletions and/or rearrangements in the AAV sequences (Giraud et al., 1995).

Site-specific integration of AAV into the human genome requires a target sequence that is similar to a virus replication origin. The experiments with the MVM system required a cloned origin sequence for integration. Could cloned AeDNV origin sequences be introduced into mosquito cells and act as target sites for densovirus integration? Mosquito cells transformed with high copy numbers of the plasmid containing the p7hyg transducing genome were created in the experiment described above. Could the virus ends of the cloned p7hyg transducing genome act as target sites for integration? In an attempt to answer this question, the p7NS1–GFP transducing genome was modified to include a kanamycin resistance gene with a bacterial promoter (Figure 8.8). The ampicillin-resistance gene was also deleted from the plasmid vector carrying the modified transducing genome. One of the hygromycin-resistant cell lines that was found to contain the p7hyg double minute chromosomes was transfected with this new construct, p7NS1–GFP–Kan–Δamp. Efficiency of transfection was assessed by detection of GFP expression by fluorescence microscopy. High-molecular-weight DNA was isolated from transfected cells. This DNA was digested with $XmnI$, which cuts only once in the ampicillin gene of the p7hyg construct, and not at all in p7NS1–GFP–Kan–Δamp. This digested DNA was then ligated at low concentration to encourage formation of circles, and then used to transform $E.\ coli$. Transformed cells were plated on medium containing both ampicillin and kanamycin, thereby selecting for events where the p7NS1–GFP–Kan–Δamp virus genome had integrated into the p7hyg plasmid array. Although three kanamycin- and ampicillin-resistant colonies were isolated, and sequenced, all three integration events appear consistent with either homologous recombination between the pUC19 sequences found in each construct, or perhaps with precise site-specific integration into the densovirus end sequences. Attempts to minimize the possibility of homologous recombination between plasmids are currently under way.

In summary, there is no evidence at this time for site-specific integration of AeDNV-based constructs in the genome of C6/36 $Ae.\ albopictus$ cells, and it seems unlikely that densoviruses integrate into mosquito genomes under natural circumstances. Extrapolating from the AAV and MVM results described above, site-specific integration of densovirus transducing genomes is likely to require a site that resembles a densovirus replication origin in the host genome. If such a site could be introduced into the mosquito genome by artificial means such as transposon-mediated transformation, this might be a target for NS1-mediated site-specific recombination into the genome. Thus, densoviral integration seems potentially equivalent to the Cre-loxP and FLP-FRT site-specific recombination systems (Carlson et al., 1995), and it may be useful for getting repeatable site-specific integration events, thereby reducing position effects seen with random integration. Further development of a densovirus-based site-specific integration system will require substantially increased knowledge regarding the mechanism of action of the NS1 protein and its potential cytotoxic effects on cells.

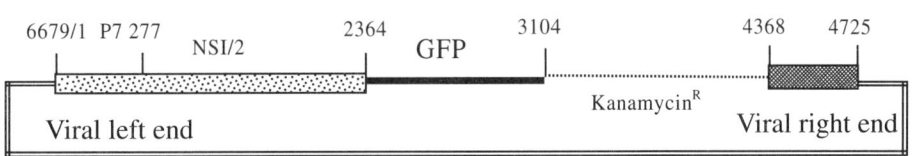

FIGURE 8.8 The plasmid p7NS1–GFP–Kan–Δamp. The p7NS1–GFP–Kan transducing genome extends from nucleotide 1 to 4725 with the various components described in the text shaded differently. The pUC19–Δamp plasmid vector is depicted as a double line.

8.5.2 POTENTIAL FOR EPISOMAL TRANSFORMATION

Could autonomously replicating vectors based on the relatively well characterized parvovirus replication machinery be used to transform insects without integration? Several instances of persistent inapparent infections of cultured mosquito cells by densoviruses have been reported. O'Neill et al. (1995) have isolated putative densoviruses from a *Toxorhynchites amboinensis* cell line, a *Culex theileri* line, and a *Haemagogus equinus* line. The *Ae. albopictus* parvovirus (AaPV), was isolated from a stock of persistently infected C6/36 cells. Although it is pathogenic for *Aedes* mosquitos, this virus had maintained itself in C6/36 cells for at least 4 years without causing apparent cytopathic effect (Jousset et al., 1993).

To test the ability of AeDNV to establish persistent infections, C6/36 cells were infected with AeDNV, and then examined by immunofluorescence at various passage intervals with antiserum to AeDNV. Initially, approximately 0.5 to 2% of the cells fluoresced upon staining with anti-AeDNV antibody. After several passages, these numbers rose to 5 to 10%, depending on the experiment. One experimental population of cells was carried through 27 1:10 passages and yielded 3% stained cells, and another, which has been passed 1:10 for 1 year, yielded 5% stained cells. Thus, the percentage of stained cells remains fairly constant for an indefinite period of time. The numbers of infected cells are comparable to the 2% reported for cells persistently infected with AaPV (Jousset et al., 1993).

The maintenance of persistent infections in cell culture by densoviruses could be explained by either of two models. It could be due to the productive lytic infection of a small percentage of the cells, which provides fresh virus to infect new cells continuously. This requires that only a small fraction of cells be susceptible to infection; otherwise all of the cells in the culture would rapidly become infected. The fraction of susceptible cells must also be continually renewed, or otherwise it would be rapidly depleted because of infection and lysis. The alternative explanation is that all cells maintain the viral genome; however, it is only productively expressed in a small but relatively constant percentage of the cells. In most cells the viral genome would remain in some sort of latent state. It has been possible to "cure" persistently infected C6/36 cells by dilution and cloning of individual cells to establish sublines. Some of these cloned sublines no longer contained cells that express AeDNV antigen (J. Corsini, unpublished results). This favors the first model.

To test the ability of transducing genomes to persist in cell culture, we have transfected C6/36 cells with β-galactosidase or GFP transducing constructs that are predicted to be self-replicating because they contain the *NS1* gene and the hairpin ends of the genome. After repeated passage of these transfected cultures, the proportion of cells containing the reporter gene eventually drops below the limit of detectability. The transduced cells therefore appear to have some growth disadvantage, and since the transducing genomes lack the ability to package themselves, they cannot be spread to new cells by the infectious process. We have isolated enriched populations of GFP-positive cells by fluorescence-activated cell sorting after transfection with p7NS1–GFP. However, the GFP-positive cells did not seem to proliferate after sorting. Whether this is an artifact caused by the rigors of transfection and cell sorting or whether the cells are unable to proliferate because of the presence of transducing genomes is unknown at this time. Cells transduced by virus particles carrying the p7NS1–GFP genome are capable of cell division, since clusters of two, four, and eight cells are readily observed up to 48 h after transduction, suggesting that they arose from a single transduced cell. This suggests that presence of transducing genomes does not preclude cell division.

Cell cycle analysis of GFP positive cells by flow cytometry shows a pronounced shift of cells to the G_2 phase at the expense of the G_1 phase when compared with GFP-negative cells. Thus the cell cycle appears to be affected by the presence of the NS1–GFP transducing genome, although the GFP protein or the presence of episomal DNA may have a similar effect. The cells do not show evidence of apoptosis using a flow cytometry assay, and appear viable. The ability of densovirus transducing genomes to persist in cells may be limited by perturbation of the cell cycle and the cytotoxic effects of the NS1 protein as is the case with mammalian parvoviruses (Winocour et al., 1988; Oleksiewski and Alexandersen, 1997). Thus, the evidence to date suggests that the presence of a transducing genome places dividing cells at a competitive disadvantage relative to untransduced cells and they are not effectively maintained in a constantly dividing population. The effect of a transducing genome on cells *in vivo,* where constant cell division does not occur, is not yet known.

8.6 SUMMARY

Densonucleosis viruses seem to have considerable potential as transducing vectors for basic research on insects and perhaps for insect control. Infectious clones have been constructed for two viruses, and these are easily manipulated to carry genes of interest. This should be possible for other densoviruses as well. Virus particles are relatively stable in the environment, and at least in the case of AeDNV will readily infect larvae. The viruses are limited in their host ranges and therefore allow targeting of specific insect pests. There are, however, limitations to densoviruses as gene-transfer vectors that must be considered. One such limitation is the relatively small genome size and the packaging constraints of an icosahedral particle. This limits the size of the DNA that can be delivered to 4 to 6 kb depending on the virus. A second limitation is the current necessity of using transfection to generate transducing particles. Optimization of transducing particle production is being continually pursued, and perhaps efficiency of particle production will improve significantly in the future.

At the present time the AeDNV seems well suited to deliver genes of interest by natural infectious mechanisms to the larval stages. Transient expression from self-replicating genomes is easily demonstrated, and delivery of gene products to the mosquito larval hemolymph through infection of the anal papillae seems highly feasible. Generation of stable transformants through genomic integration or episomal maintenance seems less likely, although more definitive experiments need to be done. Perhaps, the viruses can be used to deliver transposon vectors that integrate. Perhaps, less cytotoxic forms of NS1 can be developed.

Large gaps exist in our knowledge of these viruses. Little is known about the maintenance of virus in adult insects, and the mechanisms of transovarial transmission of the virus. Questions regarding the determinants of host range and the nature of barriers to infection and replication in nonpermissive host species must be answered if they are to be used in insect control. As with all molecular biological vector systems, the most effective ways to utilize these viruses as biological tools will only become apparent through the careful elucidation of their biology.

ACKNOWLEDGMENTS

This work was supported by grants from the National Institutes of Health (NIAID AI25629 and AI28781) and the John D. and Catherine T. MacArthur Foundation.

REFERENCES

Afanasiev BN, Galyov EE, Buchatsky LP, Kozlov YV (1991) Nucleotide sequence and genomic organization of *Aedes* densonucleosis virus. Virology 185:323–336

Afanasiev B, Kozlov Y, Carlson J, Beaty B (1994) Densovirus of *Aedes aegypti* as an expression vector in mosquito cells. Exp Parasitol 79:322–339

Afanasiev BN, Ward TW, Beaty BJ, Carlson JO (1999) Transduction of *Aedes aegypti* mosquitoes with vectors derived from Aedes densovirus. Virology 257:62–72

Allen-Miura TM, Afanasiev BN, Olson KE, Beaty BJ, Carlson JO (1999) Packaging of AeDNV-GFP transducing viruses by expression of viral proteins from a Sindbis virus expression system. Virology 257:54–61

Barreau C, Jousset F-X, Cornet M (1994) An efficient and easy method of infection of mosquito larvae from virus-contaminated cell cultures. J Virol Methods 49:153–156

Barreau C, Jousset F-X, Bergoin M (1997) Venereal and vertical transmission of the *Aedes albopictus* parvovirus in *Aedes aegypti* mosquitoes. Am J Trop Med Hyg 57:126–131

Berns KI (1996) *Parvoviridae:* the viruses and their replication. In Fields BN, Knipe DM, Howley PM (eds) Fundamental Virology, 3rd ed. Lipincott-Raven, Philadelphia, pp 1017–1041

Boublik Y, Jousset F-X, Bergoin, M (1994) Complete nucleotide sequence and genomic organization of the *Aedes albopictus* parvovirus (AaPV) pathogenic for *Aedes aegypti* larvae. Virology 200:752–763

Bredenbeek PJ, Frolov I, Rice CM, Schlesinger S (1993) Sindbis virus expression vectors: packaging of RNA replicons by using defective helper RNAs. J Virol 67:6439–6446

Buchatsky LP (1989) Densonucleosis of bloodsucking mosquitoes. Dis Aquat Org 6:145–150

Carlson J, Higgs S, Olson K, Beaty B (1995) Molecular manipulation of mosquitoes. Annu Rev Entomol 40:359–388

Cherbas L, Cherbas P (1993) The arthropod initiator: the capsite consensus plays an important role in transcription. Insect Biochem Mol Biol 23:81–90

Corsini J, Afanasiev B, Maxwell IH, Carlson JO (1996a) Autonomous parvovirus and densovirus gene vectors. Adv Virus Res 47:303–351

Corsini J, Maxwell IH, Maxwell F, Carlson JO (1996b) Expression of parvovirus LaIII NS1 from a Sindbis replica for production of LuIII-luciferase transducing virus. Virus Res 46:95–104

Corsini J, Tal J, Winocour E (1997) Directed integration of minute virus of mice DNA into episomes. J Virol 71:9008–9015

Dumas B, Jourdan M, Pascaud A, Bergoin M (1992) Complete nucleotide sequence of the cloned infectious genome of *Junonia coenia* densovirus reveals an organization unique among parvoviruses. Virology 191:202–222

Garrett MA, Bradley TJ (1984) Structure of osmoregulatory organs in larvae of the brackish-water mosquito, *Culiseta inornata* (Williston). J Morphol 182:257–277

Giraud C, Devauchelle G, Bergoin M (1992) The densovirus of *Junonia coenia* (JcDNV) as an insect cell expression vector. Virology 186:207–218

Giraud C, Winocour E, Berns K (1995) Recombinant junctions formed by site-specific integration of adeno-associated virus into an episome. J Virol 69:6917–6924

Gorziglia M, Botero L, Gil F, Esparza J (1980) Preliminary characterization of virus-like particles in a mosquito (*Aedes pseudoscutellaris*) cell line (Mos.61). Intervirology 13:232–240

Higgs S, Powers AM, Olson KE (1993) Alphavirus expression systems: applications to mosquito vector studies. Parasitol Today 9:444–452

Jousset F-X, Barreau C, Boublik Y, Cornet M (1993) A parvo-like virus persistently infecting a C6/36 clone of *Aedes albopictus* mosquito cell line and pathogenic for *Aedes aegypti* larvae. Virus Res 29:99–114

Kimmick MW (1998) Gene Expression and Regulation from the P7 Promoter of *Aedes densonucleosis* Virus. MS thesis, Colorado State University, Fort Collins

Kimmick MW, Afanasiev BN, Beaty BJ, Carlson JO (1998) Gene expression and regulation from the p7 promoter of the *Aedes* densonucleosis virus. J Virol 72:4364–4370

Kotin RM (1994) Prospects for the use of adeno-associated virus as a vector for human gene therapy. Hum Gene Ther 5:793–801

Linden RM, Ward P, Giraud C, Winocour E, Berns KI (1996a) Site-specific integration by adeno-associated virus. Proc Natl Acad Sci USA 93:11288–11294

Linden RM, Winocour E, Berns KI (1996b) The recombination signals for adeno-associated virus site-specific recombination. Proc Natl Acad Sci USA 93:7966–7972

Monroe TJ, Muhlman-Diaz MC, Kovach MJ, Carlson JO, Bedford JS, Beaty BJ (1992) Stable transformation of a mosquito cell line results in extraordinarily high copy numbers of the plasmid. Proc Natl Acad Sci USA 89:5725–5729

Oleksiewski MB, Alexandersen S (1997) S-Phase-dependent cell cycle disturbances caused by Aleutian mink disease parvovirus. J Virol 71:1386–1396

O'Neill SL, Kittayapong P, Braig HR, Andreadis TG, Gonzalez JP, Tesh RB (1995) Insect densoviruses may be widespread in mosquito cell lines. J Gen Virol 76:2067–2074

Tal J, Attathom T (1993) Insecticidal potential of the insect parvovirus GmDNV. Arch Insect Biochem Physiol 22:345–356

Winocour E, Callaham MF, Huberman E (1988) Perturbation of the cell cycle by adeno-associated virus. Virology 167:393–399

Xiong C, Levis R, Shen P, Schlesinger S, Rice CM, Huang HV (1989) Sindbis virus: an efficient, broad host range vector for gene expression in animal cells. Science 243:1188–1191

Yang CC, Xiao X, Zhu X, Ansardi DC, Epstein ND, Frey MR, Matera AG, Samulski RJ (1997) Cellular recombination pathways and viral terminal repeat hairpin structures are sufficient for adeno-associated integration *in vivo* and *in vitro*. J Virol 71:9231–9247

9 Sindbis Virus Expression Systems in Mosquitoes: Background, Methods, and Applications

Ken E. Olson

CONTENTS

9.1	Introduction 162
9.2	Sindbis Virus: Background Information 163
	9.2.1 Sindbis Virus 163
	9.2.2 Molecular Biology of Sindbis Virus..163
	9.2.3 Virogenesis of Sindbis in Cultured Cells 163
	9.2.4 Sindbis Virus/Mosquito Interactions..165
9.3	Sindbis Virus Expression Systems..166
	9.3.1 Infectious Clone Technology..166
	9.3.2 Double Subgenomic Sindbis Viruses...166
	9.3.3 Sindbis Replicon 168
9.4	Expression of Genes of Interest in Mosquito Cells and Mosquitoes by the Double Subgenomic Sindbis Virus Expression System 168
	9.4.1 Double Subgenomic Sindbis Viruses: Methodology.....................................168
	9.4.1.1 Subcloning of Genes into the Double Subgenomic Sindbis Plasmid..168
	9.4.1.2 *In Vitro* Transcription 168
	9.4.1.3 Transfection..170
	9.4.1.4 Infecting Mosquitoes with Double Subgenomic Sindbis Viruses............170
	9.4.1.4.1 Intrathoracic Inoculation of Double Subgenomic Sindbis Viruses into Mosquitoes 170
	9.4.1.4.2 *Per os* Infection of Mosquitoes with Double Subgenomic Sindbis Viruses...170
	9.4.2 Applications of Double Subgenomic Sindbis Viruses in Mosquitoes 171
	9.4.2.1 Protein Expression in Cultured Mosquito Cells............................171
	9.4.2.2 Protein Expression in Mosquitoes...171
	9.4.2.3 Protein Expression in Saliva of *Culex pipiens* 174
	9.4.2.4 Protein Expression in the Midgut of *Aedes aegypti*..................................174
	9.4.2.5 Expression of Antiviral RNAs in *Aedes aegypti* and *Aedes triseriatus*..176
	9.4.2.5.1 Expression of an Antisense RNA to LaCrosse Virus in *Aedes triseriatus* 176
	9.4.2.5.2 RNA-Mediated Interference with Flavivirus Replication in C6/36 Cells and *Aedes aegypti*..176

9.4.3 Advantages and Disadvantages of Double Subgenomic Sindbis Viruses 178
   9.4.3.1 Advantages 178
   9.4.3.2 Disadvantages 178
9.5 Expression of Genes of Interest in Mosquito Cells and Mosquitoes
    by the Sindbis Virus Replicon 180
   9.5.1 Sindbis Replicon: Methodology 180
   9.5.2 Applications of Gene Expression in Mosquito Cells and Mosquitoes Using
         the Sindbis Replicon 180
      9.5.2.1 Protein Expression from Sindbis Replicon RNA in C6/36 Cells 180
      9.5.2.2 Protein Expression from Packaged Sindbis Replicons
              in Mosquito Cells 180
      9.5.2.3 Protein Expression from Packaged Sindbis Replicons
              in Mosquitoes 181
   9.5.3 Advantages and Disadvantages of Sindbis Replicon Expression 181
9.6 Alphavirus Expression Systems: Future Applications 181
   9.6.1 Rapid Determination of Function and Utility of Effector Molecules
         in Mosquitoes Prior to Transformation 181
   9.6.2 Double Subgenomic Sindbis Viruses: Important Molecular Tools
         for Observing Virus-Specific Interactions with the Arthropod Vector 183
   9.6.3 Double Subgenomic Sindbis Viruses: Potential Tools for Down-Regulating
         Gene Expression in the Mosquito 183
   9.6.4 Development of Alphavirus Expression Systems That Are Tissue
         and Species Specific in Mosquitoes 183
   9.6.5 DNA-Based Alphavirus Replicons as an Efficient Means
         of Overexpressing Antipathogen Genes in a Heritable Fashion 184
9.7 Summary 185
References 185

9.1 INTRODUCTION

Transformation systems based on *mariner* and *Hermes* transposable elements have recently been shown to integrate heterologous genes into the *Aedes aegypti* genome in a predictable manner (Coates et al., 1998; Jasinskiene et al., 1998). This is an important accomplishment that should greatly facilitate gene manipulation in mosquitoes. However, transformation remains an intensely laborious and time-consuming procedure for characterization, mutagenesis, and expression of genes in mosquitoes. The propagation and maintenance of multiple transformed mosquito lines will be difficult in even the most spacious, well-equipped insectaries. Additionally, the complex life cycle of some medically important mosquitoes make routine transgenesis difficult, whereas other transient expression systems may more easily and rapidly answer biological questions posed by researchers.

In this chapter, RNA virus expression systems based on Sindbis (SIN; genus: *Alphavirus*, family: Togaviridae) virus are described that efficiently transduce mosquito cells and allow stable, long-term expression of heterologous genes in *Ae. aegypti, Ae. triseriatus, Culex pipiens,* and *Anopheles gambiae* (Carlson et al., 1995; Olson et al., 1998). This chapter will begin with a review of SIN virus molecular biology, virogenesis in cells, and alphavirus infection patterns in mosquitoes. A discussion of current and future uses of SIN virus expression systems as important tools for studying basic virus–vector interactions, determining biological functions of mosquito genes, and identifying molecular strategies for interfering with the transmission of vector-borne diseases will then follow. SIN virus expression systems by no means lessen the importance of arthropod vector transformation; rather they complement transformation studies by allowing researchers an opportunity to decide rationally what gene products or effector molecules merit further analysis in transgenic mosquitoes.

9.2 SINDBIS VIRUS: BACKGROUND INFORMATION

9.2.1 SINDBIS VIRUS

The AR339 SIN virus is the prototype virus of the genus, *Alphavirus*, family, Togaviridae. SIN viruses are arthropod-borne viruses that are cycled principally between *Culex* species of mosquitoes and avian vertebrate hosts (Taylor et al., 1955). However, SIN viruses are infectious to a number of other vertebrate (including humans) and arthropod species (Hurlbut and Thomas, 1960; Malherbe and Strickland-Cholmley, 1963). SIN viruses have primarily an Old World distribution and can readily be separated into distinct European–African and Oriental–Australian genetic groups (Strauss and Strauss, 1994).

9.2.2 MOLECULAR BIOLOGY OF SINDBIS VIRUS

SIN viruses are enveloped RNA viruses that are approximately 70 nm in diameter and replicate exclusively in the cytoplasm of infected cells (Strauss and Strauss, 1994). The nucleocapsid of the virus is composed of a genomic RNA and 240 copies of a single virus-encoded capsid protein, arranged in an icosahedral lattice with $T = 4$ symmetry (Paredes et al., 1993). The host-derived, virus envelope contains 80 spikes composed of trimers of E1/E2 heterodimers. The SIN virus genome is a positive-sense, single-stranded, nonsegmented RNA of about 11.7 kb (Strauss et al., 1984; Strauss and Strauss, 1994; Figure 9.1). The 5′ two thirds of the genome is translated directly into nonstructural proteins that form the viral replicase. Initially, a full-length, negative-sense RNA is transcribed from the genome. Transcription occurs in the presence of a replication complex that includes a virus-encoded RNA-dependent RNA polymerase. Transcription of antigenomic RNA shuts down early in infection as the replication complex is modified to promote positive-sense RNA transcription. The antigenome serves as a template for transcription of more genomic RNA and a subgenomic mRNA. The subgenomic (26S) mRNA, colinear with the 3′ one third of the genome, is translated into a structural polyprotein. Capsid (C) protein subunits are autocatalytically cleaved from the structural polyprotein cotranslationally; the remaining polyprotein molecule is translocated into the endoplasmic reticulum (ER) of the cell. Cleavage and modification of the envelope glycoproteins (E1 and PE2) is a multistep process that takes place during vesicular transport through the ER/Golgi complex. Glycoprotein spikes and nucleocapsids assemble at the plasma membrane to produce progeny virions. Two smaller, unpackaged polypeptides (E3 and 6K) are produced as cleavage products during glycoprotein processing. A noncoding region (NCR) at the 3′ end of genomic and subgenomic RNAs, contiguous with a poly(A) tail, contains characteristic repeated sequence elements (Strauss and Strauss, 1994) and may play a role in host specificity, possibly through interactions with cellular proteins (Kuhn et al., 1990).

9.2.3 VIROGENESIS OF SINDBIS IN CULTURED CELLS

SIN and other alphaviruses gain entry into vertebrate and invertebrate cells by a receptor-mediated endocytotic pathway (Wang et al., 1992; Strauss and Strauss, 1994; Ludwig et al., 1996). The E2 glycoprotein of the alphavirus, Venezuelan equine encephalitis (VEE), can bind a membrane-associated, laminin-binding protein of mosquito (C6/36) cells to initiate infection (Ludwig et al., 1996). However, several studies suggest that the broad host range of the virus is achieved in part by utilizing more than one protein receptor (Strauss and Strauss, 1994). The major cell receptor binding activity of SIN virus has been localized to residues 170 to 220 of the E2 glycoprotein (Strauss and Strauss, 1994). Several studies have identified dramatic changes in alphavirus replication and virulence due to single amino acid changes in this domain (Tucker and Griffin, 1991; Woodward et al., 1991; Kerr et al., 1993). After binding to the membrane receptor, the virus enters the cell by endocytosis. Once inside vesicles, SIN viruses require a low pH (5 to 6) to initiate conformational changes in the E1/E2 heterodimer that lead to fusogenic

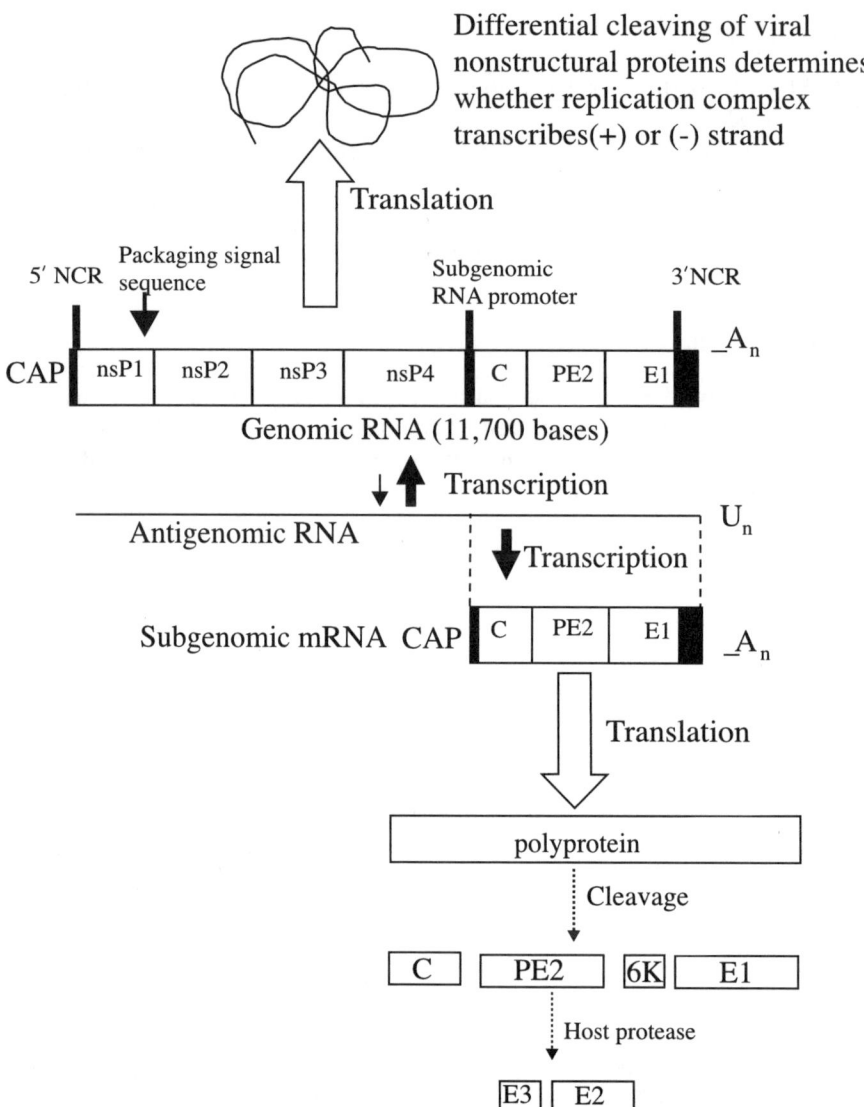

FIGURE 9.1 Genomic organization of SIN virus. The black boxes on genomic RNA contain conserved sequence elements. The 5′ two thirds of the genomic RNA is translated in the infected cell to produce the active replicase forms for plus and minus strand synthesis. The subgenomic mRNA is transcribed from minus strand viral complementary RNA. The subgenomic mRNA is translated to form a polyprotein that encodes the viral structural proteins. The polyprotein is cleaved using a capsid autoprotease and host cell proteases to form the individual structural proteins. Capsid, E2, and E1 are found in the virion.

activity of the glycoproteins and the viral nucleocapsid into the cytoplasm (Strauss and Strauss, 1994). The virus RNA genome within the nucleocapsid is then uncoated through interactions between the capsid protein and with a putative ribosome-binding site (Wengler and Wengler, 1984; Wengler et al., 1992). The naked viral genome is immediately translated to begin virus replication and expression of viral proteins. Genomic RNA and capsid protein subunits self-assemble into icosahedral nucleocapsids in the cytoplasm and are transported to the plasma membrane. The viral envelope proteins are glycosylated (N-linked glycosylation Asn-Val-Thr) as they are processed through the ER/Golgi apparatus and are subsequently transported to the plasma membrane. An important step in the maturation of SIN virus glycoproteins is the cleavage of the precursor glycoprotein PE2 into E3 and E2 by furin or a furin-like host cell protease. PE2 cleavage appears to be an obligatory event for SIN virus maturation in C6/36 cells (Heidner et al., 1996). Hyperglycosylation of PE2 can effect this cleavage and restrict virus growth in mosquito cells. In vertebrate cells, the virion envelope forms by budding of the nucleocapsid through the modified plasma membrane (Strauss and Strauss, 1994). However, in cultured mosquito cells, alphavirus maturation occurs almost exclusively within membrane-bound structures inside the cells, and virions are released by reverse phagocytosis (Miller and Brown, 1992). SIN virus infection of vertebrate cells is acute and induces apoptosis, leading to the destruction of infected cells (Griffin and Hardwick, 1997). However, SIN viruses usually establish a persistent, noncytolytic infection in many invertebrate cells. Pathology has been described in some clonal lines of mosquito cells during SIN virus infections, but apoptosis does not appear to be the cause of cell destruction (Karpf and Brown, 1998).

9.2.4 SINDBIS VIRUS/MOSQUITO INTERACTIONS

Alphavirus interactions with mosquitoes have been the subject of a number of studies (Scott et al., 1984; Weaver, 1986; 1996; Bowers et al., 1995; Woodring et al., 1996; Seabaugh et al., 1998). The sequence of events beginning with the mosquito's intake of an infected blood meal and ending with the mosquito transmitting a pathogen is the extrinsic incubation period (Hardy, 1988; Woodring et al., 1996). The extrinsic incubation period of alphaviruses usually ranges from 3 to 10 days (Weaver, 1996). During the extrinsic incubation period, the virus must interact with receptors on mosquito midgut epithelial cells, replicate within the midgut cells, disseminate through these cells, enter the hemocoel, and disseminate to other secondary target organs. Infection of salivary glands permits horizontal transmission of the virus to the next susceptible vertebrate host, when the female feeds again (Woodring et al., 1996).

Virus genetics plays an important role in determining whether a mosquito will acquire and disseminate a SIN virus infection. The dose of AR339 SIN virus required to orally infect *Ae. aegypti* is approximately 10^5 tissue culture infectious dose 50% end points ($TCID_{50}$) per mosquito (Jackson et al., 1993; Seabaugh et al., 1998). The MRE16 strain of SIN virus has a 10-fold lower infectious dose than AR339 (Seabaugh, 1997; Seabaugh et al., 1998). In addition, different SIN virus strains show different dissemination efficiencies in mosquitoes from the same rearing colony. *Aedes aegypti* (RexvilleD) orally infected with AR339, TE/3'2J/Δ2SGP, MRE1001, and MRE16 SIN viruses produce disseminated infections in 39%, 19%, 91%, and 96% of the mosquitoes after 10 days postinfection, respectively (Seabaugh, 1997; Seabaugh et al., 1998). MRE1001 virus is a chimeric virus that contains the nonstructural and *cis*-acting sequences of AR339 virus and the structural genes of MRE16 virus.

Interspecific and intraspecific genetic variability among arthropod vectors has long been recognized to have profound effects on vector competence (Grimstad et al., 1977; Tabachnick et al., 1985; Hardy, 1988). Vectorial capacity, of which vector competence is one parameter, has been analyzed among different species of mosquitoes for *per os* infection and transmission of the same alphavirus strain of SIN, VEE, and Eastern equine encephalitis (EEE). Significant differences were observed in the ability of each mosquito species to transmit the viruses (Turell et al., 1994; Dohm et al., 1995; Vaidyanathan et al., 1997). When the MRE16 strain of SIN virus is used to infect *Ae. aegypti*, *Ae. albopictus*, *Ae. triseriatus*, and *C. pipiens,* dissemination occurs in 96, 100, 35, and 0%, respectively (Seabaugh et al., 1997). When rosy-eye mutants and wild-type strains of *Ae. aegypti* are infected with the alphavirus Chikungunya, significant differences are observed in their ability to support infections (Mourya et al., 1998). Finally, the effect of environmental factors such as nutritional status, density of larvae in rearing pans, and temperature all may have an effect on the ability of alphaviruses to infect laboratory-reared mosquitoes (Turell, 1993; Nasci and Mitchell, 1994).

In the laboratory, SIN viruses can infect a number of different mosquito species that have midgut infection barriers by simply intrathoracically injecting the mosquitoes with the virus, thus bypassing the midgut. Bowers et al. (1995) have described the temporal and spatial progression of AR339 SIN virus following intrathoracic inoculation of *Ae. albopictus*. The gut-associated musculature and respiratory tracheoles are infected by 48 h post-infection but not the epithelial cell layer that lines the midgut lumen (Bowers et al., 1995). At 48 to 72 h post-infection SIN virus can be found in fat bodies, hemolymph, salivary glands, thoracic muscle, and neural tissues. Approximately 72 h post-infection the mosquito becomes persistently infected, accompanied by clearance of viral antigen from many of the infected tissues. Subsequently, viral antigen is restricted to the fat body, hindgut visceral muscles, and tracheoblasts from which the persistent infection was purportedly supported for the life of the mosquito (Bowers et al., 1995).

Several studies have followed the virogenesis of alphaviruses in vectors following oral infection (Weaver, 1996). EEE virus first infects the posterior midguts of *Culiseta melanura* mosquitoes (Scott et al., 1984). Within several days, virus disseminates from midguts and infects fat body, nerves, muscle tissue, hindgut, oviducts, and many other tissues. Salivary glands become infected between 2 and 3 days post-infection. We also have observed MRE16 SIN virus in salivary glands as early as 3 days post-infection. Researchers have demonstrated that both EEE and Western equine encephalitis (WEE) can cause cytopathic effects (CPE) in midgut cells of natural vectors, but the significance of this remains to be determined (Weaver et al., 1988; Weaver, 1996; Scott et al., 1998). However, electron microscopy studies have shown no CPE in salivary glands of *Culex theileri* or *Ae. aegypti* infected with AR86 SIN virus (Jupp and Phillips, 1998).

9.3 SINDBIS VIRUS EXPRESSION SYSTEMS

9.3.1 INFECTIOUS CLONE TECHNOLOGY

The development of infectious cDNA clone technology has been important for manipulating the genomes of positive-sense RNA viruses. Full-length infectious cDNA clones of viruses within the family Togaviridae, including SIN, Semliki Forest (SF), Ross River (RR), and VEE viruses, have been described (Rice et al., 1987; Davis et al., 1989; Kuhn et al., 1991; Liljestrom et al., 1991; Seabaugh et al., 1998). This technology has allowed genetic manipulation of RNA genomes and identification of viral determinants of host range, virulence, replication, assembly, and packaging (Lustig et al., 1988; Tucker and Griffin, 1991; Heidner et al., 1994; Strauss and Strauss, 1994; Duffus et al., 1995; Wang et al., 1996). Infectious clone technology also has led to the development of SIN virus expression systems for use in both vertebrate and invertebrate organisms (Xiong et al., 1989; Hahn et al., 1992; Carlson et al., 1995; Dubensky et al., 1996; Agapov et al., 1998; Olson et al., 1998). At the Arthropod-borne and Infectious Diseases Laboratory (AIDL; Colorado State University, Fort Collins, CO), we have pioneered the use of SIN virus–based gene expression systems in mosquito cells and in the larval and adult stages of several mosquito vectors.

9.3.2 DOUBLE SUBGENOMIC SINDBIS VIRUSES

The SIN-based transducing system most frequently used at AIDL is termed the double subgenomic SIN (dsSIN) expression system. Two dsSIN expression systems, pTE3'2J and pMRE/3'2J, are currently in use. dsSIN plasmids consist of an SP6 bacteriophage promoter followed by the 5' NCR of SIN virus, the nonstructural genes nsP1-nsP4, the capsid gene, and the envelope genes (PE2 and E1) (Figure 9.2). The origin of the glycoproteins of pTE/3'2J and pMRE/3'2J are a mouse neurovirulent strain and a Malaysian strain of SIN, respectively (Hahn et al., 1992; Seabaugh et al., 1998; Olson et al., 2000). A second internal initiation site (−109 to +49; numbered relative to the nucleotide in the genome-length positive strand, which corresponds to the first nucleotide of the subgenomic mRNA) has been placed downstream of the E1 gene. Although

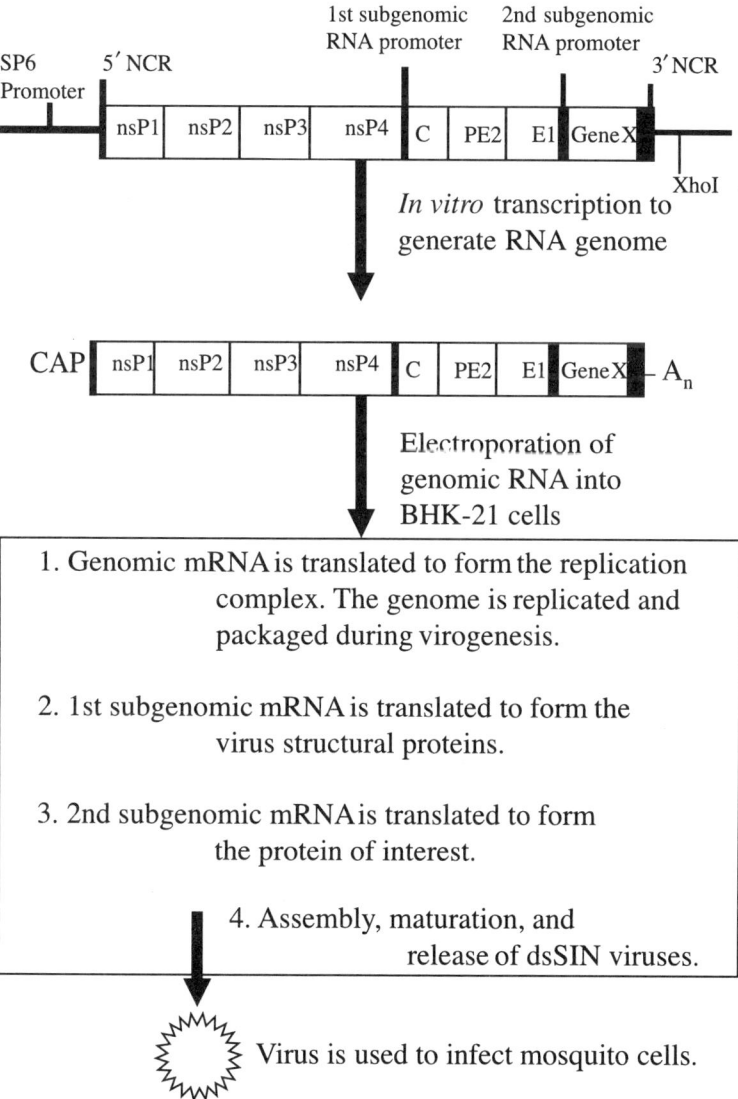

FIGURE 9.2 Outline of the protocol for producing infectious virus from dsSIN plasmids. nsP1-4 = nonstructural proteins 1-4, NCR = noncoding region, C = capsid, PE2 = precursor envelope glycoprotein E2, E1 = envelope glycoprotein E1, GeneX = gene of interest.

the core promoter for SIN subgenomic mRNA transcription maps only from position −19 to +5 relative to the subgenomic mRNA start, a sixfold increase in virus promoter activity has been observed by using the sequence −98 to +14 in the constructs (Raju and Huang, 1991). The exogenous gene is inserted 3′ to the second subgenomic promoter and is followed by the 3′ NCR of SIN virus. Full-length RNA is transcribed *in vitro* from the SP6 promoter and the RNA is transfected into the appropriate cells where virogenesis begins. Within the transfected cell three dsSIN RNA species occur, the genomic RNA, the first subgenomic mRNA that is translated to form the virus structural proteins, and a second subgenomic mRNA that can be translated to form the heterologous protein (see Figure 9.2). dsSIN viruses are infectious for multiple replicative cycles. In the mosquito, dsSIN virus infections become persistent and gene expression can be observed throughout the life of the infected mosquito. dsSIN viruses, like the parent viruses, are not vertically transmitted to the mosquito progeny.

9.3.3 SINDBIS REPLICON

Replicon expression systems have been engineered from a number of RNA viruses (Liljestrom and Garoff, 1991; Bredenbeek and Rice, 1992; Frolov et al., 1996). Alphavirus replicons are currently being developed for veterinary and medical applications as delivery vehicles of vaccines and therapeutic agents (Pushko et al., 1997; Agapov et al., 1998; Berglund et al., 1998). The SIN virus replicon expression system also is a proven means of efficiently expressing heterologous genes in mosquito cells (Olson et al., 1992; Kamrud et al., 1995). This expression system is based on a self-replicating RNA, or replicon, derived from an infectious cDNA clone of SIN virus (Xiong et al., 1989; Bredenbeek et al., 1993). The replicon (pSINrep5; Figure 9.3) consists of the 5' NCR of the virus genome, the SIN nonstructural genes (nsp1-nsp4), a SIN viral subgenomic promoter sequence located upstream of a multiple cloning site that replaces the SIN structural genes, and the 3' NCR. Heterologous genes can be cloned downstream of the SIN subgenomic promoter, and RNA transcribed *in vitro* from recombinant pSINrep5 DNA expresses the heterologous protein when transfected into cells. The SINrep5 replicon RNA contains a nucleotide sequence in the nsP1 gene that acts as a packaging signal when the structural proteins of the virus is supplied in *trans* (Bredenbeek et al., 1993; Kamrud et al., 1995). Packaging is achieved by cotransfection of the replicon with a helper RNA that provides the viral structural proteins (Bredenbeek et al., 1993). Helper constructs are available with and without packaging signals. Helper constructs with packaging signals may be important for providing continuous cell-to-cell infections of replicons with large inserts.

9.4 EXPRESSION OF GENES OF INTEREST IN MOSQUITO CELLS AND MOSQUITOES BY THE DOUBLE SUBGENOMIC SINDBIS VIRUS EXPRESSION SYSTEM

9.4.1 DOUBLE SUBGENOMIC SINDBIS VIRUSES: METHODOLOGY

The TE/3'2J dsSIN system can be obtained directly from C. M. Rice at the Washington University, St. Louis, MO. The MRE/3'2J dsSIN virus is available through the Arthropod-borne and Infectious Diseases Laboratory (AIDL). At AIDL, we produce and amplify dsSIN viruses in vertebrate and invertebrate tissue culture cells using biosafety level 2 (BSL-2) protocols and procedures. However, mosquitoes infected with these agents are maintained at BSL-3 conditions.

9.4.1.1 Subcloning of Genes into the Double Subgenomic Sindbis Plasmid

The *Xba*I restriction endonuclease (RE) site in pTE/3'2J is the most convenient RE site of ligation. Heterologous cDNAs are usually subcloned or TA-cloned into an intermediate plasmid that will flank the insert with RE sites compatible with *Xba*I (Gaines et al., 1996; Olson et al., 1996). In addition, the gene of interest can be amplified directly from cDNA by polymerase chain reaction (PCR) using forward and reverse primers that have *Xba*I RE sequences at their 5' ends. Recombinant dsSIN plasmids are then transformed into SURE bacteria (Stratagene, La Jolla, CA) and bacterial colonies containing the dsSIN plasmid plus the cDNA insert are screened by PCR using primers that flank the *Xba*I site in pTE/3'2J. Orientation of the insert relative to the second SIN internal initiation site are analyzed by PCR using primers based on TE/3'2J and insert sequences. The placement of the gene in the dsSIN plasmid is then confirmed by sequence analysis.

9.4.1.2 *In Vitro* Transcription

Approximately 50 μg of dsSIN plasmid is digested to completion with *Xho*I RE and the digestion is extracted twice with an equal volume of a phenol/chloroform/isoamyl alcohol (25:24:1) mixture. To minimize RNase activity, the aqueous phase is digested with 10 μg of proteinase K and extracted again with the phenol mixture. The template then is extracted with an equal volume

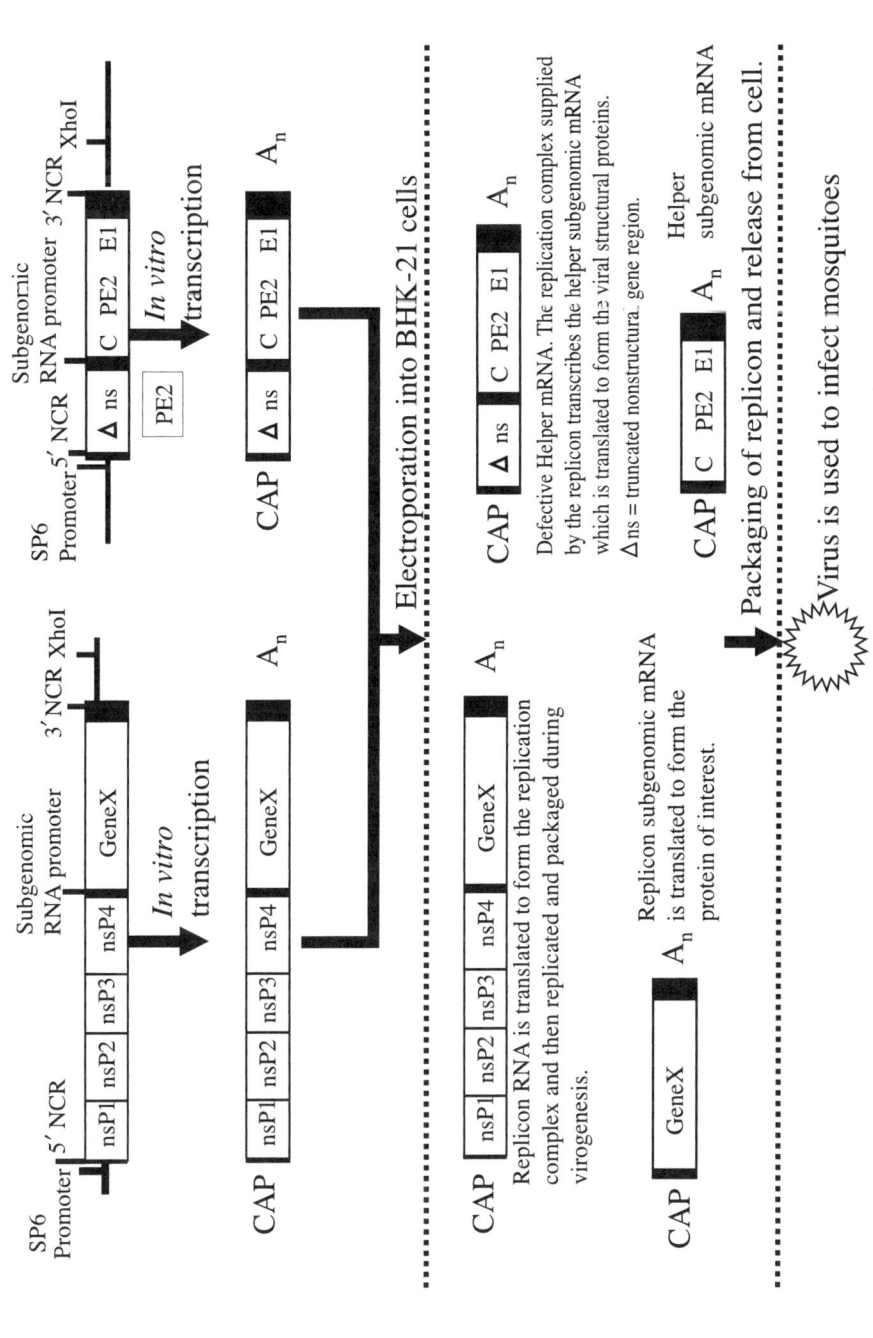

FIGURE 9.3 Outline of the protocol for producing packaged SIN replicon. nsP1-4 = nonstructural proteins 1-4, NCR = noncoding region, C = capsid, PE2 = precursor envelope glycoprotein E2, E1 = envelope glycoprotein E1, GeneX = gene of interest, ↓ = packaging signal.

of chloroform/isoamyl alcohol (24:1) and ethanol precipitated. The pellet is resuspended in the appropriate volume of sterile RNase-free water to a final concentration of 1 µg/ml of DNA. RNA is transcribed from the template DNA using the SP6 bacteriophage promoter. RNA is transcribed *in vitro* in reactions containing the following ingredients:

1. 1 to 2 µg of template DNA,
2. 1 × transcription buffer (Ambion, Inc., Austin, TX),
3. 1 mM ribonucleotides (UTP, rCTP, rGTP, and rATP),
4. 1 mM 7-methyl-guanidine triphosphate capping analogue (Ambion, Inc., Austin, TX),
5. 5 mM dithiothreitol,
6. 1 µg/µl acetylated BSA,
7. 1 unit/µl RNAsin (Promega, Madison, WI), and
8. 20 units of SP6 polymerase in a 50-µl reaction.

The reaction is incubated at 39°C for 1 h prior to transfection. The RNA is then transfected into susceptible cells, but can be stored at −70°C for up to 6 months.

9.4.1.3 Transfection

Approximately 1/5 of the transcribed RNA genome is electroporated (BTX, Inc. San Diego, CA) into 10^7/ml BHK-21 cells at 450 V, 100 µF, and 720 Ω for 0.8 ms (Powers et al., 1994; Olson et al., 1996). The BHK-21 cells used in electroporation should be of low passage and harvested when approximately 80% confluent. Immediately after electroporation, the cells are seeded into 25 cm² cell culture flasks with 5 ml of L-15 medium (10% fetal bovine serum) and incubated at 37°C for 18 to 24 h. dsSIN viruses are then harvested from the medium and titrated in BHK-21 or Vero cells using an end-point assay or plaque assay, respectively (Miller and Mitchell, 1986; Higgs et al., 1997).

9.4.1.4 Infecting Mosquitoes with Double Subgenomic Sindbis Viruses

9.4.1.4.1 Intrathoracic Inoculation of Double Subgenomic Sindbis Viruses into Mosquitoes

For many applications, dsSIN viruses can be intrathoracically inoculated into mosquitoes. This method of infecting mosquitoes has several advantages:

1. Intrathoracic inoculations ensure that 100% of the surviving mosquitoes are infected and express the gene of interest. With practice, injections become rapid, routine, and survival rates exceed 90%. Several hundred mosquitoes can be inoculated in 1 day.
2. Since intrathoracic inoculations bypass the midgut barrier to infection, gene expression from a given dsSIN virus is possible in numerous mosquito species.

Protocols for intrathoracic injections have already been described in detail (Rosen and Gubler, 1974; Higgs et al., 1997).

9.4.1.4.2 Per os Infection of Mosquitoes with Double Subgenomic Sindbis Viruses

This technique has two important applications for gene expression in midgut epithelial cells and expression of antibacterial peptides. Since midgut cells are the first mosquito cells an arthropod-borne pathogen infects following uptake of a blood meal, it is an important organ for molecular biology studies. Also, the act of intrathoracically inoculating mosquitoes results in induction of antibacterial peptides such as defensins and cecropins (Gwadz et al., 1989; Lowenberger et al.,

1995). This makes it difficult to study the role of any individual peptide or associated transcription factors expressed by the dsSIN virus. However, the *per os* route of infection does not induce this immune response in mosquitoes and is the preferred technique for these types of studies. Protocols for infecting mosquitoes by the *per os* route have been described in detail (Higgs et al., 1997; Olson et al., 2000).

9.4.2 Applications of Double Subgenomic Sindbis Viruses in Mosquitoes

9.4.2.1 Protein Expression in Cultured Mosquito Cells

Recombinant dsSIN can be used to express genes of interest in cultured mosquito cells. This has been clearly demonstrated by expression of the reporter gene encoding chloramphenicol acetyltransferase (CAT) in C6/36 (*Ae. albopictus*) cells. When C6/36 cells are infected with TE/3'2J/CAT virus at a multiplicity of infection (MOI) > 20, 100% of the cells express CAT (8.3×10^5 CAT polypeptides per cell) at 24 h post-infection (Olson et al., 1994). CAT expression and dsSIN virus titers peaked at 24 and 48 h, repectively (Figure 9.4). Characterization of dsSIN RNAs and expressed proteins can be performed routinely and conveniently in cultured C6/36 cells (Higgs et al., 1997). Other mosquito cell lines that are susceptible to dsSIN virus infection include AP-61 (*Ae. pseudoscutelaris*) and ATC15 (*Ae. aegypti*) cell lines. The *Ae. albopictus* cell line C7-10 is not only susceptible to dsSIN virus infections, but also displays distinct cytopathology as a consequence of infection.

9.4.2.2 Protein Expression in Mosquitoes

dsSIN viruses have been used to express proteins of interest in *Ae. triseriatus*, *Ae. aegypti*, *Ae. albopictus*, *Culex pipiens*, and *Anopheles gambiae* (Higgs et al., 1993; Carlson et al., 1995; Olson et al., 1998). *Ae. triseriatus* intrathoracically inoculated with 4.0 $\log_{10}TCID_{50}$ of TE/3'2J/CAT virus generate virus titers >6.0 $\log_{10}TCID_{50}$ per mosquito within 4 days. CAT activity is detected within 2 days (8×10^{-5} units of CAT/mosquito or 1.4×10^{10} CAT polypeptides), peaks at day 6 (4×10^{-3} units of CAT/mosquito or 7.2×10^{11} CAT polypeptides), and remains at peak levels to day 20 (Olson et al., 1994; Figure 9.5A). Immunofluorescence and CAT activity assays have been used to localize CAT expression in infected mosquitoes and in neural and salivary gland tissues (Figure 9.5B).

A dsSIN virus has been engineered that expresses green fluorescent protein (GFP) in larvae, pupae, and adult *Ae. aegypti*, *An. gambiae*, and *Cx. pipiens* when the recombinant virus is injected into these life stages (Higgs et al., 1996; Olson et al., 1998). GFP expression in mosquitoes is the topic of Chapter 5 by Higgs and Lewis in this volume and will not be described in detail here. Additionally, the use of dsSIN viruses to express GFP in other species of insects is discussed in that chapter.

An insect-specific scorpion toxin gene (scotox) also has been expressed in *Ae. aegypti* and *Cx. pipiens* infected with the dsSIN virus, TE/3'2J/Scotox (Higgs et al., 1995). The expression of scotox in mosquitoes demonstrates two important principles of heterologous gene expression from dsSIN viruses:

1. dsSIN viruses efficiently express neuroactive proteins in mosquitoes.
2. Gene expression from dsSIN viruses is not stable after repeatedly passing the virus in tissue culture.

First and second passage TE/3'2J/Scotox virus actively expresses the neurotoxin and kills 100% of injected mosquitoes. However, when TE/3'2J/Scotox virus was passaged five times in Vero cell culture and intrathoracically inoculated into mosquitoes, the virus was no longer lethal to mosqui-

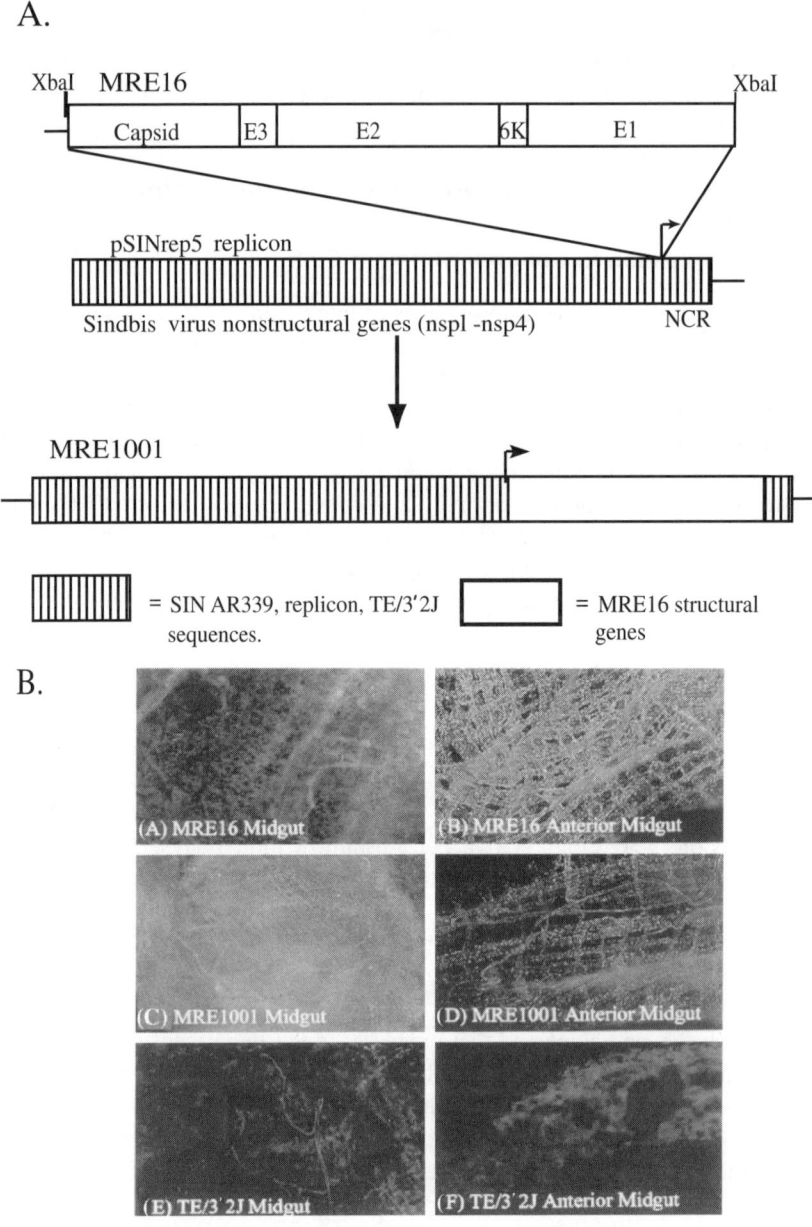

FIGURE 9.4 (A) Assembly of chimeric infectious cDNA clone, designated MRE1001. Clones pMCAP, pME2, and pME1 were sequenced and assembled utilizing unique REs to form the complete structural gene region of the MRE16 genome. The assembled structural gene coding region was amplified by PCR using primers containing an *Xba*I sequence at the extreme 5′ end. MRE16 structural gene cDNA was then ligated into the *Xba*I site of the pSINrep5 replicon. producing pMRE1001 chimeric SIN infectious cDNA. (From Seabaugh, R. C. et al., Virology, 243, 99–112, 1998. With permission.) (B) Detection of SIN E1 antigen in *Ae. aegypti* midgut and salivary gland tissues after ingestion of blood meals containing TE/3′2J/CAT and MRE16 viruses. Midgut epithelial cells of mosquitoes infected with (A) TE/3′2J/CAT and (B) MRE16 viruses at day 3 post-infection (magnification 40×). Midgut epithelial cells of mosquitoes infected with (C) TE/3′2J/CAT and (D) MRE16 viruses at day 9 post-infection (magnification 40×). Salivary gland tissues of mosquitoes infected with (E) TE/3′2J/CAT and (F) MRE16 viruses at day 9 post-infection (magnification 20×). SIN E1 was detected by IFA using monoclone 30.11a as the primary antibody.

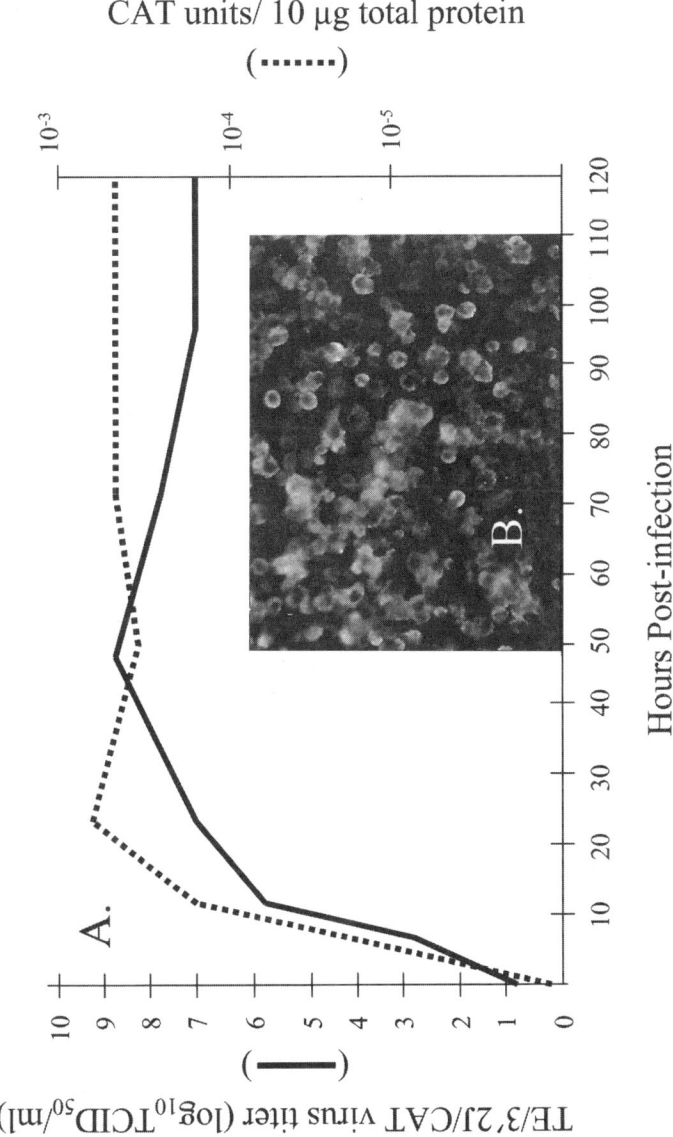

FIGURE 9.5 (A) Determination of the titer of TE/3′2J/CAT virus and CAT activity in infected C6/36 cells. (B) Detection of CAT expression in approximately 100% of the cells at day 2 post-infection using an immunofluorescence technique. (From Olson, K. E. et al., *Insect Biochem. Mol. Biol.*, 24, 39–48, 1994. With permission.)

toes (Higgs et al., 1995). The reduced toxicity with increasing passage resulted from deletions in the genome of TE/3'2J/Scotox virus. Similarly, molecular analysis of the genome of TE/3'2J/CAT virus following four *per os* passages in *Ae. aegypti* revealed that a 500-base deletion occurred that included 127 bases of the second subgenomic RNA promoter and 473 bases of the CAT gene (Seabaugh et al., 1997). These data demonstrate the importance of infecting mosquito cells with dsSIN viruses obtained either directly from the transfection or following a single passage in mosquito cells to get efficient expression of heterologous genes.

9.4.2.3 Protein Expression in Saliva of *Culex pipiens*

Gene expression in mosquitoes is a powerful approach to characterize salivary gland genes and reveal important vector determinants in pathogen transmission. TE/3'2J/CAT virus has been used to express CAT in the saliva of transduced female *Cx. pipiens* (Kamrud et al., 1997; Olson et al., 1998). Indirect immunofluorescence analysis revealed that salivary glands of mosquitoes intrathoracically infected with TE/3'2J/CAT virus were positive for both SIN E1 antigen and CAT protein 4 days post-infection. Saliva collected from mosquitoes transduced with TE/3'2J/CAT virus contained a unique 25-kDa protein that corresponded to the size of CAT protein. CAT activity assays revealed that saliva collected from mosquitoes transduced with TE/3'2J/CAT virus contained more than 5.0×10^{-5} units of CAT enzyme (3.0×10^6 CAT trimers).

9.4.2.4 Protein Expression in the Midgut of *Aedes aegypti*

As previously shown, TE/3'2J viruses efficiently infect nerve and salivary gland tissues and poorly infect midgut tissues of mosquitoes when the route of infection is by intrathoracic inoculation (Higgs et al., 1993; 1998; Olson et al., 1994; 1996; Rayms-Keller et al., 1995). Additionally, TE/3'2J viruses poorly infect midgut tissues when delivered by the *per os* route (Seabaugh et al., 1998). A Malaysian SIN virus isolate (MRE16) has been identified that efficiently infects *Ae. aegypti* midgut tissues after ingestion, and >95% of these mosquitoes also develop disseminated infections within 14 days (Figure 9.6A and B). The entire subgenomic RNA of MRE16 virus has been sequenced and a chimeric SIN cDNA infectious clone, designated MRE1001, has been developed that contains sequence elements of TE/3'2J and MRE16 virus. MRE1001 virus efficiently infects midgut cells, and greater than 90% of infected mosquitoes develop disseminated infections after 14 days extrinsic incubation (Figure 9.6A and B).

A new dsSIN infectious cDNA clone (MRE/3'2J) has been generated from the MRE1001 chimeric cDNA (Olson et al., 2000). This was accomplished by replacing the 3' NCR of MRE1001 with the second subgenomic promoter, GOI, and 3' NCR of a recombinant TE/3'2J plasmid. Using this technique, we have developed MRE/3'2J viruses that express CAT. CAT expression in *Ae. aegypti* (RexD) is now being used to characterize the tropism of MRE/3'2J dsSIN viruses. *Ae. aegypti* have been infected with either MRE/3'2J/CAT or TE/3'2J/CAT virus by the *per os* route using virus concentrations of approximately 10^7 TCID$_{50}$/ml and 10^9 TCID$_{50}$/ml, respectively. At 14 days post-infection, the experimental mosquitoes were triturated and CAT reporter gene expression was quantified by assaying for CAT activity. We have compared CAT expression in whole *Ae. aegypti* infected with MRE/3'2J/CAT and TE/3'2J/CAT by the *per os* and intrathoracic inoculation routes of infection. The results clearly demonstrated that CAT was expressed in five of the six mosquitoes orally infected with MRE/3'2J/CAT virus (Figure 9.7). In contrast, only one of six mosquitoes infected with TE/3'2J/CAT virus efficiently expressed CAT in orally infected mosquitoes. Additionally, both MRE/3'2J/CAT and TE/3'2J/CAT expressed CAT virus efficiently if the viruses were injected intrathoracically into mosquitoes bypassing the midgut. We have developed an MRE/3'2J virus that expresses GFP and have used this virus to observe enhanced expression of GFP in the midgut and dissemination of GFP expression following *per os* infection in *Ae. aegypti* (Olson et al., 2000).

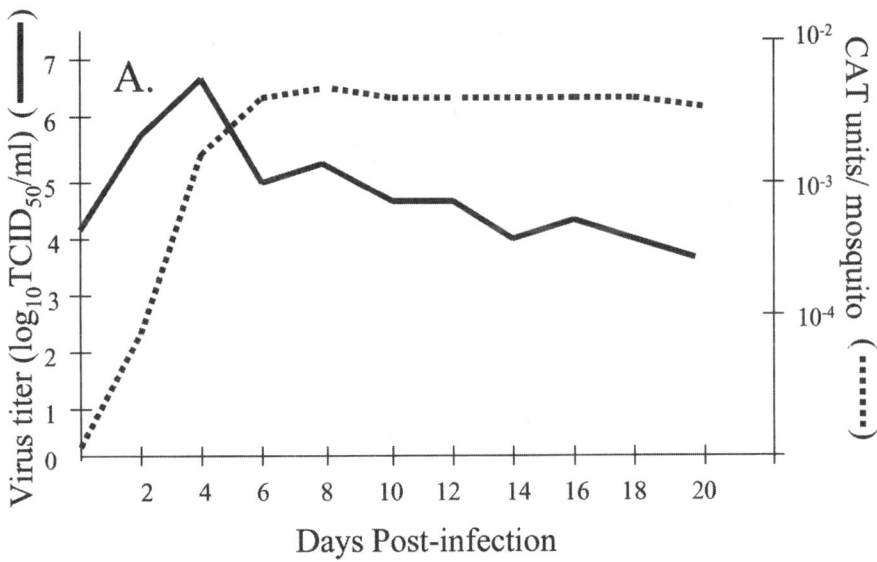

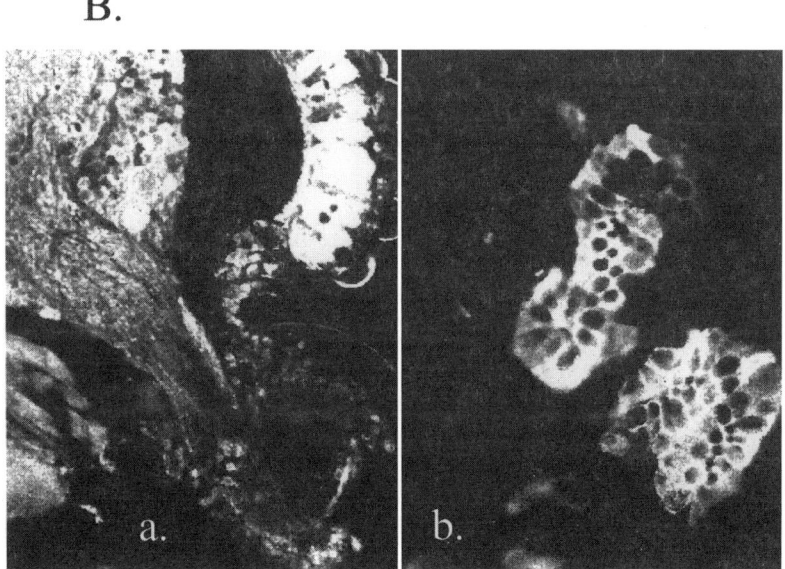

FIGURE 9.6 (A) Determination of TE/3'2J/CAT virus titer and CAT activity in intrathoracically inoculated *Ae. triseriatus*. (B) Detection of CAT expression in (a) head and (b) salivary gland tissue sections at 20 days post-infection using an immunofluorescence technique. (From Olson, K. E. et al., *Insect Biochem. Mol. Biol.*, 24, 39–48, 1994. With permission.)

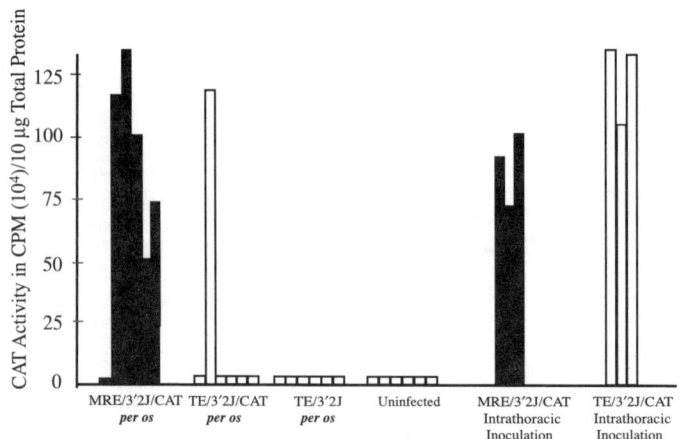

FIGURE 9.7 CAT expression in *Ae. aegypti* (Rex D) 14 days following ingestion of TE/3'2J/CAT or MRE/3'2J/CAT virus. See Olson et al. (1994) or Kamrud et al. (1995) for a description of CAT detection protocols in mosquitoes. In this figure, CAT activity is reported as total counts/min per mosquito. Each virus was also intrathoracically inoculated into mosquitoes as a positive control. TE/3'2J dsSIN virus containing no insert was fed as a negative control.

9.4.2.5 Expression of Antiviral RNAs in *Aedes aegypti* and *Aedes triseriatus*

9.4.2.5.1 *Expression of an Antisense RNA to LaCrosse Virus in* Aedes triseriatus

The temporal and spatial expression of an antisense RNA has been described in detail within *Ae. triseriatus* intrathoracically inoculated with a recombinant dsSIN virus (Rayms-Keller et al., 1995; Olson et al., 1998). In addition, the antisense RNA, which is complementary to the nucleocapsid mRNA of La Crosse virus (LAC; Bunyaviridae), profoundly inhibits LAC virus infection in mosquito cells (Powers et al., 1994; 1996).

9.4.2.5.2 *RNA-Mediated Interference with Flavivirus Replication in C6/36 Cells and* Aedes aegypti

dsSIN viruses also have been used to express effector molecules targeted to dengue and yellow fever viruses (Flaviviridae). For more information on the natural history of flaviviruses see the following reviews: Gubler et al. (1998), Rice (1996), Robertson et al. (1996). The full-length premembrane (prM) coding regions of the dengue type 2 (DEN-2; Jamaica) virus genomes have been expressed in C6/36 cells from dsSIN viruses in either the sense (D2prMs) or the antisense (D2prMa) orientation (Gaines et al., 1996). Additional dsSIN viruses have been generated containing prM antisense or NS5 RNA derived from YFV (17-D) virus (Higgs et al., 1998). C6/36 cells infected with each dsSIN virus (50 MOI) and challenged 48 h later with either DEN-2 (0.1 MOI; New Guinea C strain) or YF (0.1 MOI; Asibi) viruses. C6/36 cells infected with a control dsSIN virus supported high levels of DEN-2 (Figure 9.8) or YF virus replication. C6/36 cells infected with the dsSIN virus expressing DEN-2 or YF prM effector RNAs were completely resistant to DEN-2 or YF challenge virus, respectively (Figure 9.8). Additionally, cells expressing DEN-2 prM protein or untranslatable DEN-2 prM sense RNA also were resistant to DEN-2 challenge. Cells expressing prM protein demonstrated some breakthrough of DEN-2 virus when challenged at 10 MOI. However, expressed untranslatable sense prM RNA conferred complete protection to challenge at the high MOI (Gaines et al., 1996).

Approximately 5.0 $\log_{10}TCID_{50}$ of D2prMa or D2prMs and 3.0 $\log_{10}TCID_{50}$ of DEN-2 or YF virus were coinjected (intrathoracically) into *Ae. aegypti* females (Olson et al., 1996; Higgs et al., 1998). After 11 days the mosquito heads, salivary glands, and midguts were dissected and removed

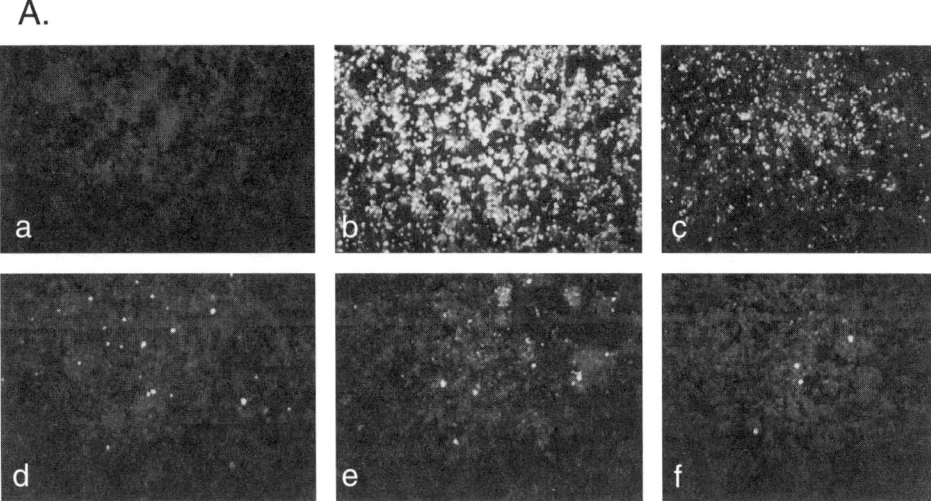

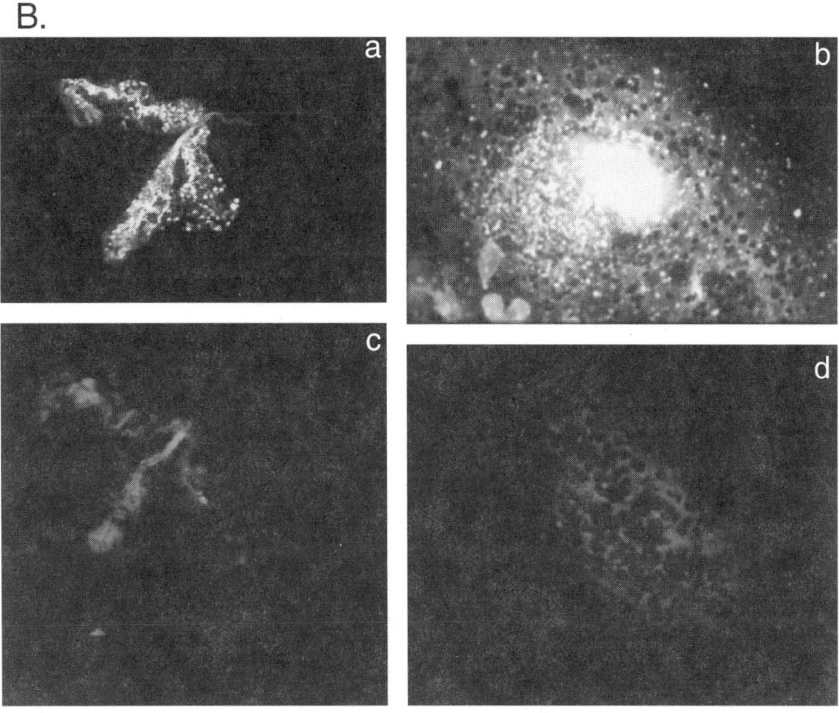

FIGURE 9.8 RNA-mediated interference with DEN-2 virus in C6/36 cells (A) and *Ae. aegypti* (B). (A) Analysis of resistance to DEN-2 virus (New Guinea C) replication in C6/36 cells. C636 cells were infected with no dsSIN virus (a and b), TE/3'2J virus (c), D2prMa virus (d), or D2prMs virus (e) at an MOI of 50 and challenged 48 h later with DEN-2 virus at an MOI of 0.1. The cells in panel (a) were not infected with DEN-2 virus. DEN-2 virus replication was determined 5 days postchallenge by IFA using an anti-E primary antibody. (From Gaines, P. et al., *J. Virol.*, 70, 2132–2137, 1996. With permission.) (B) Detection of DEN-2 E antigen in mosquito head and salivary gland tissues by IIFA 14 days after co-inoculation. Mosquitoes coinoculated with TE/3'2J and DEN-2 virus (16681) were analyzed for the presence of E antigen in salivary gland (a) and head (b) tissues. Mosquitoes co-inoculated with D2prM.np and DEN-2 virus were analyzed for the presence of DEN E antigen in salivary gland (c) and head (d) tissues. Magnifications were 20× for the salivary glands and 40× for the head tissues.

for immunofluoresence analysis. Tissues were assayed for SIN E1 protein and DEN-2 or YF E protein. Results of coinjecting D2prMs and DEN-2 viruses are shown in Figure 9.8B. Mosquitoes injected with the control dsSIN virus supported DEN-2 and YF virus replication in all tissues analyzed; the heads and salivary glands of mosquitoes injected with D2prMa, D2prMs, or YFprMa dsSIN viruses were not permissive for either DEN-2 virus or YF, respectively. Saliva was collected from mosquitoes 14 days post-infection and intrathoracically inoculated into groups of *Ae. aegypti*. After 7 days, mosquito heads were examined for DEN-2 or YF virus by immunofluorescence. D2prMa, D2prMs, and YFprMa viruses efficiently inhibited the biological transmission of the virus from which the effector RNA was derived (Olson et al., 1996; Higgs et al., 1998). In general, effector RNAs derived from DEN-2 virus profoundly inhibited replication of DEN-2 virus in mosquito cells, but were ineffective against other DEN virus serotypes and other flaviviruses (Gaines et al., 1996). This was also true for inhibition of virus transmission by *Ae. aegypti*. Conversely, effector RNAs derived from YF virus completely inhibited replication of West African strains of YF, inhibited South American strains of YF less well, and were completely ineffective against DEN viruses (Higgs et al., 1998). The dsSIN expression system has been an invaluable tool for testing in mosquito cells the interference potential of a number of different antisense RNAs of various sizes targeted to different flavivirus genes and distinct genetic virus strains. A summary of antisense RNA effects on flavivirus transmission is shown in Table 9.1.

9.4.3 ADVANTAGES AND DISADVANTAGES OF DOUBLE SUBGENOMIC SINDBIS VIRUSES

9.4.3.1 Advantages

1. The insertion of heterologous genes into dsSIN viruses is relatively simple and infectious dsSIN viruses can be generated within 24 h.
2. SIN virus has a wide host range and can be used to infect a number of different insect cell types. In tissue culture cells that are susceptible to SIN virus infections, virtually 100% of the cells can express the gene of interest within 18 h. Virus production can occur in BSL-2 conditions.
3. Relatively large numbers of effector molecules, such as antisense RNAs, can be conveniently screened for activity in mosquito cells.
4. Heterologous RNAs are transcribed to high efficiency in the cytoplasm of insect cells, and protein expression levels of the heterologous protein also are high.
5. Since dsSIN viruses replicate in the cytoplasm of the cell using their own RNA-dependent RNA polymerase, problems associated with nuclear DNA expression such as mRNA splicing, mRNA transport, and expression from poorly characterized DNA-based regulatory elements are obviated.
6. Mosquito cells often become persistently infected and long-term expression in these cells can be evaluated.
7. Because the molecular biology of SIN viruses is so well understood, it may be possible to manipulate the SIN genome to design tissue-specific and species-specific expression systems.

9.4.3.2 Disadvantages

1. The dsSIN viruses are clearly not suited for DNA promoter characterization and induction of gene expression.
2. SIN viruses do not vertically infect mosquitoes, so expression of the gene of interest in mosquitoes is limited to a single generation.
3. Gene expression is limited to cells that are susceptible to SIN virus infection unless trafficking signal sequences (e.g., secretory signals) are fused to the gene of interest to

TABLE 9.1
Antisense RNAs Targeted to YF or DEN Viruses

dsSIN Virus[a]	Antisense RNA Size No. Bases (source)[b]	Major Target Sequence	Challenge Virus(es)	Interference Potential[c]
D1prM	249 (DEN-1)	prM	DEN-1	+
D1GDD	243 (DEN-1)	NS5	DEN-1	+
D1GDDFsh	160 (DEN-1)	NS5	DEN-1	+
D1GDDRsh	183 (DEN-1)	NS5	DEN-1	+
D1GDDFshRsh	105 (DEN-1)	NS5	DEN-1	−
D1 5′-NS5	180 (DEN-1)	NS5	DEN-1	−
D2prM	540 (DEN-2)	prM	DEN-2	+
D2prM(290)	290 (DEN-2)	prM	DEN-2	+
D2NS2b-3	800 (DEN-2)	NS2b-NS3	DEN-2	−
D3prM	286 (DEN-3)	prM	DEN-3	+
D3GDD	243 (DEN-3)	NS5	DEN-3	+
D3 5′NS3	216 (DEN-3)	NS3	DEN-3	−
D3cap	223 (DEN-3)	Capsid	DEN-3	+
D4cap	240 (DEN-4)	Capsid	DEN-4	+
D4cap 5′	135 (DEN-4)	Capsid	DEN-4	−
D4cap 3′	105 (DEN-4)	Capsid	DEN-4	−
D4cap-D1GDD	240+243 (hybrid)	Capsid	DEN-4	−
		NS5	DEN-1	+
D4cap-D2prM(290)	240+290 (hybrid)	Capsid	DEN-4	−
		prM	DEN-2	+
D3cap-D4cap	223+240 (hybrid)	Capsid	DEN-1	+
		Capsid	DEN-3	+
D1GDD-D3GDD	243+243 (hybrid)	NS5	DEN-1	+
		NS5	DEN-3	+
D1prM-D3prM	290+286 (hybrid)	prM	DEN-1	+
		prM	DEN-3	+
YfprM	750 (YFV-17D)	prM	YFV (Asibi)	+
YfGDD	1009 (YFV-17D)	NS5	YFV (Asibi)	+

[a] Designated name of the dsSIN virus. All dsSIN viruses used in these studies were engineered from the TE/3′2J dsSIN plasmid.

[b] Source refers to the virus strain from which the effector RNA sequence was derived. DEN-1 (Hawaii), DEN-2 (Jamaica), DEN-3 (H-87), DEN-4 (H-241), and YFV (17D).

[c] Effector RNAs showing (+) interference potential show >95% reduction in cells capable of supporting a challenge virus infection.

move the gene product in and out of cells. Some cell types such as those found in Malpighian tubules and ovaries appear refractory to SIN virus infection.

4. While a number of different mosquito species can express genes of interest when the virus is delivered by intrathoracic inoculation, the midgut cells are usually not infected. In addition, successful gene expression in midgut tissues by the *per os* route is possible using existing SIN virus expression systems, but expression is mosquito species specific.
5. For safety considerations, insects capable of transmitting SIN viruses should be kept at BSL-3 conditions.
6. Heterologous genes can be unstable in dsSIN viruses with repeated passages of the virus.
7. The size of the heterologous gene should be less than 1.5 kb.

9.5 EXPRESSION OF GENES OF INTEREST IN MOSQUITO CELLS AND MOSQUITOES BY THE SINDBIS VIRUS REPLICON

9.5.1 SINDBIS REPLICON: METHODOLOGY

The replicon plasmid, pSINrep5, and helper plasmids DH-EB and DH(26S)5'SIN are now commercially available from Invitrogen Corp. (Carlsbad, CA). Many of the protocols used to express genes of interest from SIN replicon RNA are described in the previous section. Additional detailed methodology for recombinant DNA manipulation of pSINrep5, *in vitro* transcription of replicon and helper RNAs, transfection of RNAs into cells, and packaging of replicon RNA are available through Invitrogen. We have found that SIN replicon packaged using the SIN structural genes from DH(26S)5'SIN RNA produces the best expression of reporter genes in C6/36 cells (Kamrud et al., 1995). Estimations of virus titer can be obtained by determining the number of cells expressing the gene of interest (e.g., GFP) per unit volume of virus (Kamrud et al., 1995). Additionally, estimations of the titers of packaged replicons can be determined using a transformed BHK-21 cell line that expresses luciferase in response to replicon expression (ATCC, Rockville, MD; Olivo et al., 1994).

9.5.2 APPLICATIONS OF GENE EXPRESSION IN MOSQUITO CELLS AND MOSQUITOES USING THE SINDBIS REPLICON

9.5.2.1 Protein Expression from Sindbis Replicon RNA in C6/36 Cells

SIN Replicon RNA containing the CAT gene has been transfected into C6/36 cells (Xiong et al., 1989; Olson et al., 1992). Although transfection occurs in <1% of the C6/36 cells, CAT enzyme activity in the cells can be detected at 8 h post-transfection, peaks at 24 h, and is still be detected at 7 days post-transfection. At 24 h post-transfection, each transfected C6/36 cell expresses $>1 \times 10^7$ CAT polypeptides. Thus, high levels of expression are possible in cells that are transfected with the RNA.

9.5.2.2 Protein Expression from Packaged Sindbis Replicons in Mosquito Cells

BHK-21 cells cotransfected with pSINrep/CAT replicon RNA and either DH-EB or DH(26S)5'SIN SIN defective helper RNAs generate high-titer viruses, designated rep5/CAT/EB and rep5/CAT/26S (Kamrud et al., 1995). C6/36 cells infected with rep5/CAT/EB or rep5/CAT/26S virus at 3 MOI express high levels of CAT (1.1×10^7 and 1.8×10^7 CAT polypeptides per cell, respectively) at 2 days post-infection. Significantly, greater than 25% of the cells expressed the reporter gene, indicating that packaging increases the efficiency of delivery of the replicon. However, infecting C6/36 cells at MOIs of 10 or greater increases the percentage of cells expressing the gene of interest to greater than 50%.

Packaged SIN replicons also have been used to express and localize viral envelope proteins of LAC virus in mosquito cell culture (Kamrud et al., 1998). The expressed LAC proteins had correct molecular mass (Mr) and were antigenically similar to wild-type LAC envelope proteins. In addition, LAC G_1 and G_2 proteins colocalized when expressed from separate replicons in both mammalian and mosquito cells, suggesting that they were trafficked through the cell similarly to wild-type LAC proteins. A truncated form of the G_1 protein was secreted from mosquito cells when expressed alone. The truncated G_1 protein was also secreted from mosquito cells when expressed with the G_2 protein, but to a lesser extent than when expressed alone, suggesting that the G_2 protein sequestered G_1 protein intracellularly. Additionally, we have used the SIN replicon to express the polymerase gene, NS5, of YFV in both mammalian and mosquito cells and have demonstrated that the expressed NS5 gene localizes similarly to wild-type YFV (unpublished data). The SIN replicon system may be a powerful tool for analyzing protein maturation of pathogens within mosquito cells and mosquito organs and may allow genetic manipulation of these genes (e.g., site-directed mutagenesis) to analyze gene function.

9.5.2.3 Protein Expression from Packaged Sindbis Replicons in Mosquitoes

Aedes triseriatus mosquitoes were intrathoracically inoculated with 7×10^4 infectious focal IFU rep5/CAT/EB or 1×10^5 IFU rep5/CAT/26S virus. Virus titers remained at approximately 10^5 IFU/ml through day 2 post-infection and decreased roughly 1 log by 10 days post-infection. CAT enzyme activity was detected 2 days post-infection (rep5/CAT/EB: 1.49×10^{-4} units CAT/10 μg protein; rep5/CAT/26S: 2.03×10^{-5} units CAT/10 μg protein) and remained near these levels through 10 days post-infection (Figure 9.9A). CAT was detected in the head and salivary glands of inoculated mosquitoes by indirect immunofluorescence or CAT activity assays (Figure 9.9B). Additionally, each mosquito infected with rep5/CAT/26S CAT virus secreted approximately 1×10^6 CAT molecules into their saliva (Kamrud et al., 1997). These results suggest that packaged replicon viruses may be important tools for analyzing the function of mosquito salivary gland effector molecules. Packaged replicon viruses may also be ideal tools for expressing potentially immunogenic proteins in the saliva of mosquitoes that can be delivered to vertebrates during intake of a blood meal.

9.5.3 ADVANTAGES AND DISADVANTAGES OF SINDBIS REPLICON EXPRESSION

Many of the same advantages and disadvantages listed for dsSIN viruses as tools of gene expression in mosquitoes also apply to the SIN replicon. The SIN replicon allows expression of cDNAs of up to 5 kb in length in cells infected with the packaged replicon and greater than 5 kb in cells transfected with the RNA only. An additional advantage of SIN replicons is that new packaging systems for alphavirus replicons have been developed that eliminate recombination events and prevent production of infectious viruses, thereby reducing biosafety concerns (Pushko et al., 1997). However, several disadvantages should be noted. Determining the titers of packaged replicons is more difficult than for dsSIN because packaged replicons contain defective genomes and do not disseminate from cells initially infected with the virus. Additionally, because both *in vitro* transcription of the replicon and replication of the replicon in transfected cells are dependent on error-prone polymerases, expression of the heterologous gene can be unstable (Agapov et al., 1998). This can be overcome in cultured cells by analyzing expression soon after transfection with the SIN replicon, by selecting cells specifically expressing the gene of interest, and by expressing replicons from DNA transfection. This latter approach would transcribe replicon from a eukaryotic RNA polymerase II, thus increasing fidelity of the replicon (Agapov et al., 1998).

9.6 ALPHAVIRUS EXPRESSION SYSTEMS: FUTURE APPLICATIONS

9.6.1 RAPID DETERMINATION OF FUNCTION AND UTILITY OF EFFECTOR MOLECULES IN MOSQUITOES PRIOR TO TRANSFORMATION

SIN virus expression systems provide unique opportunities to characterize genes in juvenile and adult insect stages rapidly. TE/3'2J-based viruses can be intrathoracically inoculated into larvae or adults to evaluate expression of scorpion toxin (Higgs et al., 1995), juvenile hormone esterase (Hammock et al., 1990), and *Ae. aegypti* head peptide (Brown et al., 1994). The MRE/3'2J-based viruses are currently being used to express *Aedes* defensins in orally infected adult *Ae. aegypti*. Another use of MRE/3'2J-based viruses could include expression of serine proteases that initiate the prophenol oxidase cascade to study parasite melanization and antimicrobial responses within mosquitoes (Christensen and Severson, 1993). dsSIN or packaged SIN replicon viruses also will be important tools for analyzing whether or not an isolated mosquito gene is involved in conditioning vector competence. In addition, SIN virus expression systems have already been shown to be effective delivery vehicles of single-chain antibodies and antiviral RNA effector molecules (Jiang

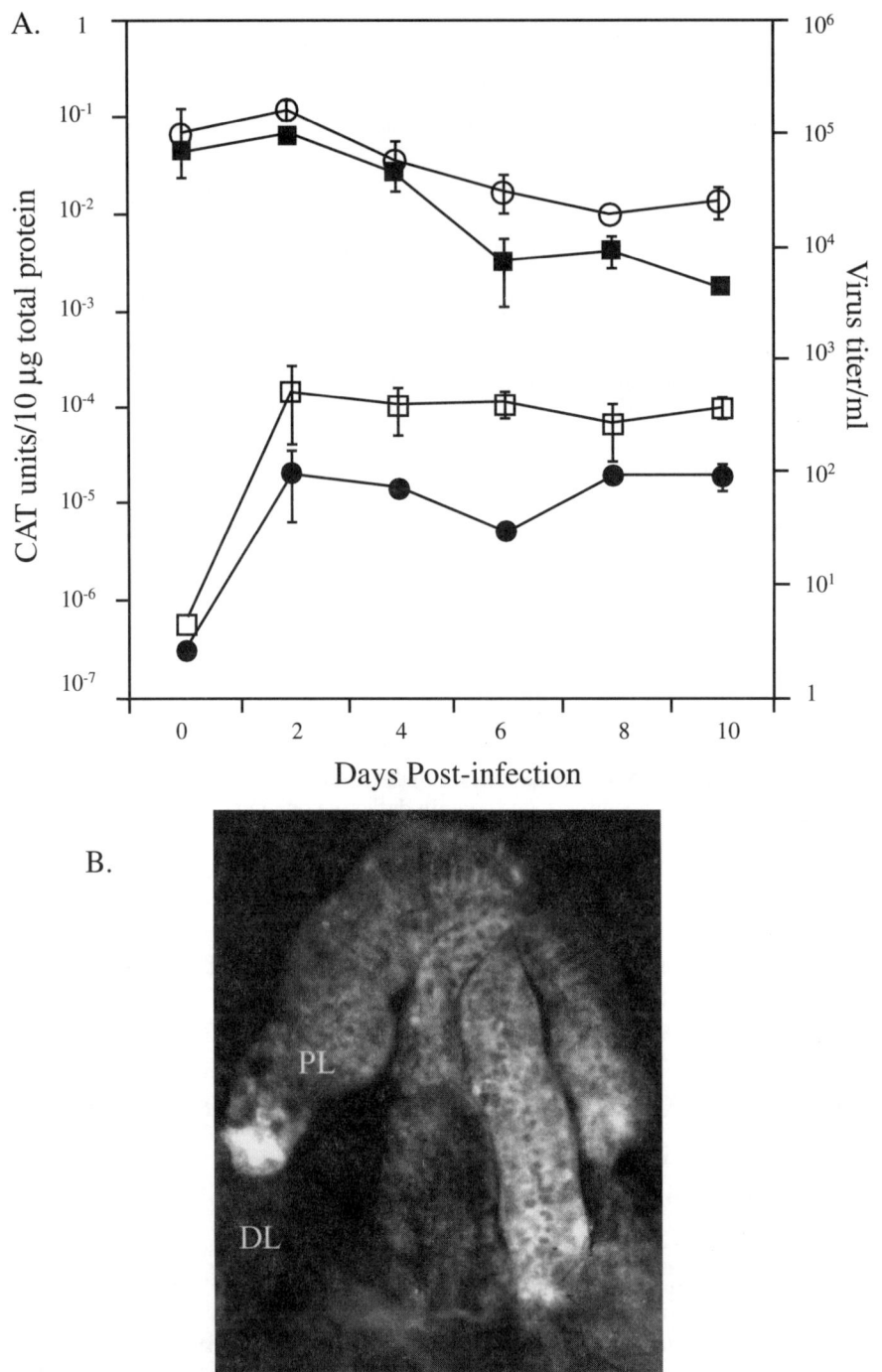

FIGURE 9.9 (A) CAT expression and corresponding titer of packaged replicons rep5/CAT/EB or rep5/CAT/26S following intrathoracic inoculation of 1×10^5 infectious particles/ml into *Ae. triseriatus*. (■) rep5/CAT/EB virus titer, (○) rep5/CAT/26S virus titer, (□) rep5/CAT/EB CAT activity, (●) rep5/CAT/26S CAT activity. (B) Immunofluorescence analysis of CAT expression in salivary glands 3 days after inoculating *Ae. triseratus* with rep5/CAT/26S virus. (From Kamrud, K. I. et al., *Exp. Parasitol.*, 81, 394–403, 1995. With permission.)

et al., 1995; Gaines et al., 1996). At AIDL, we have currently evaluated over 50 different dsSIN viruses expressing translatable sense, untranslatable sense, and antisense RNAs for their ability to inhibit the replication of DEN, YF, and LAC viruses in mosquitoes. dsSIN viruses have greatly facilitated the identification of antiviral effector RNAs that function in mosquitoes, and that would have been an extremely difficult task using transgenic strategies.

9.6.2 DOUBLE SUBGENOMIC SINDBIS VIRUSES: IMPORTANT MOLECULAR TOOLS FOR OBSERVING VIRUS-SPECIFIC INTERACTIONS WITH THE ARTHROPOD VECTOR

Future applications of dsSIN viruses include expression of marker genes such as GFP in infected mosquitoes to analyze the temporal and spatial patterns of SIN virus distribution during the extrinsic incubation period. This could be especially important in identifying cell types within midgut or salivary gland tissues that support virus infection and that might be targets of gene expression. The effects of altering specific regions of the dsSIN genome by well-established mutagenesis protocols could be rapidly evaluated by observing changes in GFP expression in infected tissues. Similarly, SIN expression systems may be important tools for facilitating the genetic manipulation, complementation, and expression of genes derived from arthropod-borne RNA viruses not readily adaptable to infectious clone technology.

9.6.3 DOUBLE SUBGENOMIC SINDBIS VIRUSES: POTENTIAL TOOLS FOR DOWN-REGULATING GENE EXPRESSION IN THE MOSQUITO

Because of the robust transcription of SIN viruses in infected cells, we have hypothesized that dsSIN viruses may be able to express antisense RNAs that can downregulate endogenous genes of mosquitoes. To test this hypothesis we have engineered a dsSIN virus that expresses the 5' 595 bases of the firefly luciferase gene in an antisense orientation. We then infected transformed mosquito cells (Klimowski et al., 1996) and mosquitoes (Coates et al., 1999) that express luciferase with the dsSIN virus, TE/3'2J-αLuc, to evaluate the effect of the antisense RNA on luciferase expression. Preliminary results show that luciferase activity in TE/3'2J-αLuc infected cells decreases >90% within 48 h post-infection (M. Bennett, personal communication). Transformed mosquitoes expressing luciferase show > 10-fold decrease in luciferase activity when infected with TE/3'2J/αLuc virus (Johnson et al., 1999). This preliminary evidence is encouraging and indicates that this approach may allow researchers to study gene function by inhibiting its expression.

9.6.4 DEVELOPMENT OF ALPHAVIRUS EXPRESSION SYSTEMS THAT ARE TISSUE AND SPECIES SPECIFIC IN MOSQUITOES

The studies of Woodward et al. (1991) and Seabaugh et al. (1998) have demonstrated that alterations in the structural genes of alphaviruses dramatically alter the ability of the virus to infect midgut cells and suggest two strategies for improving SIN virus expression systems. The first strategy takes advantage of the high genetic variability found in RNA virus populations (Holland et al., 1982). Specific genetic variants of the virus can be enriched by multiply passaging a dsSIN virus in specific mosquito tissues. We have used this approach to identify a passage 4 virus with enhanced infection of *Ae. aegypti* midguts (Seabaugh, 1997). Although the passage 4 virus lost its insert, amino acid changes were noted in the structural genes of the virus that could be engineered into dsSIN plasmids. A second strategy is to identify naturally occurring SIN virus strains that efficiently infect mosquitoes orally and then disseminate to other tissues. This strategy has led to the engineering of MRE/3'2J virus for gene expression in *Ae. aegypti* midguts (Olson et al., 2000). These strategies also can be applied to enhancing tissue-specific gene expression by dsSIN viruses in other species of mosquitoes.

A third strategy is to develop chimeric dsSIN expression viruses using random insertions of specific receptor ligands into the PE2 coding region to alter the host range of a full-length SIN viral cDNA clone (London et al., 1992; Dubuisson and Rice, 1993). This technique has already been used to develop new SIN virus expression systems that can incorporate an immunologically active neutralizing epitope of Rift Valley fever virus (Bunyaviridae) into the E2 glycoprotein of SIN virus (London et al., 1992). Others have altered the host-range specificity of SIN viruses by insertion of ligands (alpha- and beta-hCG genes) and targeting human choriocarcinoma cells (Sawai and Meruelo, 1998). Additionally, it should be possible to engineer expression systems based on alphaviruses other than SIN virus. Of particular interest would be a double subgenomic or replicon alphavirus expression system based on O'nyong nyong virus (Johnson et al., 1988). This virus naturally infects anopheline mosquitoes and could be engineered for gene expression in these mosquitoes.

A new SIN replicon has recently been described with decreased toxicity effects in vertebrate cells (Agapov et al., 1998). However, it is unclear whether this replicon has similar effects in mosquito cells. It would be interesting to compare replication and development of mosquito cells or mosquito embryos containing this replicon with that of the replicon currently in use. In addition, it may be possible to deliver SIN replicon RNAs to specific cell types. A SIN replicon transfection system has been described that results in high levels of transient protein expression in specific types of cancer cells. This system uses a mixture of streptavidin–protein A (ST-PA) fusion protein, receptor-specific mABs, and biotinylated SIN replicon in the presence of cationic liposomes (Sawai et al., 1998). This complex is able to transfect a reporter gene to specific cancer cells in an mAB dose-dependent manner. This approach may eventually allow targeting of SIN replicons to specific cell types in mosquitoes. Finally, SIN replicon RNAs packaged with structural proteins from MRE16 virus may facilitate introduction of the replicon into midguts of *Ae. aegypti* by the *per os* route (Seabaugh et al., 1998).

9.6.5 DNA-Based Alphavirus Replicons as an Efficient Means of Overexpressing Antipathogen Genes in a Heritable Fashion

Investigators have reported successful gene expression from SIN and SF replicons transcribed by eukaryotic promoters (Dubensky et al., 1996; DiCiommo and Bremner, 1998). To potentially enhance the effect of DNA-based antiviral strategies in mosquito cells, we are currently expressing reporter genes from integrated SIN replicons. We have initially constructed a transcriptional unit that can express a SIN replicon containing GFP from a constitutive baculovirus promoter hr5/IE1 that is active in insect cells (Jarvis and Finn, 1996). If this approach allows efficient expression of GFP, we can design similar transcriptional units that express gene-based antiviral agents instead of GFP. Several advantages are evident with this approach:

1. Replicon RNA transcribed from the DNA promoter should be self-replicating and large copy numbers of effector molecules can be rapidly produced.
2. Identifying eukaryotic promoters with high expression potential in mosquitoes becomes less important as a factor in generating the antiviral agent within mosquito cells.
3. This approach also minimizes adverse positional effects of stable integration on efficient expression of the transcriptional unit.
4. Since the SIN replicon replicates in the cytoplasm of cells and in similar cellular compartments as the targeted arbovirus, this approach may allow proper trafficking of the antiviral agent (particularly effector RNAs) and increase the efficiency of inhibition.
5. Increased stability of the replicon transcripts. If this approach leads to the development of cell lines that are highly refractory to LAC or DEN-2 virus replication, we will incorporate the replicon transcriptional unit into transposable elements such as the *Hermes* element used in mosquito transformation (Jasinskiene et al., 1998). Salivary gland, adult, and sex-specific promoters such as the apyrase promoter will be used to express

the replicon, which in turn will express the antiviral gene (such as antisense RNA targeted to the prM gene of DEN or YF viruses).

9.7 SUMMARY

As this chapter has demonstrated, SIN expression systems are important tools for molecularly manipulating mosquitoes. The SIN systems allow timely and ready assessment of constructs and genes in comparison with conventional transgenesis approaches. The SIN expression systems are proving to be invaluable tools in efforts to understand vector–pathogen interactions, vector competence, and other components of vector–pathogen amplification and maintenance cycles in nature. These SIN virus–based expression systems should facilitate the researcher's ability to decide which gene-based disease control strategies merit a further investment in time and resources in transgenic mosquitoes or in antimosquito control agents such as densoviruses (Parvoviridae; see Chapter 8 by Carlson). Additionally, the potential to engineer SIN-based and other alphavirus expression systems to express efficiently genes of interest in specific tissues or a particular mosquito species is a particularly attractive feature of these virus transduction systems. SIN expression systems will also undoubtedly provide considerable information about gene regulation and expression in mosquitoes and consequently a much better understanding of the biology and molecular biology of vectors. Such knowledge is essential for developing effective control strategies for vector-borne pathogens of humans and animals.

REFERENCES

Agapov EV, Frolov I, Lindenbach BD, Pragai BM, Schlesinger S, Rice CM (1998) Noncytopathic sindbis virus RNA vectors for heterologous gene expression. Proc Natl Acad Sci USA 95:12989–12994

Berglund P, Smerdou C, Fleeton MN, Tubulekas I, Liljestrom P (1998) Enhancing immune responses using suicidal DNA vaccines. Nat Biotechnol 16:562–565

Bredenbeek PJ, Rice CM (1992) Animal RNA virus expression systems. Sem Virol 3: 297–310

Bredenbeek PJ, Frolov I, Rice CM, Schlesinger S (1993) Sindbis virus expression vectors: packaging of RNA replicons by using defective helper RNAs. J Virol 67:6439–6446

Bowers DF, Abell BA, Brown DT (1995) Replication and tissue tropism of the alphavirus Sindbis in the mosquito *Aedes albopictus*. Virology 212:1–12

Brown MR, Klowden MJ, Crim JW, Young L, Shrouder LA, Lea AO (1994) Endogenous regulation of mosquito host-seeking behavior by a neuropeptide. J Insect Physiol 40:399–406

Carlson JO, Olson KE, Higgs S, Beaty BJ (1995) Molecular genetic manipulation of mosquito vectors. Annu Rev Entomol 40:359–388

Christensen BM, Severson DW (1993) Biochemical and molecular basis of mosquito susceptibility to Plasmodium and filaroid nematode. In Beckage NE, Thompson SN, Frederici BA (eds) Parasites and Pathogens of Insects, Vol 1. Academic Press, Orlando, FL, pp 245–266

Coates CJ, Jasinskiene N, Miyashiro L, James AA (1998) Mariner transposition and transformation of the yellow fever mosquito, *Aedes aegypti*. Proc Natl Acad Sci USA 95:3748–3751

Coates CJ, Jasinskiene N, Pott GB, James AA (1999) Promoter-directed expression of recombinant fire-fly luciferase in the salivary glands of Hermes-transformed *Aedes aegypti*. Gene 226:317–325

Davis NL, Willis LV, Smith JF, Johnston RE (1989) *In vitro* synthesis of infectious Venezuelan equine encephalitis virus RNA from a cDNA clone: analysis of a viable deletion mutant. Virology 171:189–204

DiCiommo DP, Bremner R (1998) Rapid, high level protein production using DNA-based Semliki Forest virus vectors. J Biol Chem 273:18060–18066

Dohm DJ, Logan TM, Barth JF, Turell MJ (1995) Laboratory transmission of Sindbis virus by *Aedes albopictus, Ae. aegypti*, and *Culex pipiens* (Diptera: Culicidae). J Med Entomol 32:818–821

Dubensky TW, Driver, DA, Polo JM, Belli BA, Latham EM, Ibanez CE, Chada S, Brumm D, Banks TA, Mento SJ, Jolly DJ, Chang SMW (1996) Sindbis virus DNA-based expression vectors: utility for *in vitro* and *in vivo* gene transfer. J Virol 70:508–519

Dubuisson J, Rice CM (1993) Sindbis virus attachment: isolation and characterization of mutants with impaired binding to vertebrate cells. Virology 67:3363–3374

Duffus WA, Levy-Mintz P, Klimjack MR, Kielian M (1995) Mutations in the putative fusion peptide of Semliki Forest virus affect spike protein oligomerization and virus assembly. J Virol 69:2471–2479

Frolov I, Hoffman TA, Pragai BM, Dryga SA, Huang HV, Schlesinger S, Rice CM (1996) Alphavirus-based expression vectors: strategies and applications. Proc Natl Acad Sci USA 93:11371–11377

Gaines P, Olson KE, Higgs S, Beaty BJ, Blair CD (1996) Pathogen derived resistance to dengue-2 virus in mosquito cells by expression of the premembrane coding region of the viral genome. J Virol 70:2132–2137

Griffin DE, Hardwick JM (1997) Regulators of apoptosis on the road to persistent alphavirus infection. Annu Rev Microbiol 51:565–592

Grimstad PR, Craig GB, Ross QE, Yuill TM (1977) *Aedes triseriatus* and La Crosse virus: geographic variation in vector susceptibility and ability to transmit. Am J Trop Med Hyg 26:990–996

Gubler DJ (1998) Dengue and dengue hemorrhagic fever. Clin Microbiol Rev 1:480–496.

Gwadz RW, Kaslow D, Lee JY, Maloy WL, Zasloff M, Miller LH (1989) Effects of magainins and cecropins on the sporogonic development of malaria parasites in mosquitoes. Infect Immun 57:2628–2633

Hahn CS, Hahn YS, Braciale TJ, Rice CM (1992) Infectious Sindbis virus transient expression vectors for studying antigen processing and presentation, Proc Natl Acad Sci USA 89:2679–2683

Hammock BD, McCutchen BF, Beetham J, Choudary PV, Fowler E, Ichinose R, Ward VK, Vickers JM, Bonning BC, Harshman LG (1993) Development of recombinant viral insecticides by expression of an insect-specific toxin and insect-specific enzyme in nuclear polyhedrosis viruses. Arch Insect Biochem Physiol 22:315–344

Hardy JL (1988) Susceptibility and resistance of vector mosquitoes. In Monath TP (ed) The Arboviruses: Epidemiology and Ecology. CRC Press, Boca Raton, FL, pp 87–126

Heidner HW, McKnight KL, Davis NL, Johnston RE (1994) Lethality of PE2 incorporation into Sindbis virus can be suppressed by second-site mutations in E3 and E2. J Virol 68:2683–2692

Heidner HW, Knott TA, and Johnston RE (1996) Differential processing of Sindbis virus glycoprotein PE2 in cultured vertebrate and arthropod cells. J Virol 70:2069–2073

Higgs S, Powers AM, Olson KE (1993) Alphavirus expression systems: applications to mosquito vector studies. Parasitol Today 9:444–452

Higgs S, Olson KE, Klimowski L, Powers AM, Carlson JO, Possee RD, Beaty BJ (1995) Mosquito sensitivity to a scorpion neurotoxin expressed using an infectious Sindbis virus vector. Insect Mol Biol 4:97–103

Higgs S, Olson KE, Kamrud KI, Powers AM, Beaty BJ (1997) Viral expression systems and viral infections in insects. In Crampton JM, Beard CB, Louis C (eds) The Molecular Biology of Disease Vectors: A Methods Manual. Chapman & Hall, London, pp 459–483

Higgs S, Traul D, Davis B, Wilcox B, Beaty B (1996) Green fluorescent protein expressed in living mosquitoes without the requirement for transformation. BioTechniques 21:660–664

Higgs S, Rayner JO, Olson KE, Davis BS, Beaty BJ, Blair CD (1998) Engineered resistance in *Aedes aegypti* to a West African and a South American strain of yellow fever virus. Am J Trop Med Hyg 58:663–670

Holland J, Spindler K, Horodysky F, Grabav E, Nichol S, Vanderpol S (1982) Rapid evolution of RNA genomes. Science 215:735–744

Hurlbut H, Thomas J (1960) The experimental host range of the arthropod-borne animal viruses in arthropods. Virology 12:391–407

Jackson AC, Bowen JC, Downe AER (1993) Experimental infection of *Aedes aegypti* (Diptera: Culicidaae) by the oral route with Sindbis virus. J Med Entomol 30:332–337

Jarvis DL, Finn EE (1996) Modifying the insect cell N-glycosylation pathway with immediate early baculovirus expression vectors. Nat Biotechnol 14:1288–1292

Jasinskiene N, Coates CJ, Benedict MQ, Cornel AJ, Rafferty CS, James AA, Collins FH (1998) Stable transformation of the yellow fever mosquito, *Aedes aegypti*, with the *Hermes* element from the housefly. Proc Natl Acad Sci USA 95:3743–3747

Jiang W, Venugopal K, Gould EA (1995) Intracellular interference of tick-borne flavivirus infection by using a single-chain antibody fragment delivered by recombinant Sindbis virus. J Virol 69:1044–1049

Johnson BK (1988) O'nyong-nyong virus disease. In Monath TP (ed) The Arboviruses: Epidemiology and Ecology. CRC Press, Boca Raton, FL, pp 217–223

Johnson BW, Olson KE, Allen-Miura T, Rayms-Keller A, Carlson JO, Coates CJ, Jasinskiene N, James AA, Beaty BJ, and Higgs S (1999). Inhibition of luciferase expression in transgenic *Aedes aegypti* mosquitoes by Sindbis virus expression of antisense luciferase RNA. Proc. Natl. Acad. Sci. USA 96:13399–13403

Jupp PG, Phillips JI (1998) An electron microscopical study of Rift Valley fever and Sindbis viral infection in mosquito salivary glands (Diptera: Culicidae). Afr Entomol 6:75–81

Kamrud KI, Powers AM, Higgs S, Olson KE, Blair CD, Carlson JO, Beaty BJ (1995) The expression of chloramphenicol acetyltransferase in mosquitoes and mosquito cells using a packaged Sindbis replicon system. Exp Parasitol 81:394–403

Kamrud KI, Olson KE, Higgs S, Powers AM, Carlson JO, Beaty BJ (1997) Detection of expressed chloramphenicol acetyltransferase in the saliva of *Culex pipiens* mosquitoes. Insect Biochem Mol Biol 27:423–429

Kamrud K, Olson K, Higgs S, Carlson J, Beaty B (1998) Use of the Sindbis replicon system for expression of LaCrosse virus envelope proteins in mosquito cells. Arch Virol 143:1365–1377

Karpf AR, Brown DT (1998) Comparison of Sindbis virus–induced pathology in mosquito and vertebrate cell cultures. Virology 240:193–201

Kerr PJ, Weir RC, Dalgarno L (1993) Ross River virus variants selected during passage in chick embryo fibroblasts: serological, genetic, and biological changes. Virology 193:446–449

Klimowski L, Rayms-Keller A, Olson KE, Tessari J, Yang R, Carlson J, Beaty B (1996) Heavy metals induce a molecular bioreporter system in mosquito cells. Environ Toxicol Chem 15:85–91

Kuhn RJ, Hong Z, Strauss JH (1990) Mutagenesis of the 3′ nontranslated region of Sindbis virus RNA. J Virol 64:1465–1476

Kuhn RJ, Niesters HG, Hong Z, Strauss JH (1991) Infectious RNA transcripts from Ross River virus cDNA clones and the construction and characterization of defined chimeras with Sindbis virus. Virology 182:430–441

Liljestrom P, Garoff H (1991) A new generation of animal cell expression vectors based on the Semliki Forest virus replicon. BioTechnology 9:1356–1361

Liljestrom P, Lusa S, Huylebroeck D, Garoff H (1991) *In vitro* mutagenesis of a full-length cDNA clone of Semliki Forest virus: the small 6,000-molecular-weight membrane protein modulates virus release. J Virol 65:4107–4113

London SD, Schmaljohn AL, Dalrymple JM, Rice CM (1992) Infectious enveloped RNA virus antigenic chimeras. Proc Natl Acad Sci USA 89:207–211

Lowenberger C, Bulet P, Charlet M, Hetru C, Hodgemean B, Christensen BM, Hoffman JA (1995) Insect immunity: isolation of three novel inducible antibacterial defensins from the vector mosquito, *Aedes aegypti*. Insect Biochem Mol Biol 25:867–873

Ludwig GV, Kondig JP, Smith JF (1996) A putative receptor for Venezuelan equine encephalitis virus from mosquito cells. J Virol 70:5592–5599

Lustig S, Jackson AC, Hahn CS, Griffin DE, Strauss EG, Strauss JH (1988) Molecular basis of Sindbis virus neurovirulence in mice. J Virol 62:2329–2336

Malherbe H, Strickland-Cholmley M (1963) Sindbis virus infection in man report of a case with recovery of virus from skin lesions. S Afr Med J 37:547–552

Miller ML, Brown DT (1992) Morphogenesis of Sindbis virus in three subclones of *Aedes albopictus* (mosquito) cells. J Virol 66:4180-4190

Miller BR, Mitchell CJ (1986) Passage of yellow fever virus: its effect on infection and transmission rates in *Aedes aegypti*. Am J Trop Med Hyg 35:1302–1309

Mourya DT, Ranadive SN, Gokhale MD, Barde PV, Padbidri VS, Banerjee K (1998) Putative chikungunya virus-specific receptor proteins on the midgut brush border membrane of *Aedes aegypti* mosquito. Indian J Med Res 107:10–14

Nasci RS, Mitchell CJ (1994) Larval diet, adult size, and susceptibility of *Aedes aegypti* (Diptera: Culicidae) to infection with Ross River virus. J Med Entomol 31:123–126

Olivo PD, Frolov I, Schlesinger S (1994) A cell line that expresses a reporter gene in response to infection by Sindbis virus: a prototype for detection of positive strand RNA viruses. Virology 198:381–384

Olson KE, Carlson JO, Beaty BJ (1992) Expression of the chloramphenicol acetyltransferase gene in *Aedes albopictus* (C6/36) cells using a non-infectious Sindbis virus expression vector. Insect Mol Biol 1:49–52

Olson KE, Higgs S, Hahn CS, Rice CM, Carlson JO, Beaty BJ (1994) The expression of chloramphenicol acetyltransferase in *Aedes albopictus*, C6/36 cells and *Aedes triseriatus* mosquitoes using a double subgenomic recombinant Sindbis virus. Insect Biochem Mol Biol 24:39–48

Olson K, Higgs S, Powers A, Davis BS, Carlson JO, Blair CD, Beaty BJ (1996) Genetically engineered resistance in mosquitoes to dengue virus transmission. Science 272: 884–886

Olson K, Beaty B, Higgs S (1998) Sindbis virus expression systems for the manipulation of insect vectors. In Miller LK, Ball LA (eds) The Insect Viruses. Plenum Press, New York, pp 371–404

Olson KE, Myles KM, Seabaugh RC, Higgs S, Carlson JO, Beaty BJ (2000) Development of a Sindbis virus expression system that efficiently expresses green fluorescent protein in midguts of *Aedes aegypti* following *per os* infection. Insect Mol Biol 9:57–65

Paredes AM, Simon M, Brown DT (1993) The mass of the Sindbis virus nucleocapsid suggests it has T = 4 icosahedral symmetry. Virology 187:324–332

Powers AM, Olson KE, Higgs S, Carlson JO, Beaty BJ (1994) Intracellular immunization of mosquito cells to La Crosse virus using a recombinant Sindbis virus vector. Virus Res 32:57–67

Powers A, Kamrud K, Olson K, Higgs S, Carlson J, Beaty B (1996) Molecularly engineered resistance to California serogroup virus replication in mosquito cells and mosquitoes. Proc Natl Acad Sci USA 93:4187–4191

Pushko P, Parker M, Ludwig GV, Davis NL, Johnston RE, Smith JF (1997) Replicon helper systems from attenuated Venezuelan equine encephalitis virus: expression of heterologous genes *in vitro* and immunization against heterologous pathogens *in vivo*. Virology 239:389–401

Raju R, Huang HV (1991) Analysis of Sindbis virus promoter recognition in vivo, using novel vectors with two subgenomic mRNA promoters. J Virol 65:2501–2510

Rayms-Keller A, Powers AM, Higgs S, Olson KE, Kamrud KI, Carlson JO, Beaty BJ (1995) Replication and expression of a recombinant Sindbis virus in mosquitoes. Insect Mol Biol 4:245–251

Rice CM (1996) Flaviviridae: the viruses and their replication. In Fields BN, Knipe DM, Howley PM (eds) Fields Virology, 3rd ed. Lippincott-Raven, Philadelphia, pp 961–1034

Rice CM, Levis R, Strauss JH, Huang HV (1987) Production of infectious RNA transcripts from Sindbis virus cDNA clones: mapping of lethal mutations, rescue of a temperature-sensitive marker, and *in vitro* mutagenesis to generate defined mutants. J Virol 61:3809–3819

Robertson SE, Hull BP, Tomori O, Bele O, LeDuc JW, Esteves K (1996) Yellow fever: a decade of reemergence. J Am Med Assoc 276:1157–1162

Rosen L, Gubler D (1974) The use of mosquitoes to detect and propagate dengue viruses. Am J Trop Med Hyg 23:1153–1160

Sawai K, Meruelo D (1998) Cell-specific transfection of choriocarcinoma cells by using Sindbis virus hCG expressing chimeric vector. Biochem Biophys Res Commun 248:315–323

Sawai K, Ohno K, Iijima Y, Levin B, Meruelo D (1998) A novel method of cell-specific mRNA transfection. Mol Genet Metab 64:44–51

Scott TW, Lorenz LH (1998) Reduction of *Culiseta melanura* fitness by Eastern equine encephalomyelitis virus. Am J Trop Med Hyg 59:341–346

Scott TW, Hildreth SW, Beaty BJ (1984) The distribution and development of Eastern equine encephalitis virus in its enzootic mosquito vector, *Culiseta melanura*. Am J Trop Med Hyg 33:300–310

Seabaugh RC (1997) Genetic Determinants of Sindbis Oral Infectivity in *Aedes aegypti* mosquitoes. Dissertation. Colorado State University, Fort Collins, CO

Seabaugh RC, Olson KE, Higgs S, Carlson JO, Beaty BJ (1998) Development of a chimeric Sindbis virus with enhanced *per os* infection of *Aedes aegypti*. Virology 243:99–112

Strauss EG, Rice CM, Strauss JH (1984) Complete nucleotide sequence of the genomic RNA of Sindbis virus. Virology 133:92–110

Strauss JH, Strauss EG (1994) The alphaviruses: gene expression, replication and evolution. Microbiol Rev 58:491–562

Tabachnick WJ, Wallis GP, Aitken TH, Miller BR, Amato GD, Lorenz L, Powell JR, Beaty BJ (1985) Oral infection of *Aedes aegpti* with yellow fever virus: geographic variation and genetic considerations. Am J Trop Med Hyg 34:1219–1224

Taylor RM, Hurlbut HS, Work TH, Klingston JR, Frothingham TE (1955) Sindbis virus: a newly recognized arthropod-transmitted virus. Am J Trop Med Hyg 4:844–862

Tucker PC, Griffin DE (1991) Mechanism of altered Sindbis virus neurovirulence associated with a single amino acid change in the E2 glycoprotein. J Virol 65:1551–1557

Turell MJ (1993) Effect of environmental temperature on the vector competence of *Aedes taeniorhynchus* for Rift Valley fever and Venezuelan equine encephalitis viruses. Am J Trop Med Hyg 49:672–676

Turell MJ, Beaman JR, Neely GWJ (1994) Experimental transmission of Eastern equine encephalitis virus by strains of *Aedes albopictus* and *A. taeniorhynchus* (Diptera: Culicidae). J Med Entomol 31:287–290

Vaidyanathan R, Edman JD, Cooper LA, Scott TW (1997) Vector competence of mosquitoes (Diptera: Culicidae) from Massachusetts for a sympatric isolate of Eastern equine encephalomyelitis virus. J Med Entomol 34:346–352

Wang HL, O'Rear J, Stollar V (1996) Mutagenesis of the Sindbis virus nsP1 protein: effects on methyltransferase activity and viral infectivity. Virology 217:527–531

Wang KS, Kuhn RJ, Strauss EG, Ou S, Strauss JH (1992) High affinity laminin receptor is a receptor for Sindbis virus in mammalian cells. J Virol 66:4992–5001

Weaver SC (1986) Electron microscopic analysis of infection patterns for Venezuelan equine encephalomyelitis virus in the vector mosquito *Culex taeniopus*. Am J Trop Med Hyg 35:624–631

Weaver SC (1996) Vector biology in virus pathogenesis. In Nathanson N (ed) Viral Pathogenesis. Raven Press, New York, pp 329–352

Weaver SC, Scott TW, Lorenz LH, Lerdthusnee K, Romoser WS (1988) Togavirus-associated pathologic changes in the midgut of a natural mosquito vector. J Virol 62:2083–2090

Wengler G, Wengler G (1984) Identification of a transfer of viral core protein to cellular ribosomes during the early stages of alphavirus infection. Virology 134:435–442

Wengler G, Wurkner D, Wengler G (1992) Identification of a sequence element in the alphavirus core protein which mediates interaction of cores with ribosomes and the dissassembly of cores. Virology 191:880–888

Woodring JL, Higgs S, Beaty BJ (1996) Natural cycles of vector-borne pathogens. In Beaty BJ, Marquardt WC (eds) The Biology of Disease Vectors, University of Colorado Press, Niwot, CO, pp 51–72

Woodward TM, Miller BR, Beaty BJ, Trent DW, Roehrig JT (1991) A single amino acid change in the E2 glycoprotein of Venezuelan equine encephalitis virus affects replication and dissemination in *Ae. aegypti* mosquitoes. J Gen Virol 72: 2431–2435

Xiong C, Levis R, Shen P, Schlesinger S, Rice CM, Huang HV (1989) Sindbis virus: an efficient broad host range vector for gene expression in animal cells. Science 243:1188–1191

10 Retrotransposons and Retroviruses in Insect Genomes

Christophe Terzian, Alain Pélisson, and Alain Bucheton

CONTENTS

10.1 Introduction 191
10.2 Common Structural Features of the LTR-RT, the EnRV, and the ExRV 192
    10.2.1 Long Terminal Repeat..192
    10.2.2 Primer Binding Site 192
    10.2.3 Retroelement Gene Products 193
    10.2.4 PolyPurine Tract...193
    10.2.5 Poly(A) Tract..194
10.3 Functional Similarities among LTR-RT, EnRV, and ExRV 194
10.4 Structural and Functional Differences among LTR-RT, EnRV, and ExRV 194
    10.4.1 The *env* Gene 194
    10.4.2 The *pol* Gene 194
    10.4.3 The Replication Cycle 194
10.5 The Endogenous Retroviruses of Insects 195
    10.5.1 The *Gypsy* Element..195
        10.5.1.1 Structure of the Proviral *Gypsy* 195
        10.5.1.2 Infectious Properties of *Gypsy*..196
        10.5.1.3 Control of *Gypsy* by the *Drosophila flamenco* Gene...............................196
    10.5.2 Other Insect EnRVs 197
10.6 Potential Uses and Limitations of Retroviral Vectors for Insect Transgenesis 197
10.7 Conclusion...199
References 199

10.1 INTRODUCTION

Retroelements form a large and diverse family of mobile elements that can be found in all eukaryotes. They all share a common mechanism of replication in that they propagate by reverse transcription of RNA intermediates, and integrate their genetic information into the genome of the host cell. The polymerase that copies RNA into DNA is the reverse transcriptase (RT) (Varmus and Swanstrom, 1985). Amino acid sequences of RT were used to determine the evolutionary relationships among the retroelements (Xiong and Eickbush, 1990; Springer et al., 1995). The phylogeny determined in this way fits well with the general organization of these elements, including the nucleotidic structures of the 5′ and 3′ extremities, the number of open reading frames (ORFs), and the organization of the polymerase domain. This classification permits the distinction of eight major classes of eukaryotic retroelements

TABLE 10.1
The Eight Classes of Eukaryotic Retroelements

Class of Retroelements	Example (Host)	Ref.
Group II introns	aI2 (yeast)	Moran et al., 1995
Retroplasmids	Mauriceville (*Neurospora crassa*)	Wang and Lambowitz 1993
Non-LTR retrotransposons	I factor (*Drosophila*)	Busseau et al., 1994
or poly(A) retrotransposons	Dong (*Bombyx mori*)	Xiong and Eickbush, 1993
LTR-Retrotransposons (LTR-RT)		
Ty1/copia	Ty1 (yeast); copia, 1731 (*Drosophila*)	Emori et al., 1985
Ty3/412	Ty3 (yeast); 412, Ulysses (*Drosophila*)	Evgen'ev et al., 1992; Fourcade-Peronnet et al., 1988; Yuki et al., 1986
Endogenous retroviruses EnRV	Gypsy (*Drosophila*)	Marlor et al., 1986
Exogenous retroviruses ExRV	MoMLV, HIV1, RSV (vertebrates)	Schwartz et al., 1983; Van Beveren et al., 1981; Wain-Hobson et al., 1991
Hepadnaviruses	HBV (human)	Loeb et al., 1991
Caulimoviruses	CaMV (cauliflower; restricted to plants)	Balazs et al., 1982

(Table 10.1). This chapter will focus on three of those classes: long terminal repeat-retrotransposons (LTR-RT), endogenous retroviruses (EnRV), and exogenous retroviruses (ExRV). Two of these (LTR-RT and EnRV) are present in the genomes of insects, are integrated in the genome of the germ cells, and are transmitted vertically like cellular genes. The "true" retroviruses (ExRV) propagate strictly horizontally by cell-to-cell infection, and so far have been described only in vertebrates. The ExRV class is included in this chapter for two main reasons: (1) some structural and functional properties have not yet been described for LTR-RT and EnRV, but they may be inferred given similarities with ExRV, and (2) a large number of tools for gene transfer are based on ExRV. The chapter will compare structural and functional properties of these three classes of retroelements and emphasize their similarities where possible. The characteristics of the *Gypsy* EnRV element from *Drosophila* will then be developed. Finally, the potential use of the EnRV as tools for insect transgenesis will be discuss.

10.2 COMMON STRUCTURAL FEATURES OF THE LTR-RT, THE EnRV, AND THE ExRV

This description only concerns full-length retroelements that are autonomous for their replication, although defective integrated forms of LTR-RT and EnRV are present in nearly all host genomes.

10.2.1 LONG TERMINAL REPEAT

The structural features of LTR-RT and EnRV presented here concern their integrated DNA forms, known as the provirus for the ExRV. LTR-RT and EnRV, like ExRV, possess identical or almost identical LTRs that contain *cis*-acting elements necessary for transcription. The LTRs are divided into three elements : U3, R, and U5 (Figure 10.1). The size of the LTR can vary considerably from several hundred to thousands of nucleotides. The transcription initiation site is the first nucleotide between U3 and R in the 5' LTR, and the polyadenylation site is between R and U5 in the 3' LTR. Most of the *cis*-acting elements necessary for transcription lie in the U3 sequence; hence, the U3 sequence is critical in determining the tissue and developmental-time specificity of replication.

10.2.2 PRIMER BINDING SITE

The primer binding site (PBS) region is adjacent to the 5' LTR and complementary to the 3' terminus of a specific host tRNA species. This tRNA functions as the primer for the reverse transcriptase to

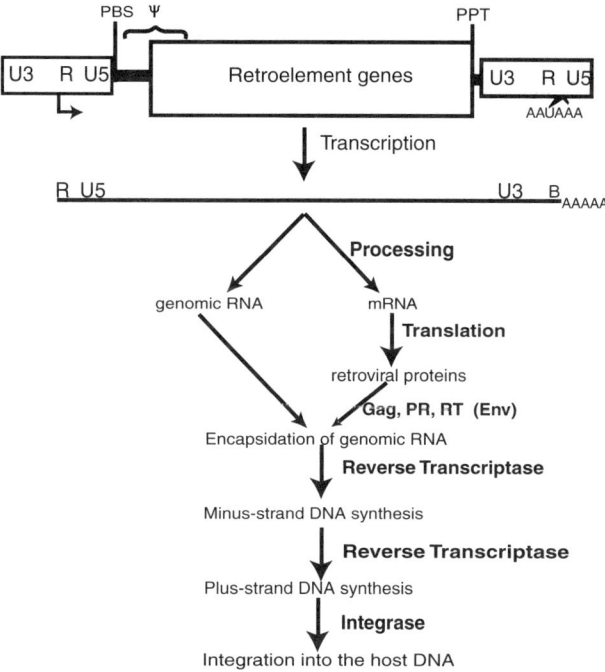

FIGURE 10.1 Common steps of replication for LTR-RT, EnRV, and ExRV. The retroviral Env protein is required only for the extracellular step of ExRV replication (see text for a description).

initiate synthesis of the first (minus) DNA strand; however, it has been shown that interactions between sequences outside the PBS and the primer tRNA play a role in the reverse transcription pathway for the LTR-RT, Ty1 (Friant et al., 1996), as well as for avian ExRV (Aiyar et al., 1992) and HIV1 (Arts et al., 1994).

10.2.3 RETROELEMENT GENE PRODUCTS

The *gag* gene encodes a polyprotein that in ExRV is usually cleaved into three major structural proteins, matrix (MA), capsid (CA), and nucleocapsid (NC). The most conserved region of Gag is the major homology region (MHR) localized in the CA domain. Another conserved region is a cysteine-rich array found in one or two copies within the NC. However, these two regions are not present in all *gag* genes, whatever the class of retroelements. The product of *pro* is a protease essential for the maturation of the polyproteins encoded by the retroelements.

The *pol* gene encodes a polyprotein that contains two main domains, the first of which, an RT domain, has two distinct enzymatic properties: a DNA polymerase activity that copies RNA and DNA templates and a ribonuclease (RNase H) activity that is specific for the RNA strand of RNA:DNA duplexes. The second is an integrase domain (IN) necessary for integration of the double-stranded DNA copy of the RNA genome into the host cell DNA.

10.2.4 POLYPURINE TRACT

A purine-rich sequence is located immediately upstream of the 3′LTR, is relatively resistant to RNase H degradation, and generates the RNA primer used for synthesis of the second DNA strand (plus strand).

10.2.5 Poly(A) Tract

A signal for polyadenylation AAUAAA is generally present in the R sequence, up to 20 nucleotides upstream of the site of polyadenylation. The poly(A) tract is added post-transcriptionally, as for most cellular transcripts.

10.3 FUNCTIONAL SIMILARITIES AMONG LTR-RT, EnRV, AND ExRV

The mechanism of replication is believed to be similar for LTR-RT, EnRV, and ExRV (see Figure 10.1). The retrotransposition cycle can be divided into the following steps: transcription of the genomic and messenger RNAs, translation of mRNAs, genomic RNA encapsidation into particles, synthesis of first and second DNA strands by the reverse transcriptase, nuclear entry in quiescent (HIV1) or dividing cells (oncoretroviruses; Lewis and Emerman, 1994), and integration. Intermediates of reverse transcription were experimentally identified for the *Drosophila* LTR-RT *copia* (Shiba and Saigo, 1983), *mdg1* and *mdg3*, and for the *Drosophila Gypsy* EnRV (Arkhipova et al., 1986).

10.4 STRUCTURAL AND FUNCTIONAL DIFFERENCES AMONG LTR-RT, EnRV, AND ExRV

10.4.1 The *env* Gene

The major difference among LTR-RT, EnRV, and ExRV is the presence of an additional open reading frame for the *env* gene, in EnRV and ExRV. *Env* has a potential for encoding a protein precursor to an envelope glycoproteins. The function of this Env protein is to mediate the adsorption and penetration of the retroviral particle into the host cell. The Env precursor polypeptide is translated from a subgenomic RNA generated by splicing of the genomic RNA. Env then is cleaved to yield the SU (surface) and TM (transmembrane) proteins that attach to each other usually by noncovalent interactions to form oligomers. The SU/TM oligomers are incorporated into the budding viral particles at the plasma membrane. The recognition of the host cell receptor by the retrovirus is mediated by the SU protein, whereas the fusion of the cellular and viral membranes is directed by TM. The core of the retrovirus is then introduced into the cytoplasm of the cell, and reverse transcription is activated (for a review, see Hunter and Swanstrom, 1990).

10.4.2 The *pol* Gene

A classification of LTR-RT, EnRV, and ExRV was proposed based on structural characteristics of the *pol* gene. Among them, the organization of the *pol* domains permits the distinction of two groups among LTR-RT, the *Ty3/412* group whose domains (RT, IN) are ordered as in the EnRV and ExRV *pol* gene, and the *Ty1/copia* group which displays the reverse gene order (IN, RT) (Figure 10.2). Furthermore, given their RT amino acid sequences, the *Ty3/412* group appears closer to the RV than the *Ty1/copia* group (Xiong and Eickbush 1990; Springer et al., 1995).

10.4.3 The Replication Cycle

Some specific peculiarities in the replication cycle can distinguish LTR-RT, EnRV, and ExRV. First, unlike known EnRV and ExRV, the priming of the first-strand DNA (minus-strand synthesis) by a methionine tRNA has been described for some LTR-RT. In addition, integration of LTR-RT and EnRV into the host germline DNA displays various degrees of target site specificity. This specificity is certainly linked to the extent of potential damage to the host genome, as every proviral integration is potentially mutagenic.

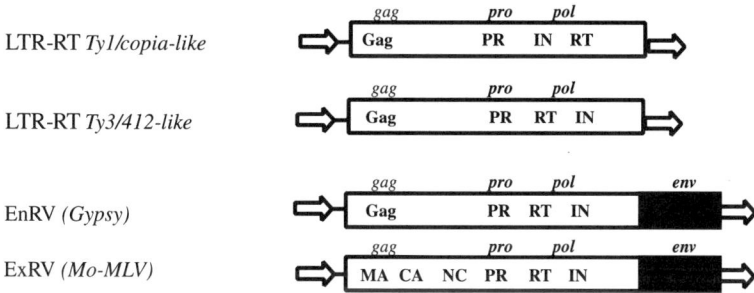

FIGURE 10.2 Conserved structural domains of LTR-RT, EnRV, and ExRV (not to scale). PR: protease; RT: reverse transcriptase; IN: integrase; MA: matrix; CA: capsid; NC: nucleocapsid.

This overview of the structural and functional similarities between LTR-RT, EnRV, and ExRV indicates that the basic properties that make ExRV efficient gene vectors are also observed in the EnRV. Hence, the insect EnRV might be considered as tools for transgenesis. Among them, the *Gypsy* retrovirus of *D. melanogaster* is the only one in invertebrates for which infectious properties have been demonstrated.

10.5 THE ENDOGENOUS RETROVIRUSES OF INSECTS

10.5.1 THE *GYPSY* ELEMENT

Gypsy (also called *mdg4*) was first cloned and described as a transposable element present in the genome of *D. melanogaster* (Bayev et al., 1984). All *D. melanogaster* stocks contain defective *Gypsy* elements located in the centromere, and many strains contain only a few active *Gypsy* proviruses (fewer than five copies; Vieira and Biemont, 1996). However, a few strains with a high copy number of *Gypsy* proviruses have been described, and high mutabilities were observed in these stocks due to high frequencies of transposition of *Gypsy* (Mevel-Ninio et al., 1989; Kim et al., 1990).

10.5.1.1 Structure of the Proviral *Gypsy*

Three independent *Gypsy* proviruses from *D. melanogaster* have been completely sequenced, and they display differences in coding and noncoding primary sequences (Marlor et al., 1986; Liubomirskaia et al., 1998). The primary structure of one of them (Accession number M12927) is described here (Figure 10.3). The two 482 bp LTRs are identical and are composed of the U3 (238 bp), R (53 bp), and U5 (190 bp) regions. Their termini are composed of 5'-AG...$^T/_C$T-3' rather than the TG...CA observed for other LTR-RT, EnRV, and ExRV. The PBS site extends from nt 482 to nt 493 and is complementary to tRNA^{Lys}. Interestingly, the last nucleotide of the 5' LTR is the first nucleotide of the PBS, which should theoretically lead to the formation of linear DNA lacking one terminal base pair, but this has never been observed. Analysis of the reverse transcription intermediates are necessary to understand how the missing base pair is regenerated. A single PolyPurine tract (PPT) is identified from nt 6978 to nt 6987. The consensus sequence of *Gypsy* insertion sites is YRYRYR (where Y = pyrimidine and R = purine), which indicates that *Gypsy* may have a higher degree of target specificity than most ExRV (Dej et al., 1998).

The *gag* gene does not exhibit identifiable primary sequence similarity to other retroviral *gag* genes; however, an arginine-rich region in the C-terminus was described as a putative RNA-binding motif (Alberola and de Frutos, 1996). The *pro* and *pol* genes are potentially expressed as a Gag-Pro-Pol polyprotein that is produced by a −1 frameshift near the C-terminus of Gag.

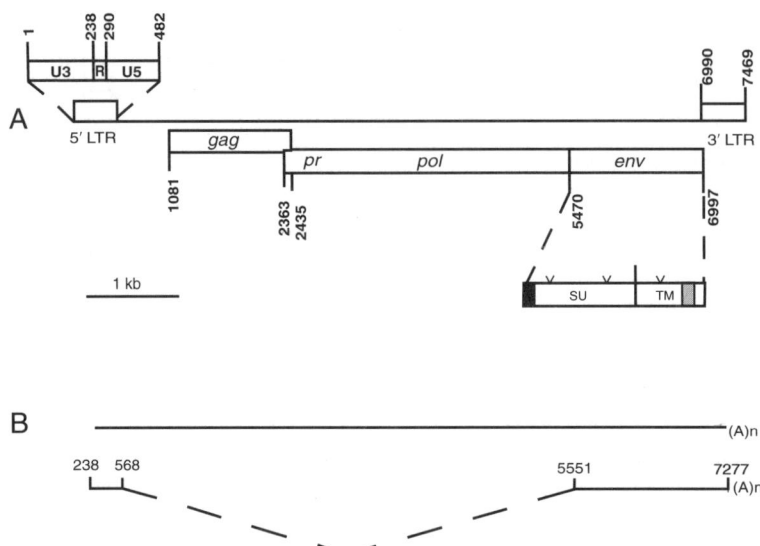

FIGURE 10.3 Structure of the *Gypsy* element. (A) Organization of the proviral *Gypsy* element. The predicted 54-kDa protein encoded by *env* is represented with the signal peptide (black box), potential N-glycosylation sites (Y), the dibasic cleavage site (arrow), and the transmembrane domain (gray box); SU: surface protein; TM: transmembrane protein. (B) Structure of genomic and *env* subgenomic transcripts of the *Gypsy* element, with numbers referring to the nucleotides at the transcription start site, the splicing donor and acceptor sites, and the transcription termination site.

The *env* gene is expressed from a spliced 2.1 kb mRNA and potentially encodes a protein of 54 kDa; the C-terminus of Env lies within the U3 region of the 3'LTR. *Env* produces *in vivo* a 66 kDa N-linked glycosylated polyprotein, which is processed into putative SU and TM proteins (Song et al., 1994) at a cleavage site identified by homology with retroviral consensus sites (Pelisson et al., 1994). Two hydrophobic domains characteristic of the signal peptide of SU and the membrane-spanning domain of TM have also been identified (Pelisson et al., 1994).

10.5.1.2 Infectious Properties of *Gypsy*

Infectious properties of *Gypsy* were demonstrated by experiments in which crude extracts of pupae from a *flamenco* permissive strain (see below) containing a high copy number of actively transposing *Gypsy* elements were put in contact with permissive larvae lacking active *Gypsy* elements (Kim et al., 1994). A number of the exposed flies were shown to contain *Gypsy* proviral copies integrated into the germline, and the frequency of infected flies appeared to be rather high in some experiments (Kim et al., 1994). This experiment was reproduced with an additional step of partial purification of *Gypsy* particles using sucrose gradient centrifugation (Song et al., 1994). Direct evidence that *Gypsy* Env products are responsible for infectious properties was obtained using a Moloney murine leukemia virus–based retroviral vector pseudotyped by the *Gypsy* Env protein: such particles, produced in the 293GP human cell line, can infect *Drosophila* cells (Teysset et al., 1998; see Chapter 7 by Burns).

10.5.1.3 Control of *Gypsy* by the *Drosophila flamenco* Gene

The transposition and infectious properties of *Gypsy* are controlled by a host gene called *flamenco* (Prud'homme et al., 1995). Restrictive alleles repress *Gypsy* transcription and transposition, whereas permissive alleles allow high rates of transposition and the expression of its infectious properties (Pelisson et al., 1994). Transposition only occurs in the progeny of females homozygous for permissive *flamenco* alleles, indicating that this gene has a maternal effect on *Gypsy* amplification. The regulation of *Gypsy* by *flamenco* is tissue specific, the RNAs and Env proteins accumulate in the ovaries of *flamenco* permissive females (Pelisson et al., 1994). This accumulation takes place near the apical membrane of the somatic follicle cells that surround the oocyte where many *Gypsy* viruslike particles (VLP) can be observed (Lecher et al., 1997). However, transposition involves transfer to the germline since integration occurs in the progeny. This has led to the hypothesis that *Gypsy* VLPs could enter the oocyte due to their infectious properties.

10.5.2 OTHER INSECT ENRVS

A significant number of insect retroelements share structural characteristics with *Gypsy* (Table 10.2). Among them, an *env*-like subgenomic transcript has been detected for the *tom* (Tanda et al., 1988), *ZAM* (Leblanc et al., 1997) and *nomad* (Whalen and Grigliatti, 1998) elements. The *tom env*-like RNA was shown to be translated into glycosylated Env-like polypeptides (Tanda et al., 1988). The TED ORF3 was also shown to produce a membrane glycoprotein with properties characteristic of retroviral Env proteins (Ozers and Friesen, 1996). However, in contrast to *Gypsy*, the infectious properties of these elements remain to be demonstrated.

TABLE 10.2
Eleven Insect EnRV That Are Now Cloned and Sequenced

Insect EnRV	Accession No.	Host
Gypsy	M12927	*D. melanogaster*
ZAM	AJ000387	*D. melanogaster*
17.6	X01472	*D. melanogaster*
297	X03431	*D. melanogaster*
Idefix	AJ009736	*D. melanogaster*
nomad	AF039416	*D. melanogaster*
TOM	Z24451	*D. ananassae*
Osvaldo	AJ133521	*D. buzzatii*
Tv1	AF056940	*D. virilis*
yoyo	U60529	*Ceratitis capitata*
TED	M32662	*Trichoplusia ni*

10.6 POTENTIAL USES AND LIMITATIONS OF RETROVIRAL VECTORS FOR INSECT TRANSGENESIS

The ability of retroviruses to integrate their genome into chromosomal DNA efficiently and the fact that retroviral proteins can act in *trans* are two characteristics that have permitted the design of gene vectors that are extensively used in vertebrates (see Chapter 7 by Burns). In such vectors, retroviral genes are replaced by equal amounts of foreign DNA and propagate in the presence of "helper" elements providing Gag, PR, RT, IN, and Env proteins. Retroviral packaging cell lines have been designed to provide the viral proteins in *trans*, but not permitting the packaging of the RNAs that encode these functions (Figure 10.4). However, critical *cis*-acting elements necessary for replication of the vector must be retained. These include (1) promoter and enhancer signals located in the 5'LTR; (2) PBS, PTT, and R regions necessary for reverse transcription; (3) a complex packaging signal (Ψ) localized between the end of the 5'LTR and *gag*, which directs incorporation of the vector RNA into the virion; and (4) a polyadenylation signal. These sequences are well identified in many avian and mammalian ExRVs, and most of them may be recognized in insect EnRVs given similarities with ExRV, except for the Ψ packaging signal which can only be experimentally determined. Such experiments remain difficult to perform with

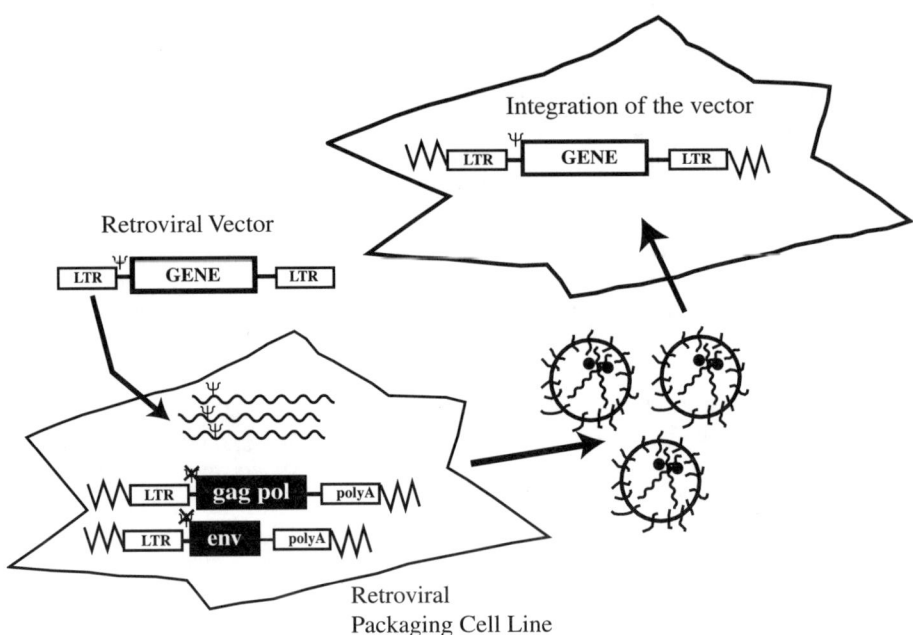

FIGURE 10.4 Packaging cell lines provide all the viral proteins required for capsid production and the virion maturation of the vector. They have been designed in such a way that they express the Gag, Pol, and Env functions. The retroviral DNA vector is introduced by transfection into the packaging cell line. Viral particles are harvested, and used to infect recipient cells into which the vector can integrate. Wavy line: vector RNA; zigzag line: cellular DNA.

insect EnRVs because of the lack of permissive cell cultures. Hence, it could be hazardous to design a true insect retroviral vector without detailed knowledge of the minimal *cis*-acting sequences required for replication.

A different approach involves extending the host range of existing vertebrate retroviral vectors by replacing the vertebrate retroviral Env proteins with Env from other viruses. One successful approach has the Moloney murine leukemia virus–based retroviral vectors pseudotyped with the vesicular virus envelope protein (VSV-G), which have been shown to infect, integrate, and mediate foreign gene expression in insect cell cultures and in adults (see Chapter 7 by Burns). Another approach is to produce pseudotyped Moloney murine leukemia virus–based retroviral vectors with the *Gypsy* Env protein (Teysset et al., 1998). Immunostaining has shown that the *Gypsy* Env protein can be produced in human 293GP/LNhsp70lucL.3 packaging cells and vector particles collected from these cells can infect *Drosophila* cultured cells. These data also suggest that the insect *Gypsy* envelope is correctly processed in human cells. The proteolytic cleavage of the retroviral envelope protein is a necessary step for its incorporation into infectious particles and budding. Among the candidate endoproteases, it has been shown that the furin protease has a role in cleavage of the human immunodeficiency virus envelope proteins (Decroly et al., 1994). *Dfur1* and *Dfur2* are two *D. melanogaster* genes exhibiting homology with the human *furin*, suggesting that *furin*-like functions are conserved among taxa (De Bie et al., 1995). This might explain why the *Gypsy* envelope protein is processed in mammalian cells. However, the infection efficiency of these pseudotyped particles is much lower (1:100) than that of the VSV-G-pseudotyped vectors. Moreover, infection efficiencies for *D. hydei* and *Anopheles gambiae* cultured cells are even lower. Hence, these *Gypsy* Env-pseudotyped particles are not at present useful as tools for transgenesis. However, it could be worthwhile testing different insect EnRV Env proteins instead of *Gypsy* Env. Another way to improve infection efficiency would be to design chimeric envelopes, where the SU protein, which contains the domain interacting with the infected cell, has an EnRV insect origin, whereas the TM protein, which may interact with the core particle, has the same origin as the vector. This might increase the efficiency of SU/TM incorporation into the core particle, and then the virus titer.

10.7 CONCLUSION

The development of retroviral gene transfer tools based on insect EnRV should be considered in the light of the major advances in vertebrate retroviral vector technology. A greater understanding of the functions of the insect EnRV products, and especially the Env protein, will certainly improve the development of methodologies for gene transfer in insects.

REFERENCES

Aiyar A, Cobrinik D, Ge Z, Kung H J, Leis J (1992) Interaction between retroviral U5 RNA and the T psi C loop of the tRNA(Trp) primer is required for efficient initiation of reverse transcription. J Virol 66:2464–2472

Alberola TM, de Frutos R (1996) Molecular structure of a *Gypsy* element of *Drosophila subobscura* (*gypsyDs*) constituting a degenerate form of insect retroviruses. Nucl Acids Res 24:914–923

Arkhipova IR, Mazo AM, Cherkasova VA, Gorelova TV, Schuppe NG, Llyin, YV (1986) The steps of reverse transcription of *Drosophila* mobile dispersed genetic elements and U3-R-U5 structure of their LTRs. Cell 44:555–563

Arts EJ, Li X, Gu Z, Kleiman L, Parniak MA, Wainberg MA (1994) Comparison of deoxyoligonucleotide and tRNA(Lys-3) as primers in an endogenous human immunodeficiency virus-1 in vitro reverse transcription/template-switching reaction. J Biol Chem 269:14672–14680

Balazs E, Guilley H, Jonard G, Richards K (1982) Nucleotide sequence of DNA from an altered-virulence isolate D/H of the cauliflower mosaic virus. Gene 19:239–249

Bayev AA Jr, Lyubomirskaya NV, Dzhumagaliev EB, Ananiev EV, Amiantova IG, Ilyin YV (1984) Structural organization of transposable element mdg4 from *Drosophila melanogaster* and a nucleotide sequence of its long terminal repeats. Nucl Acids Res 12:3707–3723

Busseau I, Chaboissier MC, Pelisson A, Bucheton A (1994) I factors in *Drosophila melanogaster*: transposition under control. Genetica 93:101–116

De Bie I, Savaria D, Roebroek AJ, Day R, Lazure C, Van de Ven WJ, Seidah, NG (1995) Processing specificity and biosynthesis of the *Drosophila melanogaster* convertases dfurin1, dfurin1-CRR, dfurin1-X, and dfurin2. J Biol Chem 270:1020–1028

Decroly E, Vandenbranden M, Ruysschaert JM, Cogniaux J, Jacob GS, Howard SC, Marshall G, Kompelli A, Basak A, Jean F, et al. (1994) The convertases furin and PC1 can both cleave the human immunodeficiency virus (HIV)-1 envelope glycoprotein gp160 into gp120 (HIV-1 SU) and gp41 (HIV-I TM). J Biol Chem 269:12240–12247

Dej KJ, Gerasimova T, Corces VG, Boeke JD (1998) A hotspot for the *Drosophila Gypsy* retroelement in the ovo locus. Nucl Acids Res 26:4019–4025

Emori Y, Shiba T, Kanaya S, Inouye S, Yuki S, Saigo K. (1985) The nucleotide sequences of copia and copia-related RNA in Drosophila virus-like particles. Nature 315:773–776

Evgen'ev MB, Corces VG, Lankenau DH (1992) Ulysses transposable element of *Drosophila* shows high structural similarities to functional domains of retroviruses. J Mol Biol 225:917–924

Fourcade-Peronnet F, d'Auriol L, Becker J, Galibert F, Best-Belpomme M (1988) Primary structure and functional organization of *Drosophila* 1731 retrotransposon. Nucl Acids Res 16:6113–6125

Friant S, Heyman T, Wilhelm ML, Wilhelm FX (1996) Extended interactions between the primer tRNAi(Met) and genomic RNA of the yeast Ty1 retrotransposon. Nucl Acids Res 24:441–449

Hunter E, Swanstrom R (1990) Retrovirus envelope glycoproteins. Curr Top Microbiol Immunol 157:187–253

Kim AI, Belyaeva ES, Aslanian, MM (1990) Autonomous transposition of *Gypsy* mobile elements and genetic instability in *Drosophila melanogaster*. Mol Gen Genet 224:303–308

Kim A, Terzian C, Santamaria P, Pelisson A, Prud'homme N, Bucheton A (1994) Retroviruses in invertebrates: the *Gypsy* retrotransposon is apparently an infectious retrovirus of *Drosophila melanogaster*. Proc Natl Acad Sci USA 91:1285–1289

Leblanc P, Desset S, Dastugue B, Vaury C (1997) Invertebrate retroviruses: ZAM a new candidate in *D. melanogaster*. EMBO J 16:7521–7531

Lecher P, Bucheton A, Pelisson A (1997) Expression of the *Drosophila* retrovirus *Gypsy* as ultrastructurally detectable particles in the ovaries of flies carrying a permissive flamenco allele. J Gen Virol 78:2379–2388

Lewis PF, Emerman M (1994) Passage through mitosis is required for oncoretroviruses but not for the human immunodeficiency virus. J Virol 68:510–516

Liubomirskaia NV, Smirnova Iu B, Avedisov SN, Surkov SA, Il'in Iu V (1998) [Comparative analysis of the structure and retrotransposable activity of two variants of the *Drosophila melanogaster MDG4 (gypsy)* mobile element]. Mol Biol (Mosk) 32:823–829

Loeb DD, Hirsch RC, Ganem D (1991) Sequence-independent RNA cleavages generate the primers for plus strand DNA synthesis in hepatitis B viruses: implications for other reverse transcribing elements. EMBO J 10:3533–3540

Marlor RL, Parkhurst SM, Corces VG (1986) The *Drosophila melanogaster gypsy* transposable element encodes putative gene products homologous to retroviral proteins. Mol Cell Biol 6:1129–1134

Mevel-Ninio M, Mariol MC, Gans M (1989) Mobilization of the *gypsy* and *copia* retrotransposons in *Drosophila melanogaster* induces reversion of the *ovoD* dominant female-sterile mutations: molecular analysis of revertant alleles. EMBO J 8:1549–1558

Moran JV, Zimmerly S, Eskes R, Kennell JC, Lambowitz AM, Butow RA, Perlman PS (1995) Mobile group II introns of yeast mitochondrial DNA are novel site-specific retroelements. Mol Cell Biol 15:2828–2838

Ozers MS, Friesen PD (1996) The Env-like open reading frame of the baculovirus-integrated retrotransposon TED encodes a retrovirus-like envelope protein. Virology 226:252–259

Pelisson A, Song SU, Prud'homme N, Smith PA, Bucheton A, Corces VG (1994) Gypsy transposition correlates with the production of a retroviral envelope-like protein under the tissue-specific control of the *Drosophila flamenco* gene. EMBO J 13:4401–4411

Prud'homme N, Gans M, Masson M, Terzian C, Bucheton A (1995) Flamenco, a gene controlling the *Gypsy* retrovirus of *Drosophila melanogaster*. Genetics 139:697–711

Schwartz DE, Tizard R, Gilbert W (1983) Nucleotide sequence of Rous sarcoma virus. Cell 32:853–869

Shiba T, Saigo K (1983) Retrovirus-like particles containing RNA homologous to the transposable element copia in *Drosophila melanogaster*. Nature 302:119–124

Song SU, Gerasimova T, Kurkulos M, Boeke JD, Corces VG (1994) An env-like protein encoded by a *Drosophila* retroelement: evidence that *Gypsy* is an infectious retrovirus. Genes Dev 8:2046–2057

Springer MS, Tusneem NA, Davidson EH, Britten RJ (1995) Phylogeny, rates of evolution, and patterns of codon usage among sea urchin retroviral-like elements, with implications for the recognition of horizontal transfer. Mol Biol Evol 12:219–230

Tanda S, Shrimpton AE, Chueh LL, Itayama H, Matsubayashi H, Saigo K, Tobari YN, Langley CH (1988) Retrovirus-like features and site-specific insertions of a transposable element, tom, in *Drosophila ananassae*. Mol Gen Genet 214:405–411

Teysset L, Burns JC, Shike H, Sullivan BL, Bucheton A, Terzian C (1998) A Moloney murine leukemia virus–based retroviral vector pseudotyped by the insect retroviral *Gypsy* envelope can infect *Drosophila* cells. J Virol 72:853–856

Van Beveren C, van Straaten F, Galleshaw JA, Verma IM (1981) Nucleotide sequence of the genome of a murine sarcoma virus. Cell 27:97–108

Varmus H, Swanstrom R (1985) Replication of retroviruses. In Weiss R, Teich N, Varmus H, Coffin J (eds) RNA Tumor Viruses. Cold Spring Harbor Laboratory Press, Cold Spring Harbor, NY, pp 369–512

Vieira C, Biemont C (1996) Selection against transposable elements in *D. simulans* and *D. melanogaster*. Genet Res 68:9–15

Wain-Hobson S, Vartanian JP, Henry M, Chenciner N, Cheynier R, Delassus S, Martins LP, Sala M, Nugeyre MT, Guetard D (1991) LAV revisited: origins of the early HIV-1 isolates from Institut Pasteur. Science 252:961–965

Wang H, Lambowitz AM (1993) The Mauriceville plasmid reverse transcriptase can initiate cDNA synthesis de novo and may be related to reverse transcriptase and DNA polymerase progenitor. Cell 75:1071–1081

Whalen JH, Grigliatti TA (1998) Molecular characterization of a retrotransposon in *Drosophila melanogaster*, nomad, and its relationship to other retrovirus-like mobile elements. Mol Gen Genet 260:401–409

Xiong Y, Eickbush TH (1990) Origin and evolution of retroelements based upon their reverse transcriptase sequences. EMBO J 9:3353–3362

Xiong Y, Eickbush TH (1993) Dong, a non-long terminal repeat (non-LTR) retrotransposable element from *Bombyx mori*. Nucl Acids Res 21:1318

Yuki S, Inouye S, Ishimaru S, Saigo K (1986) Nucleotide sequence characterization of a *Drosophila* retrotransposon, 412. Eur J Biochem 158:403–410

11 Polydnaviruses and Insect Transgenic Research

Bruce A. Webb

CONTENTS

11.1 Introduction 203
11.2 Polydnavirus Biology...204
11.3 Polydnavirus and Parasitoid Phylogeny 205
11.4 Polydnavirus Genome Organization 206
    11.4.1 Segment Integration Sites 207
    11.4.2 Coding and Noncoding Sequences in the *Campoletis sonorensis*
           Ichnovirus Genome 207
    11.4.3 Genes Encoding Polydnavirus Structural Proteins..........................208
11.5 Relationships Between Genome Organization and Viral Gene Function.........209
11.6 Summary 212
Acknowledgments 213
References 213

11.1 INTRODUCTION

There is considerable evidence that genes have moved across species boundaries over evolutionary time with several genetic elements such as plasmids, viruses, and transposable elements implicated in gene transfer (e.g., Andersson and Kurland, 1998a). The impact of these mobile DNA elements is evident both within individual genomes and at the population and species levels. Much of transgenic research has been directed toward understanding these naturally occurring mobile elements to exploit preexisting mechanisms for gene delivery and integration. The results of this approach to transgenesis have been spectacularly successful and continue to have great potential for application in the future.

The interest of the author's laboratory in insect transgenesis stems from analysis of an unusual insect virus, the *Campoletis sonorensis* polydnaviruses (CsIV). Polydnaviruses have similarities to two types of genetic elements used extensively for transgenic research, transposable elements and viruses. Moreover, the role of polydnaviruses in their somewhat complex life cycle is to deliver genetic information from the genome of one insect to an infected insect cell for transient expression of viral genes. Therefore, analysis of the mechanisms through which polydnavirus genomes excise, replicate, and are delivered for expression in infected cells is relevant to transgenic research.

Conventional viruses, the presumed evolutionary antecedents of polydnaviruses, can be defined as infectious particles that contain protein and nucleic acid (either RNA or DNA) and replicate in infected cells. Polydnavirus genomes are nucleic acid (DNA), are encapsidated by a protein coat, and enter host cells where viral gene expression causes pathological effects (reviewed in Fleming, 1992; Stoltz, 1993; Webb, 1997). However, polydnavirus replication does not occur in newly infected cells, but only in cells containing a preexisting integrated viral genome. In this sense polydnaviruses may be considered defective or degenerate viruses (Stoltz and Whitfield, 1992). Interestingly, polydnaviruses have conceptual similarities to experimental viruses engineered to replicate only in "helper" cell lines expressing an essential viral gene and helper-dependent transposable elements (e.g., *Dissociation* elements in maize). There are, of course, many viruses that integrate as a part of their normal life cycle. However, unlike conventional viruses, polydnaviruses do not establish viral infections capable of supporting subsequent rounds of virus replication. As a result, polydnavirus replication is restricted to cells inheriting the proviral DNA "infection."

Like transposable elements, polydnavirus genome segments are integrated at multiple sites in the genome and undergo excision from genomic DNA. Although analyses of polydnavirus integration sites is limited, some have similarity to bacterial *Hin* recombination sites (Gruber et al., 1996; Savary et al., 1997). Hin recombination causes a phase-dependent shift in gene expression by inverting an enhancer element in *Salmonella typhimurium* (Scott and Simon 1982) and altering promoter sequences. Similar elements appear to be widespread in prokaryotic organisms. Other polydnavirus integration sites are characterized by target site duplications flanking direct terminal repeats of varying size and complexity (Fleming and Summers, 1991; Cui and Webb, 1997). However, polydnaviruses do not normally integrate in their life cycles, but excise in a developmental-stage and tissue-dependent manner to initiate virus replication. In replicative and infected tissues episomal polydnavirus segments exist with little evidence of reinsertion into chromosomal DNA (Theilmann and Summers, 1986; 1987; Strand et al., 1992). After at least 100 generations in a laboratory colony the known viral integration site of the *C. sonorensis* ichnovirus segment B has not changed (Fleming and Summers, 1991; B. Webb, unpublished). Integration of polydnavirus segments into the genome of infected cells has been shown to occur in tissue culture infections (Kim et al., 1996; McKelvey et al., 1996) but its biological significance is unclear.

Because polydnavirus replication occurs only from an integrated provirus, the viral genomes are resident within the cellular genomes of the wasp host. Andersson and Kurland (1998b) suggest that resident genomes have evolutionary characteristics analogous to those described by Muller (1964) for small asexual populations. That is, resident genomes are subject to reductive evolutionary forces operating through the action of Muller's ratchet to reduce the fitness, size, and capability of the resident genome over evolutionary time. As a result, resident mutualistic or parasitic genomes (e.g., mitochondrial genomes and *Mycoplasma genitalium*) are greatly reduced in size and functional capability in the absence of mechanisms for maintaining genetic variability. Like other resident genomes, polydnaviruses evolve only within the genome of their mutualistic wasp host, presumably because they are dependent upon the nuclear genome for genes required during virus replication. However, resident genomes characteristically show sweeping effects of reductive evolution with large decreases in genome size while polydnavirus genomes are among the largest known viral genomes, have many duplicated genes, and contain significant amounts of noncoding sequences. Although no polydnavirus genome has been fully sequenced, they are clearly among the most structurally complex viruses in both sequence complexity and organization. Thus, polydnaviruses are not conventional viruses, nor are polydnaviruses transposable elements as they are conventionally understood. Polydnaviruses are unique genetic elements having an unusual and intriguing mix of genetic attributes. In the spirit of considering the biological properties of potential utility in transgenic research this chapter will first describe the life cycle of polydnaviruses and then discuss the genetic specializations evident from analysis of the organization of the *C. sonorensis* ichnovirus (CsIV) genome. The chapter closes with a somewhat speculative consideration of the potential utility of polydnaviruses in transgenic research.

11.2 POLYDNAVIRUS BIOLOGY

Polydnaviruses are obligate symbionts of some parasitic hymenoptera that replicate from integrated proviral DNA only in specialized "calyx" cells of the female reproductive tract (Norton and Vinson, 1983; Stoltz, 1993; Albrecht et al., 1994; Stoltz et al., 1995). The polydnavirus life cycle is characterized by asymptomatic virus replication in the parasitic wasp followed by pathogenic infection without virus replication in the larval host of the wasp. Thus, polydnavirus replication and function is inextricably linked to the life cycle of the parasitic wasp (Stoltz and Whitfield, 1992).

Virus is introduced from the female reproductive tract into parasitized insects (mostly Lepidoptera) during oviposition. A species-specific subset of viral genes is expressed in parasitized insects in the absence of viral DNA replication. There is no indication that viral gene expression is *trans*-activated in parasitized insects (Theilmann and Summers, 1987). In some species, viral genes are expressed rapidly and at a constant level throughout endoparasite development (Theilmann and Summers, 1988; Strand et al., 1992; Harwood and Beckage, 1994). In others, expression is restricted to a very limited number of viral genes over a short temporal period (Hayakawa et al., 1994; Asgari et al., 1997). Viral gene expression disrupts some host physiological systems and is required for successful development of the endoparasitic wasp (Edson et al., 1981; Beckage et al., 1994; Strand and Pech, 1995b; Lavine and Beckage, 1996).

Thus, the polydnavirus /wasp parasite/lepidopteran host system provides an unusual example of an obligate mutualistic association between a virus and a parasitic wasp that functions to the extreme detriment of the lepidopteran host of the parasite. As a result of the mutually obligate associations between polydnaviruses and their wasp hosts, polydnaviruses are found in every individual of an infected species, but do not replicate outside of their associated wasp host (Stoltz et al., 1986; Stoltz, 1990; Fleming, 1992). To emphasize this point, polydnaviruses do not infect then replicate in host cells. Rather, polydnaviruses replicate only in cells carrying the proviral DNA and persist, evolutionarily, on the basis of their incorporation in the wasp genome.

Polydnaviruses accumulate to high densities in the wasp oviduct with virus particles reaching concentrations capable of diffracting light and imparting an opalescent blue color upon the entire oviduct. Viral replication begins in the wasp oviduct during the late pupal period (Norton and Vinson, 1983), possibly in response to changes in ecdysteroid titers that drive adult metamorphosis (Webb and Summers, 1992; Gruber et al., 1996). After injection into their host insect, polydnaviruses infect a variety of host tissues, predominantly hemocytes (Strand et al., 1992; Soldevila et al., 1996). In infected cells the viral nucleocapsid releases DNA into the nucleus where viral genes are expressed. Viral gene expression is required for survival of the wasp egg (Edson et al., 1981). In the absence of virus, wasp eggs are usually encapsulated and killed by the lepidopteran immune system. Along with their effects on host immunity, polydnaviruses may disrupt other aspects of host physiology. In *C. sonorensis*, host growth is often suppressed with synthesis of several growth-associated proteins inhibited by virus infection (Shelby and Webb, 1994; 1997). Polydnaviruses alter immunity and development of parasitized hosts by expressing a species-dependent subset of viral genes (Theilmann and Summers, 1987). Viral gene products may be targeted intracellularly (Strand, 1994), to the cell membrane or secreted into the hemolymph of the parasitized host (Li and Webb, 1994; Cui et al., 1997). Secreted proteins are known to bind to hemocytes, although other tissues may also be targeted. The effect of individual viral proteins on host physiology is not well understood. Virus infection of *Microplitis demolitor* granulocytes induces apoptosis (Strand and Pech, 1995b), while secreted CsIV proteins bind to host hemocytes and inhibit but do not totally block encapsulation (Li and Webb, 1994; Cui et al., 1997). The functional analysis of polydnavirus proteins is complicated by the difficulty in studying viruses that are not amenable to genetic manipulation, act intracellularly in a labile host cells (i.e., hemocytes), and express several related viral genes to cause their pathogenic affects.

11.3 POLYDNAVIRUS AND PARASITOID PHYLOGENY

Phylogenetic classification of viruses is initially based on the viral nucleic acid (RNA or DNA) and the organization of the genome (Murphy et al., 1995). Polydnaviruses, as the only group of viruses having profusely segmented DNA genomes, are easily characterized by these criteria, although other characteristics such as their obligate association with parasitic wasps are equally distinctive. Two polydnavirus genera are recognized, the ichnoviruses found in two ichneumonid subfamilies, Campoplegineae and Banchinae, and the bracoviruses known from three braconid subfamilies, Microgastrinae, Cheloninae, and Cardiochilinae. The ichnoviruses, with CsIV being the type species, are phylogenetically described from 25 species of parasitic wasps (Stoltz et al., 1995) with as many as 14,000 species predicted to exist based on the estimated size of the ichnovirus-carrying groups. The bracoviruses are described from 30 braconid species with approximately 17,500 bracovirus-containing species estimated from the size of the polydnavirus-containing bracovirus clades (Whitfield, 1997; Webb, 1998). The two polydnavirus-containing wasp lineages are nonoverlapping, suggesting that this mutualism arose independently in bracoviruses and ichnoviruses. Moreover, the ichnoviruses and bracoviruses are morphologically distinct, do not share nucleic acid similarities, are not antigenically related, and have distinct physiological effects. Thus, it is unlikely that ichnoviruses and bracoviruses are derived from the same ancestral virus. Assuming that polydnaviruses are derived from two different conventional viruses, their viral ancestors would, presumably, have had an unsegmented DNA genome. As a result, segmentation in the two polydnavirus genera must be a convergently derived character state. In other words, polydnavirus genomes may be segmented as a result of their mutualistic and pathogenic roles in their associations with parasitic wasps.

Polydnaviruses are found in two of the most speciose and economically important groups of parasitic wasps, the braconids and ichneumonids. Relative to sister-groups, the polydnavirus-containing subfamilies are much more species rich. Although much of the host range data are anecdotal, groups of polydnavirus-carrying wasps appear to have (in general) broader host ranges (M. Sharkey and G. Wahl, personal communication). This observation has led to the suggestion that polydnaviruses may have allowed the successful parasitization of a wider host range, and that the accompanying physiological advantages improved the relative evolutionary success of the polydnavirus-containing groups (Stoltz and Whitfield, 1992)

The most recent phylogenetic analysis of the known polydnavirus-containing subfamilies of Ichneumonidae suggests that the ichnovirus genera do not form a monophyletic group (Wahl, 1991), but this is not yet supported by molecular analyses of the virus. The existing phylogeny suggests either lateral transfer of ichnoviruses or, alternatively, independent colonization events in the parasitic wasps. In contrast, all bracoviruses are found within a single large clade of the Braconidae and appear to be derived from a single wasp–virus association (Whitfield, 1997). The possibility of lateral transfer or multiple colonizations of ichnoviruses is intriguing, potentially important, and directly relevant to transgenic research. It suggests that it may be possible to identify evolutionary events that fixed the virus–wasp associations in different phylogenetic lineages and/or the evolutionary events that liberated virus from species-specific host associations.

11.4 POLYDNAVIRUS GENOME ORGANIZATION

The unique role of polydnaviruses in the life cycle of parasitic Hymenoptera provides unusual opportunities and constraints for viral genome evolution. Because polydnavirus transmission is not dependent upon the infectious phase of the viral life cycle, viral genes required for replication (e.g., structural proteins, replicative enzymes) may not be packaged in the virion. Conversely, the essential function of the virus, the alteration of physiology in the parasitized insect, could favor diversification of viral genes and/or enhancement of viral gene expression.

Polydnavirus genomes comprise multiple, closed circular DNA segments ranging in size from 2.0 to over 30 kb (Stoltz et al., 1995). Genome segment number varies from 6 to over 30 with segments present in highly variable molar ratios. Conventionally, polydnavirus segments are identified alphabetically from the smallest to the largest with comigrating DNA segments identified by a numerical suffix (e.g., O^1, O^2, etc.). Polydnavirus genome size estimates are complicated by comigrating segments (Theilmann and Summers, 1987) and sequence homologies between segments (Xu and Stoltz, 1993; Cui and Webb, 1997). Relative to segmented RNA viruses, segmentation of polydnavirus genomes is extreme. The more complex segmented RNA viruses have about 12 segments with an aggregate genome size of about 20 kb (Murphy et al., 1995). The *C. sonorensis* polydnavirus has at least 28 DNA segments with genome size estimates in excess of 250 kb (Krell et al., 1982; Krell, 1991). Although a segmented genome is a hallmark of polydnaviruses, there is little understanding of the evolutionary pressures that have driven or, alternatively, allowed segmentation of the viral genome. One hypothesis ventured is that genome segmentation (and the presence of repeated DNA and viral gene families) enhances the capability of the virus to generate genetic variability through recombination (Dib-Hajj et al., 1993; Summers and Dib-Hajj, 1995). While the importance of mechanisms for generating diversity is unquestionable, it does not account for the existence of segments with extensive sequence identity. Specifically, mapping of cross-hybridizing segments suggests that some smaller DNA segments are excised from larger DNA segments, an organizational pattern known as "segment nesting" (Xu and Stoltz, 1993; Cui and Webb, 1997). Segment nesting increases the number of polydnavirus segments without significantly increasing the sequence complexity of the virus. However, nesting does increase the representation of nested sequences within the packaged viral genome (Fleming and Krell, 1993; Webb and Cui, 1998).

11.4.1 SEGMENT INTEGRATION SITES

Polydnavirus segment integration sites have been analyzed from two braconid species (*C. inanitus* and *C. congregatus*; Gruber et al., 1996, and Savary et al., 1997, respectively). These two braconid integration sites have recombination site similarity to the Hin recombination site of *S. typhimurium*. In both species segments excise from integration sites during replication and an "empty site" can be amplified after replication. Interestingly, analysis of the *C. congregatus* site indicates that two viral segments are integrated in tandem array. Excision of the more abundant EP1 locus is described, whereas the excision site of the adjacent, less abundant locus A is not. A comparison of the integration sites of the two differentially replicated tandem segments would be interesting.

The CsIV genome comprises segments that are excised from proviral DNA to produce an individual viral DNA segment (unique or orphan segments) and segments that undergo multiple reductive recombination events to produce segment families (Webb, 1998). Fleming and Summers (1991) analyzed the integration site of the orphan segment B. Segment B is a 6.6-kb segment that is integrated at an imperfect 59-bp direct repeat. Segment B does not hybridize to other segments at high stringency and is stably integrated at a single locus. Two variants of extrachromosomal segment B were isolated, with one containing each copy of the imperfect repeat. Based on these data, a model was developed in which the repeats were alternatively excised during recombination.

Cui and Webb (1997) analyzed the integration site of segment W. Segment W is an abundant 15.8-kb segment that is the parental member of a segment family that includes segments R, M, and C2. The integration site of segment W is markedly different from segment B and the bracovirus integration sites. Segment W is integrated at a 1186-bp perfect direct repeat with a high degree of internal sequence complexity. Of particular interest, the long direct repeat contains a smaller repeat of 350 bp found at two other places within segment W. Based on the association of the 350-bp repeated sequence with the intramolecular events that produce the nested segments R and M, this has been designated the recombination repeat. The differences in the homology, length, and internal complexity of the segment B and W integration sites may be related to the differences in abundance of the two segments. It is suspected that the segment W repeat may promote more efficient excision and/or replication than integrated repeats of the B segment type containing shorter, less highly conserved integration repeats.

11.4.2 CODING AND NONCODING SEQUENCES IN THE *CAMPOLETIS SONORENSIS* ICHNOVIRUS GENOME

A basic premise of reductive evolution is that genes that become nonessential as a result of parasitism are subject to elimination. Reductive evolution may be indicated by an increased number of pseudogenes and noncoding sequence in microbial genomes (Andersson and Kurland, 1998b), but it is the norm for these nonessential sequences to be eliminated over evolutionary time. Viruses in particular have limited noncoding sequence and gene duplications are rare.

In polydnaviruses, viral gene families have been described from both bracoviruses and ichnoviruses. In spite of the complexity and large size of polydnavirus genomes, the number of transcripts expressed in parasitized insects is not great. An estimated 20 mRNAs are detected in CsIV, whereas baculoviruses with a somewhat smaller genome size encode in excess of 100 genes (Volkman et al., 1995). It is possible that many virally encoded genes are expressed only for virus replication in the wasp host where less work has been done but sequence analysis of CsIV segments B (W. Rattanedechakul and B. Webb, unpublished) and W (Cui and Webb, 1997) do not support this possibility. Rather, the sequence analysis of these segments indicates that the majority of the polydnavirus genome is noncoding sequence.

Segment B encodes two genes, one of which is expressed in parasitized insects (BHv0.9; Theilmann and Summers, 1988) and the other expressed in the oviduct during virus replication (WCs1.0; W. Rattandechakul and B. Webb, unpublished). The coding sequence of the two genes represents approximately 29% of the 6.6-kb segment B. Neither gene contains introns so the inescapable conclusion is that the majority of the segment is noncoding sequence. Similarly, the 15.8-kb segment encodes four genes with an aggregate size of less than 6 kb, less than 40% of the segment (Cui and Webb, 1997). Rather than being streamlined and reduced, these CsIV segments appear to have excess DNA.

Sequence analysis of segment W and a segment with similar properties, segment V, provides an indication of the source of the excess sequence. The nested segments V and W encode related members of the cys-motif gene family suggesting that they result from gene duplication (Blissard et al., 1989; Dib-Hajj et al., 1993; Cui and Webb, 1996). Noncoding sequences from the nested segments are also indicative of sequence duplication (Cui and Webb, 1997). The 15.8-kb segment W appears to result from the threefold duplication of an approximately 5-kb segment while the 15.2-kb segment V has a 3-kb region that is now present in five variants on the segment. The duplications on segment W have produced pseudogenes. Segment W encodes four expressed genes (WHv1.0, WHv1.6, WCs1 and WCs2; Blissard et al., 1989), as well as two pseudogenes with homology to the expressed sequences on segment W (Cui and Webb, 1997; segment W Genbank Accession No. AF004378, pseudogenes present in bp 10076 to 10514,WCs-3 pseudogene, and 7096 to 7448 cys-motif gene). The relatively low density of expressed sequences in the CsIV genome appears to be pervasive. Of the 46 kb of DNA from the four segments that have been entirely sequenced, approximately 12 kb encode proteins suggesting that almost 75% of the sequence in the CsIV genome will be noncoding.

11.4.3 Genes Encoding Polydnavirus Structural Proteins

Reductive evolution should target genes for removal that are no longer essential in the resident genome. Genes encoding viral structural proteins could be subject to reductive evolution through gene transfer to the nuclear genome. Analysis of two CsIV structural proteins strongly indicates that gene transfer from the viral genome to the wasp genome has occurred.

CsIV is structurally complex, having two membranes separated by a matrix region that surrounds a large DNA-containing nucleocapsid. This structural complexity is reflected in the protein profile of SDS-PAGE analyses of purified virions with at least 25 proteins detected in varying degrees of abundance (Krell et al., 1982). The first two genes encoding polydnavirus virion proteins were recently isolated and sequenced (Deng and Webb, 1999; Deng et al., 2000). Interestingly, one of the two genes encoding viral structural proteins is encoded by the virus (Deng and Webb, 1999). By contrast, the second is encoded only in the genome of the mutualistic wasp (Deng et al., 2000). These data suggest that associations between ichnoviruses and wasp hosts may have become fixed through transfer of one or more essential viral genes to the wasp genome.

The cDNA encoding an abundant polydnavirus structural protein was isolated by determining the N-terminal amino acid sequence, designing degenerate oligonucleotide primers encoding the known amino acid sequence, and amplifying the cDNA by reverse transcriptase-PCR (rPCR). The cDNA encoded a predicted protein of ~50 kDa (p44) that lacked a signal peptide and had regions of high positive charge reminiscent of nuclear localization signals (Deng et al., 2000). The cDNA sequence had no significant homology to nucleic acid or protein sequence in the Genbank and SwissProt databases. Immuno-EM with antiserum raised to recombinant p44 protein expressed from bacteria localized the virion protein to either the inner membrane or matrix region of the virion. In genomic Southerns with viral and wasp genomic DNA, the cDNA clone encoding this virion protein did not hybridize to viral DNA but hybridized only to wasp DNA, indicating that the p44 protein was not present within the packaged viral genome (Deng et al., 1999). The relative abundance of the p44 protein in the virion and the tissue-specific expression of the gene indicate that the association of the p44 protein with the virion is unlikely to be fortuitous. Rather, the data suggest that the gene encoding an abundant viral structural protein resides in the genome of its associated host where it is expressed only in the female oviduct for assembly into virions.

A 12-kDa viral structural protein was isolated from a bacteriophage expression oviduct cDNA library using an antiserum raised to whole virions. The p12 cDNA is encoded on viral segment Y, and is expressed only in the oviduct during virus replication (Deng and Webb, 1999). Antisera raised to bacterially expressed recombinant p12 localized the protein in the virion and demonstrated that p12 was not expressed in other wasp tissues.

Interestingly, both the p12 and the p44 genes are amplified during virus replication. That p44 is amplified for virus replication is additional evidence indicating p44 is a viral structural gene that retains a mechanism (at this point undetermined) for amplification in association with virus replication (Deng et al., 2000). It is suspected that the p44 gene is excised and amplified during virus replication but has lost signals required for packaging into the virion for export.

Stoltz (1993) hypothesized that the polydnavirus genome may be a mosaic of host and viral genes selected for inclusion in the viral capsid. Genes required only for virus replication may lose the capacity to excise from the parasitoid genome or be packaged into the virion for export during parasitization. Because polydnaviruses replicate only in cells containing the integrated provirus, it is not essential that the virus encode all of the proteins required for virus replication if genes expressing these essential "viral" proteins reside in the genome of their associated hymenopteran hosts and are expressed in tissues supporting virus replication. Conversely, "wasp" genes providing functions important for parasite survival in the lepidopteran host (e.g., immune suppression) may be selected for inclusion in the viral genome (Webb and Summers, 1990). Few polydnavirus genomes have been described in detail, but even the limited analyses available indicate that polydnavirus genomes are unusually fluid. Segment number varies significantly in closely related species (Stoltz et al., 1986). Segment polymorphisms have been described within geographic populations of a single species (Stoltz, 1990). Asgari et al. (1996) have reported that only a single gene is expressed after infection with the *C. rubecula* polydnavirus. Thus, in the context of polydnavirus genomes, transfer of "viral" genes to the nuclear genome of the host is one of several features indicating that polydnavirus genomes are subject to significant modification. In many ways transfer of "viral" genes to the wasp nuclear genome is analogous to the transfer of mitochondrial genes to the nuclear genome, and may be similarly indicative of plasticity and specialization within polydnavirus genomes.

11.5 RELATIONSHIPS BETWEEN GENOME ORGANIZATION AND VIRAL GENE FUNCTION

The CsIV genome comprises unique or "orphan" segments that are excised directly from proviral DNA and produce a single viral DNA segment and segments that belong to segment families. In a segment family a master or "parental" segment is excised from a single genomic locus but produces multiple viral segments by undergoing multiple intramolecular recombination events to produce several smaller derivatives of the parental segment. The 15.8-kb segment W is a nested segment that undergoes sequential recombination to produce three smaller segments (R, 13.1 kb; M, 10.4 kb; and C^2, 6.6 kb; Cui and Webb, 1997). All four members of this segment family are packaged in the viral genome and exist in nonequimolar amounts with the parental segment, W, predominating. Examination of the genes encoded by parental and daughter segments reveals that the copy number of cysteine-rich genes (WHv1.0 and WHv1.6) encoded on segment W is altered by segment nesting. The copy number of the WHv1.6 gene increases relative to WHv1.0 because WHv1.6 is present on all members of the W segment family (W, R, M, and C^2), whereas the WHv1.0 gene is only on the parental segment W. Segment V (15.2 kb) is similarly nested, giving rise to segments T, L^2, K, and C^1 (B. Webb et al., unpublished). Segment V also encodes two highly expressed cys-motif proteins, VHv1.1 and VHv1.4 (Cui and Webb, 1996). The VHv1.1 gene is present only on the parental segment V while VHv1.4 is present on all the V-family segments. Thus, segment nesting provides for specific amplification of the VHv1.4 and WHv1.6 genes by increasing the copy number of these genes in the packaged viral genome relative to the VHv1.1 and WHv1.0 genes, which are found only on the parental segment (Cui and Webb, 1997).

Polydnavirus genes have been divided into three groups on the basis of their expression (Blissard et al., 1987; Theilmann and Summers, 1987). Polydnavirus genes expressed only in the parasitic wasp are designated as class 1, genes expressed only in parasitized lepidopteran larvae as class 2, and genes that are expressed in both the wasp and the parasitized host are class 3. Characterization of class 1 and class 3 genes has been limited, but demonstrates that at least some mRNAs expressed in the oviduct are virally encoded (Blissard et al., 1986; 1987; 1989; Theilmann and Summers, 1988).

Class 2 genes have been characterized in the most detail because these viral genes alter the physiology of their host in dramatic and interesting ways. Class 2 genes from two families have been described from CsIV. The "cys-motif" gene family has been characterized on the basis of a cysteine-rich motif that is found in a single copy on segment W (WHv1.0 and WHv1.6), but is present in two copies on segment V genes (VHv1.1 and VHv1.4; Dib-Hajj et al., 1993). Members of the cys-motif gene family encode secreted proteins with similar gene structures that are abundantly expressed in parasitized insects. The segment V–encoded proteins bind to host hemocytes and have been implicated in suppressing the host immune system (Li and Webb, 1994). The other known CsIV class 2 gene family is the *"rep"* gene family. *Rep* genes are identified on the basis of a 540-bp repeat nucleotide sequence that is ubiquitously distributed within the viral genome and present in at least one copy on most viral DNA segments (Theilmann and Summers, 1987; 1988). The *rep* sequence is unrelated to the segment W recombination repeat described above. The BHv0.9 gene is the only *rep* gene sequence that has been reported (Theilmann and Summers, 1988) and it is unrelated to other sequences in the protein databases. BHv0.9 does not encode a signal peptide and its mRNA is less abundant than the cys-motif mRNAs. More recent analysis of the *rep* gene family indicates that this is the most diverse gene family in CsIV with at least seven expressed genes having significantly conserved peptide as well as nucleotide sequence (R. Hilgarth and B. Webb, unpublished).

From expression studies of WHv1.0, WHv1.6, VHv1.1, VHv1.4, and BHv0.9, a consistent pattern of CsIV gene expression emerges. In parasitized lepidopteran hosts class 2 genes are expressed rapidly and persistently throughout endoparasite development with little variation in mRNA levels (Theilmann and Summers, 1987; Li and Webb, 1994; Cui and Webb, 1996). This type of expression is similar to viral "early" genes that require only host transcription factors for expression. Indeed, all three of the class 2 polydnavirus promoters that have been tested function as "early" promoters in recombinant baculoviruses (Soldevila and Webb, 1996). Although some variability in the apparent levels of gene expression was noted in one study (Theilmann and Summers, 1988), there is no indication that CsIV exploits *trans*-activating factors that alter viral gene expression in a temporally dependent manner. Rather, class 2 genes are expressed rapidly and constitutively. Under this type of regulatory system, the level of gene expression is completely dependent upon the *cis*-dependent promoter activity and the number of copies of the gene (gene dosage). Clearly, segment nesting has the capability to alter gene dosage.

Segment nesting appears to be a widespread phenomenon in polydnaviruses. The *Hyposoter fugitivus* ichnovirus encodes nested segments U and L (Xu and Stoltz, 1993). Preliminary evidence also suggests that segment nesting occurs in the *M. croceipes* bracovirus (B. Webb, unpublished) and *M. demolitor* (M. Strand, unpublished). Segmentation makes virus replication and genome packaging a more complex process, suggesting that there must be selective advantages offsetting the negative effects associated with replication and packaging of a complex segmented genome. Viral genome segmentation confers unique opportunities for virus reassortment through exchange of genome segments between related viruses, as has been demonstrated for bunyaviruses (Elliot et al., 1990; Bouloy, 1991) and flaviviruses (Chambers et al., 1990). Recombination between RNA viruses can also alter biological properties of the virus such as host range and pathogenicity. Segmentation also allows modification of viral gene expression by promoting changes in viral "gene dosage" or copy number. Polydnavirus genome segmentation may also enhance the capability of the virus to generate genetic variability through recombination (Dib-Hajj et al., 1993; Summers and Dib-Hajj, 1995). The presence of repeated DNA sequences as described for segments V and W and viral gene families would similarly promote recombination and enhance genetic variation.

Campoletis sonorensis is a generalist parasite attacking over 30 known lepidopteran hosts with varying degrees of success (Lingren et al., 1970). In a study of the relationship between polydnavirus infectious dose and host range, Cui et al. (2000) evaluated the effects of manipulating virus dose in fully permissive, semipermissive, and nonpermissive hosts. In fully permissive hosts the VHv1.4 protein titers rise rapidly and remain at high levels, while in semipermissive and nonpermissive hosts the VHv1.4 protein is transiently expressed at levels that positively correlate with permissive host status and the number of parasitization events. Increasing hemolymph titers of immunosuppressive polydnavirus proteins via multiple parasitizations improves the success of parasitization in semipermissive hosts. For example, a semipermissive host that is multiply parasitized is more likely to be successfully parasitized than if singly parasitized. The titer of the VHv1.4 CsIV protein in the hemolymph of semipermissive hosts is also correlated with the number of parasitization events. Thus, more heavily parasitized larvae have a higher titer of immunosuppressive proteins. If not all parasitizations are successful and success is correlated with the level of polydnavirus gene expression, selection would favor increased levels of viral gene expression (and biological activity) in parasitized insects. Taken together, the data suggest that the success of an insect parasite is dependent upon the level of viral gene expression, which is, in turn, dependent upon gene dosage. Webb and Cui (1998) suggest that the level of polydnavirus gene expression is directly correlated with the number of gene copies introduced. The success of parasitization in semipermissive hosts is increased by increasing copy number, and this has favored the abundant production of virions and the amplification of some viral genes by segment nesting.

The general strategy of increasing gene copy number to increase expression levels is well known in other systems. Protozoans, insects, and vertebrates exploit variations on this theme by producing minisatellite chromosomes, polyploid, and polytene cells (Spradling and Mahowald, 1980; Stark and Wahl, 1984). Inducible gene amplification in mammalian cultured cells is also associated with the development of antibiotic resistance (Schimke, 1984). All of these systems allow for the selective amplification of genes and are correlated with higher levels of expression of the amplified gene (Long and Dawid, 1980). Because polydnavirus genome segments exist in nonequimolar ratios and viral genome segments are nested, some polydnavirus genes are introduced and remain at relatively low copy numbers, while other genes have consistently higher copy numbers. Presumably, the diverse expression levels of CsIV genes reflect the different physiological targets of the viral genes.

11.6 SUMMARY

Polydnavirus evolution may be driven, in part, by a requirement for a high copy number of functionally active genes. The level of viral protein in the hemolymph is critical to the ability of the virus to suppress the immune system and inhibit host growth. In the absence of an infectious process or a regulatory cascade that results in greatly increased mRNA transcription, the gene dosage determines the level of viral protein in the hemolymph. Nonequimolar segmentation of the viral genome and segment nesting may have evolved, in part, to increase the copy number of essential genes. Finally, the loss of viral structural proteins to the wasp genome reduces the viral genome size without affecting viral replication in the wasp oviduct, thereby increasing the available capacity for delivery of multiple copies of viral genes that are functionally important in parasitized insects.

This chapter has focused on the physical organization of the CsIV genome as a means to elucidate the relationships between the genomic organization of polydnaviruses and their distinctive evolutionary and functional roles. The data suggest that segmentation in polydnavirus genomes may have evolved, at least in part, to increase the copy number of essential viral genes. While this hypothesis is strongly supported by the preliminary data from the *C. sonorensis* system, broader evolutionary issues cannot be effectively addressed through analysis of a single polydnavirus. We would like to know if gene duplication is associated with speciation rates, if divergence of cys-motif genes enhances host range, if evolutionary forces progressively reduce viral genomes, or alternatively, if evolution counters reductive evolutionary pressures and actually increases genetic complexity. These issues are fundamental not only to polydnavirus evolution, but also to the evolution of diverse parasitic and mutualistic associations ranging from transposable elements to the higher eukaryotic parasites. The polydnaviruses with their species-rich phylogenetic lineages, relatively small genomes, and varying host specificity (from obligate monophagy to extreme polyphagy) may serve as a useful model for the effects of parasitism on resident genomes.

The CsIV genome is replete with evidence of duplication and divergence of genes. Summers and Dib-Hajj (1995) suggest that the extant polydnavirus genomes largely reflect selection for genetic variability. Ongoing molecular analyses clearly show that CsIV genes and segments have undergone multiple duplication events. CsIV segments also undergo multiple, intramolecular recombination events as a normal part of virus replication, and these intramolecular recombinations produce nucleotide sequence alterations that are not predicted from the parental molecule (Cui and Webb, 1997). Analyses of the CsIV cys-motif genes indicates that they are evolving at rates greater than that predicted by the neutral theory of evolution (Hughes and Nei, 1989; Fitch et al., 1991) indicating that mutations in these genes are usually advantageous. Of the 11 CsIV genes expressed in parasitized insects that have been identified, all belong to two gene families, the cys-motif and the *rep* genes. The data suggest that most if not all genes expressed in parasitized insects are either cys-motif or *rep* genes. If so, the seemingly inordinate complexity and unusual organization of the CsIV genome, and by extension other polydnavirus genomes, may be evolving to support the expression and diversification of only two or three ancestral viral genes.

The biology of polydnaviruses could interface directly with transgenic research in several areas. Viral DNA appears to infect and integrate in lepidopteran cell cultures (Kim et al., 1996; McKelvey et al., 1996), leading to the suggestion that polydnaviruses have potential for gene delivery in transgenic systems. However, the inability to engineer polydnaviruses genetically limits the direct utility of this system. Indirectly, polydnavirus research suggests several research avenues that could have more general utility in transgenic research. For example, regulation of tranposition may be accomplished through developmentally regulated expression of a tranposase. Expression of a transposed gene may be regulated by altering gene copy number and or integration state. Transposition and expression of transposed genes can be separated and in the case of polydnaviruses occur in separate insect species. Without the ability to engineer polydnaviruses genetically in cultured cells or in a similar system, the potential for direct utility of polydnavirus infection in transforming insects or even transforming insect cells seems problematic than other approaches. Thus, polydnavirus research may initially impact transgenic research by suggesting novel ways in which existing transgenic systems may be regulated or elaborated. Given the ability to engineer polydnaviruses, the opportunities to exploit polydnavirus and parasitoid biology for transgenic research proliferate and could include, as an example, the use of polydnaviruses delivered via parasitization to transform target insects in the field.

ACKNOWLEDGMENTS

I would like to acknowledge support of grants from the USDA–NRI (98-35302-6822), NIH (AI 331140), and the NSF (MCB-9603504) which make our polydnavirus research possible. This is publication 99-08-108 of the University of Kentucky Agricultural Experiment Station.

REFERENCES

Albrecht A, Wyler T, Pfister-Wilhelm R, Heiniger P, Hurt E, Gruber A, Schumperli D, Lanzrein B (1994) Polydnavirus of the parasitic wasp *Chelonus inanitus* (Braconidae): characterization, genome organization and time point of replication. J Gen Virol 75:3353–3363

Andersson SG, Kurland CG (1998a) Ancient and recent horizontal transfer events: the origins of mitochondria. APMIS Suppl. 84:5–14

Andersson SG, Kurland CG (1998b) Reductive evolution of resident genomes. Trends Microbiol. 267:263–268

Andersson SG, Zomorodipour A, Andersson JO, Sicheritz-Ponten T, Alsmark UC, Podowski RM, Naslund AK, Eriksson AS, Winkler HH, Kurland CG (1998) The genome sequence of *Rickettsia prowazekii* and the origin of mitochondria. Nature 39:133–140

Asgari S, Hellers M, Schmidt O (1997) Host hemocyte inactivation be an insect parasitoid: transient expression of a polydnavirus gene. J Gen Virol 77:3061–3070

Beckage NE, Tan FE, Schleifer KW, Lane RD, Churubin LL (1994) Characterization and biological effects of *Cotesia congregata* polydnavirus on host larvae of the tobacco hornworm, *Manduca sexta*. Arch Insect Biochem Physiol 26:165–195

Blissard GW, Vinson SB, Summers MD (1986) Identification, mapping and in vitro translation of *Campoletis sonorensis* virus mRNAs from parasitized *Heliothis virescens* larvae. J Virol 57:318–327

Blissard GW, Smith OP, Summers MD (1987) Two related viral genes are located on a single superhelical DNA segment of the multipartite *Campoletis sonorensis* virus genome. Virology 160:120–134

Blissard GW, Theilmann DA, Summers MD (1989) Segment W of the *Campoletis sonorensis* virus: expression, gene products, and organization. Virology 169:78–79

Bouloy M (1991) Bunyaviridae: genome organization. Adv Virus Res 40:235–266

Carlson RW (1979) Family Ichneumonidae. In Krombein KV (ed) Catalog of Hymenoptera in America north of Mexico. Smithsonian Institution Press. Washington, D.C.

Chambers T, Hahn C, Galler R, Rice C (1990) Flavivirus genome organization, expression and replication. Annu Rev Microbiol 44:649–688

Cui L, Webb BA (1996) Isolation and characterization of a member of the cysteine-rich gene family from *Campoletis sonorensis* polydnavirus. J Gen Virol 77:797–809

Cui L, Webb BA (1997) *Campoletis sonorensis* polydnavirus W segment family: integration in the wasp genome and segment nesting. J Virol 71:8504–8513

Cui L, Soldevila AI, Webb BA (1997) Expression and hemocyte-targeting of a *Campoletis sonorensis* polydnavirus cysteine-rich gene in *Heliothis virescens* larvae. Arch Insect Physiol Biochem 36:251–271

Cui L, Soldevila AI, Webb BA (2000) Relationships between polydnavirus gene expression and host range of the parasitoid wasp, *Campoletis sonorensis*. J Insect Physiol 46:000:000

Deng L, Webb BA (2000) Cloning and expression of a gene encoding a *Campoletis sonorensis* polydnavirus structural protein. Arch Insect Physiol Biochem 40:30–40

Deng L, Stoltz D, Webb BA (2000) The gene encoding a polydnavirus structural polypeptide is not encapsidated. Virology 268:000–000

Dib-Hajj SD, Webb BA, Summers MD (1993) Structure and evolutionary implications of a "cysteine-rich" *Campoletis sonorensis* polydnavirus gene family. Proc Natl Acad Sci USA 90:3765–3769

Edson KM, Vinson SB, Stoltz DB, Summers MD (1981) Virus in a parasitoid wasp: suppression of the cellular immune response in the parasitoid's host. Science 211:582–583

Elliot R, Schmaljohn C, Collet M (1990) Bunyaviridae genome structure and gene expression. In Current Topics in Microbiology and Immunology. Springer-Verlag, Berlin, pp 91–42

Fitch WM, Leiter JME, Li X (1991) Positive Darwinian evolution in human influenza A viruses. Proc Natl Acad Sci USA 88:4270–4274

Fleming JGW (1992) Polydnaviruses: mutualists and pathogens. Annu Rev Entomol 37:401–425

Fleming JGW, Krell PJ (1993) Polydnavirus genome organization. In: Beckage NE, Thompson SN, Federici BA (eds) Parasites and Pathogens of Insects, Vol. 1 Parasites. Academic Press, New York, pp 189–225

Fleming JGW, Summers MD (1991) Polydnavirus DNA is integrated in the DNA of its parasitoid wasp host. Proc Natl Acad Sci USA 88:9770–9774

Gruber A, Heiniger P, Schumperli D, Lanzrein B (1996) Polydnavirus DNA of the braconid wasp *Chelonus inanitus* is integrated into the wasp genome and excised in the females at late stages of pupal stages of the female. J Gen Virol 77:2873–2879

Harwood SH, Beckage NE (1994) An abundantly expressed hemolymph glycoprotein isolated from newly parasitized *Manduca sexta* larvae is a polydnavirus gene product. Virology 205:381–392

Hayakawa Y, Yazaki K, Yamanaka A, Tanaka T (1994) Expression of polydnavirus genes from the parasitoid wasp *Cotesia kariyai* in two noctuid hosts. Insect Mol Biol 3:97–103

Hughes AL, Nei M (1989) Nucleotide substitution at major histocompatibility complex class II loci: evidence for overdominant selection. Proc Natl Acad Sci USA 86:958–962

Kim M, Sisson G, Stoltz DB (1996) Ichnovirus infection of an established gypsy moth cell line. J Gen Virol 77:2321–2328

Krell PJ (1991) The polydnaviruses: multipartite DNA viruses from parasitic hymenoptera. In Kurstak E (ed) Viruses of Invertebrates. Marcel Dekker, New York, pp 141–177

Krell PJ, Summers MD, Vinson SB (1982) Virus with a multipartite superhelical DNA genome from the ichneumonid parasitoid *Campoletis sonorensis*. J Virol 43:859–870

Lavine MD, Beckage NE (1996) Polydnaviruses: potent mediators of host insect immune dysfunction. Parasitol Today 11:368–378

Li X, Webb BA (1994) Apparent functional role for a cysteine-rich polydnavirus protein in suppression of the insect cellular immune response. J Virol 68:7482–7489

Lingren PD, Guerra RJ, Nickelsen JW, White C (1970) Hosts and host age preference of *Campoletis perdistinctus*. J Econ Entomol 63:518–522

Long EO, Dawid IB (1980) Repeated genes in eukaryotes. Annu Rev Biochem 49:727–764

McKelvey TA, Lynn DW, Gundersen-Rindal D, Guzo D, Stoltz D, Guthrie KP, Taylor PB, Dougherty EM (1996) Transformation of gypsy moth (*Lymantria dispar*) cell lines by infection with *Glyptapanteles indiensis* polydnavirus, Biochem Biophys Res Commun 225:764–779

Muller HJ (1964) The relation of recombination to mutational advance. Mutat Res 1:2–9

Murphy FA, Fauquet CM, Bishop DHL, Ghabrial SA, Jarvis AW, Martelli GP, Mayo MA, Summers MD (1995) Virus Taxonomy. Springer-Verlag, New York

Nilsen TW (1994) Unusual strategies of gene expression and control in parasites. Science 264:1868–1867

Norton WN, Vinson SB (1983) Correlating the initiation of virus replication with a specific pupal developmental phase of an ichneumonid parasitoid. Cell Tissue Res 231:387–389

Norton WN, Vinson SB, Stoltz DB (1975) Nuclear secretory particles associated with the calyx cells of the ichneumonid parasitoid *Campoletis sonorensis* (Cameron). Cell Tissue Res 162:195–208

Savary S, Beckage NE, Tan F, Periquet G, Drezen J-M (1997) Excision of the polydnavirus chromosomal integrated EP1 sequence of the parasitiod wasp *Cotesia congregata* (Braconidae, Microgastrinae) at potential recombinase binding sites. J Gen Virol 78:3125–3134

Schimke RT (1984) Gene amplification in cultured animal cells. Cell 37:705–713

Scott TN, Simon MI (1982) Genetic analysis of the mechanism of the Salmonella phase variation site specific recombination system. Mol Gen Genet 188:313–321

Sharkey MJ, Wahl DB (1992) Cladistics of the Ichneumonoidea (Hymenoptera). J Hymenoptera Res 1:15–24

Shelby KS, Webb BA (1994) Polydnavirus infection inhibits synthesis of an insect plasma protein, arylphorin. J Gen Virol 75:2285–2292

Shelby KS, Webb BA (1997) PDV infection inhibits translation of specific growth-associated host proteins. Insect Biochem Mol Biol 27:263–270

Soldevila AI, Webb BA (1996) Expression of polydnavirus genes under polydnavirus-promoter regulation in baculovirus recombinants. J Gen Virol 27:201–211

Soldevila AI, Heuston S, Webb BA (1996) Purification and analysis of a polydnavirus gene product expressed using a poly-histidine baculovirus vector. Insect Biochem Mol Biol 27:201–211

Spradling AC, Mahowald AP (1980) Amplification of genes for chorion proteins during oogenesis in *Drosophila melanogaster*. Proc Natl Acad Sci USA 77:1069–1074

Stark GR, Wahl GM (1984) Gene amplification. Annu Rev Biochem 53:447–491

Stoltz DB (1990) Evidence for chromosomal transmission of polydnavirus DNA. J Gen Virol 71:1051–1060

Stoltz DB (1993) The polydnavirus life cycle. In Beckage NE, Thompson SN, Federici BA (eds) Parasites and Pathogens of Insects, Vol. 1 Parasites. Academic Press, New York, pp 167–187

Stoltz DB, Whitfield JB (1992) Viruses and virus-like entities in the parasitic Hymenoptera. J Hymenoptera Res 1:125–139

Stoltz DB, Guzo D, Cook D (1986) Studies on polydnavirus transmission. Virology 155: 120–131

Stoltz DB, Beckage NE, Blissard GW, Fleming JGW, Krell PJ, Theilmann DA, Summers MD, Webb BA (1995) Polydnaviridae. In: Murphy FA et al. (eds) Virus Taxonomy. Springer-Verlag, New York, pp 143–147

Strand MR (1994) *Microplitis demolitor* polydnavirus infects and expresses in specific morphotypes of *Pseudoplusia includens* haemocytes, J Gen Virol 75:3007–3020

Strand MR, Pech LL (1995a) Immunological basis for compatibility in parasitoid–host relationships. Annu Rev Entomol 40:31–56

Strand MR, Pech LL (1995b) *Microplitis demolitor* polydnavirus induces apoptosis of a specific hemocyte morphotype in *Pseudoplusia includens*. J Gen Virol 76:283–291

Strand MR, Mckenzie DI, Grassl V, Dover BA, Aiken JM (1992) Persistence and expression of *Microplitis demolitor* polydnavirus in *Pseudoplusia includens*. J Gen Virol 73:1627–1635

Summers MD, Dib-Hajj S (1995) Polydnavirus facilitated endoparasite protection against host immune defenses. Proc Natl Acad Sci USA 92:29–36

Theilmann DA, Summers MD (1986) Molecular analysis of *Campoletis sonorensis* polydnavirus DNA in the lepidopteran host *Heliothis virescens*. J Gen Virol 67:1961–1969

Theilmann DA, Summers MD (1987) Physical analysis of the *Campoletis sonorensis* virus multipartite genome and identification of a family of tandemly repeated elements. J Virol 61:2589–2598

Theilmann DA, Summers MD (1988) Identification and comparison of *Campoletis sonorensis* virus transcripts expressed from four genomic segments in the insect hosts *Campoletis sonorensis* and *Heliothis virescens*. Virology 167:329–341

Volkman LE, Blissard GW, Friesen P, Keddie BA, Posser R, Theilmann DA (1995) Baculoviridae. In Murphy FA et al (eds) Virus Taxonomy. Springer-Verlag, New York, pp 104–113

Wahl DB (1991) The status of *Rhimphoctona*, with special reference to the higher categories within Campopleginae and relationships of the subfamily. Trans Am Entomol Soc 117:193–213

Webb BA (1998) Polydnavirus biology, genome structure and evolution. In Miller LK, Ball A (eds) The Insect Viruses. Plenum Press, New York, pp 105–139

Webb BA, Cui L (1998) Relationships between polydnavirus genomes and viral gene expression. J Insect Physiol 44:785–793

Webb BA, Summers MD (1990) Venom and viral expression products of the endoparasitic wasp *Campoletis sonorensis* share epitopes and related sequences. Proc Natl Acad Sci USA 87:4961–4965

Webb BA, Summers MD (1992) Stimulation of polydnavirus replication by 20-hydroxyecdysone. Experientia 48:1018–1022

Whitfield JB (1997) Molecular and morphological data suggest a single origin of the polydnaviruses among braconid wasps. Naturwissenschaften 84:502–507

Xu D, Stoltz DB (1991) Evidence for chromosomal location of polydnavirus DNA in the ichneumonid parasitoid *Hyposoter fugitivis*. J Virol 65:6693–6704

Xu D, Stoltz D (1993) Polydnavirus genome segment families in the ichneumonid parasitoid *Hyposoter fugivitus*. J Virol 67: 1340–1349

Yamanaka A, Hayakawa Y, Noda H, Nakashima N, Watanabe H (1996) Characterization of polydnavirus-encoded RNA in parasitized armyworm larvae. Insect Biochem Mol Biol 5:529–536

Section V

Transposable Element Vectors

12 *Hermes* and Other *hAT* Elements as Gene Vectors in Insects

Peter W. Atkinson and David A. O'Brochta

CONTENTS

12.1 Introduction 219
12.2 Transposable Element Mobility Assays and the Development of Gene Transfer
     Vectors in Insects 220
12.3 Mobility Properties of Prototypical *hAT* Elements 223
     12.3.1 *Ac* and *Tam3* 223
     12.3.2 *hobo* Mobility in *D. melanogaster* 223
     12.3.3 Mobility in Non-Drosophilid Species 226
12.4 Non-Drosophilid *hAT* Elements 226
     12.4.1 Discovery 226
         12.4.1.1 The *Hermes* Element of *M. domestica* 226
         12.4.1.2 The *Homer* Element of *B. tryoni* 227
         12.4.1.3 The *hermit* Element of *L. cuprina* 228
         12.4.1.4 The *hopper* Element of *B. dorsalis* 228
         12.4.1.5 The *huni* Element of *Anopheles gambia* 229
     12.4.2 Mobility Properties of the *Hermes* Element 229
         12.4.2.1 Range of Species 229
         12.4.2.2 Germline Integration 229
         12.4.2.3 Analysis of Chromosomal Integration Sites 230
     12.4.3 Mobility Properties of the *Homer* Element 230
12.5 Issues Relating to the use of *Hermes* and Other *hAT* Elements in Insects 230
References 232

12.1 INTRODUCTION

Three reasons can be advanced for the emergence of a successful non-drosophilid transformation technology over the past 4 years. One has been the availability of cloned marker genes and corresponding mutations in the target species, and the more recent use of the green fluorescent protein as a dominant selectable marker gene in insect species in which there is a paucity or absence of genetic mutations (Chalfie et al., 1994; Zwiebel et al., 1995; Cornel et al., 1997). The second has been the recognition and acceptance that the *P* transposable element of *Drosophila melanogaster* cannot be used to transform other insect species. This directly led to a search for other class II transposable elements (i.e., elements that transpose via a DNA intermediate) that might be useful gene vectors in insects. The discovery of the *Minos* (Franz and Savakis, 1991), *Hermes* (Atkinson et al., 1993; Warren et al., 1994), *mariner* (Medhora et al., 1991), and *piggyBac* (Fraser et al., 1995) elements was a consequence of this. The third, and final, reason has been the development and implementation of transposable element mobility assays that quickly enable the mobility properties of a particular element in a given species to be determined. Indeed, these assays provided critical information showing that the *P* element was incapable of proper excision in insects other than *D. melanogaster* and some closely related drosophilids (O'Brochta and Handler, 1988; O'Brochta et al., 1991). These results prompted many investigators to abandon *P* as a non-drosophilid vector and to seek alternate transposable elements.

This chapter focuses on the development of transposable element mobility assays and the role they played in the isolation and characterization of *hAT* elements, most particularly the *Hermes* element of *Musca domesctica*. Also discussed are some important issues that will arise from the use of *hAT*, and other transposable elements, as gene vectors in insects.

12.2 TRANSPOSABLE ELEMENT MOBILITY ASSAYS AND THE DEVELOPMENT OF GENE TRANSFER VECTORS IN INSECTS

Transposable element mobility assays can be used to measure two separate events: element excision from a donor site and element integration into a target site. Both types of assays typically require the microinjection of plasmid molecules into preblastoderm insect embryos and the subsequent recovery, usually within 24 h, of these plasmids from developed embryos or larvae (Figure 12.1). One of these plasmids, the donor or indicator plasmid, possesses a transposable element containing bacterial genetic markers, such as genes conferring drug resistance upon a host, the *Escherichia coli β-galactosidase* gene, or the ColE1 origin of replication sequence. Appropriate genetic strains of *E. coli* are then transformed using these plasmids and a phenotype consistent with transposable element excision and/or integration is sought. Plasmid DNA is prepared from candidate colonies and then mapped with appropriate restriction enzymes to determine whether a size difference predicted from excision or integration of the transposable element exists. If so, the DNA sequences of the excision or integration breakpoints are determined to establish if they bear the hallmarks of the excision or integration of the transposable element.

A variation of these assays has been to perform them in tissue culture cells and examine the transposition of the transposable element from the chromosomes to a baculovirus infecting the cell. This type of approach led to the identification of the *piggyBac* transposable element through the production of a mutant baculovirus genotype that could be detected during routine passaging of cells (Fraser et al., 1995). Subsequently, the ability of *piggyBac* to transpose into baculoviruses was used to define further the mobility properties of this element (Elick et al., 1996; 1997). Zhao and Eggleston (1998) transfected *Anopheles gambiae* cells with a donor plasmid containing the *Hermes* element together with a helper plasmid that provided a source of *Hermes* transposase. Integration of the *Hermes* element into the *An. gambiae* chromosomes was detected and DNA sequencing of the integration breakpoints confirmed that, in the majority of cases, integration had occurred by transpositional recombination.

Excision assays measure the ability of the element to excise from a donor site in a plasmid molecule. Excision is initially detected either through the loss of a genetic marker located with the transposable element or by the restoration of a genetic marker following the excision of the transposable element originally located within it. Excision assays for transposable elements were originally developed to monitor the movement of bacterial transposable elements (Berg et al., 1983) and then were applied to the study of plant transposable elements (Baker et al., 1987). They were first utilized in insects as a means to examine the mobility properties of the *P* element of *D. melanogaster* (Rio et al., 1986). In these experiments, indicator plasmids were designed that could be used to monitor the production of *P* element transposase in *D. melanogaster* cell culture. These assays were gain-of-function assays. The *P* element was inserted into the α-peptide-coding region

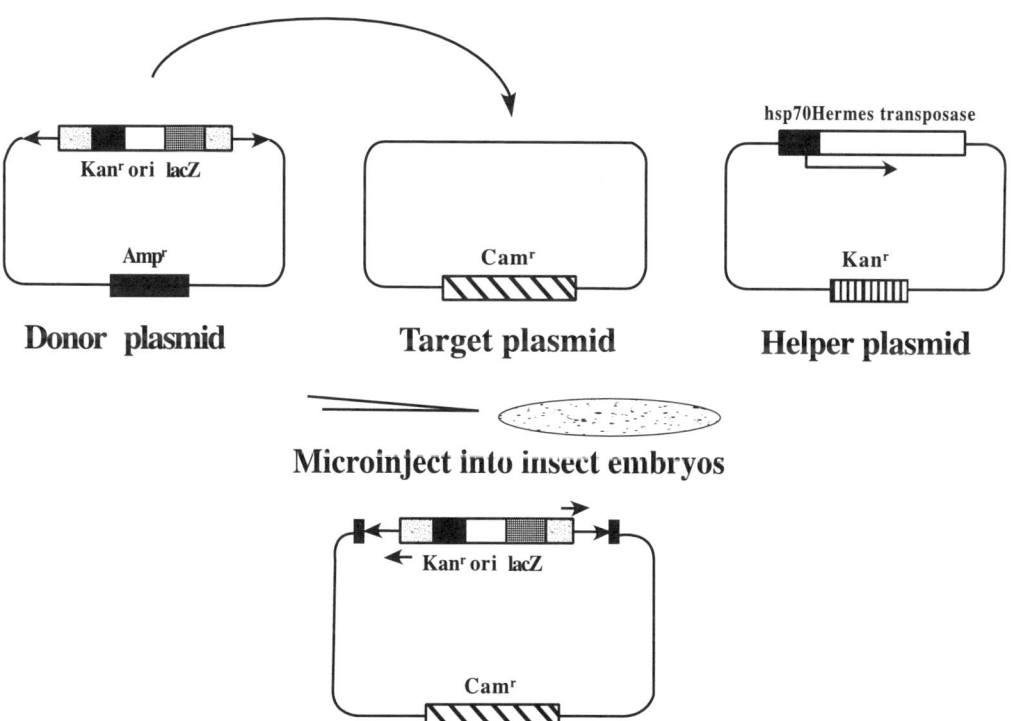

FIGURE 12.1 Interplasmid transposition assays in insect embroyos. For interplasmid transposition assays, three plasmids consisting of a donor plasmid, a target plasmid, and a helper plasmid are microinjected into preblastoderm insect embryos. During embryonic development, the assay measures the ability of the transposase enzyme encoded by the helper plasmid to mediate the transpositon of the transposable element located on the donor plasmid to the target plasmid. The result is the plasmid shown in which only transposable element sequences have inserted, by transpositional recombination, into the target plasmid. Plasmids are recovered from developed embryos ~1 day after injection and transformed by electroporation into appropriate strains of *E. coli* that are then plated on defined media. Plasmid DNA is prepared from colonies displaying the desired phenotype and then analyzed for the presence of the element in the target plasmid. The breakpoints of transposition are determined by DNA sequencing. *Key:* Kan^r, Cam^r, Amp^r each refer to *E. coli* genes conferring resistance to kanamycin, chloramphenicol, and ampicillin; lacZ refers to the gene encoding β-galactosidase; ori refers to the *E. coli* ColE1 origin of replication. In the example shown, the target plasmid is derived from *Bacillus subtilis* and cannot replicate in *E. coli*. Arrows above and below the transposed element represent sequencing primer sites.

of the *E. coli β-galactosidase* gene and excision of *P* was detected by the restoration of the open reading frame (ORF) of the α-peptide which, in turn, led to the production of β-galactosidase. This could be detected upon the subsequent introduction of the indicator plasmids into appropriate strains of *E. coli* and then plating in appropriate defined medium.

These *P* element assays were utilized by O'Brochta and Handler (1988) to examine the mobility of *P* in species other than *D. melanogaster*. At the time of these experiments a number of research groups were becoming frustrated at the apparent failure of *P* to be a robust transformation vector in non-drosophilid insects. While some success in mosquitoes had been obtained (Miller et al., 1987; McGrane et al., 1988; Morris et al., 1989) the transformation frequencies were low and integration of the *P* element apparently had not occurred through transpositional recombination mediated by the *P* element transposase. O'Brochta and Handler (1988) found that the ability of *P* to excise in drosophilid species other than *D. melanogaster* was severely limited and was, in fact, nonexistent in the few non-drosophilid species they examined. Modifications to the helper plasmid subsequently showed that correct splicing of the *P* element transposase gene did not occur in these non-drosophilid species. However, even when a *P* transposase cDNA gene was supplied, no excision in non-drosophilids occurred (O'Brochta and Handler, 1988).

These experiments provided unequivocal data showing that *P* could not move in non-drosophilid species. This was in stark contrast to many unsuccessful transformation experiments in a range of insect species, none of which supplied any data explaining why the *P* element was failing as a vector in these species. Furthermore, these experiments illustrated the ease and speed with which the indicator or helper plasmids could be altered and then retested. Questions could be asked and answered in a short period of time without the need to generate transgenic lines of insects.

Transposable element excision and transposition assays have been utilized for three of the four elements that can transform non-drosophild insects. In the case of *hAT* elements, *hobo* excision assays performed in the nonhost species, *M. domestica*, first indicated the presence of an endogenous transposable element called *Hermes* (Atkinson et al., 1993). *Hermes* has since been developed as a genetic transformation vector in *D. melanogaster* (O'Brochta et al., 1996), the yellow fever mosquito, *Aedes aegypti* (Jasinskiene et al., 1998; Pinkerton et al., 2000; D. O'Brochta and P. Atkinson, unpublished data), and the Mediterranean fruit fly, *Ceratitis capitata* (K. Michel, D. O'Brochta, A. Pinkerton, A. Stanerova, G. Franz, A. Robinson, and P. Atkinson, unpublished data). In addition, transposition assays have shown that *Hermes* is mobile in at least 12 insect species (Sarkar et al., 1997a, b; M. Lehane, P. Atkinson, and D. O'Brochta, unpublished data) including the mosquitoes, *An. gambiae* (F. Collins, P. Atkinson, and D. O'Brochta, unpublished data) and *Culex quinquefasciatus* (M. Allen, D. O'Brochta, and P. Atkinson, unpublished data), and the noctuid, *Helicoverpa armigera* (Pinkerton et al., 1996). Transposition assays performed with the *hobo* element also have shown that it is mobile in at least three nonhost species (O'Brochta et al., 1994), while excision assays using this element in *D. melanogaster* strains either lacking or containing endogenous *hobo* elements both have revealed the unique structure of empty sites remaining after the excision of the *hobo* element (and thereby provided information concerning the mechanism of *hobo* element excision) and also have shown that *hobo* most likely has somatic, as well as germline activity (Atkinson et al., 1993). Excision assays performed with the *Mos1* element of the *mariner* element family demonstrated that this element was mobile in *D. melanogaster*, the Australian sheep blowfly, *Lucilia cuprina,* and the Queensland fruit fly, *Bactrocera tryoni* (Coates et al., 1995). This was also subsequently confirmed using transposition assays in the same species, as well as in *Ae. aegypti* (Coates et al., 1997; 1998). The *Mos1* element was then used to generate transgenic lines of *Ae. aegypti* (Coates et al., 1998). Different forms of excision and transposition assays have been used to illustrate the mobility of the *piggyBac* element from *Trichoplusia n*i in cell culture (Fraser et al., 1995, Elick et al., 1996; 1997). Indeed, the ability of *piggyBac* to transpose from the chromosome into a baculovirus genome resident in infected cells was used to identify initially and isolate the *piggyBac* element (Fraser et al., 1995). More recently, Thibault et al. (1999) used plasmid-based transposition assays to show that *piggyBac* could transpose in developing embryos of the pink bollworm, *Pectinophora gossypiella*.

These assays have made a significant contribution to the development of transposable elements as gene vectors in non-drosophilid insects. They continue to provide information concerning the mobility properties of *Hermes, hobo, piggyBac,* and *Mos1* in insects and will most likely become a standard assay procedure for testing new transposable elements that are sought for use as gene vectors in insects. For example, transposition assays performed with the *Homer* element (see below) from the Queensland fruit fly, *B. tryoni,* have shown that this element is mobile in *D. melanogaster* and *B. tryoni*, but at a very low frequency relative to both *hobo* and *Hermes*, indicating that this element, in its current form, will most likely be of limited use as a gene vector in at least these species of insects (A. Pinkerton, D. O'Brochta, S. Whyard, and P. Atkinson, unpublished).

Excision and transposition assays will also enable the cross-mobility of existing transposable elements to be determined. This will be important in establishing whether a given element will be stable in the presence of a transposase expressed by another transposable element. As gene tagging and enhancer trap strategies using two different transposable element systems are developed for non-drosophilid insects, information about cross-mobility will become critical. Excision assays in particular have proved to be sensitive indicators of transposable element cross-mobility. When applied to the *hAT* element system, *hobo* excision assays detected the presence of the *Hermes* element (Atkinson et al., 1993) and, when applied to *D. melanogaster* strains containing *hobo* elements, can also detect the presence of the autonomous *hobo* element contained in the plasmid pHFL1 (Handler and Gomez, 1995). Excision assays may therefore provide information concerning the presence of endogenous recombination systems that may remobilize genetically engineered transposable elements introduced into the genome. To date, transposition assays have not provided this degree of sensitivity.

12.3 MOBILITY PROPERTIES OF PROTOTYPICAL *hAT* ELEMENTS

12.3.1 Ac AND TAM3

The *Ac* element is one of the most well-studied eukaryotic transposable elements and has mobility properties characteristic of elements within the *hAT* family that make them useful tools for gene discovery and analysis. *Ac*, like many bacterial transposable elements, transposes by a cut-and-paste mechanism (Federoff, 1989). Analysis of the *Ac* transposase promoter and *Ac* mRNA levels shows that this promoter is not cell-cycle regulated; however, it is expressed at higher levels in dividing cells, leading to an increase in *Ac* copy number in cells (Fridlender et al., 1998). In addition, *Ac* transposition has a strong tendency to be local. That is, *Ac* elements that have just transposed are not usually found more than 200 kb from the site of the original insertion site (Jones et al., 1990). The high levels of *Ac* activity in maize have been used in gene tagging experiments in this species (Balcells et al., 1991) and, because few plant species contain any characterized endogenous transposable elements, elements such as *Ac* and *Tam3* were considered for use in heterologous plant species. Both *Ac* and *Tam3* have the notable ability to undergo transposition and excision in foreign cellular environments (Hehl, 1994). The ability of *Ac* to function in heterologous hosts has, and continues to be, exploited as a gene-finding or -targeting tool. *Ac*-tagged genes have been found in a number of species such as *Petunia* (Chuck et al., 1993) and *Arabidopsis thaliana* (Altmann et al., 1995).

The data demonstrating the potential of *Tam3* and *Ac* to function over a broad host range were significant to insect biologists because they clearly demonstrated the potential for some transposable elements to be host unrestricted in their mobility. This was in stark contrast to the lessons being learned by insect biologists attempting to use the *D. melanogaster P* element in insects such as mosquitoes, fruit flies, and moths. Despite its remarkable success as a tool for creating transgenic *D. melanogaster*, and for executing various gene-discovery strategies such as gene tagging and enhancer trapping, *P* element utility remains completely confined to drosophilids. The mobility properties of *P* elements with respect to interspecific mobility are unique and, to a large extent, enigmatic. Virtually every short-inverted repeat-type insect transposable element system discovered and analyzed since the discovery of *P* elements has been capable of functioning in diverse nonhost species. These enigmatic mobility properties of *P* elements begin to raise questions concerning the usefulness of *P* as a model for future studies of transposable element behavior in insect populations. These questions pertain to creating transgenic insects, spreading transgenes through wild insect populations using transposable elements as driving agents, and to the development of repression systems that inhibit transposable element movement. That *hobo* is a member of a family of elements that includes *Ac* and *Tam3* begs the question of the ability of this element to function in species other than *D. melanogaster*.

12.3.2 HOBO MOBILITY IN D. MELANOGASTER

The *hobo* element can be active in *D. melanogaster*, and certain parallels have been drawn between the mobility properties of *hobo* and *P* elements in this species. For example, genetic studies of the *hobo* system by a number of investigators reported a phenomenon resembling hybrid dysgenesis, in which crossing certain strains of *D. melanogaster* resulted in sterility arising from gonadal dysgenesis and chromosomal mutations such as deletions and rearrangements (Lim, 1979; Blackman et al., 1987; Yannaopoulos et al., 1987). The presence of *hobo* elements was strongly correlated with the occurrence of dysgenic traits; however, there were features of *hobo* dysgenesis that distinguished it from *P*-element-mediated dysgenesis. For example, unlike *P* element hybrid dysgenesis, *hobo*-mediated dysgenesis did not show a strong dependence on the direction of the test crosses to detect dysgenesis (Blackman et al., 1987). *P*-element-mediated hybrid dysgenesis is induced almost exclusively when certain *P* element-containing males (P) are crossed to females lacking any or most *P* elements (M). The reciprocal cross of P females mated with M males does not result in a strong dysgenic response. This was not the case with the *hobo* system in which dysgenesis was observed in crosses beginning with either *hobo*-containing (H strain) males or females (Blackman et al., 1987; Bazin and Higuet, 1996). Upon further analysis, the dysgenesis associated with the *hobo* system has proved somewhat unusual. Bazin and Huguet (1996) showed that there was no correlation between the different dysgenic parameters (such as gonadal dysgenesis and element mobilization) nor between the dysgenic parameters and the molecular characteristics of the strain being tested (such as the presence of complete or deleted forms of the *hobo* element). More recently, Bazin et al. (1999) showed that *D. melanogaster* E strains can differ in their ability to permit the movement of *hobo* elements introduced from an H strain. Furthermore, they showed that this ability, or permissiveness, altered with increasing age of the E females. The *hobo* system thus appears to be fundamentally different from the P-M and I-R systems of *D. melanogaster* (Finnegan, 1989).

The distribution of *hobo* both within *D. melanogaster* and among the *Drosophila* species has led to the proposal that, like the *P* element, *hobo* has undergone a recent horizontal transfer into *D. melanogaster* (Periquet et al., 1994). In old lines of *D. melanogaster* established from individual wild-type females more than 50 years ago, *hobo* elements are absent, whereas more recently established lines contain *hobo* elements. In addition, the *hobo* elements present within *D. melanogaster*, *D. simulans,* and *D. mauritiana* show remarkably low levels of sequence divergence, which is consistent with the idea that *hobo* has recently transferred into these species (Periquet et al., 1994). More recent analysis suggests that multiple transfers into the *D. melanogaster* species complex may have occurred from a common source in the recent past. Investigations into the dynamics of *hobo* elements in natural populations, based on the geographic distribution of *hobo* activity as measured in a variety of ways, have not revealed any clear patterns that would suggest an origin of the introduction (Bonnivard et al., 1997). In fact, the different *hobo* activities measured, including gonadal dysgenic activity and element mobilization in dysgenic crosses, are unequally distributed and show no geographic gradient. This is distinctly different from our understanding of the *P* element system and indicates that the dynamics of *hobo* elements in natural populations may be quite different from that of the *P* element.

The spreading abilities of transposable elements have been of interest particularly to those who envision using transposable elements as tools for genetically manipulating wild insect populations as a way, for example, to reduce insect-borne disease transmission rates. Studies using the *P* element system have demonstrated that *P* elements can spread through cage populations of *D. melanogaster* rather quickly under some conditions (Kidwell et al., 1988). Given some of the known differences between the *hobo* and the *P* element systems, it has been relevant to address the question of whether *hobo* can serve as a similar genetic drive agent and to question whether the spreading characteristics of these elements are significantly different. Few experiments looking at the ability of *hobo* to spread through a population of *D. melanogaster* have been reported, but those that have suggest that this element has the potential to serve as a genetic driver (Galindo et al., 1995; Ladeveze et al., 1998). Galindo et al. (1995) attempted to make a direct comparison of the *hobo* and *P* systems and found that, despite the overall genetic similarities of these systems, they differed significantly in their abilities to spread. An interesting observation arising from these studies was that *hobo* activity and the subsequent development of *hobo*-repression potential increased at 25°C compared with 21°C. Temperature has also been reported to affect the dynamics of invasion by *P* elements, but in the reciprocal manner (Kidwell et al., 1988). At 20°C *P* element invasion was more effective than at 25°C. These studies clearly demonstrated that the dynamics of *P* and *hobo* element spreading were different. As an autonomous element (*P* or *hobo*) spreads through a population, a repression or regulatory system usually develops. In the case of *P* elements the appearance of this repression system is acute and is thought to correlate with the appearance of a specific internally deleted *P* element or factor producing a modified form of the transposase that acts as a repressor. The *hobo* repression system that develops during the spread of autonomous *hobo* elements does not abruptly appear and seems to have no correlation with the presence of internally deleted *hobo* elements since these forms of elements did not appear during these experiments. This suggests that the autoregulation of *P* and *hobo* elements is quite different. Thus, the propensity of *P* elements to evolve deleted forms contrasts to what is observed with *hobo* during spreading experiments, and may indicate that *hAT* elements will be more prone to retain a transgene they might be carrying. The copy number of *hobo* elements following a spreading experiment remained low (from 2 to 7) which, again, is in contrast to the *P* element in which successful spreading resulted in the accumulation of many copies (from 20 to 60) of the element (Galindo et al., 1995).

Another aspect of *hobo* mobility in *D. melanogaster* that distinguishes it to some extent from the *P* element system is the potential of the *hobo* element to display somatic activity. A number of lines of evidence support the conclusion that *hobo* mobility is not as strictly limited to the germline of *D. melanogaster* as is the case with *P* elements. First, Yannopoulos et al. (1983) observed chromosomal rearrangements in single salivary nuclei that were apparently mediated by *hobo* elements as suggested by the presence of these elements at the breakpoints. Second, using a plasmid-based *hobo* excision assay, Handler and Gomez (1995) observed element excisions in *hobo*-containing strains (Oregon-R), as well as a strain containing a single copy of the *hobo* transposase gene under the control of its native promoter. The plasmid-based assay used by these investigators measured somatic transposase activity. Third, Calvi and Gelbart (1994) reported evidence for low levels of *hobo* promoter activity in the soma, although they did not report any evidence of somatic mobilization of the element. Finally, we have been able to detect *hobo* transcription in the somatic tissue of *D. melanogaster* (D. O'Brochta and P. Atkinson, unpublished). The strict limitation of *P* element movement to the germline of *D. melanogaster* was a hallmark of that system, and the extent to which the *hobo* and *P* element systems are similar will be reflected, in part, by similarities in their regulation. Based on the current understanding of *hobo* mobility in *D. melanogaster*, the regulation of these two elements appears distinctly different. The degree to which an element is regulated within a species may serve as an indicator of the ability of the element to function in other species. Elements that show relatively little tissue specificity, like *mariner*, and to a lesser extent *hobo*, may be more widely suitable as gene vectors than *P* elements. To date, this generalization appears to be true.

hobo was first used as a gene vector in *D. melanogaster* in 1989 (Blackman et al., 1989). The original report demonstrating the utility of *hobo* as a gene vector employed a germline transformation strategy identical to that used to transform *D. melanogaster* with *P* element vectors. A binary gene vector system was created consisting of a vector and a helper element. The vector element contained a nonautonomous, internally deleted *hobo* element with intact inverted terminal repeats into which the *D. melanogaster rosy* gene had been inserted to serve as a genetic marker to permit the recognition of transgenic individuals. The helper element was a plasmid containing the *hobo* ORF under the control of its endogenous promoter. Of the fertile adults arising from injected embryos, 25% gave rise to transgenic offspring (Blackman et al., 1989). This rate of integration is comparable with that observed with the *P* element system.

The ability of *hobo* to be remobilized following its chromosomal integration has been investigated and exploited for the purposes of developing an enhancer trapping system. Smith et al. (1993) constructed an enhancer-trap system employing a reporter gene and genetic marker system similar to that employed for *P*-element-based enhancer-trapping systems (Wilson et al., 1989). The motivation for developing a second enhancer trap system arose from the improbability that a complete sampling of the genome of *D. melanogaster* could be achieved using only *P* elements. This was based both on the known insertion site preferences of *P* elements and the distinctly nonrandom distribution of *P* insertions into the *D. melanogaster* genome. The genome of *D. melanogaster* shows differential susceptibility to *P* element insertion, with some sites being hot spots for element integration and others being cold spots. The results of Smith et al. (1993) demonstrate a number of important points about the mobility properties of *hobo*. The observation that *hobo* resulted in a higher rate of recessive lethal phenotypes (17% on the second chromosome and 28% on the third chromosome) than did *P* (10 and 11%, respectively) indicates that the insertion site preferences between *P* and *hobo* elements may be distinctly different. This difference was further reflected in a comparison of the distribution of *P* and *hobo* elements along the third chromosome that revealed that there was no association between *P* and *hobo* element integration sites. A more recent comparison of *P* and *hobo* elements further supports the observed differences in integration site preference of these two transposable elements. In an experiment designed to test the ability of *hobo* elements to transpose preferentially over small intrachromosomal distances, Newfeld and Takaesu (1999) found that *hobo* was capable of integrating into a region of the *decapentaplegic* (*dpp*) locus that is refractory to *P* element integration. The data of Newfeld and Takaesu (1999) also suggested that *hobo* undergoes local hopping, a phenomenon described originally in the *Ac* system and subsequently found to be a characteristic of the *P* element.

The mobility properties of *hobo* in *D. melanogaster* demonstrate that this element can potentially form the basis of a useful genetic system in this species. *hobo* transposes at rates that are of practical use and it undergoes local hopping and thus can be used as an enhancer trap and gene-tagging agent. In addition, *hobo* appears to recognize a rather distinct part of the genome relative to *P* elements, making it a complementary tool for gene targeting in this species.

12.3.3 MOBILITY IN NON-DROSOPHILID SPECIES

DNA and amino acid sequence similarities, together with the structure of empty excision sites between *hobo* and *Ac*, suggest they may be related phylogenetically. That these elements may be members of the same family of elements further suggests that they may share other properties as well. In particular, *hobo* and other insect *hAT* elements might share the unrestricted ability of *Ac* to transpose when introduced into nonhost species. This idea was tested directly by Atkinson et al. (1993), who found that *hobo* could indeed undergo excision when introduced into developing embryos of *M. domestica*. A further test was conducted by O'Brochta et al. (1994) who showed that *hobo* could also transpose in *M. domestica* as well as *B. tryoni*. Lozovskaya et al. (1996) along with Gomez and Handler (1997) independently demonstrated the ability of *hobo* to serve as a germline transformation vector in *D. virilis*. Handler and Gomez (1995; 1997) also demonstrated the ability of *hobo* to undergo excision in the drosophilids, *D. melanica*, *D. repleta*, *D. virilis*, *D. saltans*, *D. simulans*, *D. willistoni*, and *Chymomyza procnemis*, and the tephritids, *Anastrepha suspensa*, *B. cucurbitae*, *B. dorsalis*, *Ceratitis capitata*, and *Toxotrypana curvicauda*. Pinkerton et al. (1996) reported the ability of *hobo* to transpose in embryos of the Old World cotton bollworm, *H. armigera*. DeVault et al. (1996) subsequently reported the stable transformation of the corn earworm, *Helicoverpa zea*, further indicating that *hobo* mobility is not confined to Diptera.

12.4 NON-DROSOPHILID *hAT* ELEMENTS

12.4.1 DISCOVERY

12.4.1.1 The *Hermes* Element of *M. domestica*

The *Hermes* element was discovered on the basis of its mobility properties. Plasmid-based excision assays using the *hobo* element from *D. melanogaster* were performed in *M. domestica* embryos, and imprecise excision of the *hobo* element was observed (Atkinson et al., 1993). Excision did not occur if the inverted terminal repeats (ITRs) of the *hobo* element were deleted, or if the entire *hobo* element was deleted, indicating that excision was dependent on the presence of *hobo* sequences on the indicator plasmid. Most significantly, *hobo* excisions occurred in the absence of helper plasmid containing the *hobo* transposase suggesting that an endogenous *M. domestica* activity was responsible for the excision of the *hobo* element. By comparison, when these assays were performed in *D. melanogaster* E strains in which endogenous *hobo* elements are absent, no excision of the *hobo* element from the indicator plasmid occurred in the absence of helper plasmid (Atkinson et al., 1993; Handler and Gomez, 1995). Assays performed in *D. melanogaster* H strains, in which endogenous *hobo* elements are present, showed that excision of the *hobo* element from the indicator plasmid occurred in the absence of helper plasmid (Handler and Gomez, 1995).

To test the hypothesis that the excision of *hobo* elements in *M. domestica* was caused by an endogeous *hAT* element transposase, a PCR-based approach was undertaken to determine if a *hAT* element was present in the *M. domestica* genome. Three regions of nucleic acid and amino acid similarity identified in the transposases from the *hobo* and *Ac* elements were used to design degenerate oligonucleotide primers (Feldmar and Kunze, 1991; Calvi et al., 1991). Amplified fragments of the predicted sizes were obtained from the *M. domestica* genome. These were cloned and their DNA sequences determined (Atkinson et al., 1993). The level of nucleic acid and amino acid similarity across the amplified region to the *hobo* element and the *hobo* transposase strongly suggested that the amplified fragment did contain a region of a *hAT* element, which was called *Hermes*. Subsequently, an inverse PCR strategy was used to clone the regions flanking this original clone and the ITRs of the *Hermes* element identified (Warren et al., 1994).

Hermes is 2749 bp in size and contains 17 bp ITRs. The ITRs are conserved — 10/12 of the distal 12 bp of the *Hermes* ITRs are identical to the 12-bp *hobo* ITRs. The *Hermes* transposase is approximately 72 kDa in size and is 55% identical and 71% similar to the *hobo* transposase. Overall, the nucleic acid identity between *Hermes* and *hobo* is 55%. *Hermes* is present in the *M. domestica* genome in multiple copies, and full-length copies of *Hermes* are widely dispersed in wild populations of *M. domestica*. Full-length copies of *Hermes* have so far been identified in *M. domestica* strains collected from Maryland, Florida, Texas, Missouri, Georgia, California, Australia, Thailand, Senegal, Uruguay, Panama, Canada, and Zimbabwe. Full-length copies of *Hermes* were not detected in strains of *M. domestica* collected from Korea (L. Cathcart, E. Krafsur, P. Atkinson, and D. O'Brochta, unpublished).

12.4.1.2 The *Homer* Element of *B. tryoni*

The same degenerate primers used to amplify regions of the *Hermes* element were applied to the genome of the Queensland fruit fly, *B. tryoni*. As a member of the family tephritidae, *B. tryoni* is a member of one of the most significant insect pest families known to humankind. Two different fragments of expected size were amplified, cloned, and their DNA sequence determined (Pinkerton et al., 1999). These were called *Homer* and the *Homer*-like element (*HLE*). As done for *Hermes*, inverse PCR was used to clone the complete *Homer* element.

The *Homer* element is 3789 bp in size and, as such, is the longest insect *hAT* element yet characterized. However, the *Homer* transposase, at 621 amino acids in length, is in the same size range as the *hobo* and *Hermes* transposases. The extra sequence in *Homer* is located downstream from the termination codon of the *Homer* transposase. Analysis of this 1 kb of extra sequence does not reveal the presence of any extra-long ORFs.

The *Homer* element is 53% identical to the *Hermes* element and 54% identical to the *hobo* element along its length. The *Homer* transposase is 53% identical and 70% similar to the *Hermes* transposase and 54% identical and 71% similar to the *hobo* transposase. The *Homer* ITRs are 12 bp in length and are 10/12 identical with the 12 bp ITRs of the *hobo* element and 10/12 bp identical with the distal 12 bp of the 17 bp of the *Hermes* ITRs.

Homer elements coexist in the *B. tryoni* genome with *Homer*-like elements. The copy number of *Homer* elements is approximately 6 to 8 per genome while the copy number of the *HLE*s in each genome is approximately 5 (Pinkerton et al., 1999). This is the only case so far discovered in which two types of *hAT* element are present in the same genome of one insect species. While the 3' ITR of the *HLE* remains to be discovered, over 4 kb of *HLE* sequence has been determined, including the entire reading frame of the putative *HLE* transposase. A conceptual translation shows that this transposase is as related to the *Homer* transposase (48% identity, 66% similarity) as either is to the *hobo* or *Hermes* transposases. The same is true when nucleic acid sequence comparisons are made. The *HLE* transposase is, however, inactive due to a number of frameshift mutations along its length.

Several *Homer* long ORFs encoding slight variants of the *Homer* transposase have been isolated and examined for function using interplasmid transposition assays performed in *D. melanogaster* and *B. tryoni* (see below). As described below, all but one of these appear to be functional.

12.4.1.3 The *hermit* Element of *L. cuprina*

The *hermit* element is a nonfunctional *hAT* element from the Australian sheep blowfly, *L. cuprina* (Coates et al., 1996). It was initially found by low stringency hybridization screening of an *L. cuprina* genomic library using a DNA probe from the *hobo*108 element. The entire *hermit* element was cloned and its DNA sequence determined. The *hermit* element is 2716 bp long and contains perfect 15 bp ITRs, the distal 12 of which are identical to the *hobo* ITRs at 10/12 positions. At the nucleic acid level *hermit* is 49% identical to the *hobo* element and 51% identical to the *Hermes* element. Several frameshift mutations interrupt the long ORF of the *hermit* transposase, rendering it inactive. Conceptual translation of the *hermit* transposase shows that it is 42% identical and 64% similar to the *hobo* transposase (Coates et al., 1996).

One interesting feature of the *hermit* element is that it is present only once in the *L. cuprina* genome. No other copies, full length or deleted, are present. This stands in contrast to the distribution of *hobo*, *Hermes*, and *Homer* elements in the genomes of their respective host species. The *hermit* element is clearly inactive — the frameshift mutations that interrupt the reading frame that would have encoded the hermit transposase clearly preclude function. However, it is clear that *hermit* was once active. Flanking the 15 bp ITRs are 8 bp duplications, the sequence of which, 5' GTTGCAAC 3', conforms with the consensus sequence of target site duplications resulting from the insertion of other *hAT* elements into insect genomes. The *hermit* element appears to have been inactivated soon after integrating into the *L. cuprina* genome.

12.4.1.4 The *hopper* Element of *B. dorsalis*

The *hopper* element is another *hAT* element isolated from a tephritid species — in this case the Oriental fruit fly, *B. dorsalis* (Handler and Gomez, 1997). Similar to the PCR-based approach used for *Hermes* and *Homer*, degenerate oligonucleotide primers were used based on regions of amino acids conserved between the *hobo* and *Ac* elements. An amplified DNA fragment of expected size was cloned from *B. dorsalis* and its DNA sequence determined. In addition, other tephritid species such as *B. cucurbitae*, *C. capitata* and *Anastrapha suspensa* were examined, using the same strategy, for the presence of *hAT*-like sequences in their genomes (Handler and Gomez, 1996). Sequences homologous to *hobo* and *Hermes* were found in all species, although there was a significant difference in the relatedness of these tephritid elements to each other and to *hobo* and *Hermes*. The *hopper* element was chosen for further study and a full-length clone of this element obtained by screening a *B. dorsalis* genomic library with the original PCR fragment. Two full-length *hopper* clones were obtained. *Hopper* is 3120 bp in size and contains 19-bp imperfect ITRs which, in the terminal 12 bp, are 5/12 identical to the ITRs of the *hobo* element and 6/12 identical to the ITRs of the *Homer* element from *B. tryoni*. A consensus sequence for the putative *hopper* transposase was constructed and compared with the transposase sequence of other *hAT* transposases. Surprisingly, the *hopper* transposase was found to be equally related to the *hobo* and *Ac* transposases. Amino acid identity and similarities were approximately 24 and 44% for both *hopper–hobo* and *hopper–Ac* comparisons, respectively (Handler and Gomez, 1997). Furthermore, the nucleic acid identity between *hopper* and *Homer* is only 40%, while the corresponding amino acid identity and similarity values are 23 and 45%.

The difference in amino acid similarity between *hopper* and *Homer* is intriguing given that both come from insects from the same genus. This, together with the low level of nucleic acid identity between *hopper* and the other insect *hAT* elements, suggests that *hopper* may have entered the bactrocerids by horizontal transfer.

12.4.1.5 The *huni* Element of *Anopheles gambiae*

The *huni* element is the only *hAT* element so far isolated from a nematoceran species. It was amplified from *An. gambiae* genomic DNA using degenerate oligonucleotide primers based on regions of conservation between insect *hAT* elements (D. O'Brochta, F. Collins, and P. Atkinson, unpublished). An 800-bp fragment was amplified, cloned and its sequence found to contain part of an ORF of a putative *hAT* element transposase. Conceptual translation revealed that this region of the *huni* transposase was 25% identical and 40% similar to the corresponding region of the *Hermes* transposase, 22% identical and 38% similar to the corresponding region of the *hobo* transposase, and 25% identical and 45% similar to the corresponding region of the *Ac* transposase.

12.4.2 MOBILITY PROPERTIES OF THE *HERMES* ELEMENT

12.4.2.1 Range of Species

Hermes has a remarkably wide host range. To date it has been shown to transpose accurately in 12 species of insects: 11 dipterans and 1 lepidopteran (Pinkerton et al., 1996; Sarkar et al., 1997a; 1997b; D. O'Brochta, P. Atkinson, and M. Lehare, unpublished; M. Allen, C. Levesque, D. O'Brochta, P. Atkinson, unpublished data). Demonstration of transposition was achieved using plasmid-based transposition assays in which three plasmids, a *Hermes* donor, a *Hermes* helper, and a target plasmid, are microinjected into preblastoderm embryos. The ability of *Hermes* to transpose accurately in three mosquito species and a lepidopteran is particularly significant given the medical and agricultural impact of these species. *Hermes* transposition has also been demonstrated in a cell culture line of *An. gambiae*; however, in this case, the target for *Hermes* integration was the *Anopheles* chromosome, not a target plasmid (Zhao and Eggleston, 1998).

As techniques for plasmid delivery and recovery are developed for other insect species, the known host range of *Hermes* is likely to increase. In those species in which *Hermes* is known to transpose, transgenic lines have subsequently been generated in three species: *D. melanogaster* (O'Brochta et al., 1996), *Ae. aegypti* (Jasinskiene et al., 1998; Pinkerton et al., 2000), and *C. capitata* (K. Michel, D. O'Brochta, A. Pinkerton, A. Stanenova, G. Franz, A. Robinson and P. Atkinson, unpublished data), suggesting that plasmid-borne transposition assays are reliable indicators of the ability for *Hermes* to be used as a transformation vector in these species.

12.4.2.2 Germline Integration

The frequency of transformation of *D. melanogaster* is typically on the order of 20 to 40%, making *Hermes* as efficient a transformation vector as the *P* element. Genetic markers used to recognize *Hermes*-mediated *Drosophila* transgenics have been the mini-*white* gene, the *C. capitata white* cDNA gene placed under hsp70 control, and the actin5C::EGFP gene. The relationship between insert size and *Hermes* transformation frequency is largely unexplored. *Hermes* transformation frequency of *Ae. aegypti* is in the order of 2 to 5% (Jasinskiene et al., 1998; Pinkerton et al., 2000). *Hermes* has been used to generate several transgenic lines of *Ae. aegypti* in several laboratories using either the *D.*

melanogaster cinnabar gene or the enhanced green or yellow fluorescent protein gene as the genetic marker (Jasinskiene et al., 1998; Pinkerton et al., 2000). Coates et al. (1999) used *Hermes*-mediated transformation to demonstrate that approximately 1.5 kb of the *Ae. aegypti apyrase* promoter and approximately 1.5 kb of the *Ae. aegypti maltase* promoter were sufficient to drive sex-specific expression of the firefly luciferase gene in transgenic mosquitoes. More recently K. Michel, D. O'Brochta, A. Pinkerton, A. Stanenova, G. Franz, A. Robinson and P. Atkinson (unpublished data) used the *Hermes* element to transform genetically the medfly, *C. capitata*. Transformation frequency was low since only one transgenic line was generated; however, at least four independent integrations of the *Hermes* element were obtained from this single line.

12.4.2.3 Analysis of Chromosomal Integration Sites

In both *D. melanogaster* and *C. capitata* transgenic lines, *Hermes* elements create target site duplications upon transposition into target sites (O'Brochta et al., 1996; K. Michel, D. O'Brochta, A. Pinkerton, A. Stanenova, G. Franz, A. Robinson and P. Atkinson, unpublished data). These duplications are 8 bp long and conform with the consensus sequence 5' GTNNNNAC 3'. This consensus sequence is conserved whether transposition is into a target sequence residing in either a plasmid or chromosome, and is species independent. Moreover, this 8-bp target site consensus sequence is not unique to the *Hermes* element. Other insect *hAT* elements are also flanked by 8-bp target site duplications that conform with this sequence. This sequence is thus a property of insect *hAT* elements and is most likely caused by the *hAT* element transposase recognizing these sequences and making staggered single-strand cuts 8 bp apart. Sequences flanking the donor *Hermes* element do not appear to participate in transposition of *Hermes*, just as they apparently play no detectable role in *Hermes* excision.

Integration of *Hermes* into the *Ae. aegypti* genome is accompanied by DNA sequences flanking the *Hermes* transposable element, including, in some cases, plasmid DNA (Pinkerton et al., 2000; Jasinskiene et al., 1999). However, this appears to be *Hermes* transposase-dependent, since injections of vector plasmid in the absence of *Hermes* transposase do not result in the generation of transgenic mosquitoes (Jasinskiene et al., 1999). Interplasmid transposition assays performed in *Ae. aegypti* clearly show that *Hermes* can transpose accurately in this species. The basis for the difference in types of integration events between plasmid-based transposition assays and chromosomal integrations is unknown. One possibility is that endogenous *hAT* elements that may be present in the *Ae. aegypti* genome could be affecting the ability of *Hermes* elements to integrate precisely into the *Ae. aegypti* chromosomes. Nevertheless, *Hermes* elements appear to be stable once integrated. The original transgenic lines have been constantly maintained for over 2 years now with no loss of eye color, while transgenic lines containing green fluorescent protein as the genetic marker have bred true for this phenotype for eight generations.

12.4.3 MOBILITY PROPERTIES OF THE *HOMER* ELEMENT

Homer interplasmid transposition assays performed in *D. melanogaster* and *B. tryoni* show that, while *Homer* is capable of transposition, the frequency is extremely low in both species (A. Pinkerton, D. O'Brochta, S. Whyard, and P. Atkinson, unpublished). These data suggest that, in its present form, *Homer* will not be an efficient gene vector in at least these two insect species.

12.5 ISSUES RELATING TO THE USE OF *HERMES* AND OTHER *hAT* ELEMENTS IN INSECTS

Within the last 5 years the community of insect molecular geneticists has seen the availability of gene transformation tools increase from zero to at least four functional systems. Interested researchers are no longer faced with the question of whether or not there is a transformation tool available that may work in their insect system, but instead are now faced with the difficult problem of deciding which tool they will employ. Because insect transformation demands a large commitment of time and resources, the problem of vector choice will be critical. On what basis should such a decision be made? To what extent is enough known about the existing elements and their relative performance characteristics to enable a good decision to be made? Perhaps of primary concern is knowing whether a particular element will function (with any efficiency) in the species of interest. The number of species in which insect *hAT* elements have been tested is growing. All species of Diptera examined have been able to support *hobo* and/or *Hermes* movement. To a more limited extent, it is known that both *hobo* and *Hermes* can function in two species of *Helicoverpa* but efforts to test in other lepidoptera have been very limited. Clearly, both the *hobo* and *Hermes hAT* elements are potentially useful gene vectors in Diptera. It is important, however, to understand more fully the range of species in which these elements will function. Not only is this a practical question of interest to those looking for genetic tools, but it is a question that we can anticipate being important in any sort of risk assessment effort. While gene vectors with broad host ranges will be particularly valuable in the laboratory, they may pose special risks if used in the field. For example, it has become clear from the study of transposable elements that they can be transferred among species. The natural history of many of the transposable elements being investigated as gene vectors indicates that they have been transferred horizontally in nature, and, in some cases, this transfer has been very recent. The multiple instances of recent horizontal transfer (within the last 75 years) begs the question of how common this event is in nature. If the situation observed in *D. melanogaster* involving the recent introduction of *P* and *hobo* elements is typical, then perhaps these types of events are more common than was initially estimated. The frequency and mechanism of horizontal transfer are aspects of these genetic systems that require additional study. While experimentally addressing the issue of horizontal transfer may be difficult, we are in a better position to address the related question of host range. The likelihood of interspecies transfer will depend upon the mobility properties of the element in its new host. Such questions can be approached experimentally with existing tools. We anticipate that questions regarding the ability of *hAT* and other transposable elements to function in a wide range of insect and noninsect species, such as vertebrates, will become increasingly important as efforts to bring this technology to the field advance.

Another issue that has emerged as a result of investigation of insect *hAT* elements is the potential for related transposable elements to interact. Transposases appear to bind either terminal or subterminal sequences within the element with a degree of specificity high enough to prevent widely divergent elements such as *P* and *hobo*, for example, from interacting. The ability of closely related elements to interact and to result in cross-mobilization has rarely been tested directly, although it has been suggested by ourselves and others. For example, Atkinson et al. (1993) reported that *hobo* elements present on plasmids introduced into the developing embryos of the housefly, *M. domestica*, could excise in the absence of *hobo*-encoded transposase whereas, under similar conditions in *D. melanogaster*, *hobo* excision would absolutely require *hobo*-encoded transposase. We concluded at the time that *M. domestica* embryos contained a *hobo* transposase-like activity capable of interacting with and mobilizing *hobo*. The *Hermes* element was subsequently found in *M. domestica*, lending support to the cross-mobilization hypothesis.

Recently, Sundararajan et al. (1999) directly tested the ability of the *hobo* and *Hermes* transposases, which are 55% identical at the amino acid level, to cross-mobilize the *Hermes* and *hobo* elements, respectively. They found that *hobo* transposase could cause the excision of *Hermes* elements in both plasmid-based excision assays and chromosome-based excision assays in *D. melanogaster*. In fact, *hobo* transposase was almost as good at mobilizing *Hermes* as was *Hermes* transposase. The reciprocal interaction, however, was not as effective — *Hermes* transposase was unable to mobilize *hobo* elements.

Handler and Gomez (1995) reported *hobo* excision from plasmid substrates that were introduced into the embryos of a number of *Drosophila* species in the absence of any experimentally supplied *hobo* transposase and suggested that this also represented cross-mobilization by endogenous *hobo*-like elements as was reported in *Musca*. Fraser et al. (1995) reported what also appeared to be transposase-independent movement of the *piggyBac* element in various cell lines and concluded that the cell lines likely contained a *piggyBac*-related transposase activity. The unique retrotransposon *Penelope* from *D. virilis,* when injected into developing *D. virilis* embryos, resulted in the induction of visible mutations (Evgen'ev et al., 1997), which were due to the mobilization of two unrelated elements, *Ulysses* and *Paris*. The authors proposed that *Penelope* was directly responsible for this cross-mobilization. A member of the *P* element family was isolated from *Scaptomyza pallida* that was 76% identical to the canonical *P* element of *D. melanogaster* yet was still capable of interacting with and mobilizing nonautonomous *P* elements in *D. melanogaster* (Simonelig and Anxolabéhère, 1991). Finally, the *Ubiquitous* and *Ac* transposable element systems of *Zea mays* are related and appear to be capable of interaction (Pisbarro et al., 1991).

The phenomenon of element cross-mobilization has important implications for future work on transgenic insects and the development of insect gene vectors. Perhaps the most immediate implication is for the stability of integrated transgenes. The presence of related elements within the host could destabilize an integrated transgene. Gene vector choice may therefore be influenced by the potential of the vector to interact with endogenous transposable elements. The phenomenon of element interaction/cross-mobilization should stimulate at least a cursory examination of the host genome for transposable elements belonging to the family of elements from which the vector was constructed. In addition, element interaction/cross-mobilization justifies the development of multiple independent gene vector systems to permit a system to be chosen that will minimize the possibility of interactions with endogenous elements. Finally, element interaction/cross-mobilization provides a target for future efforts to modify and improve transposable element-based gene vectors to increase their specificity and to curtail their ability to be cross-mobilized. While the past 4 years have witnessed tremendous strides in the ability to genetically engineer pest insects using transposable elements, we are now faced with the interesting and important problem of understanding the molecular basis of the transposition of these elements in these new hosts.

REFERENCES

Altmann T, Felix G, Jessop A, Kauschmann A, Uwer U, Pena-Cortes H, Willmitzer L (1995) *Ac/Ds* transposon mutagenesis in *Arabidopsis thaliana*: mutant spectrum and frequency of *Ds* insertion mutants. Mol Gen Genet 247:646–652

Atkinson PW, Warren WD, O'Brochta DA (1993) The *hobo* transposable element of *Drosophila* can be cross-mobilized in houseflies and excises like the *Ac* element of maize. Proc Natl Acad Sci USA 83:9693–9697

Baker B, Coupland G, Federoff N, Starlinger P, Schell J (1987) Phenotypic assay for excision of the maize controlling element *Ac* in tobacco. EMBO J 6:1547–1554

Balcells L, Swinburn J, Coupland G (1991) Transposons as tools for the isolation of plant genes. Trends Biotechnol 9:31–37

Bazin C, Higuet D (1996) Lack of correlation between dysgenic traits in the *hobo* system of hybrid dysgenesis in *Drosophila melanogaster*. Genet Res 67:219- 226

Bazin C, Denis B, Capy P, Bonnivard E, Higuet D (1999) Characterization of permissivity for *hobo*-mediated gonadal dysgenesis in *Drosophila melanogaster*. Mol Gen Genet 261:480–486

Berg DE, Schmandt MA, Lowe JB (1983) Specificity of transposon Tn5 insertion. Genetics 105:813–828

Blackman RK, Grimaila R, Koehler MD, Gelbart WM (1987) Mobilization of *hobo* elements residing within the *decapentaplegic* gene complex: suggestion of a new hybrid dysgenesis system in *Drosophila melanogaster*. Cell 49:497–505

Blackman RK, Macy M, Koehler D, Grimaila R, Gelbart WM (1989) Identification of a fully functional *hobo* transposable element and its use for germ-line transformation of *Drosophila*. EMBO J 8:211–217

Bonnivard E, Higuet D, Bazin C (1997) Characterization of natural populations of *Drosophila melanogaster* with regard to the *hobo* system: a newhypothesis for the invasion. Genet Res 69:197–208

Calvi B, Gelbart WM (1994) The basis for germline specificity of the *hobo* transposable element in *Drosophila melanogaster*. EMBO J 13:1636–1640

Calvi BR, Hong TJ, Findley SD, Gelbart WM (1991) Evidence for a common evolutionary origin of inverted terminal repeat transposons in *Drosophila* and plants: *hobo*, *Activator* and *Tam3*. Cell 66:465–471

Chalfie M, Tu Y, Euskirchin G, Ward WW, Prasher DC (1994) Green fluorescent protein as a marker for gene expression. Science 263:802–806

Chuck G, Robbins T, Nijja RC, Ralston E, Courtney-Gutterson N, Dooner HK (1993) Tagging and cloning of a *Petunia* flower color gene with the maize transposable element *Activator*. Plant Cell 5:317–328

Coates CJ, Turney CL, Frommer M, O'Brochta DA, Atkinson PW (1995) The transposable element *mariner* can excise in non-drosophild insects. Mol Gen Genet 249:246–252

Coates CJ, Johnson KN, Perkins HD, Howells AJ, O'Brochta DA, Atkinson PW (1996) The *hermit* transposable element of the Australian sheep blowfly, *Lucilia cuprina*, belongs to the *hAT* family of transposable elements. Genetica 97:23-31

Coates CJ, Turney CL, Frommer M, O'Brochta DA, Atkinson PW (1997) Interplasmid transposition of the *mariner* transposable element in non- drosophild insects. Mol Gen Genet 253:728–733

Coates CJ, Jasinskiene N, Miyashiro L, James AA (1998) *Mariner* transposition and transformation of the yellow fever mosquito, *Aedes aegypti*. Proc Natl Acad Sci USA 95:3748–3751

Coates CJ, Jasinskiene N, Pott GB, James AA (1999) Promoter-directed expression of recombinant fire-fly luciferase in the salivary glands of *Hermes*-transformed *Aedes aegypti*. Gene 226:317–325

Cornel AJ, Benedict MQ, Rafferty CS, Howells AJ, Collins FH (1997) Transient expression of the *Drosophila melanogaster cinnabar* gene rescues eye color in the *white* eye (WE) strain of *Aedes aegypti*. Insect Biochem Mol Biol 27:993- 997

DeVault JD, Hughes KJ, Leopold RA, Johnson OA, Narang SK (1996) Gene transfer into corn earworm (*Helicoverpa zea*) embryos. Genome Res 6:571- 529

Elick TA, Bauser CA, Fraser MJ (1996) Excision of the *piggyBac* transposable element *in vitro* is a precise event that is enhanced by the expression of its encoded transposase. Genetica 98:33–41

Elick TA, Lobo N, Fraser MJ (1997) Analysis of the *cis*-acting DNA elements required for *piggyBac* transposable element excision. Mol Gen Genet 255:605–610

Evgen'ev MB, Zelentsova H, Shostak N, Kozitsina M, Barskyi V, Lankenau DH, Corces VG (1997) *Penelope*, a new family of transposable elements and its possible role in hybrid dysgenesis in *Drosophila virilis*. Proc Natl Acad Sci USA 94:196–201

Federoff NV (1989) Maize transposable elements. In Berg DE, Howe MM (eds) Mobile DNA. American Society for Microbiology, Washington, D.C., pp 375–412

Feldmar S, Kunze R (1991) The ORFa protein, the putative transposase of maize transposable element *Ac*, has a basic DNA binding domain. EMBO J 10:4003–4010

Finnegan DJ (1989) The I-factor and I-R hybrid dysgenesis in *Drosophila melanogaster*. In Berg DE, Howe NM (eds) Mobile DNA. American Society for Microbiology, Washington, D.C., pp 503–517

Franz G, Savakis C (1991) *Minos*, a new transposable element from *Drosophila hydei*, is a member of the Tc1-like family of transposons. Nucl Acids Res 19:6646

Fraser MJ, Cary L, Boonvisudhi K, Wang H-GH (1995) Assay for movement of lepidopteran transposon IFP2 in insect cells using a baculovirus genome as target DNA. Virology 211:397–407

Fridlender M, Sitrit Y, Shaul O, Gileadi O, Lvey AA (1998) Analysis of the *Ac* promoter: structure and regulation. Mol Gen Genet 258:306–314

Galindo I, Ladeveze V, Lemeunier F, Kalmes R, Periquet G, Pascual L (1995) Spread of an autonomous transposable element *hobo* in the genome of *Drosophila melanogaster*. Mol Biol Evol 12:723–734

Gomez SP, Handler AM (1997) A *Drosophila melanogaster hobo-white*(+) vector mediates low frequency gene transfer in *D. virilis* with full interspecific *white*(+) complementation. Insect Mol Biol 6:165–171

Handler AM, Gomez SP (1995) The *hobo* transposable element has transposase- dependent and –independent excision activity in drosophild insects. Mol Gen Genet 247:399–408

Handler AM, Gomez SP (1996) The *hobo* transposable element excises and has related elements in tephritid species. Genetics 143:1339–1347

Handler AM, Gomez SP (1997) A new *hobo, Ac, Tam3* transposable element, hopper, from *Bactrocera dorsalis* is distantly related to *hobo* and *Ac*. Gene 185:133–135

Hehl R (1994) Transposon tagging in heterologous host plants. Trends Genet 10:385–386

Jasinskiene N, Coates CJ, Benedict MQ, Cornel AJ, Salazar-Rafferty C, James AA, Collins FH (1998) Stable, transposon-mediated transformation of the yellow fever mosquito, *Aedes aegypti*, using the *Hermes* element from the house fly. Proc Natl Acad Sci USA 95:3743–3747

Jasinskiene N, Coates CJ, James AA Structure of *Hermes* integrations in the germ line of the yellow fever mosquito, *Aedes aegypti*, Insect Mol Biol, 9:11–18

Jones JDG, Carland F, Lim E, Ralston E, Dooner HK (1990) Preferential transposition of the maize element *Activator* to linked chromosomal locations in tobacco. Plant Cell 2:701–707

Kidwell MG, Kimura K, Black D (1988) Evolution of hybrid dysgenesis potential following *P* element contamination in *Drosophila melanogaster*. Genetics 119:815–828

Ladeveze V, Galindo I, Chaminade N, Pascual L, Periquet G, Lemeunier F (1998) Transmission pattern of *hobo* transposable elements in transgenic lines of *Drosophila melanogaster*. Genet Res 71:97–107

Lim JK (1979) Site-specific instability in *Drosophila melanogaster*: the origin of mutation and cytogenetic evidence for site specificity. Genetics 93:681–701

Lozovskaya ER, Nurminsky DI., Hartl DL, Sullivan DT (1996) Germline transformation of *Drosophila virilis* by the transposable element *hobo*. Genetics 142:173–177

McGrane V, Carlson JO, Miller BR, Beaty BJ (1988) Microinjection of DNA into *Aedes triseriatus* ova and detection of integration. Am J Trop Med Hyg 39:502–510

Medhora M, Maruyama K, Hartl DL (1991) Molecular and functional analysis of the mariner mutator element *Mos1* in *Drosophila*. Genetics 128:311–318

Miller LH, Sakai RK, Romans P, Gwadz RW, Kantoff P, Coon HG (1987) Stable integration and expression of a bacterial gene in the mosquito, *Anopheles gambiae*. Science 237:779–781

Morris AC, Eggleston P, Crampton JM (1989) Genetic transformation of the mosquito, *Aedes aegypti*, by micro-injection of DNA. Med Vet Entomol 3:1–7

Newfeld SJ, Takaesu NT (1999) Local transposition of a *hobo* element within the *decapentaplegic* locus of *Drosophila*. Genetics 151:177–187

O'Brochta DA, Handler AM (1988) Mobility of *P* elements in drosophilids and non-drosophilids. Proc Natl Acad Sci USA 85:6052–6056

O'Brochta DA, Gomez SP, Handler AM (1991) *P* element excision in *Drosophila melanogaster* and related drosophilids. Mol Gen Genet 225:3878–3894

O'Brochta DA, Warren WD, Saville KJ, Atkinson PW (1994) Interplasmid transposition of *Drosophila hobo* elements in non-drosophilid insects. Mol Gen Genet 244:9–14

O'Brochta DA, Warren WD, Saville KJ, Atkinson PW (1996) *Hermes*, a functional non-drosophilid gene vector from *Musca domestica*. Genetics 142:907–914

Periquet G, Lemeunier F, Bigot Y, Hamelin MH, Bazin C, Ladeveze V, Eeken J, Galindo MI, Pascual L, Boussy I (1994) The evolutionary genetics of the *hobo* transposable element in the *Drosophila melanogaster* complex. Genetica 93:79–90

Pinkerton AC, O'Brochta DA, Atkinson PW (1996) Mobility of *hAT* transposable elements in the Old World American bollworm, *Helicoverpa armigera*. Insect Mol Biol 5:223–227

Pinkerton AC, Michel K, O'Brochta DA, Atkinson PA (2000) Green fluorescent protein as a genetic marker in transgenic *Aedes aegypti*. Insect Mol Biol 9:1–10

Pinkerton AC, Whyard S, Mende HM, Coates CJ, O'Brochta DA, Atkinson PW (1999) The Queensland fruit fly, *Bactrocera tryoni*, contains multiple members of the *hAT* family of transposable elements. Insect Mol Biol 8:423–434

Pisabarro AG, Martine WF, Peterson PA, Saedler H, Gierl A (1991) Molecular analysis of the *Ubiquitous* (*Uq*) transposable element system of *Zea mays*. Mol Gen Genet 230:201–208

Rio DC, Laski FA, Rubin GM (1986) Identification and immunochemical analysis of biologically active *Drosophila P* element transposase. Cell 44:21–32

Sarkar A, Yardley K, Atkinson PW, James AA, O'Brochta DA (1997a) Transposition of the *Hermes* element in embryos of the vector mosquito, *Aedes aegypti*. Insect Biochem Mol Biol 27:359–363

Sarkar A, Coates CJ, Whyard S, Willhoeft U, Atkinson PW, O'Brochta DA (1997b) The *Hermes* element from *Musca domestica* can transpose in four families of cylorrhaphan flies. Genetica 99:15–29

Simonelig M, Anxolabéhére D (1991) A *P* element of *Scaptomyza pallida* is active in *Drosophila melanogaster*. Proc Natl Acad Sci USA 88:6102–6106

Smith D, Wohlgemuth J, Calvi BR, Franklin I, Gelbart WM (1993) *Hobo* enhancer trapping mutagenesis in *Drosophila* reveals an insertion specificity different from *P* elements. Genetics 135:1063–1076

Sundararajan P, Atkinson PW, O'Brochta DA (1999) Transposable element interactions in insects: cross-mobilization of *hobo* and *Hermes*. Insect Mol Biol 8:359–368

Thibault ST, Nhak V, Miller TA (1999) Precise excision and transposition of *piggyBac* in pink bollworm embryos. Insect Mol Biol 8:119–123

Warren WD, Atkinson PW, O'Brochta DA (1994) The *Hermes* transposable element from the housefly, *Musca domestica*, is a short inverted repeat-type element of the *hobo*, *Ac*, and *Tam3* (*hAT*) element family. Genet Res Camb 64:87–97

Wilson C, Pearson, RK, Bellen HJ, O'Kane CJ, Grossniklaus U, Gehring WJ (1989) *P*-element-mediated enhancer detection: an efficient method for isolating and characterizing developmentally regulated genes in *Drosophila*. Genes Dev 9:1301–1313

Yannaopoulos G, Stamatis N, Zacharopoulou A, Pelecanos M (1983) Site specific breaks induced by the male recombination factor 23.5MRF in *Drosophila melanogaster*. Mutat Res 108:185–202

Yannopoulos G, Stamatis N, Monastirioti M, Hatzopoulos P, Louis C (1987) *Hobo* is responsible for the induction of hybrid dysgenesis by strains of *Drosophila melanogaster* bearing the male recombination factor *23.5MRF*. Cell 49:487–495

Zhao Y, Eggleston P (1998) Stable transformation of an *Anopheles gambiae* cell line mediated by the *Hermes* mobile element. Insect Biochem Mol Biol 28:219–231

Zwiebel LJ, Saccone G, Zacharopolou A, Besansky NJ, Favia G, Collins FH, Louis C, Kafatos FC (1995) The *white* gene of *Ceratitis capitata*: a phenotypic marker for germline transformation. Science 270:2005–2008

13 Genetic Engineering of Insects with *mariner* Transposons

David J. Lampe, Kimberly K. O. Walden, John M. Sherwood, and Hugh M. Robertson

CONTENTS

13.1 Introduction ... 237
13.2 Background ... 237
13.3 Diverse *mariners* Are Found in Diverse Hosts ... 238
13.4 *mariners* Persist through Repeated Transfers into New Hosts ... 239
13.5 *mariner* Transposase Functions without Host Factors ... 240
13.6 *mariners* Can Function in Diverse Hosts ... 241
13.7 *Himar1* Is Only Marginally Active in *D. melanogaster* ... 242
13.8 Hyperactive Mutants of *Himar1* ... 243
13.9 *mariners* from Different Subfamilies Do Not Interact ... 244
13.10 *mariners* as Genetic Tools in Insects ... 244
References ... 246

13.1 INTRODUCTION

The *mariner* family of transposons holds considerable promise for development as genetic tools in insects. Exploitation of this family of transposons generally follows the model of the *P* element in *Drosophila melanogaster*; however, there are several features of the *mariner* family that make its members particularly appropriate as genetic tools for insects. First, they are extremely widespread and diverse in animal genomes. Second, they persist primarily by repeated horizontal transfers into new host genomes. Third, their transposases are capable of functioning autonomously of host proteins. Fourth, they are capable of functioning in diverse host environments. Fifth, their functioning in bacteria such as *Escherichia coli* allows manipulation of their transposase and inverted terminal repeats to generate improved versions. Sixth, *mariner*s from different subfamilies, and perhaps divergent lineages within subfamilies, do not interact. This chapter reviews some consequences of their prospects as genetic tools for insects.

13.2 BACKGROUND

The type member of the family is the canonical *mariner* element in *D. mauritiana* first identified as an insertion in the *white eye* locus that caused an unstable mutation called w^{peach} (Jacobson and Hartl, 1985; Haymer and Marsh, 1986). It was cloned using the *w* gene of *D. melanogaster* as a probe, and is a 1286-bp element with 28-bp imperfect inverted terminal repeats (ITRs) that generates a TA target duplication upon integration (Jacobson et al., 1986). Subsequently, other versions of this element in *D. mauritiana* and *D. simulans* called "Mosaic factors" that differ by one or more amino acids in their encoded transposase were shown to be active versions (Medhora et al., 1991; Maruyama et al., 1991; Capy et al., 1991). An active element called *Mos1* was capable of mobilizing the original *w*^{peach} insert when both were introduced into the germline of *D. melanogaster* (Garza et al., 1991) and has been demonstrated as a transformation system for *D. melanogaster* (Lidholm et al., 1993) and the yellow fever mosquito, *Aedes aegypti* (Coates et al., 1998). Autonomous versions are also active in *D. virilis* (Lohe and Hartl, 1996a). Similar elements have been identified in other drosophilids (Maruyama and Hartl, 1991a; Capy et al., 1992), including an instance of horizontal transfer between a *Drosophila* and a *Zaprionus* species (Maruyama and Hartl, 1991b). Hartl's group has recently explored the possible regulatory mechanisms of this *mariner* (Lohe and Hartl, 1996b; Lohe et al., 1996; Hartl et al., 1997b), examined the functionality of its transposase (Lohe et al., 1997), and published reviews of it (Hartl, 1989; Hartl et al., 1997b).

The initial evidence of the activity of the *Mos1* version of *mariner* would be enough to encourage its testing in other insects and, indeed, other organisms (see below), but in the past decade a lot more has been learned about the large family of related transposons that give it and these other *mariners* even more potential. We emphasize that we are using the word *family* in the sense that it is used for families of proteins, in this case the transposases that *mariners* encode (Robertson, 1995). (Previously the term *family* has been used to describe the many copies of a particular transposon in a species, e.g., the P family in *D. melanogaster*.) We use the name *mariner* to describe all members of this family, and employ a naming system that uses the two-letter abbreviation of the host species name followed by *mar* and a number to describe particular *mariners* in a species. Thus, the canonical *mariner* would be *Dmmar1*, and the *Mos1* copy would be *Dmmar1.Mos1* (Robertson and Asplund, 1996; other authors have preferred to refer to all the related transposons as *mariner*-like elements or MLEs). We have developed a taxonomy for the family that gives subfamily names to major subdivisions (Robertson and MacLeod, 1993). These subfamily names are generally the species name of the first host in which a member of the subfamily was recognized; thus the mauritiana subfamily includes *Dmmar1* and other closely related *mariner* elements.

13.3 DIVERSE *MARINERS* ARE FOUND IN DIVERSE HOSTS

We and others have described an enormous diversity of *mariners* found in a wide range of animals. The initial discovery was that of Lidholm et al. (1991), who found a divergent *mariner* in an intron of a cecropin gene in the giant silkmoth, *Hyalophora cecropia*. Comparison of the encoded transposases of the fly and moth elements allowed the design of fully degenerate PCR primers to conserved stretches of five to six amino acids that amplified a ±450-bp fragment representing the central half of the ±1000-bp transposase gene. Our initial survey of 404 diverse insect taxa revealed the presence of a wide diversity of related transposons with some unusual characteristics (Robertson, 1993; Robertson and MacLeod, 1993). Most striking was the complete incongruence of the phylogenetic relationships of these *mariner* sequences and the phylogenetic relationships of their host insects, from which we inferred that horizontal transfers, some fairly recent and between orders of insects, must be invoked to explain the distribution of *mariner* transposons. In addition, it was possible to recognize several distinct subfamilies of *mariners*, and many species contained multiple *mariners* belonging to different and sometimes the same subfamily in their genomes. The sequences of these PCR fragments showed that many of these were no longer functional *mariners* because their transposase genes had suffered various mutations, including stop codons and insertions/deletions (indels) causing frameshifts.

Subsequent studies too numerous to detail here have confirmed this pattern, with *mariners* being found widely in animals, ranging from flatworms (Garcia-Fernàndez et al., 1995; Robertson, 1997), nematodes (Sedensky et al., 1994; Wiley et al., 1997; Grenier et al., 1999), and hydras (Robertson, 1997) to humans and other primates (see Robertson and Zumpano, 1997;

Robertson and Martos, 1997). Many others have been found in diverse insects (Blanchetot and Gooding, 1995; Ebert et al., 1995; Robertson and Lampe, 1995; Robertson and Asplund, 1996; Gomulski et al., 1997; Russell and Shukle, 1997) and a mite (Jeyaprakash and Hoy, 1995). We continued our homology-PCR survey of insects and related arthropods, revealing additional instances of likely recent horizontal transfers as well as sequences likely to represent additional smaller subfamilies (Robertson et al., 1998). A highly divergent *mariner* has even been found in soybeans (Jarvik and Lark, 1998).

We summarize the phylogenetic relationships and subfamily classification of most of the known full-length *mariner*s in Figure 13.1. In addition to the two known active *mariner*s from the mauritiana (*Mos1*) and the irritans subfamilies (*Himar1*), there are *mariner*s representative of each of the major subfamilies for which active versions could be obtained (names in bold in Figure 13.1). In most cases this is because consensus sequences could be relatively easily generated from the known genomic copies, as we did for *Himar1* (Lampe et al., 1996), and more recently we have identified several *mariner*s in the nematode *Caenorhabditis elegans* genome where individual copies still reflect the presumably active consensus sequences. In addition to the seven subfamilies shown in Figure 13.1, the existence of as many as 10 additional subfamilies is known from PCR fragments (Robertson, 1993; 1997; Robertson et al., 1998), and these subfamilies might also eventually yield divergent active *mariner*s.

13.4 *MARINER*S PERSIST THROUGH REPEATED TRANSFERS INTO NEW HOSTS

The evidence for recent and ancient horizontal transfers of *mariner*s between animal hosts is overwhelming (Maruyama and Hartl, 1991; Robertson, 1993; 1997; Robertson and MacLeod, 1993; Lohe et al., 1995; Robertson and Lampe, 1995; Garcia-Fernàndez et al., 1995; Robertson et al., 1998). Indeed, taken to its logical conclusion, the evidence suggests that every *mariner* discovered results from a horizontal transfer into the current host lineage, the age of this event being deduced from the level of divergence of the copies within the host. Thus, most known *mariner*s represent old horizontal transfers because the extant copies are quite divergent from their consensus and they exist in related species through vertical inheritance within the host clade, but there are a few relatively recent events known.

There is no known mechanism whereby particular copies of a *mariner* transposon within a particular host species might be selected for continued activity. Because the transposase is synthesized in the cytoplasm and returns to the nucleus, it can interact with the ITRs of any cognate *mariner* in the genome, thus mobilizing defective elements in *trans*. We and others have therefore predicted that within host genomes copies of a particular *mariner* should evolve neutrally and eventually become defective. The overwhelming evidence for defective copies in most genomes is strong support for this notion, and this neutral pattern of evolution has been observed in some instances (Robertson and Lampe, 1995); we summarize others in Robertson et al. (1998). Analysis of the entire genome complement of several *mariner*s in the nematode *C. elegans* also supports this conclusion (H. Robertson, unpublished results).

In contrast, comparisons of *mariner* lineages across host species indicate that the invading copies were active elements; that is, they were selected for activity during the horizontal transfer event. Thus, even when all the extant copies within a genome are defective with multiple frameshifting indels and in-frame stop codons in their transposase genes, for example, the 80-Myr-old *Hsmar2* copies in the human genome (Robertson and Martos, 1997), their consensus sequence encodes a potentially active *mariner* transposase, which presumably represents the original invading copy. There are a few exceptions to this rule in the nematode *C. elegans* genome, but they appear to be the result of multiple cross-interacting different types of *mariner*s that this genome has experienced.

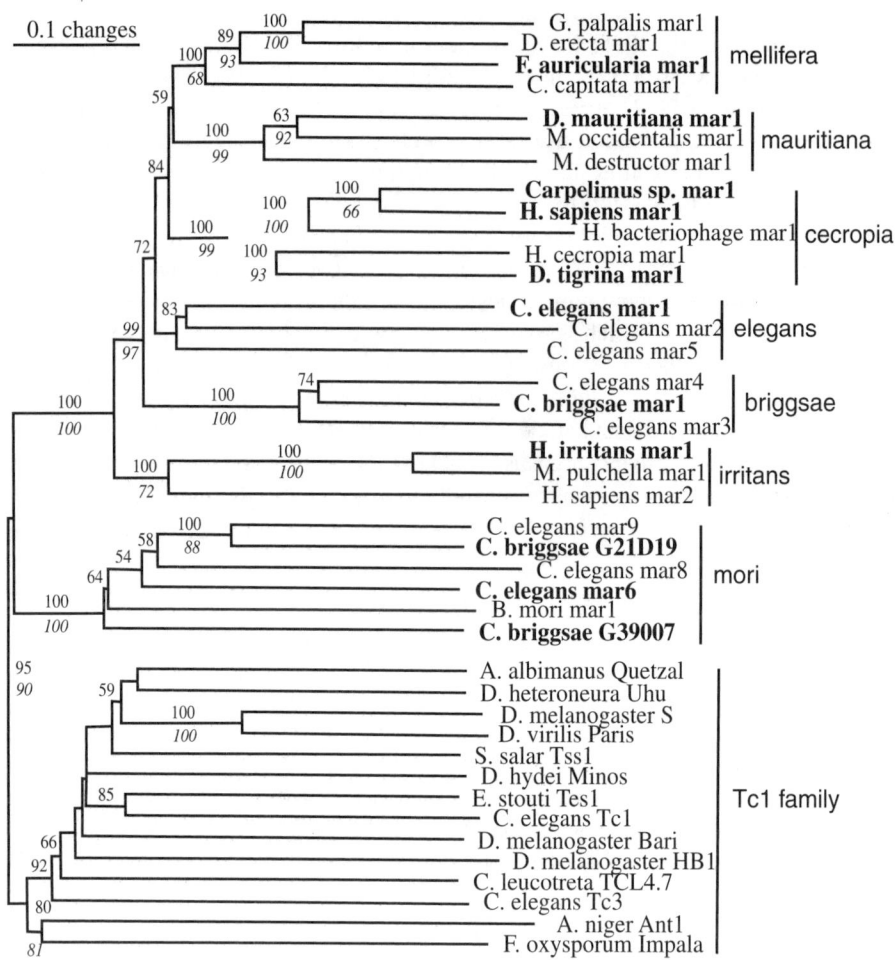

FIGURE 13.1 Phylogenetic relationships of *mariner*s based on their encoded transposase sequences. Most available full-length elements are shown, aligned as in Robertson and Asplund (1996) and rooted with the *Tc1* family. This is the single tree obtained with the neighbor-joining heuristic search algorithm of PAUP4*(4.0b2a). Multiple equally parsimonious trees obtained using maximum parsimony do not differ significantly in the relationships of the *mariner* family, although they do for the poorly supported relationships of the *Tc1* family. Bootstrap support for particular branches was evaluated with 1000 replications of both algorithms, with the neighbor-joining values shown above branches supporting nodes present in more than 50% of the replications, and the generally lower support values obtained with maximum parsimony below branches in italics. The *mariner* subfamilies are separated for clarity, and two new subfamilies of consensus sequences from the nematodes *C. elegans* and *C. briggsae* are named. Those *mariner*s for which active versions are available or could readily be obtained are shown in bold (the earwig *Forficula auricularia* contains copies of a nearly intact mellifera subfamily *mariner* similar to those from the honeybee *Apis mellifera*). The naming system for the *C. elegans mariner*s has been updated to accommodate additional newly recognized members in that genome.

13.5 *MARINER* TRANSPOSASE FUNCTIONS WITHOUT HOST FACTORS

Evidence of horizontal transfer of *mariner*s across orders of insects and phyla of animals led us to predict that *mariner* transposases should be capable of autonomous activity. This prediction was tested using the *Himar1* transposase of a *mariner* that is present in 17,000 copies in the genome of the horn fly *Haematobia irritans*. We were able to reconstruct the consensus sequence of *Himar1*, all examined copies of which in the *H. irritans* genome are highly defective (Robertson and Lampe, 1995), by first engineering the consensus sequence of the closely related *Cpmar1* from the green lacewing *Chrysoperla plorabunda* using a particular copy that differs from its consensus by just one amino acid, and then making another amino acid change to the *Himar1* consensus (Lampe et al., 1996). This transposase was expressed in *E. coli*, purified, and tested for activity in several *in vitro* assays. First, it binds the cognate 30-bp ITRs of the *Himar1* consensus sequence. Second, it cleaves these termini on both strands. Third, it is capable of catalyzing the transposition of a marked *Himar1* construct from one plasmid into another. The distantly related *Tc1* transposase is similarly capable of autonomous transposition activity in vitro (Vos et al., 1996). It therefore seems reasonable to conclude that all active *Tc1–mariner* superfamily elements will prove capable of autonomous activity. It remains to be seen whether even more distantly related transposases, such as those of the *pogo* superfamily (Smit and Riggs, 1996; Robertson, 1996; Wang et al., 1999), are similarly autonomous.

13.6 *MARINERS* CAN FUNCTION IN DIVERSE HOSTS

The first demonstration of function in a heterologous host was particularly dramatic, that of *Mos1* in a kinetoplastid protist, *Leishmania major* (Gueiros-Filho and Beverley, 1997). Subsequently *Mos1* has been shown to be active in zebrafish (Fadool et al., 1998) and is a candidate transforming agent for chicken embryos (Sherman et al., 1998). Our collaborators have demonstrated activity of *Himar1* in human cells (Zhang et al., 1998). More recently, activity of *Himar1* has been demonstrated in bacteria such as *Haemophilus influenzae* and *E. coli* (Rubin et al., 1999). Similarly widespread activity has been reported for members of the *Tc1* family, including activity of *Minos* in the medfly (Loukeris et al., 1995a), *Tc1* in human cells (Shouten et al., 1998), *Tc3* in zebrafish embryos (Raz et al., 1997), and the *Sleeping Beauty* element (reconstructed from fish elements) in human and mouse cells (Ivics et al., 1997; Luo et al., 1998). It seems reasonable to generalize that any *Tc1–mariner* superfamily element will be capable of activity in any host organism; however several more critical tests, for example, in plants and fungi, are warranted.

Most importantly for prospects in insects, Coates et al. (1998) have achieved remarkable success with *Mos1* in transformation of *Ae. aegypti* mosquitoes, obtaining rates of 4% of G_0 animals yielding transformed progeny. This achievement is particularly encouraging given the low level of transformation obtained in *D. melanogaster* (Lidholm et al., 1993) with this same *mariner*, suggesting that activity in *D. melanogaster* is not a reliable guide to activity in other insects. *Drosophila melanogaster* is clearly an excellent host species for some transposons, in particular the *P* element, the *hAT* superfamily elements *hobo* and *Hermes* (O'Brochta et al., 1996), and some members of the *Tc1* family such as *Minos* (Loukeris et al., 1995b; Arcà et al., 1997). Activity of *Mos1* was fairly readily demonstrated when it was introduced in a *P* element and its activity detected through mobilization of the peach copy from a white eye gene (Garza et al., 1991); however, efforts to use *Mos1* as a transformation vector in analogous experiments to the *P*, *hobo*, *Hermes*, and *Minos* systems produced low rates of transformation (Lidholm et al., 1993), and the transformants were subsequently stable in the presence of *Mos1*.

There are also some reports of failure to obtain activity in heterologous systems, in particular failure of *Tc1* to mobilize in *D. melanogaster* (Szekely et al., 1994). Indeed, it seems naive to assume that significant activity will always be obtained in any host. Theory predicts that hosts should evolve diverse regulatory mechanisms to suppress the activity of invading transposons, and many instances of such mechanisms are known for transposons in various plants and fungi. Indeed, several lines of evidence suggest that various transposons in *D. melanogaster* are limited in their activity by host factors, for example, mutations of the *flamenco* gene allow activity of *gypsy* retrotransposons (Bucheton, 1995). In addition, many transposons have their own built-in regulatory or repressive systems. These are particularly well characterized for bacterial transposons, but are also reasonably well known for the *P* element in *D. melanogaster* (Roche et al., 1995), and are proposed for many other Class II transposons, including the canonical *mariner* (Lohe and Hartl, 1996b; Lohe et al., 1996; Hartl et al., 1997).

13.7 HIMAR1 IS ONLY MARGINALLY ACTIVE IN *D. MELANOGASTER*

We have undertaken extensive and determined efforts to demonstrate activity of *Himar1* in *D. melanogaster* with only minimal success. First, we attempted to use *Himar1* to transform *D. melanogaster* following protocols that included standard coinjection of a marked version and a "helper" or "marooned" construct that encodes the transposase but has no ITRs. Various marker genes, specifically a mini-*white* gene and a *yellow* gene, have been employed, and various promoters for the transposase have been employed, including *hsp70* with and without heat shock, *hsp82*, and *hsp26*. We have also tried coinjection of purified transposase with marked *Himar1* constructs, following the demonstration of this approach for *P* elements (Kaufman and Rio, 1991). Together, these experiments have employed injection of at least 5000 embryos, and examination of the G_1 progeny of at least 500 G_0 flies, but no transformants have been recovered. Our proficiency at transformation procedures is not the difficulty because we have repeatedly transformed in various *Himar1* constructs using *P* element vectors (see below).

As a second approach to this issue we attempted to demonstrate activity of *Himar1* transposase in *D. melanogaster* by using *P* elements to integrate the transposase gene driven by various promoters and then asking whether it could mobilize a marked *Himar1* separately transformed using *P* element vectors. The marked element has a mini-*white* gene and was ~5 kb in size; inserts in two locations on the X chromosome were employed. Insertions of the *Himar1* transposase construct on the second and third chromosomes were used to mobilize the target insertions. The two elements were genetically combined by crosses, and mobilization of the marked *Himar1* constructs scored in the next generation, generally by crossing of males with the marked construct on the X chromosome to white-eyed attached-X females. Transpositions of the marked *Himar1* element to an autosome would be detected as red-eyed female progeny. Despite repeated experiments involving tests of at least 500 such males, no convincing examples of such transpositions were recovered (some false positives involving unusual chromosome dynamics were occasionally recovered). Indeed, no convincingly white-eyed males were recovered, suggesting that the marked construct was not even excising from the X chromosome. Immunolabeling with a polyclonal antibody to *Himar1* transposase confirmed that the protein was being produced in the male germline and was localized to the nucleus, so it remains unclear why it was not active in these experiments.

As a third approach, we engineered a construct in which an autonomous copy of *Himar1* is inserted into the promoter region of a mini-*white* gene. This construct was transformed into *D. melanogaster* using *P* element vectors and the *yellow* gene as a marker. Two insertions on the X chromosome that had yellow pigmentation of the eyes, indicating minimal expression of this mutant white-eye gene, were employed in crosses of males to white-eyed attached-X females, looking for red-eyed male progeny that would indicate excision of the autonomous *Himar1* copy from the mini-*white* gene, thereby restoring function. In addition, all males were examined for somatic eye color mosaicism. Little indication of somatic mosaicism was seen in thousands of males examined, and only a few red-eyed males were recovered (fewer than 1/1000). Molecular characterization of these few red-eyed males by PCR amplification across the site of the *Himar1* insertion in the mini-*white* gene promoter revealed that the *Himar1* had indeed excised, and sequencing of nine excision sites revealed the expected three, or more rarely two, base-pair footprint of *Himar1* excision (Lampe et al., 1996). Furthermore, examination of genomic DNA from single yellow-eyed males revealed the presence of two bands, one appropriate for the intact construct and one for the shorter excision product. Because the excision product was much shorter (1.2 vs. 2.5 kb), it would be amplified preferentially; therefore, we cannot estimate the frequency of these excisions, but the levels of somatic mosaicism and germline excision suggest that it is low.

We therefore conclude that while *Himar1* is capable of activity in the *D. melanogaster* genome and nuclear and cellular environments, this activity is at a very low level. Presumably this level is so low that it does not allow detection of transformation events in normal-size experiments. It is tantalizing to suggest that something in the host environment of *D. melanogaster* is less than hospitable to *mariner* family transposons, and that this is responsible for the very low activity of *Himar1*, the low activity of *Mos1*, and the near absence of endogenous *mariners*. What exactly this inhibitory activity might be is a mystery, and how it might inhibit *mariner* family transposons and not some of the related *Tc1* family is unclear. Indeed, it remains an open question whether all *mariners* are inhibited and why the *Tc1* element itself did not function when introduced into *D. melanogaster* (Szekely et al., 1994).

Several collaborators have attempted to achieve transformation of some other insects with *Himar1*, as well as detect activity of *Himar1* in several insects and other organisms without success. We cannot provide details of these attempts here, and some were on a small scale; however, they confirm that, while *Himar1* might not require host factors for activity, it might nevertheless be prevented from functioning well in many hosts. It is also unclear how many similar attempts have been made with *Mos1*, and whether similar difficulties will be encountered with other *mariners* when active versions become available.

13.8 HYPERACTIVE MUTANTS OF *HIMAR1*

Demonstration of activity of *Himar1* in *E. coli* (Rubin et al., 1999) opens the door to using this extraordinarily well-developed molecular genetic system to try to obtain improved versions of *Himar1* that might be more effective genetic tools in insects. In collaboration with E. Rubin and B. Akerley in J. Mekalanos' laboratory at Harvard Medical School, we have developed a papillation assay for the activity of *Himar1* in *E. coli* and have used it to screen for hyperactive transposase mutants (Lampe et al., 1999). This assay is essentially the same as that developed for *Tn5* and other transposons (Krebs and Reznikoff, 1988). Essentially, a *Himar1* construct was generated that contains a *lacZ* gene fused 72 bp from the 5′ ITR such that there is an open reading frame through the ITR to the *lacZ* gene. Transposition of this construct from an F plasmid into the host genome would occasionally yield an in-frame fusion with an expressed gene, thus yielding a *lacZ*⁺ cell, which on MacConkey plates containing lactose would have a growth advantage and form red-colored papillae (the color difference is due to a pH change indicated by neutral red). Transformation of this *lacZ*-bearing strain with a plasmid expressing *Himar1* transposase yielded approximately 20 papillae per large colony after 2 to 3 days growth (nontransformed bacteria were eliminated by selection with an antibiotic and the presence of the F plasmid was similarly ensured). We then mutagenized the *Himar1* transposase gene using error-prone PCR, cloned the products into the expression plasmid, and transformed pools of them into the *lacZ*-bearing strain. Under conditions that lead to high levels of mutagenesis, almost every colony contained a null or hypomorphic mutant, and only one hypertransposing mutant was obtained out of 2500 examined. Under conditions producing lower levels of mutagenesis, we obtained 10 potential hypertransposers out of 2300 colonies/plasmids examined. Of these 11, only 2 survived subsequent validation tests, and when sequenced both revealed two amino acid changes. In one case both changes contribute to the hyperactivity, while in the other only one does. These mutants were tested as expressed and purified proteins in an *in vitro* assay, and although their levels of hyperactivity dropped from 10- to 50-fold to 5- to 7-fold, they are still more active than wild type, demonstrating that they are not simply hyperactive due to an accommodation to the *E. coli* environment. We are undertaking additional characterization of these mutants, as well as screens for additional mutants and superimposed mutants. Initial tests of hyperactive transposases in *D. melanogaster* have not revealed any increased ability to mobilize a marked *Himar1* construct, so it might be necessary to generate combinations of mutants that have 100-fold increases in activity to see an improvement in *D. melanogaster*. Alternatively, it might be necessary to obtain mutants in *D. melanogaster* directly, ones that might overcome the apparent host inhibition, although none was detected in a screen for mutants of the *mariner* transposase (Lohe et al., 1997; however, this approach to mutagenesis of the *mariner* transposase is far less efficient than the *E. coli* papillation screens).

This ability to manipulate the *Himar1* transposase, and potentially also the ITRs, in mutagenic screens in *E. coli* will surely be useful in some contexts, even if not immediately in improved genetic tools for insects. For example, the hyperactive mutants we already have are considerably more efficient at mutagenesis *in vitro* (Lampe et al., 1999).

13.9 *MARINERS* FROM DIFFERENT SUBFAMILIES DO NOT INTERACT

The amino acid divergence of *mariners* in different subfamilies is high, with mauritiana, mellifera, cecropia, elegans, and briggsae subfamily members sharing at most 50% encoded amino acid identity. The irritans and mori subfamilies are even more divergent from the others (less than 40 and 30% identity with other subfamily members, respectively). This divergence, together with the divergence of their ITRs (Figure 13.2), suggests that *mariners* from different subfamilies, and perhaps even divergent *mariners* within subfamilies, are unlikely to interact. There has been little direct work on this hypothesis, however, and our initial attempts to evaluate it have involved testing for ability of purified *Himar1* transposase to bind the ITRs of *mariners* representing the other major subfamilies. We employed footprinting assays similar to those described in Lampe et al. (1996) and found that there was no binding of *Himar I* transposase to the ITRs of *mariners* from other subfamilies, and only weak binding to the ITR of the *Mantispa pulchella mar1*, another irritans subfamily *mariner* (see Figure 13.1). Additional assays for the ability to cleave the ends of the ITRs or catalyze transposition *in vitro* agreed with these results. In addition, van Leunen et al. (1993) have shown that the *Tc3* transposase does not cross-mobilize copies of *Tc1* in the nematode *C. elegans* genome. *Tc1* and *Tc3* can be viewed as representing very different subfamilies of the *Tc1* family (see Figure 13.1). Additional support for this hypothesis comes from the observation that, when insects and other animals contain multiple different kinds of *mariners*, they are usually from different subfamilies or are highly divergent members of the same subfamily.

13.10 *MARINERS* AS GENETIC TOOLS IN INSECTS

Perhaps the primary advantage of using *mariners* as genetic tools for insects is our considerable understanding of their diversity and molecular evolution. Members of the major subfamilies (meaning the most commonly found in insects, that is, mauritiana, mellifera, cecropia, and irritans subfamilies) can easily be detected using degenerate PCR primers designed from conserved regions of the transposase (Robertson 1993; 1997), allowing ready determination of the endogenous complement of *mariners* for any targeted host insect. Simple sequencing of a pool of PCR clones reveals the diversity of *mariners* in the host genome (limited only by great differences in copy number), and whether or not they are likely to encode functional transposases. An appropriate active *mariner* from a subfamily not already present in the host genome can then be selected as a genetic tool. This strategy has not yet been necessary for most attempted transformation experiments, because most insects do not have representatives of both the mauritiana and irritans subfamilies, but it will become more important as this genetic technology spreads throughout molecular entomology and beyond. The subsequent stability of transgenes is usually of great importance and can be fairly readily assured with sufficient knowledge of the host *mariner* complement.

A seeming disadvantage of *mariners* has been their apparently low rates of transposition, as evidenced by the low levels of transformation of *D. melanogaster* by *Mos1* (Lidholm et al., 1993)

```
Position number          1........11........21........31...
G.palpalis.mar1          TTAGTTTGGGGAAAAAGAAATCCATTATTTTT
D.erecta.mar1            WTAGKTTGGCAAATATCTCCCTTCCGCYTTTTTG
A.mellifera.mar1         TTGGGTTGCCCAAAAAGTAATTGCGGATTTTT

D.mauritiana.mar1        YCAGGTGTACAAGTAKGGAATGTCGKTT
M.occidentalis.mar1      YCAGGGGTGTAAMTATGAAACGAGAAA
M.destructor.mar1        TTGGGTGTACAACTTAAAAACCGGAATT

H.sapiens.mar1           TTAGGTTGGTGCAAAAGTAATTGCGGTTTT
H.cecropia.mar1          TTRGGTCCTTACATATGAAATTRGCGTTTTGT
D.tigrina.mar1           TTAGGTTGTTCGATATGAAACGGGTCAAAMTT

C.elegans.mar1           TCAGGTTGTCCCATAAGTTTTTGTACT
C.elegans.mar2           TTAGGTTGGTCGAAAAGTCTTTGCAAAATTT
C.elegans.mar5           TTAGGCTGTGAAAAAAGTTTTCTCCGAAATT

C.briggsae.mar1          TTAGGTTGGCGGGAAAGTCTTTGTCACTCTGT
C.elegans.mar3           TTAGGTTGAACCGGAAGTCTTTGTGACT
C.elegans.mar4           TTAGGTTGGCGGGAAAGTCCTGTCCTA

C.plorabunda.mar1        WYAGGTTGGCTGATAAGTCCCCGGTCTRACA
M.pulchella.mar1         ACAGGCGRGTRGATAATTCCCCGRTCTSMCA
H.sapiens.mar2           YAAGGGGTCTTCAAAAAGTTCATGGAAAATG

C.elegans.mar6           CAGGGTGAGTCAAAATTATGGTAAGT
C.elegans.mar8           CAGGGTGGGGAATAAATATAGCGCCGCA
C.elegans.mar9           CAGGGTGGGCCAAAAGTATGGTAACA
B.mori.mar1              CTTAGTCTGGCCATAAATACTGTTACAA
```

FIGURE 13.2 The inverted terminal repeats of *mariners*. Representatives of all the major subfamilies in Figure 13.1 are shown. In each case a duplicated direct TA repeat of the host DNA flanks these ITRs. IUPAC degenerate nucleotide codes are used when the 5′ and 3′ ITRs differ or when there is no clear-cut consensus among copies from a particular species. The subsets of *mariners* from the different major subfamilies are separated by spaces for clarity.

and our failure to demonstrate transformation of this species with *Himar1*. This problem extends farther for more advanced techniques such as enhancer trapping, which requires remobilization of already transformed *mariner* constructs, because the data so far suggest that most *mariner* integrations are highly stable. Of course, this cannot always be the case; otherwise, *mariners* would cease to persist, and could never generate the thousands of copies seen in some host genomes. Nevertheless, our lack of understanding of what host processes might limit *mariner* transposition is a disadvantage. Mobility assays like those of Coates et al. (1997; 1998) can be critical in deciding whether a particular *mariner* will be an effective tool in a particular species (see Chapter 12 by Atkinson and O'Brochta for a discussion of transposition assays).

Another possible disadvantage is that there is evidence that transposition efficiency of *mariners* is sensitive to size. The initial transformation experiments of Lidholm et al. (1993) used a large 13-kb construct, which might have been responsible for the low efficiency. Lohe and Hartl (1996c) reported remarkable stability of transformed constructs ranging from 4.5 to 12 kb, however, their results were complicated by an inference of a requirement for sequences internal to the ITRs for transposition, something not seen in any of the *Himar1 in vitro* transposition assays that use constructs with approximately 60 bp on each terminus. Subsequent work with these *in vitro* assays has shown that transposition frequency indeed decreases exponentially with increasing size of the construct (Lampe et al., 1998). Nevertheless, our papillation experiments in *E. coli* employ a 13-kbp construct that mobilizes efficiently (Lampe et al., 1999) and the successful transformation of *Ae. aegypti* with *Mos1* used a 4.7-kb *cinnabar* marker gene (Coates et al., 1998), so the size limitations on *mariner* transposition remain unclear, and are unlikely to prevent development of genetic tools.

Some might feel that the evidence for horizontal transfers of *mariners* into new hosts is disadvantageous, which might be considered a risk for release of transgenic insects created with this technology (Hoy, 1995). We do not consider this to be a serious concern for several reasons. First, when nonautonomous *mariners* are used for transformation using a transient transposase source provided in *trans*, the resultant transformants are entirely stable. The only way such a transgene could be mobilized would be in *trans* by the *mariner* transposase from a closely related same-subfamily *mariner* already present in the host genome, a possibility that should be excluded before a particular *mariner* is used as a vector for transgenics to be released. Second, even if autonomous *mariners* are to be used to drive transgenes into host insect populations, the risk of subsequent horizontal transfer of such a gene to other insects or other animals remains extremely small on a human timescale. Our evidence for horizontal transfers of *mariners* involves evolutionary timescales with the most recent events occuring at least 100,000 years ago (Robertson and Lampe, 1995) and probably much longer ago. Prospects for horizontal transfers into our own species are extraordinarily remote given that only two *mariners* have invaded our genome in the past 100 million years (Robertson and Zumpano, 1997; Robertson and Martos, 1997).

There are many reasons to be optimistic about the future utility of *mariners* as genetic tools for insects. First, their functioning in *E. coli* provides an unparalleled environment for improvement, and similar approaches in *E. coli* or other model systems such as yeast might usefully be employed with other transposons. Second, their ability to function in diverse hosts will lead, we hope, to their being employed as genetic tools in a diverse array of organisms beyond insects, and improvements of their technological aspects in these systems by diverse researchers will, we hope, eventually benefit their usefulness in insects. Third, with seven major subfamilies and 10 minor subfamilies, *mariners* provide a large number of potentially useful noninteracting transposons that might be exploited, if necessary, for repeated introductions of transgenes in wild populations of a particular host insect. Many of these features apply to the sister *Tc1* family (Robertson, 1995; Plasterk, 1996), which includes the *Minos* element already utilized to transform the medfly (Loukeris et al., 1995a), and so the *Tc1–mariner* superfamily provides an almost limitless supply of raw material for development of genetic tools for insects.

REFERENCES

Arcà B, Zabalou S, Loukeris TG, Savakis C (1997) Mobilization of a *Minos* transposon in *Drosophila melanogaster* chromosomes and chromatid repair by heteroduplex formation. Genetics 145:267–279

Blanchetot A, Gooding RH (1995) Identification of a *mariner* element from the tsetse fly, *Glossina palpalis palpalis*. Insect Mol Biol 4:89–96

Bucheton A (1995) The relationship between the *flamenco* gene and *gypsy* in *Drosophila*: how to tame a retrovirus. Trends Genet 11:349–353

Capy P, Maruyama K, David JR, Hartl DL (1991) Insertion sites of the transposable element *mariner* are fixed in the genome of *Drosophila sechellia*. J Mol Evol 33:450–456

Capy P, Koga A, David JR, Hartl DL (1992) Sequence analysis of active *mariner* elements in natural populations of *Drosophila simulans*. Genetics 130:499–506

Coates CJ, Turney CL, Frommer M, O'Brochta DA, Atkinson PW (1997) Interplasmid transposition of the *mariner* transposable element in non-drosophilid insects. Mol Gen Genet 253:728–733

Coates CJ, Jasinskiene N, Miyashiro L, James AA (1998) *Mariner* transposition and transformation of the yellow fever mosquito, *Aedes aegypti*. Proc Natl Acad Sci USA 95:3748–3751

Ebert PR, Hileman IV JP, Nguyen HT (1995) Primary sequence, copy number, and distribution of *mariner* transposons in the honey bee. Insect Mol Biol 4:69–78

Fadool JM, Hartl DL, Dowling JE (1998) Transposition of the *mariner* element from *Drosophila mauritiana* in zebrafish. Proc Natl Acad Sci USA 95:5182–5186

Garcia-Fernàndez J, Bayascas-Ramírez JR, Marfany G, Muñoz-Mármol AM, Casali A, Baguñà J, Saló E (1995) High copy number of highly similar *mariner*-like transposons in planarian (Platyhelminthe): evidence for a trans-phyla horizontal transfer. Mol Biol Evol 12:421–431

Garza D, Medhora M, Koga A, Hartl DL (1991) Introduction of the transposable element *mariner* into the germline of *Drosophila melanogaster*. Genetics 128:303–310

Gomulski LM, Torti C, Malacrida AR, Gasperi G (1997) *Ccmar1*, a full-length *mariner* element from the Mediterranean fruit fly, *Ceratitis capitata*. Insect Mol Biol 6:241–253

Grenier E, Abdaon M, Brunet F, Capy P, Abad P (1999) A *mariner*-like transposable element in the insect parasite nematode *Heterorhabditis bacteriophora*. J Mol Evol 48:328–336

Gueiros-Filho FJ, Beverley SM (1997) Trans-kingdom transposition of the *Drosophila* element *mariner* within the protozoan *Leishmania*. Science 276:1716–1719

Hartl DL (1989) Transposable element *mariner* in *Drosophila* species. In Berg DE, Howe MM (eds) Mobile DNA. American Society for Microbiology, Washington, D.C., pp 5531–5536
Hartl DL, Lohe AR, Lozovskaya ER (1997a) Modern thoughts on an ancyent *marinere*: function, evolution and regulation. Annu Rev Genet 31:337–358
Hartl DL, Lozovskaya ER, Nurminsky DI, Lohe AR (1997b) What restricts the activity of *mariner*-like transposable elements? Trends Genet 13:197–201
Haymer DS, Marsh JL (1986) Germ line and somatic instability of a *white* mutation in *Drosophila mauritiana* due to a transposable element. Dev Genet 6:281–291
Hoy MA (1995) Impact of risk analyses on pest-management programs employing transgenic arthropods. Parasitol Today 11:229–232
Ivics Z, Hackett PB, Plasterk RH, Izsvak Z (1997) Molecular reconstruction of *Sleeping Beauty*, a Tc1-like transposon from fish, and its transposition in human cells. Cell 91:501–510
Jacobson JW, Hartl DL (1985) Coupled instability of two X-linked genes in *Drosophila mauritiana*: germinal and somatic mutability. Genetics 111:57–65
Jacobson JW, Medhora MM, Hartl DL (1986) Molecular structure of a somatically unstable transposable element in *Drosophila*. Proc Natl Acad Sci USA 83:8684–8688
Jarvik T, Lark KG (1998) Characterization of *Soymar1*, a *mariner* element in soybean. Genetics 149:1569–1574
Jeyaprakash A, Hoy MA (1995) Complete sequence of a *mariner* transposable element from the predatory mite *Metaseiulus occidentalis* isolated by an inverse PCR approach. Insect Mol Biol 4:31–39
Kaufman PD, Rio DC (1991) Germline transformation of *Drosophila melanogaster* by purified P element transposase. Nucl Acids Res 19:6336
Krebs MP, Reznikoff WS (1988) Use of a Tn5 derivative that creates *lacZ* translational fusion to obtain a transposition mutant. Gene 63:277–285
Lampe DJ, Churchill MEA, Robertson HM (1996) A purified *mariner* transposase is sufficient to mediate transposition *in vitro*. EMBO J 15:5470–5479
Lampe DJ, Grant TE, Robertson HM (1998) Factors affecting transposition of the *Himar1 mariner* transposon *in vitro*. Genetics 149:179–187
Lampe DJ, Ackerley BJ, Rubin EJ, Mekalanos JJ, Robertson HM (1999) Hyperactive transposase mutants of the *Himar1 mariner* transposon. Proc Natl Acad Sci USA 96:11428–11433
Lidholm D-A, Gudmundsson GH, Boman HG (1991) A highly repetitive, *mariner*-like element in the genome of *Hyalophora cecropia*. J Biol Chem 266:11518–11521
Lidholm D-A, Lohe AR, Hartl DL (1993) The transposable element *mariner* mediates germline transformation in *Drosophila melanogaster*. Genetics 134: 859–868
Lohe AR, Hartl DL (1996a) Germline transformation of *Drosophila virilis* with the transposable element *mariner*. Genetics 143:365–374
Lohe AR, Hartl DL (1996b) Autoregulation of *mariner* transposase activity by overproduction and dominant-negative complementation. Mol Biol Evol 13:549–555
Lohe AR, Hartl DL (1996c) Reduced germline mobility of a *mariner* vector containing exogenous DNA: effect of size or site? Genetics 143:1299–1306
Lohe AR, Moriyama EN, Lidholm D-A, Hartl DL (1995) Horizontal transmission, vertical inactivation, and stochastic loss of *mariner*-like transposable elements. Mol Biol Evol 12:62–72
Lohe AR, Sullivan DT, Hartl DL (1996) Subunit interactions in the *mariner* transposase. Genetics 144:1087–1095
Lohe AR, de Aguiar D, Hartl DL (1997) Mutations in the *mariner* transposase: the D,D(35)E consensus sequence is nonfunctional. Proc Natl Acad Sci USA 94:1293–1297
Loukeris TG, Livadaras I, Arcà B, Zabalou S, Savakis C (1995a) Gene transfer into the medfly, *Ceratitis capitata*, using a *Drosophila hydei* transposable element. Science 170:2002–2005
Loukeris TG, Arcà B, Livadaras I, Dialektaki G, Savakis C (1995b) Introduction of the transposable element *minos* into the germ line of *Drosophila melanogaster*. Proc Natl Acad Sci USA 92:9485–9489
Luo G, Ivics Z, Izsvák Z, Bradley A (1998) Chromosomal transposition of a Tc1/mariner-like element in mouse embryonic stem cells. Proc Natl Acad Sci USA 95:10796–10773
Maruyama K, Hartl DL (1991a) Evolution of the transposable element *mariner* in *Drosophila* species. Genetics 128:319–329
Maruyama K, Hartl DL (1991b) Evidence for interspecific transfer of the transposable element *mariner* between *Drosophila* and *Zaprionus*. J Mol Evol 33:514–524

Maruyama K, Schoor KD, Hartl DL (1991) Identification of nucleotide substitutions necessary for *trans*-activation of *mariner* transposable elements in Drosophila: analysis of naturally occurring elements. Genetics 128:777–784

Medhora M, Maruyama K, Hartl DL (1991) Molecular and functional analysis of the *mariner* mutator element *Mos1* in Drosophila. Genetics 128:311–318

O'Brochta DA, Warren WD, Saville KJ, Atkinson PW (1996) *Hermes*, a functional non-drosophilid insect gene vector from *Musca domestica*. Genetics 142:907–914

Plasterk RHA (1996) The Tc1/mariner transposon family. Curr Top Microbiol Immunol 204:125–143

Raz E, van Leunen HGAM, Schaerringer B, Plasterk RHA, Driever W (1997) Transposition of the nematode *Caenorhabditis elegans Tc3* element in the zebrafish *Danio rerio*. Curr Biol 8:82–88

Robertson HM (1993) The *mariner* element is widespread in insects. Nature 362:241–245

Robertson HM (1995) The *Tc1-mariner* superfamily of transposons in animals. J Insect Physiol 41:99–105

Robertson HM (1996) Members of the pogo superfamily of DNA-mediated transposons in the human genome. Mol Gen Genet 252:761–766

Robertson HM (1997) Multiple *mariners* in flatworms and hydras are related to those of insects. J Hered 88:195–201

Robertson HM, Asplund M (1996) *Bmmar1*: a basal lineage of the *mariner* family of transposable elements in the silkmoth, *Bombyx mori*. Insect Biochem Mol Biol 26:945–954

Robertson HM, Lampe DJ (1995) Recent horizontal transfer of a *mariner* element between Diptera and Neuroptera. Mol Biol Evol 12:850–862

Robertson HM, MacLeod EG (1993) Five major subfamilies of *mariner* transposable elements in insects, including the Mediterranean fruit fly, and related arthropods. Insect Mol Biol 2:125–139

Robertson HM, Martos R (1997) Molecular evolution of the second ancient human *mariner* transposon, *Hsmar2*, illustrates patterns of neutral evolution in the human genome lineage. Gene 205:219–228

Robertson HM, Zumpano KL (1997) Molecular evolution of an ancient *mariner* transposon, *Hsmar1*, in the human genome. Gene 205:203–217

Robertson HM, Soto-Adames FN, Walden KKO, Avancini RMP, Lampe DJ (1998) The *mariner* transposons of animals: horizontally jumping genes. In Syvanen M, Kado C (eds) Horizontal Gene Transfer. Chapman & Hall, London, pp 268–284

Roche SE, Schiff M, Rio DC (1995) *P* element repressor autoregulation involves germ-line transcriptional repression and reduction of third intron splicing. Genes Dev 9:1278–1288

Rubin EJ, Akerley BJ, Novik VN, Lampe DJ, Husson RN, Mekalanos JJ (1999) *In vivo* transposition of *mariner*-based elements in enteric bacteria and mycobacteria. Proc Natl Acad Sci USA 96:1645–1650

Russell VW, Shukle RH (1997) Molecular and cytological analysis of a *mariner* transposon from Hessian fly. J Hered 88:72–76

Sedensky MM, Hudson SJ, Everson B, Morgan PG (1994) Identification of a *mariner*-like repetitive sequence in *C. elegans*. Nucl Acids Res 22:1719–1723

Sherman A, Dawson A, Mather C, Gilhooley H, Li Y, Mitchell R, Finnegan D, Sang H (1998) Transposition of the *Drosophila* element *mariner* into the chicken germ line. Nat Biotech 16:1050–1053

Shouten GJ, van Leunen HGAM, Verra NCV, Valerio D, Plasterk RHA (1998) Transposon Tc1 of the nematode *Caenorhabditis elegans* jumps in human cells. Nucl Acids Res 26:3013–3017

Smit AFA, Riggs AD (1996) *Tiggers* and other DNA transposon fossils in the human genome. Proc Natl Acad Sci USA 93:1443–1448

Szekely AA, Woodruff RC, Mahendran R (1994) *P* element mediated germ line transformation of *Drosophila melanogaster* with the *Tc1* transposable DNA element from *Caenorhabditis elegans*. Genome 37:356–366

van Leunen HGAM, Colloms SD, Plasterk RHA (1993) Mobilization of quiet endogenous Tc3 transposons of *Caenorhabditis elegans* by forced expression of Tc3 transposase. EMBO J 12:2513–2520

Vos JC, De Baere I, Plasterk RHA (1996) Transposase is the only nematode protein required for *in vitro* transposition of Tc1. Genes Dev 10:755–761

Wang H, Hartswood E, Finnegan DJ (1999) *Pogo* transposase contains a putative helix-turn-helix DNA binding domain that recognises a 12 bp sequence within the terminal inverted repeats. Nucl Acids Res 27:455–461

Wiley LJ, Riley LG, Sangster NC, Weiss AS (1997) *mle-1*, a mariner-like transposable element in the nematode *Trichostrongylus colubriformis*. Gene 188:235–237

Zhang L, Sankar U, Lampe DJ, Robertson HM, Graham FL (1998) The *Himar1 mariner* transposon cloned in a recombinant adenovirus vector is functional in mammalian cells. Nucl Acids Res 26:3687–3692

14 The TTAA-Specific Family of Transposable Elements: Identification, Functional Characterization, and Utility for Transformation of Insects

Malcolm J. Fraser, Jr.

CONTENTS

14.1 Introduction 250
14.2 Origin of Lepidopteran TTAA-Specific Transposable Elements 250
    14.2.1 The FP Plaque Morphology Mutation of Baculoviruses 250
    14.2.2 Transposon-Induced Mutations of Baculoviruses 251
14.3 The *tagalong* (TFP3) Element 252
    14.3.1 The *tagalong* Target Site Preference Is Not Limited to Virus Insertions 252
    14.3.2 Excision of tagalong Involves a Site-Specific Recombination Event and Is Precise 252
14.4 The *piggyBac* Element 253
    14.4.1 Structural Features of *piggyBac*..253
    14.4.2 Insertion of *piggyBac* at TTAA Target Sites Is Not Limited to Viral Insertions ..254
14.5 Assays for TTAA-Specific Transposon Mobility 254
    14.5.1 The *piggyBac* ORF Encodes a Functional Transposase..................................254
    14.5.2 Excision of *piggyBac* from Baculovirus Insertion Sites Involves Site-Specific Recombination and Is Precise...256
    14.5.3 Plasmid-Based Excision and Transposition Assays for *piggyBac* Mobilization ...256
        14.5.3.1 Plasmid-Based Excision Assays Demonstrate Excision of *piggyBac* Is Dependent on a Helper-Expressed Transposase 256
        14.5.3.2 Excision Assays with Terminal Duplications Demonstrate the Significance of Terminal Sequences in Excision.............................256
        14.5.3.3 Excision of *piggyBac* Involves a Site-Specific Cut-and-Paste Recombination Event..257
        14.5.3.4 Interplasmid Transposition Assays Demonstrate Mobility of *piggyBac* in a Variety of Insect Species 258
        14.5.3.5 Transposition of *piggyBac* is Nonreplicative, Confirming a Strict Cut-and-Paste Recombination Mechanism..............................259
14.6 Host-Specific Accessory Factors?...259
14.7 Transformation of Insects with *piggyBac*...260
    14.7.1 First Attempts: Tantalizing Inconclusive Results in a Few Insect Species............260
    14.7.2 Reports of Successful *piggyBac* Transformation 261

14.8 Horizontal Transmission of *piggyBac*..262
  14.8.1 Evidence of TTAA-Specific Elements outside the Order Lepidoptera..................262
  14.8.2 General Considerations for Horizontal Transmission 262
  14.8.3 Viruses as Vectors of DNA among Species..262
    14.8.3.1 Baculoviruses as Potential Vectors of Transposons among
              Certain Species 263
14.9 Safety Concerns..263
14.10 Summary..264
References 265

14.1 INTRODUCTION

The TTAA-specific, short-repeat elements are a group of transposons that share similarity of structure and properties of movement. These elements were originally defined in the order Lepidoptera, but may be common among other animals as well. While the importance of these elements for genetic manipulation has been appreciated only recently, it is becoming increasingly apparent that they will be extremely useful tools for transformation of insects in the orders Diptera, Lepidoptera, and Coleoptera. This chapter outlines the history of these elements from their discovery to their most recent use in the genetic transformation of insects.

14.2 ORIGIN OF LEPIDOPTERAN TTAA-SPECIFIC TRANSPOSABLE ELEMENTS

The original identification of these unusual elements came through a somewhat unconventional route relative to most other Class II mobile elements (Fraser et al., 1983). The identification of the TTAA-specific transposable elements parallels the development of baculovirus genetics, and reflects a unique approach to the isolation and identification of active transposable elements in certain eukaryotic systems.

14.2.1 THE FP PLAQUE MORPHOLOGY MUTATION OF BACULOVIRUSES

Baculovirus genetics began in earnest with the establishment of invertebrate cell cultures capable of supporting the full replication of certain of these viruses in the early 1970s. Establishment of one insect cell line in particular, the TN-368 line of Hink (1970) from ovarian tissue of *Trichoplusia ni*, was a breakthrough in that it allowed exceptional growth of the *Autographa californica* nuclear polyhedrosis virus (AcNPV) in culture, leading to the eventual acceptance of this virus as the model baculovirus system. This cell line also permitted the establishment of methods for enumeration and isolation of AcNPV strains.

In the first application of a plaque assay using the TN-368 cell line and field-isolated AcNPV, two distinctive and reproducible plaque morphotypes were identified for this virus (Hink and Vail, 1973). These morphotypes were designated MP (Many Polyhedra) and FP (Few Polyhedra). Similar plaque morphologies were subsequently identified for the *T. ni* NPV (Potter et al., 1976), and later for the *Galleria mellonella* NPV (Fraser and Hink, 1982b) upon serial passage in these TN-368 cells. While all these viruses represent closely related variants of the AcNPV, the commonality of this observation between different isolates and among separate researchers suggested that the variable plaque morphology might be universal among baculoviruses. In fact, later observations of both *Helicoverpa zea* NPV (Fraser and McCarthy, 1984) and *Lymantria dispar* NPV (Slavicek et al., 1995) passaged in cell lines derived from their respective hosts confirmed the universality of spontaneous plaque morphology mutants resembling FP mutants, even though the genetic basis for these similar phenotypes may be quite different (Bischoff and Slavicek, 1997).

These serial passage-derived FP mutants have several distinguishing characteristics from mutagen-induced or temperature-sensitive, occlusion-body-deficient viruses, or the several morphogenic mutants with defective occlusions observed in long-term serial passages. These spontaneous mutants are often isolated within a couple of passages in the cell line, conditions that favor the accumulation of viruses producing more infectious extracellular budded virions than occluded virions (Potter et al., 1976; Fraser and Hink, 1982a; Fraser et al., 1983; Cary et al., 1989). More importantly, the accumulation of these same FP mutants is not restricted to cell culture, but can also occur during serial microinjection passage for the virus in insects, a procedure that also selects against the occluded virus, and favors mutants producing more extracellular budded virus (Hink and Strauss, 1976; Potter et al., 1976; Fraser and Hink, 1982a; Fraser et al., 1983).

With the advent of restriction endonuclease technology, an examination of the genetic basis for the FP mutation of baculoviruses became possible. However, the methylcellulose plaque assay procedure could not easily ensure the separation of genetically pure clonal isolates of the virus. Subsequent development of agarose-based overlay plaque assay procedures (Wood, 1977; Fraser and Hink, 1982b; Fraser, 1982) ultimately facilitated the isolation of genetically pure strains of these viruses, permitting a detailed analysis of their genomes.

14.2.2 Transposon-Induced Mutations of Baculoviruses

The fact that host cell transposons could insert and cause mutations in baculovirus genomes was established through the characterization of these spontaneous FP mutations (Miller and Miller, 1982; Fraser et al., 1983; 1985). The identification of a retrotransposon insertion, TED (Miller and Miller, 1982), in an FP mutant of AcNPV isolated following long-term passage in TN-368 cells was apparently a fortuitous association of this host cell transposon with a separate FP mutation, and has not been repeated. However, this observation was significant in establishing that transposon mutagenesis of baculovirus genomes can occur, even though it did not provide a reproducible, experimentally manipulable system for the analysis of this genetic interaction between these viruses and their hosts.

In contrast, most of the FP mutations of either AcNPV or GmNPV that are isolated from low-passage viruses are linked to the frequent insertion of host cell sequences in the FP25K protein gene (Fraser et al., 1983; Cary et al., 1989; Wang and Fraser, 1993). Examination of nine individual FP mutant strains isolated from these two related baculoviruses identified a common alteration in *Hind*III restriction fragment profiles. This common genetic alteration was the addition of variable lengths of DNA to a 4.95 kb *Hind*III fragment of both virus strains. Further examination revealed these insertions were localized within a 500 bp section of this *Hind*III fragment, and homology comparisons established that several of these insertions were related to one another (Fraser et al., 1983). Later analyses revealed that insertions within this region of the virus genome account for the majority of genome rearrangements observed among AcMNPV mutants arising from serial propagation in the TN-368 cell line (Kumar and Miller, 1987; Cary et al., 1989; Wang and Fraser, 1993).

My research has identified and characterized several mobile host DNA insertions within the FP locus of AcMNPV and GmMNPV (Fraser et al., 1983; 1985; Cary et al., 1989; Wang et al., 1989; Bauser et al., 1996). The insertions most extensively studied are those now designated as *tagalong* (formerly TFP3) and *piggyBac* (formerly IFP2). These insertions exhibit a unique preference for TTAA target sites, whether inserting within the viral FP-locus (Cary et al., 1989; Wang et al., 1989) or at other regions of the viral genome (Wang and Fraser, 1993; Fraser et al., 1995). Both of these elements are part of a larger family of TTAA-target site-specific insertion elements that includes the *T. ni*–derived *piggyBac* and *tagalong* elements, the *Spodoptera frugiperda*–derived elements IFP1.6 (Beames and Summers, 1988; 1990) and the 290 bp insertion of Carstens (1987), and the transposon-like insertion within the *Eco*RI-J,N region of *A. californica* nuclear polyhedrosis virus (Oellig et al., 1987; Schetter et al., 1990), whose origin is undefined.

Transposon-induced mutations are also associated with larval-propagated virus and are not simply an artifact of *in vitro* propagation (Jehle et al., 1995). This emphasizes the fact that the relatively high frequency of transpositional and illegitimate recombination (Xiong et al., 1991) that occurs with these viruses is an important aspect of natural baculovirus evolution. These genetic interactions almost certainly contribute to the evolution of the baculovirus genome through the acquisition of genes that enhance the virus host range or virulence (O'Reilly and Miller, 1989; O'Reilly et al., 1992; reviewed in Blissard and Rohrmann, 1990).

14.3 THE *TAGALONG* (TFP3) ELEMENT

The *tagalong* elements were originally isolated from the spontaneous mutants GmFP3 and AcFP6 (Fraser et al., 1983), and were later seen among several FP mutations isolated independently by Kumar and Miller (1987). This element was originally defined as having homology to dispersed and repetitive DNA of the host cell genome, and was one of the smaller insertion elements originally isolated in the virus mutant screen. Sequencing analysis of this insertion provided the first verification that some of these elements do transpose into the viral genome, and are not simply acquired through nonspecific recombination (Fraser et al., 1985).

All *tagalong* elements identified to date, whether from FP mutant viruses or TN-368 genomic clones, have similar restriction sites and overall length and do not exhibit significant open reading frames (ORFs) or identifiable promoter regions (Wang et al., 1989; Wang and Fraser, 1993). *tagalong* elements insert with the duplication of a target viral sequence, TTAA (Fraser et al., 1985; Wang et al., 1989; Wang and Fraser, 1993), and have 13/15 bp imperfect terminal inverted repeats (Wang et al., 1989). These elements comprise a family of dispersed, low-repeat sequences (Fraser et al., 1983; Wang et al., 1989) common among all *T. ni* genomes, suggesting this element is a well-established transposon in the *T. ni* population (Wang et al., 1989).

14.3.1 THE *TAGALONG* TARGET SITE PREFERENCE IS NOT LIMITED TO VIRUS INSERTIONS

While specificity for and duplication of TTAA target sites was a confirmed quality of the insertion of these *tagalong* elements into the baculovirus genome, there remained some question whether this phenomenon was directed as part of some virus recombination mechanism or was a distinct property of the movement of the transposons. The answer to this question could be provided by examining representatives of the element embedded in the uninfected cellular genome.

By using inverse PCR isolation protocols, a *tagalong* insertion was identified within the TN-368 cell line genome at a position that was not occupied in the larval *T. ni* genome (Wang and Fraser 1993). This genomic *tagalong* insertion was similar in size and sequence to the viral insertion elements, and had duplicated a TTAA target site that was present only once in the uninterrupted larval version of the gene, demonstrating that the element has moved in the TN-368 genome since establishment of the cell line from *T. ni* pupae (Wang and Fraser, 1993).

14.3.2 EXCISION OF *TAGALONG* INVOLVES A SITE-SPECIFIC RECOMBINATION EVENT AND IS PRECISE

Initial observations of reversions associated with *tagalong* insertions in the baculovirus genome suggested precise excision was possible for this element (Fraser et al., 1983). In the original characterization of the GmFP3 mutation, spontaneous reversions to the wild-type phenotype were identified following transfection of the GmFP3 viral DNA in the IPLB-SF21AE cell line. These spontaneous phenotype revertants had identical restriction endonuclease profiles to that of the wild-type virus, and regenerated the capacity to produce the 25-kDa protein in infected cells, a protein that was absent in all FP mutant virus infections (Fraser et al., 1983).

While these genetic analyses had seemingly demonstrated precise excision leading to restoration of the wild-type phenotype, confirmation of reproducible and preferential precise excision was obtained for the *tagalong* transposon by recombining *lacZ*-tagged transposon constructs back into the virus genome and following the fate of the tagged transposon upon passage in either TN-368 or SF21AE cells (Fraser et al., 1996). In both cell lines we were able to recover the revertant white-plaque virus, demonstrating conclusively that excision from the baculovirus genome could occur, even in the absence (SF21AE cell line) of genomic representatives of the *tagalong* element. This result was not unexpected, based upon the apparent reversion of a *tagalong*-containing virus mutant observed previously (Fraser et al., 1983).

The *tagalong* element itself contained no significant coding capability, and no coding capacity could be associated with genomic representatives in the TN-368 cell line; yet movement of this element in two lepidopteran cell lines (insertion in the TN-368 cell line, excision in the SF21AE cell line) had clearly been demonstrated. The identification of another active TTAA-specific transposable sequence in the SF21AE cell line (Carstens, 1987; Beames and Summers, 1990) provides some evidence that these TTAA-specific elements may be relatively common among lepidopteran genomes. We speculate that excision of *tagalong* in SF21AE cells might result from the activity of a *trans*-mobilizing transposase. Alternatively, there may be cell-specific factors that mediate the excision of *tagalong* and other TTAA-specific elements.

The analysis of the *tagalong* transposon provided fundamental observations on the commonality of TTAA-specific transposable element function. This element or some as-yet-unidentified related elements may eventually prove useful for genetic engineering in insects. However, the lack of an identified mobilizing function for this element currently limits its utility.

14.4 THE *PIGGYBAC* ELEMENT

The *piggyBac* elements occurred in 13% of all TN-368-derived FP mutants isolated in our hands (Cary et al., 1989). The *piggyBac* element was observed as an insertion in three of the original nine FP mutants, and was subsequently identified in two additional characterized FP mutants. Like *tagalong*, *piggyBac* elements also insert into the viral genome with the duplication of the target site, TTAA (Cary et al., 1989). All characterized *piggyBac* insertions isolated from baculovirus mutants have a similar size and physical map (Cary et al., 1989).

The element is similar to dispersed repetitive DNA in both TN-368 cells (Fraser et al., 1983) and the independently established TN-5B1 cell line (Cary et al., 1989; Elick et al., 1996a), but not to DNA isolated from the independently established cell lines TN-R^2, TN-4B, or TN-4B31 (Elick et al., 1996a). The *piggyBac* sequences in the TN-368 cell genome did not appear to be degenerate based upon the first *SacI/ClaI* digest analyses of genomic Southern blots (Fraser et al., 1983).

Similarly, no degenerate copies of *piggyBac* were recovered during direct PCR amplifications, nor from reverse-transcripterase (RT)-PCR analyses using the TN-368 cell line which harbors many more copies than does the larval genome. However, it is worth noting that these analyses would only have detected internal degeneracy, and are less likely to have identified degeneracy from one or the other end of the element. The apparent incomplete distribution of this element, coupled with its apparent lack of degeneracy (Elick et al., 1996a), suggests it is recently introduced into the *T. ni* genome, or that it is genetically detrimental to its host in large numbers.

14.4.1 STRUCTURAL FEATURES OF *PIGGYBAC*

The *piggyBac* element is 2.4 kb in length and terminates in 13-bp perfect inverted repeats that have the same 5' CCC...GGG 3' terminal trinucleotide sequence configuration as the *tagalong* element, with additional internal 19-bp inverted repeats located asymmetrically with respect to the ends (Cary et al., 1989). The initial sequence analysis of the *piggyBac* element revealed a potential RNA polymerase II promoter sequence configuration, typical Kozak translational start signal, and two apparently overlapping long open reading frames (ORFs). Primer extension analysis with polyadenylated mRNA positioned the 5' end of the *piggyBac* transcript near the identified consensus promoter region (Cary et al., 1989). However, questions remained regarding the nature of the transcript, its exact size, and whether or not splicing of the sequence occurred.

Subsequent Northern analyses, and RT-PCR and sequencing of *piggyBac*-specific RNA transcripts from TN-368 cells confirmed that the major transcript is unspliced (Elick et al., 1996a). All cDNAs were identical in size and similar in sequence with the exception that a single A residue present in the original sequence at the point of overlapping ORFs (Cary et al., 1989) was missing from the derived amplified sequences. Reexamination of additional *piggyBac* sequences amplified from the TN-368 cell genome, as well as the plasmid p3E1.2 (see Section 14.5.1), confirmed the lack of this additional A in the original sequence, and the corrected sequence could be read as a single ORF encoding a polypeptide with a predicted size of 64 kDa.

Later cloning and expression of the full-length and truncated versions of the ORF confirmed a single protein product of a size predicted for the corrected sequence (T. Elick and M. Fraser, unpublished). However, because the sequence of *piggyBac* or its encoded polypeptide does not resemble any transposases previously described, there was no initial assurance that this sequence represented a full-length transposon capable of autonomous movement, or that the polypeptide encoded had the properties of a transposase.

14.4.2 INSERTION OF *PIGGYBAC* AT TTAA TARGET SITES IS NOT LIMITED TO VIRAL INSERTIONS

Inverse PCR analyses of genomic representatives of the *piggyBac* element allowed identification of numerous *piggyBac* insertion termini, either end specific or both ends simultaneously (Elick et al., 1996a). The sequences adjacent to the termini of nearly all these amplified ends was TTAA, confirming that specificity for the TTAA target site was a transposon-specific property and not a property unique to insertions within the baculovirus genome. Interestingly, one of the termini exhibited a 5' ATAA 3' target site associated with the *Sau*09 single-terminus amplified product. This particular aberration was attributed to either a PCR amplification artifact or some allowed degeneracy in the target sites. In later mutagenesis experiments, the latter interpretation proved to be correct.

14.5 ASSAYS FOR TTAA-SPECIFIC TRANSPOSON MOBILITY

Computer-assisted analyses of the *piggyBac* DNA or predicted protein sequences do not reveal significant similarities with other known transposon sequences. To characterize the functional significance of the *piggyBac* ORF, a virus-based transposition assay, and both virus and plasmid-based excision assays were developed.

14.5.1 THE *PIGGYBAC* ORF ENCODES A FUNCTIONAL TRANSPOSASE

The mobility of the *piggyBac* element was first examined by developing an assay based upon the already demonstrated mobilization of the element into the baculovirus genome (Fraser et al., 1995). The *piggyBac* element was tagged by inserting a polyhedrin-driven *lacZ* reporter gene at the unique, internal *Pst*I site. This tagged transposon, when mobilized into the target virus genome, would produce a blue plaque phenotype. The experiment was performed in the presence or absence of an unmodified helper transposon carried within the plasmid p3E1.2 (see below) by transfecting the plasmids and target viral DNA into SF21AE cells. The SF21AE cell line lacks any endogenous *piggyBac*-homologous elements, eliminating the possibility that mobilization could occur in the absence of the helper. While numerous blue plaque viruses were recovered from this experiment, subsequent cloning and sequencing analyses established that movement of the tagged transposon into the virus genome in transfected insect cells occurred via transposition only when the helper element was cotransfected (Fraser et al., 1995).

This experiment was seminal because it confirmed several important facts of *piggyBac* movement. First, and perhaps most importantly, the transposon itself encodes a function that facilitates its own movement, and this function acts in a *trans* fashion, being supplied on a helper plasmid. Second, *piggyBac* could transpose into the baculovirus genome in these insect cells while carrying a marker gene, *polh/lacZ*. Third, movement of the *piggyBac* element could be demonstrated in cells from a lepidopteran species distantly related to the species from which it orignated. These observations proved that this transposon could be used as a helper-dependent vector for transfer of genes in insect cells, and that *piggyBac* movement was not restricted to the species of origin. These observations led directly to the current interest in the *piggyBac* transposon as a tool for genetic engineering in insects.

The original p3E1.2 helper plasmid was derived from a *SalI/HindIII* fragment containing an entire *piggyBac* insertion within the GmNPV FP locus that was ligated into the multiple cloning site of pUC18 (Cary et al., 1989; Fraser et al., 1995). However, some confusion in the sequence of the p3E1.2 plasmid has resulted following its distribution to other laboratories. We now know that this plasmid was unintentionally modified upon transfer to the laboratory of P. Shirk (USDA–ARS, Gainesville, FL) to include an intact multiple cloning site at the 5′ end of the element. Investigators who received the Shirk derivative rather than our original p3E1.2 construct note additional restriction sites not defined in the original clone. In some cases these sites have proved handy for generating deletions of the 5′ terminus, rendering the helper transposon

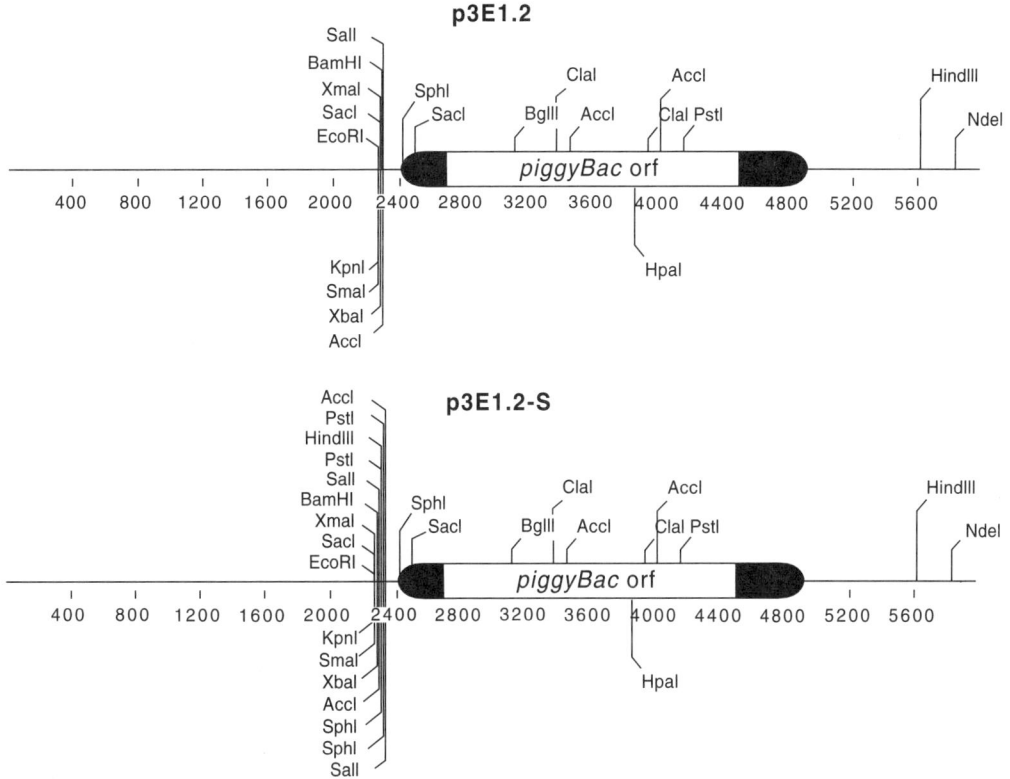

FIGURE 14.1 Comparative maps of the original p3E1.2 and the derivative p3E1.2-S plasmids. The sequence of the transposon is indicated in black, and the location of the *piggyBac* ORF within the transposon sequence is shown as a white box. An apparent expansion of a cloning artifact occurred during transfer of the plasmid p3E1.2 from the laboratory of Dr. Malcolm Fraser to that of Dr. Paul Shirk, resulting in the plasmid p3E1.2-S having reiterations of restriction enzyme sites in the multiple cloning site of the parent pUC18 plasmid. Both plasmids are capable of supplying a functional transposase.

immobile (Handler et al., 1998). Restriction site maps of both plasmids are presented in Figure 14.1 for clarification.

14.5.2 EXCISION OF piggyBac FROM BACULOVIRUS INSERTION SITES INVOLVES SITE-SPECIFIC RECOMBINATION AND IS PRECISE

One of the recombinants generated from the transposition assay was used to examine the excision properties of *piggyBac* relative to *tagalong* (Fraser et al., 1996). If mobility of the two were truly related, then we would expect that precise excision of both should occur in the absence of transposase helpers in SF21AE cells. As expected, *lacZ*-tagged *piggyBac* excised precisely from its insertion site within the baculovirus genome upon passage in either TN-368 or SF21AE cell lines, providing additional evidence of similar mechanisms for movement of both TTAA-specific transposons (Fraser et al., 1995).

14.5.3 PLASMID-BASED EXCISION AND TRANSPOSITION ASSAYS FOR piggyBac MOBILIZATION

While the baculovirus-based assays provided a foundation for analysis of the movement of the element, the assay was less than perfect due to a considerable background of illegitimate recombination events that occurred between the baculovirus target and the cotransfected plasmid leading to a *lacZ*-positive phenotype. Consequently, characterization of the resulting recombinant virus was time-consuming and labor intensive. A much better assay had been developed based upon movement of transposons from plasmid DNAs upon injection of insect embryos. Excision analyses were being conducted with other Class II mobile elements such as *P* (O'Brochta et al., 1991), *hobo*/*Hermes* (Atkinson, 1993; Handler and Gomez, 1995; 1996) with great success and relative facility in injected insect embryos. These assays were used not only to examine the movement properties of the element itself, but also to validate their mobility in a variety of alternative host insects.

14.5.3.1 Plasmid-Based Excision Assays Demonstrate Excision of *piggyBac* Is Dependent on a Helper-Expressed Transposase

We were uncertain at the time whether or not precise excision was a characteristic and invariant feature of *piggyBac* movement since excisive movement had only been observed with the baculovirus genome insertions, and excision and recombination mechanisms are quite active in a baculovirus-infected cell. The plasmid-based assay seemed the best approach to demonstrate excision properties for these elements conclusively, and to characterize these excision products with relative ease.

Our first analyses using plasmid-based strategies utilized a bacterial *supF* tRNA gene within the context of the transposon. Movement of the tagged element from the plasmid could be detected as reversion to white colony phenotype in MBL50 cells (Elick et al., 1996b). In all cases we were able to define excision products as being precise, regenerating the characteristic TTAA target site in the plasmid. This realization allowed us to detect these excision products more easily for both the tagged and unmarked helper constructs using a simple restriction digest, since each excision product generated a new *Ase*I site (ATTAAT) (Elick et al., 1996b).

14.5.3.2 Excision Assays with Terminal Duplications Demonstrate the Significance of Terminal Sequences in Excision

A second assay was easily developed from the first by constructing a vector having a duplication of the 3′ terminal repeat, and tagged with *supF* between the 5′ terminal repeat and the first 3′ repeat domain, and a kanamycin-resistance gene inserted between the two 3' terminal repeats (Elick et al., 1997). Excision of the element using the proximal 3' repeat yielded plasmids that were no longer capable of suppressing the amber mutation in the β-*galactosidase* gene of *Escherichia coli* strain MBL50, but were kanamycin resistant. Excision utilizing the distal repeat yielded plasmids lacking both the suppressor activity and kanamycin resistance. This construct allowed us to replace the proximal 3' terminus with various mutant repeats and to examine the effect of mutations on the target site or 3' terminus on the excision of the element. Mutations in the proximal repeat that were prohibitive would force utilization of the distal terminus for excision (Elick et al., 1997).

This assay was dependent on the fact that either the proximal repeat region would be favored in nonmutagenized constructs or that both distal and proximal repeats would be utilized with equal frequency. The control experiment using wild-type termini demonstrated that both terminal repeats were used with equal frequency (Elick et al., 1997). This observation was somewhat unexpected, since earlier experiments with another Class II element, the *P* element, demonstrated the proximal terminal repeat was highly favored for use in the excision reaction suggesting a scanning model for terminal recognition by the *P*-element transposase. In contrast, our results suggested that *piggyBac* transposase does not operate by scanning from an internal binding site toward the termini, but rather seems to identify termini directly.

Substitutions and deletions of bases in the 5' GGGTTAA 3' sequence revealed an essential requirement for the terminal 5' GGG 3' residues, as well as an essential requirement for the TTA of the TTAA target site (Elick et al., 1997). This latter observation was unprecedented in that no other transposon has such a requirement for the target site sequences to retain mobility. This observation also agreed nicely with the single *Sau*09 5'...ATAACCC...3' terminal sequence recovered from PCR amplifications of genomic *piggyBac* termini. The *Sau*09 genomic insertion seems to have occurred without reliance on the strict TTAA sequence configuration, and may reflect the fact that a certain amount of degeneracy is permitted in searching for target sites. Interestingly, this *Sau*09 terminus, if aligned with the mutagenized construct, would appear as 5'...GGGTTAT...3', in perfect agreement with the experimental observations of degeneracy in the terminal residue of the TTAA site. The introduction of asymmetry, however, brings up a currently unanswered question of the nature of the target sequence. Either the target site for the *Sau*09 genomic insertion was TTAT instead of TTAA or mutagenesis of the target site sequence may have occurred following insertion of the element; or the original supposition that this may represent a PCR artifact is true. This question will most likely be resolved through continued and extensive use of the transposon in gene transformation studies.

14.5.3.3 Excision of *piggyBac* Involves a Site-Specific Cut-and-Paste Recombination Event

The excision results emphasize the uniqueness of the TTAA-specific transposons. This is apparent when one considers that all current models proposed to explain the rare precise excision events seen with most transposons are inadequate in explaining the preference for precise excision exhibited by the TTAA-specific elements. Models that involve replicative deletion of the transposon sequences predict a small percentage of precise excision events among a larger percentage of imprecise events. These models include strand migration (Goldberg et al., 1990), "slippage" (Gordenin et al., 1993), or "copy choice" (d'Alençon et al., 1994). All of these models necessarily involve DNA polymerase reading across a hairpin structure that forms upon denaturation of the double-stranded DNA molecule during replication or repair. Misreading by the polymerase leads to short duplications of terminal sequences or deletions of the target sequence at the excision site. Neither of these features is observed following *piggyBac* or *tagalong* excision

The homologue-dependent gap repair model that explains precise excision of the *P* element from *Drosophila melanogaster* chromosomes (Engels et al., 1990) is also eliminated. Neither the baculovirus nor the plasmid-based excision assays can provide sequences homologous to the insertion site that might be used as repair templates.

Our research has established that excision of these TTAA-specific elements such as *piggyBac* and *tagalong* must involve direct breakage and then immediate joining of the DNA strands at the excision breakpoint (Elick et al., 1997). Most transposons resolve the initial double-strand cuts by fill-in repair at the excision breakpoint that generates deletions through exonuclease activity on the free ends, or can generate insertions of sequences related to the flanking DNA around the excision breakpoints as a result of either template switching (Saedler and Nevers, 1985) or resolution of hairpins by alternative nicking (Coen et al., 1989; Takasu-Ishikawa et al., 1992; see Rommens et al., 1993, for review). Clearly, these repair models cannot explain the extreme precision of *tagalong* and *piggyBac* excision events.

The model we have proposed for excision of these elements begins with synapsis of the transposon termini followed by double-strand cleavage at the target site (Elick et al., 1996a). While the initial process of site-specific cleavage may be similar to the process that occurs with other transposons (see Mizuuchi, 1992, for review), subsequent repair of the double-strand cleavage necessarily involves immediate, direct ligation of the ends of the double-strand break (Elick et al., 1996b). Any fill-in repair or exonuclease digestions will not regenerate the single TTAA target sites as a predominant product of the excision reaction (Saedler and Nevers, 1985; Coen et al., 1986; Rommens et al., 1993).

14.5.3.4 Interplasmid Transposition Assays Demonstrate Mobility of *piggyBac* in a Variety of Insect Species

Although there is reason to believe that excision and transposition are coupled events for most transposons, particularly if a cut-and-paste model is proposed, experimental examination of the transpositional movement of *piggyBac* in a species cannot be verified simply through the use of an excision assay. The work of O'Brochta et al. (1994) provided a strategy for examining transposition in embryos of a given species in the absence of actual transformation of the insect. The construction of the plasmid pB[KOα] by Thibault et al. (1999) permitted examination of the movement of *piggyBac* from one plasmid to another. The recipient plasmid, pGDV1, was derived from *Bacillus subtilis* and is incapable of growth in *E. coli* unless it obtains the [KOα]-tagged *piggyBac* transposon, which includes an *E. coli* origin of replication.

Thibault et al. (1999) were able to demonstrate interplasmid transposition in the pink bollworm, *Pectinophora gossypiella*, using the phspBac helper construct (Handler and Harrell, 1999). Both excision and interplasmid transposition were dependent on the presence of the helper transposon. In their hands, the frequencies of interplasmid transposition were comparable with those obtained for other elements developed as transformation vectors for insects. Distribution of the insertions within the pGDV1 plasmid was restricted to TTAA target sites, as expected, but there appeared to be some preference for only 9 of the possible 21 TTAA sites that might be recovered. The exact reason for restriction to particular TTAA sites was not apparent based solely on sequence comparisons in the vicinity of these target sites. These observations later led to successful phenotypic transformation of *P. gossypiella* (T. Miller, personal communication).

Using pB[KOα] and the pGDV1 target in combination with phspBac, similar analyses were conducted in embryos of *D. melanogaster, Aedes aegypti,* and *T. ni* (Lobo et al., 1999). Although the lack of recovery of interplasmid transposition events in the dipteran species in the absence of the helper was not surprising, the lack of events in *T. ni* was unexpected, since resident copies of the element are present in this species. The reduced efficiency of transposition in *T. ni*, coupled with the lack of observed transposition events in the absence of the helper, suggests that repression effects may be operating in this species.

Significant differences were observed both in the frequency of transposition and the distribution of target sites between the dipteran species and *T. ni*. Once again, utilization of TTAA target sites in the target plasmid, pGDV1, was restricted to 8 of the possible 21 sites in the lepidopteran species, and these sites were the same as those observed by Thibault et al. (1999) for *P. gossypiella*. Interestingly, only 2 of the possible 21 sites were utilized in the two dipteran species (Lobo et al., 1999). We have since confirmed these observations with additional assays (X. Li and M. Fraser, personal communication).

Successful interplasmid transposition results have been obtained for both *Bombyx mori* (T. Tamura, personal communication) and *Anopheles gambiae* (G. Grossman and M. Benedict, personal communication). In the former case, further work has resulted in transformation of the silkworm (T. Tamura et al., 2000), confirming the value of the interplasmid transposition assay as a predictor of successful transformation.

14.5.3.5 Transposition of *piggyBac* is Nonreplicative, Confirming a Strict Cut-and-Paste Recombination Mechanism

The possibility that only a few events were being amplified after transposition in these dipteran embryos was ruled out by replicating the assays, and by examining recovered plasmids for evidence of methylation. Methylation of plasmids does occur in *E. coli* cells, and can be detected through the use of methylation-specific restriction endonucleases like *Dpn*I. In contrast, methylation does not occur in *B. subtilis*, and the pGDV1 plasmid is not sensitive to the activity of *Dpn*I. Since the [KOα]-tagged *piggyBac* does have *Dpn*I sites that would be methylated, the only way to recover interplasmid transposition products following digestion with *Dpn*I would be if duplication of the *piggyBac* element had taken place during movement (in which case movement would be replicative) or the product plasmid had replicated following transposition. Since interplasmid transposition products cannot be recovered from the assays if they are digested with *Dpn*I prior to transformation of bacteria, we may conclude that the interplasmid transposition products are not replicated in the embryos with transposition of *piggyBac* from the *E. coli* plasmid being nonreplicative; transfer of the methylated site to the product would render the pGDV1 target sensitive to *Dpn*I digestion (Lobo et al., 1999).

14.6 HOST-SPECIFIC ACCESSORY FACTORS?

Although the combined results have demonstrated that *piggyBac* movement is not apparently species dependent, the differences observed for insertion sites between dipterans and lepidopterans suggest there may be species-specific differences in target site selection. These differences might be simply due to peculiarities of plasmid configuration in individual species, or could reflect species-specific accessory factors. The possibility of host-specific accessory factors is not without precedent, since cell-specific factors are involved in the transposition of several mobile elements, such as the *P* element of *Drosophila* (Rio and Rubin, 1988; Beall et al., 1994; Staveley et al., 1995) and bacteriophage Mu (Craigie et al., 1985; Craigie and Mizuuchi, 1987; Surette et al., 1987; Surette and Chaconas, 1989).

In the case of *piggyBac* or *tagalong*, excision from baculovirus genomes in insect cells did occur in the absence of homologous transposon-encoded protein products. This seemed to suggest that there might be some cellular factors involved in the mobility of these elements or that there might be cross-mobilizing transposases present in the SF21AE cell line. At least some of these factors might have the properties of identifying the terminal repeats of the element, at a minimum, and recognize terminal repeats of several, if not all, TTAA-specific elements.

We began a search for endogenous cellular factors using nuclear extracts from SF21AE cells, a cell line that harbors no *piggyBac* or *tagalong* elements, yet does allow movement of these elements out of the virus genome at a relatively low frequency. A fraction of the nuclear extract from SF21AE cells did contain a protein, designated TRBP (Terminal Repeat Binding Protein), that was capable of binding a *piggyBac*-specific terminal sequence as well as a *tagalong*-specific terminal sequence

(Bauser et al., 1999). This component binds the terminus of the *piggyBac* sequence in a manner that protects it from *Exo*III degradation from the TTA of the TTAA target site through the terminal inverted repeat domain. The identical fraction from SF21AE cells also bound the *tagalong* repeat primer, fulfilling the criterion of recognition of the termini of both elements.

A similar activity was present in TN-368 cells, suggesting that a ubiquitous DNA-binding protein that has affinity for the sequence configuration CCCTTA(A) may exist since this was the only sequence in common between the two primers used as probes. The exact role of this activity in the movement of these transposons remains to be determined (Bauser et al., 1999).

Analogous host cell factors in other systems do not provide transposase activity, and given the previous demonstrations of functionality for the *piggyBac* ORF in both excision and transposition, it is unlikely that this TRBP activity mediates either excision or transposition alone. Although there may be some ubiquitous excision mechanism that operates with some degree of efficiency in most eukaryotic cells, the most efficient excisions are obtained only in the presence of the transposon-specific transposase activity (Elick et al., 1996a).

14.7 TRANSFORMATION OF INSECTS WITH *PIGGYBAC*

Although assays that provide evidence for the movement of a transposon in embryos of a given species are suggestive of the possibility for transformation of that species, they do not provide the conclusive proof that a given transposon will be useful for transformation of that species. Many variables impinge on the relative utility of a given element as a vector for transgenesis. Transposon-related variables include the relative efficiency of movement into the genome, the stability of the transposon insertion in the genome, and the potential for cross-mobilizing or suppressive effects from similar resident elements. Variables related to the insect being examined include the availability of suitable selectable marker genes, the survivorship of microinjected embryos, and the survivorship and fecundity of emerged injected insects. Refinements in technique and identification of suitable marker genes have contributed at least as much as the application of new transposons have to the recent surge in the successful transgenic engineering of key species.

14.7.1 First Attempts: Tantalizing Inconclusive Results in a Few Insect Species

The utility of *piggyBac* for transformation of insects was first explored using the host species from which the element had originally been isolated, *T. ni* (P. Shirk and M. Fraser, unpublished). Injection of *T. ni* with *piggyBac* elements tagged with *lac*Z marker genes proved to be less than definitive, although tantalizing differential staining was observed in larvae that had been microinjected over those which had not. As an alternative, the organophosphate dehydrogenase gene (*opd*; Benedict et al., 1994; see Chapter 6 by ffrench-Constant and Benedict) was used as a dominant selectable marker for selection of *T. ni* or *Plodia interpunctella* larvae resistant to paraoxon. This seemed an attractive marker gene that would confer selective survivorship on transformed insects (Benedict et al., 1994). Resistant progeny were obtained in several trials with both insects, and stable lines were maintained for several generations before being lost. One of the *T. ni* lines, A1, was derived from a female G_1 that also exhibited an uncommon white-eye mutation (M. Fraser, unpublished). The G_2 progeny exhibited Southern hybridization patterns reflecting additional *piggyBac* elements distributed in the genome compared with the background wild-type hybridization pattern. However, both the *T. ni* and *P. interpunctella* lines were lost before definitive data concerning the identity of the elements could be obtained. It became apparent that working with these lepidopteran systems would prove less tractable than with other, better-characterized genetic systems.

The *opd* marker was also utilized in attempts to transform the mosquito, *Ae. aegypti* (N. Lobo and M. Fraser, unpublished). Once again, tantalizing resistance was obtained which could be maintained through several generations, while no resistance was obtained in the control insects.

However, evaluation of genomic Southerns revealed this resistance was not associated with the insertion of either the *piggyBac* vector or the *opd* gene. Eventually, work with the *opd* marker gene was discontinued in favor of more conclusive genetic markers. While this gene seemed promising as a generalized dominant selectable marker for obtaining and amplifying transformed insects, the results suggest that initial optimism for its use was apparently unwarranted.

14.7.2 Reports of Successful *piggyBac* Transformation

The definitive demonstration of genetic transformation of an insect by the *piggyBac* transposon was finally accomplished using the Mediterranean Fruit Fly as a model system (Handler et al., 1998). This system had the advantage of a recently identified white-eye mutation and complementing gene. A transposase helper plasmid was used having a deletion from the *Sac*I site within the 5′ untranslated leader region of the *piggyBac* element to a flanking *Sac*I site within the pUC18 multiple cloning site sequence. This deletion removed the 5′ inverted repeat domains, as well as the apparent endogenous *piggyBac* promoter sequence and 5′ mRNA initiation site (Cary et al., 1989). In spite of this deletion, the transposase apparently was expressed at levels sufficient to obtain a transformation frequency of 3 to 5%, suggesting an alternative promoter configuration may exist upstream of the ORF in this construct. In any event, the *piggyBac* element is transcriptionally active in this alternative host.

Insertion of the *Ceratitis capitata* white-eye marker into the *piggyBac* transposon at the unique *Hpa*I site had no deleterious effect on the movement of the transposon. Coinjection with the helper allowed rescue of the white-eyed flies to a red-eyed coloration. Variations were apparent in the intensity of the eye color phenotypes, suggesting a position effect suppression of expression for the integrated *white* gene cDNA in the medfly. Subsequent hybridization, inverse PCR amplification, and sequencing analyses confirmed the insertion of the entire transposon carrying the *white* gene into the *C. capitata* genome, with the duplication of a TTAA target site. In one strain, maintenance of the gene at its original position was observed through 15 generations, indicating the *piggyBac* is a stable transformation vector in this species.

In a more recent paper, Handler and Harrell (1999) demonstrated the utility of *piggyBac* for the transformation of *D. melanogaster* as well. In one set of transformations the helper construct was the pBΔSac, while in a separate set of transformation experiments, the heat shock promoter, *hsp*70, was substituted at the *Sac*I site, upstream of the start of translation for the *piggyBac* ORF. The pBΔSac construct yielded 1 to 3% transformation frequency, comparable with that seen for the medfly transformations. In contrast, the *hsp*70 helper, phspBac, yielded an eightfold or more increase in transformation efficiency. This report also introduced a green fluorescent protein (GFP)-tagged construct that yielded a higher efficiency for detection of transformants than the cotransfected white-eye marker gene, apparently overcoming the position silencing effects that are common for the *white* gene transformants. Once again, the insertion with specificity for and duplication of the characteristic TTAA target site was confirmed using inverse PCR and sequencing. More recent results with other dipteran species include the Carribean fruit fly, *Anastrepha suspensa* (A. Handler and R. Harrell, personal communication), the oriental fruit fly, *Bactrocera dorsalis* (A. Handler and S. McCombs, personal communication), and the yellow fever mosquito *Ae. aegypti* (N. Lobo and M. Fraser, unpublished).

Importantly, other researchers have reported successful *piggyBac* phenotypic transformations in several lepidopterans including *Bombyx mori* (Tamura et al., 2000) and *Pectinophora glossypiella* (T. Miller, personal communication), and a coleopteran, *Tribolium castaneum* (Berghammer et al., 1999). The *Tribolium* transformation occurred at an unusually high frequency of 60% (compared with 1% for *Hermes*) based on GFP marker expression, and it will be important to know the molecular nature of the integrations, as well as to see if this frequency can be repeated. At this point it is clear that the *piggyBac* transposon is established as an extremely useful, helper-dependent gene-transfer vector for transformation in a wide variety of insects.

14.8 HORIZONTAL TRANSMISSION OF *PIGGYBAC*

14.8.1 Evidence of TTAA-Specific Elements outside the Order Lepidoptera

To date, there have only been reports of mobile TTAA elements in cell lines from two species of lepidopterans, *Spodoptera frugiperda* (Carstens, 1987; Beames and Summers, 1988) and *Trichoplusia ni* (Fraser et al., 1985; Cary et al., 1989), although dispersed and repeated *piggyBac* homologous sequences have recently been found in the oriental fruit fly (A. Handler, personal communication).

Computer-assisted analysis of sequences obtained from the human genome has revealed what appear to be 100 to 500 copies of a fossil element called LOOPER, which has sequence homology to *piggyBac*, terminates in 5' CCY...GGG 3', and apparently targets TTAA insertion sites (J. Jurka, personal communication). The LOOPER consensus sequence is on average 77% similar to individual sequences identified in the human genome, indicating it is at least 60 million years old. There are two other TTAA-specific fossil repeat elements, MER75 and MER85 (estimated at 2000 copies per genome), which appear to target TTAA insertion sites and terminate in 5' CCC...GGG 3'. Evidence is now accumulating that suggests a superfamily of TTAA-specific mobile elements exists in a diversity of organisms, and that *piggyBac*-related sequences may be present in a diversity of species.

14.8.2 General Considerations for Horizontal Transmission

Several lines of investigation have verified that horizontal transmission of transposons does occur between species (Daniels et al., 1984; 1990; Anxolabéhère et al., 1985; Kidwell, 1993; McDonald, 1993; Robertson, 1993). These observations raise the perennial question of how such transmission might be effected among a diversity of plant and animal species. In this puzzle one must consider several key factors. The fact that similar elements populate the genomes of widely divergent species is well demonstrated. The argument that at least some of these elements have been recently introduced into these divergent species seems to be credible. The nagging question is, what vector could possibly transduce mobile sequences among such a diversity of species? Parasitic mites have been proposed as one vector for *P*-element movement among related species (Houck et al., 1991), but even if these were a valid means of dispersal among a limited number of species, it cannot account for the radiation of these several elements among the wealth of divergent species observed.

In considering a suitable vector for dissemination one must also consider the breaching of germline tissues by the vector, allowing deposition of the transposon in a genetically heritable manner. This added level of complexity is highly significant. Among those vectors that could potentially access germline tissue, it seems that at least some of the obvious candidates would be some sort of ubiquitous pathogen. Most investigators acknowledge that one of the most likely modes of horizontal transmission for fragments of nucleic acids could be as part of a viral genome (Miller and Miller, 1982; Syvanen, 1984; Fraser, 1986; Kidwell, 1993; McDonald, 1993). This is an obvious hypothesis that follows from the dual observations of host DNA insertions, including transposons, within virus genomes and the remarkable relatedness of certain transposons found among vastly different species (see Robertson and Lampe, 1995, for a review).

14.8.3 Viruses as Vectors of DNA among Species

Recombination with host genomes is a generalized phenomenon among viruses. The most obvious examples of vectoring of host sequences by viruses are the retroviral acquisition of oncogenes and transduction in lambda (see Amabile-Cuevas and Chicurel, 1993, for a review). Examples of recombinatorial acquisition of host sequences by eukaryotic DNA animal viruses include adenovirus and SV40 defective particles (Deuring et al., 1981; Norkin and Tirrell, 1982) and the FP mutants of baculoviruses.

In contrast to most other virus systems, baculoviruses may acquire host DNA, including transposable elements, without the concomitant loss of essential replication functions, and these acquired host sequences could conceivably be introduced back into a cellular genome, perhaps during abortive infections (Miller and Miller, 1982; Fraser, 1986). Furthermore, at least in insect systems, baculovirus infections can easily ramify throughout the insect using the system of tracheoles instead of relying on the longer process of breaching the midgut epithelium (Engelhard et al., 1994). Transmission of the virus through tracheoblasts could effectively place the virus in germline tissue quickly. If such positioning occurred in nonpermissive or semipermissive insect hosts, then horizontal transmission could be effected.

Although a direct demonstration of transposon vectoring by viruses is not yet available in an animal model, Sugimoto et al. (1994) demonstrated that a geminivirus engineered to carry a *Ds* element could serve as an effective vector for transposition of the element into the genome of rice calli, essentially demonstrating that if a virus acquired a transposon it might serve as a vector for that transposon. Nonetheless, continuity between the acquisition and transmission of a transposon has not been established in any virus/host system, and this must be established to validate the hypothesis completely.

14.8.3.1 Baculoviruses as Potential Vectors of Transposons among Certain Species

At present, baculoviruses seem to provide the only practical experimental system for direct validation of the hypothesis that viruses might act as vectors for horizontal transmission of mobile DNA. We have proved that viruses can acquire transposons that are mobile and remain nondefective. This part of the puzzle is not in doubt. The question becomes, can a baculovirus carry a transposable element into nonhost cells? The *piggyBac* transposon represents a unique opportunity to answer this question. The fact that it can be mobilized in tissues of a diversity of insect species, and at a relatively high frequency compared with other transposons like *mariner* and *Hermes/hobo*, provides a unique opportunity to examine the transmission of transposons between species by viruses.

The baculoviruses seem ideally suited as vectors for horizontal transfer of genes. These viruses have a tolerance for acquired host genes and mobile elements, a wide natural host range among insects of the order Lepidoptera, and a demonstrated ability to adsorb and penetrate cells of nonhost species, including vertebrates (Hofmann et al., 1995; Boyce and Bucher, 1996). Several lines of experimentation have demonstrated that NPV can effectively invade and initiate abortive replication in nonsusceptible host species such as *D. melanogaster, Ae. aegypti,* or poikilothermic vertebrates (Sherman and McIntosh, 1979; Carbonell et al., 1985; Brusca et al., 1986; Flipsen et al., 1995). Aborted infections of baculoviruses in nonreceptive lepidopteran hosts have also been demonstrated (Washburn et al., 1996). These abortive infections could allow transposons carried on these baculovirus genomes to escape into the cellular genome.

Because of its apparently wide adsorption and penetration capabilities, one could argue that if any virus might serve as a generally useful transfer vehicle following acquisition of transposons, the baculoviruses are among the prime candidates. The successful demonstration of transposon vectoring by baculoviruses will provide significant experimental support for the involvement of viruses in the horizontal transmission phenomenon. This is one of the most significant and exciting avenues of research afforded by this system.

14.9 SAFETY CONCERNS

If transposon vectors are to be useful for engineering environmentally released transgenic insects, the potential for horizontal transmission of any transposon must be addressed from an environmental risk standpoint. At present, the impressive lack of information on this topic makes any predictions of probability for such an event impossible. A considerable amount of research must be expended toward this issue in the near future if the use of transposons for practical engineering purposes in insects is ever to be realized.

Since horizontal transmission of transposons is apparently evolutionarily inevitable, one must consider the possible ramifications of using a transposon vector to introduce genes into insects planned for environmental release. In this case, there are several key features that must be given careful consideration. First, the gene that is to be vectored is of prime concern. This would be true even if transposons are not used for genetic engineering, since illegitimate recombination events abound in nature. Introduction of genes that are functional only in the target species or whose expression is essentially devoid of effect on the target species (such as viral antisense genes) are the most preferred genes for any environmental release of transgenic insects.

Understanding the fundamental mechanisms of the movement of an element is one of the most critical issues in the application of any transposon vector for these purposes. This information will almost certainly lead to greater control over their movement, ideally permitting restriction of the element to the target species. As understanding of the mechanism for mobilization accumulates, altering the target site specificity of an element like *piggyBac* through modification of the transposase or accessory protein may become feasible, allowing restricted movement of the element to laboratory manipulations. Tailoring the transposase function to unusual engineered terminal sequences would significantly reduce the risk of cross-mobilization by naturally occurring elements. Strategies for crippling the element once it is inserted may also prove practical with greater understanding of the molecular requirements for an element's mobility.

Since the most likely use of this technology would be to introduce genes on nonautonomous transposon vectors that rely on an added helper element or added purified transposase protein, there would be little danger of amplification of the gene of concern unless resident cross-mobilizing or homologous elements are present. Therefore, careful examination for potential *trans*-activating or ectopic recombining elements is an important consideration for both target species as well as nontarget species in the ecosystem of concern. Obviously, this will involve a serious commitment of resources, but assays do abound for the elements in question. As with any risk assessment experimentation, the difficulty in examining these questions revolves around the proof of a negative result. Experimental design must be rigorous, ensuring that if risk exists it can be detected and quantified.

14.10 SUMMARY

The TTAA-specific lepidopteran transposon *piggyBac* has already proved useful as a gene-transfer vector for efficient transformation of a wide variety of insects. This element has several unique properties of movement and few specific homologues among insect species examined. However, the probability that other TTAA-specific elements exist in a target species genome appears to be significant, and there remains a possibility that cross-mobilization might occur in certain insects. The availability of several constructs including suitable marker genes for detection in a variety of species has already stimulated investigators to attempt wider applications of this transposon in insect species with remarkable successes. As additional species are tested, the identification of endogenous TTAA-specific elements related to *piggyBac* will almost certainly follow.

There remains a significant amount of basic research that is essential to characterize the mechanism of movement for these TTAA-specific elements in general, and *piggyBac* in particular. Fine-tuning and controlling the mobility of a given transposon is of critical importance for the practical utilization of the element, and this will be true for *piggyBac* as well as other vectors currently applied for this purpose. Assays are already in place to facilitate many of these functional analyses, while others that examine the molecular mechanisms of the *piggyBac* transposase have yet to be developed. While the development of *piggyBac* has seemingly lagged behind that of other transposons identified through more traditional means, there can be no doubt now that this element is a significant transformation vector with wide application, and merits serious research consideration.

The analysis of *piggyBac* and its interaction with baculovirus genomes has provided significant insights into a probable means for horizontal transmission employing viruses as vectors. This is yet another aspect of these TTAA-specific elements in particular, and transposon evolution in general, that merits serious investigation. The observation of the relatively efficient mobility of *piggyBac* in so many different species, coupled with accumulating data suggesting the presence of fossil and potentially active TTAA-specific elements among diverse animal species, suggests this is an ideal experimental system for analysis of this and other important aspects of transposon evolution.

Strategies for ensuring environmental stability and reduced risk for horizontal transmission of vectored genes need to be developed for these TTAA-specific elements if manipulations of natural populations are to become feasible. It is not too early to begin considering these approaches. Once feasibility is demonstrated, progress to implementation will be rapid. The better one understands their unique properties for mobilization, the more effective these vectors will become for the stable and nonproliferative introduction of genes into insects destined for environmental release.

REFERENCES

Amabile-Cuevas C, Chicurel ME (1993) Horizontal gene transfer. Am Sci 81:332–341

Anxolabéhère D, Nouaud D, Periquet G, Tchen P (1985) *P*-element distribution in Eurasian populations of *Drosophila melanogaster*: A genetic and molecular analysis. Proc Natl Acad Sci USA 82:5418–5422

Atkinson PW, Warrren WD, O'Brochta DA (1993) The hobo transposable element of *Drosophila* can be cross-mobilized in houseflies and excises like the Ac element of maize. Proc Natl Acad Sci USA 90:9693–9697

Bauser CA, Elick TE, Fraser MJ (1996) Characterization of *hitchhiker*, a transposon insertion frequently associated with baculovirus FP mutants derived upon passage in the TN-368 cell line. Virology 216:235–237

Bauser CA, Elick TE, Fraser MJ (1999) Proteins from nuclear extracts of two lepidopteran cell lines recognize the ends of TTAA-pecific transposons *piggyBac* and tagalong. Insect Mol Biol 8:1–8

Beall EL, Admon A, Rio DC (1994) A *Drosophila* protein homologous to the human p70 Ku autoimmune antigen interacts with the P transposable element inverted repeats. Proc Nat Acad Sci USA 91:12681–12685

Beames B, Summers MD (1988) Comparisons of host cell DNA insertions and altered transcription at the site of insertions in few polyhedra baculovirus mutants. Virology 162:206–220

Beames B, Summers MD (1990) Sequence comparison of cellular and viral copies of host cell DNA insertions found in *Autographa californica* nuclear polyhedrosis virus. Virology 174:354–363

Benedict MQ, Scott JA, Cockburn AF (1994) High-level expression of the bacterial *opd* gene in *Drosophila melanogaster*: improved inducible insecticide resistance. Insect Mol Biol 3:247–252

Berghammer AJ, Klingler M, Wimmer EA (1999) A universal marker for transgenic insects. Nature 402:370–371

Bischoff DS, Slavicek JM (1997) Phenotypic and genetic analysis of *Lymantria dispar* nucleopolyhedrovirus few polyhedra mutants: mutations in the 25K FP gene may be caused by DNA replication errors. J Virol 71:1097–1106

Blissard GW, Rohrmann GF (1990) Baculovirus diversity and molecular biology. Annu Rev Entomol 35:127–155

Boyce FM, Bucher NLR (1996) Baculovirus-mediated gene transfer into mammalian cells. Proc Natl Acad Sci USA 93: 2348–2352

Brusca J, Summers MD, Couch J, Courtney L (1986) *Autographa californica* nuclear polyhedrosis virus efficiently enters but does not replicate in poikilothermic vertebrate cells. Intervirology 26:207–222

Carbonell LF, Klowden MJ, Miller LK (1985) Baculovirus-mediated expression of bacterial genes in dipteran and mammalian cells. J Virol 56:153-160

Carstens E (1987) Identification and nucleotide sequence of the regions of *Autographa californica* nuclear polyhedrosis virus genone carryoing insertion elements derived from *Spodoptera frugiperda*. Virology 167:8–17

Cary LC, Goebel MJ, Corsaro B, Wang HG, Rosen E, Fraser MJ (1989) Transposon mutagenesis of Baculoviruses: analysis of *Trichoplusia ni* transposon IFP2 insertions within the FP-locus of nuclear polyhedrosis viruses. Virology 172:156–169

Coen ES, Carpenter R, Martin C (1986) Transposable elements generate novel spatial patterns of gene expression in *Antirrhinum majus*. Cell 47:285–296

Coen EA, Robbins TP, Almeida J, Hudson A, Carpenter R (1989) Consequences and mechanisms of transposition in *Antirrhinum majus*. In Berg DE, Howe MM (eds) Mobile DNA. American Society for Microbiology, Washington, D.C., pp 413–436

Craigie R, Mizuuchi K (1987) Transposition of Mu DNA: joining of Mu to target DNA can be uncoupled from cleavage at the ends of Mu. Cell 51:493–501

Craigie R, Arndt-Jovin DJ, Mizuuchi K (1985) A defined system for the DNA strand-transfer reaction at the initiation of bacteriophage Mu transposition: protein and DNA substrate requirements. Proc Natl Acad Sci USA 82:7570–7574

d'Alençon E, Petranovic M, Michel B, Noirot P, Aucouturier A, Uzest M, Ehrlich SD (1994) Copy-choice illegitimate DNA recombination revisited. EMBO J 13:2725–2734

Daniels SB, Strausbaugh LD, Ehrman L, Armstrong R (1984) Sequences homologous to *P* elements occur in *Drosophila paulistorum*. Proc Natl Acad Sci USA 81:6794–6797

Daniels SB, Peterson KR, Strausbaugh LD, Kidwell MG, Chovnick A (1990) Evidence for horizontal transmission of the *P* transposable element between *Drosophila* species. Genetics 124:339–355

Deuring R, Klotz G, Doerfler W (1981) An unusual symmetric recombinant between adenovirus type 12 DNA and human cell DNA. Proc Natl Acad Sci USA 78:3142-3146

Elick TA, Bauser CA, Principe NM, Fraser MJ (1996a) PCR analysis of insertion site specificity, transcription, and structural uniformity of the lepidopteran transposable element *piggyBac* (IFP2) in the TN-368 cell genome. Genetica 97:127–139

Elick TA, Bauser CA, Fraser MJ (1996b) Excision of the *piggyBac* transposable element *in vitro* is a precise event that is enhanced by the expression of its encoded transposase. Genetica 98:33–41

Elick TA, Lobo N, Fraser MJ (1997) Analysis of the *cis*-acting DNA elements required for *piggyBac* transposable element excision. Mol Gen Genet 255:605–610

Engelhard EK, Kam-Morgan NW, Washburn JO, Volkman LE (1994) The insect tracheal system: a conduit for the systemic spread of *Autographa californica* M nuclear polyhedrosis virus. Proc Natl Acad Sci USA 91:3224–3227

Engels WR, Johnson-Schlitz DM, Eggleston WB, Sved J (1990) High-frequency *P* element loss in *Drosophila* is homolog dependent. Cell 62:515–525

Flipsen JT, Martens JW, van Oers MM, Vlak JM, van Lent JW (1995) Passage of *Autographa californica* nuclear polyhedrosis virus through the midgut epithelium of *Spodoptera exigua* larvae. Virology 208:328–335

Fraser MJ (1982) Simplified agarose overlay plaque assay for insect cell lines and insect nuclear polyhedrosis viruses. J Tissue Cult Methods 7:43–46

Fraser MJ (1986) Transposon-mediated mutagenesis of baculoviruses: transposon shuttling and implications for speciation. Ann Entomol Soc Am 79:773–783

Fraser MJ, Hink WF (1982a) The isolation and characterization of MP and FP plaque variants of *Galleria mellonella* nuclear polyhedrosis virus. Virology 117:366–378

Fraser MJ, Hink WF (1982b) Comparative sensitivity of several plaque assay techniques employing TN-368 and IPLB-SF21AE insect cell lines for plaque variants of *Galleria mellonella* nuclear polyhedrosis virus. J Invertebr Pathol 40:89–97

Fraser MJ, McCarthy WJ (1984) The detection of FP plaque variants of *Heliothis zea* nuclear polyhedrosis virus grown in the IPLB-HZ 1075 insect cell line. J Invertebr Pathol 43:427–429

Fraser MJ, Smith GE, Summers MD (1983) Acquisition of host cell DNA sequences by baculoviruses: relationship between host DNA insertions and FP mutants of *Autographa californica* and *Galleria mellonella* nuclear polyhedrosis viruses. J Virol 47:287–300

Fraser MJ, Brusca JS, Smith GE, Summers MD (1985) Transposon-mediated mutagenesis of a baculovirus. Virology 145:356-361

Fraser MJ, Cary L, Boonvisudhi K, Wang HH (1995) Assay for movement of lepidopteran transposon IFP2 in insect cells using a baculovirus genome as a target DNA. Virology 211:397–407

Fraser MJ, Ciszczon T, Elick T, Bauser C (1996) Precise excision of TTAA-specific lepidopteran transposons *piggyBac* (IFP2) and *tagalong* (TFP3) from the baculovirus genome in cell lines from two species of Lepidoptera. Insect Mol Biol 5:141–151

Goldberg JB, Won J, Ohman DE (1990) Precise excision and instability of the transposon Tn5 in *Pseudomonas aeruginosa*. J Gen Microbiol 136:789–796

Gordenin DA, Lobachev KS, Degtyareva NP, Malkova AL, Perkins E, Resnick MA (1993) Inverted DNA repeats: a source of eukaryotic genomic instability. Mol Cell Biol 13:5315–5322

Handler AM, Gomez SP (1995) The *hobo* transposable element has transposase-dependent and -independent excision activity in drosophilid species. Mol Gen Genet 247:399–408

Handler AM, Gomez SP (1996) The *hobo* transposable element excises and has related elements in Tephritid species. Genetics 143:1339–1347

Handler AM, Harrell RA (1999) Germline transformation of *Drosophila melanogaster* with the *piggyBac* transposon vector. Insect Mol Biol 8:449–457

Handler AM, McCombs S, Fraser MJ, Saul SH (1998) The lepidopteran transposon vector, *piggyBac*, mediates germ-line transformation in the Mediterranean fruit fly. Proc Natl Acad Sci USA 95:7520–7525

Hink WF (1970) Established insect cell line from the cabbage looper, *Trichoplusia ni*. Nature (London) 226:466–467

Hink WF, Strauss E (1976) Replication and passage of alfalfa looper nuclear polyhedrosis virus plaque variants in cloned cell cultures and larval stages of four host species. J Invertebr Pathol 27:49–55

Hink WF, Vail PV (1973) A plaque assay for titration of alfalfa looper nuclear polyhedrosis virus in the cabbage looper (TN-368) cell line. J Invertebr Pathol 22:168–174

Hofmann C, Sandig V, Jennings G, Rudolph M, Schlag P, Strauss M (1995) Efficient gene transfer into human hepatocytes by baculovirus vectors. Proc Natl Acad Sci USA 92:10099–10103

Houck MA, Clark JB, Peterson KR, Kidwell MG (1991) Possible horizontal transfer of *Drosophila* genes by the mite *Proctolaelaps regalis*. Science 253:1125–1128

Jehle JA, Fritsch E, Nickel A, Huber J, Backhaus H (1995) TCl4.7: a novel lepidopteran transposon found in *Cydia pomonella* granulosis virus. Virology 207:369–379

Kidwell M (1993) Voyage of an ancient mariner. Nature 362:202

Kumar S, Miller LK (1987) Effects of serial passage of *Autographa californica* nuclear polyhedrosis virus in cell culture. Virus Res 7:335–349

Lobo N, Li X, Fraser MJ (1999) Transposition of the *piggyBac* element in embryos of *Drosophila melanogaster, Aedes aegypti*, and *Trichoplusia ni*. Mol Gen Genet 261:803–810

McDonald JF (1993) Evolution and consequences of transposable elements. Curr Opin Genet Dev 3:855–864

Miller DM, Miller LK (1982) A virus mutant with an insertion of a copia-like transposable element. Nature (London) 299:562–564

Mizuuchi K (1992) Transpositional recombination: mechanistic insights from studies of Mu and other elements. Annu Rev Biochem 61:1011–1051

Norkin LC, Tirrell SM (1982) Emergence of Simian Virus 40 variants during serial passage of plaque isolates. J Virol 42:730–733

O'Brochta DA, Gomez SP, Handler AM (1991) *P*-element excision in *Drosophila melanogaster* and related drosophilids. Mol Gen Genet 225:387–394

O'Brochta DA, Warren WD, Saville KJ, Atkinson PW (1994) Interplasmid transposition of *Drosophila* hobo elements in non-drosophilid insects. Mol Gen Genet 244:9–14

Oellig C, Happ B, Muller T, Doerfler W (1987) Overlapping sets of viral RNAs reflect the array of polypeptides in the EcoRI J and N fragments (map positions 81.2 to 85.0) of the *Autographa californica* nuclear polyhedrosis virus. J Virol 61:3048–3057

O'Reilly DR, Miller LK (1989) A baculovirus blocks insect molting by producing ecdysteroid UDP-glucosyl transferase. Science 245:1110–1112

O'Reilly DR, Brown MR, Miller LK (1992) Alteration of ecdysteroid metabolism due to baculovirus infection of the fall armyworm *Spodoptera frugiperda*: host ecdysteroids are conjugated with galactose. Insect Biochem Mol Biol 22:313–320

Potter KN, Faulkner P, MacKinnon EA (1976) Strain selection during serial passage of *Trichoplusia ni* nuclear polyhedrosis virus. J Virol 18:1040–1050

Rio DC, Rubin GM (1988) Identification and purification of a *Drosophila* protein that binds to the terminal 31-base-pair inverted repeats of the P transposable element. Proc Natl Acad Sci USA 85:8929–8933

Robertson HM (1993) The *mariner* transposable element is widespread in insects. Nature 362:241–245

Robertson HM, Lampe DJ (1995) Distribution of transposable elements in arthropods. Ann Rev Entomol 40:333–357

Rommens C, Van Haaren MJ, Nikamp J, Hille J (1993) Differential repair of excision gaps generated by transposable elements of the "Ac family." BioEssays 15:507–512

Rothe M, Pan M-G, Henzel WJ, Ayres TM, Goeddel DV (1995) The TNFR2-TRAF signaling complex contains two novel proteins related to baculoviral inhibitor of apoptosis proteins. Cell 83:1243–1252

Saedler H, Nevers P (1985) Transposition in plants: a molecular model. EMBO J 4:585–590

Schetter C, Oellig C, Doerfler W (1990) An insertion of insect cell DNA in the 81-Map-Unit segment of *Autographa californica* nuclear polyhedrosis virus DNA. J Virol 64:1844–1850

Sherman KE, McIntosh AH (1979) Baculovirus replication in a mosquito (Dipteran) cell line. Infect Immunol 26:232–234

Slavicek JM, Hayes-Plazolles N, Kelly ME (1995) Rapid formation of few polhedra mutants of *Lymantria dispar* multinucleocapsid nuclear polyhedrosis virus during serial passage in cell culture. Biol Control 5:251–261

Staveley BE, Heslip TR, Hodgetts RB, Bell JB (1995) Protected *P*-element termini suggest a role for inverted-repeat-binding protein in transposase-induced gap repair in *Drosophila melanogaster*. Genetics 139:1321–1329

Sugimoto K, Otsuki Y, Saji S, Hirochika H (1994) Transposition of the *maize* Ds element from a viral vector to the rice genome. Plant J 5:863–871

Surette MG, Chaconas G (1989) A protein factor that reduces the negative supercoiling requirement in the Mu strand transfer reaction is *Escherichia coli* integration host factor. J Biol Chem 264:3028–3034

Surette MG, Buch SJ, Chaconas G (1987) Transpososomes: stable protein–DNA complexes involved in the *in vitro* transposition of bacteriophage Mu DNA. Cell 49:253–262

Syvanen M (1984) The evolutionary implication of mobile genetic elements. Annu Rev Genet 18:271–293

Takasu-Ishikawa E, Yoshihara M, Hotta Y (1992) Extra sequences found at *P* element excision sites in Drosophila melanogaster. Mol Gen Genet 232:17–23

Tamura T, Thibert T, Royer C, Kanda T, Eappen A, Kamba M, Kômoto N, Thomas J-L, Mauchamp B, Chavancy G, Shirk P, Fraser M, Prudhomme J-C, Couble P (2000) A *piggyBac* element-derived vector efficiently promotes germ-line transformation in the silkwork *Bombyx mori* L. Nat Biotechnol 18:81–84

Thibault ST, Luu HT, Vann N, Miller TA (1999) Precise excision and transposition of *piggyBac* in pink bollworm embryos. Insect Mol Biol 8:119–123

Vaughn JL, Goodwin RH, Thompkins G, McCawley P (1977) The establishment of two cell lines from the insect *Spodoptera frugiperda* (Lepidoptera: Noctuidae). *In Vitro* 13:213–217

Wang HH, Fraser MJ (1993) TTAA serves as the target site for TFP3 lepidopteran insertions in both nuclear polyhedrosis virus and *Trichoplusia ni* genomes. Insect Mol Biol 1:109–116

Wang HH, Fraser MJ, Cary LC (1989) Transposon mutagenesis of baculoviruses: analysis of TFP3 lepidopteran insertions at the FP locus of nuclear polyhedrosis viruses. Gene 81:97–108

Washburn JO, Kirkpatrick BA, Volkman LE (1996) Insect protection against viruses. Nature 383:767

Wood HA (1977) An agar overlay plaque assay method for *Autographa californica* nuclear polyhedrosis virus. J Invertebr Pathol 29:304–307

Xiong G, Schorr J, Tjia ST, Doerfler W (1991) Heterologous recombination between *Autographa californica* nuclear polyhedrosis virus DNA and foreign DNA in non-polyhedrin segments of the viral genome. Virus Res 21:65–85

Section VI

Symbiont Vectors

15 *Wolbachia* as a Vehicle to Modify Insect Populations

Steven P. Sinkins and Scott L. O'Neill

CONTENTS

15.1 Introduction 271
15.2 *Wolbachia pipientis* 272
    15.2.1 Overview 272
    15.2.2 Host Range and Phylogeny of *Wolbachia* 273
15.3 Unidirectional Cytoplasmic Incompatibility and Gene Drive 273
    15.3.1 Population Dynamics of Unidirectional Cytoplasmic Incompatibility....273
        15.3.1.1 Population Structure..276
    15.3.2 Fitness Effects of Transgenes 276
        15.3.2.1 Transgenes Compatible with Cytoplasmic Drive........................277
    15.3.3 Cytoplasmic Incompatibility between *Wolbachia*-Infected Populations....277
    15.3.4 Use of Other Maternally Inherited Elements as Expression Systems....279
        15.3.4.1 Nutritive Symbionts 279
        15.3.4.2 Transovarially Transmitted Viruses 280
15.4 Other Applications of *Wolbachia* 280
    15.4.1 Bidirectional Cytoplasmic Incompatibility for Genome Replacement....280
    15.4.2 Cytoplasmic Incompatibility and Population Suppression 281
    15.4.3 *Wolbachia* and Modification of Population Age Structure 282
15.5 *Wolbachia* Research Priorities 282
    15.5.1 *Wolbachia* Interspecific Transfer 283
    15.5.2 *Wolbachia* Transformation 283
15.6 Conclusions 284
References 284

15.1 INTRODUCTION

The goals of insect transgenesis may be divided into two major categories. The first is the transformation of target non-drosophilid species in the laboratory, as a tool to study their molecular biology and gene expression. The second is the transformation of field populations of pest insects with transgenes designed for specific applications, for example making disease vectors unable to transmit pathogens to humans, crops, or agricultural animals. This is clearly a longer-term and more technically ambitious goal than laboratory transformation. There is certainly great need for sustainable new control methods of this kind, due to the severe problems associated with insecticide resistance (compounded in the case of disease vectors by parasite drug resistance). Huge-scale mass releases of transformed insects to effect field population changes by simple dilution are unlikely to be feasible in most cases. Therefore, potential transgenic strategies are heavily dependent on the use of genetic mechanisms able to drive transgenes through natural populations from small release seedings.

Wolbachia is an obligately intracellular bacterium found in many arthropod species. It possesses the ability to spread itself rapidly through populations by means of cytoplasmic incompatibility, an induced crossing sterility that is a manipulation of host reproduction. It has therefore become one of the main candidate systems for driving beneficial genes through field populations.

15.2 WOLBACHIA PIPIENTIS

15.2.1 Overview

Wolbachia pipientis was first observed in the ovaries and testes of the mosquito *Culex pipiens* using light microscopy (Hertig and Wolbach, 1924), and named and described in the same insect by Hertig (1936). It is coccoid or bacilliform in morphology with two cell membranes, surrounded by a third membrane thought to be of host origin, and averages 0.8 to 1.5 µm in length (Figure 15.1). Sequence analyses based on 16S rRNA have confirmed that morphological similarities to the Rickettsia are based on phylogenetic relatedness (O'Neill et al., 1992; Rousset et al., 1992; Stouthamer et al., 1993), and revealed that within the alpha-Proteobacteria it is most closely related to the *Erlichia* clade. *Wolbachia* is maternally inherited through the egg cytoplasm, and horizontal infectious transfer between individuals has never been observed.

Wolbachia infections have long been thought to be associated primarily with insect reproductive tissues, although some studies did report its presence in somatic tissues, particularly Malpighian tubules (e.g., Binnington and Hoffman, 1989; Louis and Nigro, 1989). More recent work, however, has revealed that in some host/*Wolbachia* associations there are heavy somatic infections and *Wolbachia* tissue tropism is generally much wider than has been appreciated (Min and Benzer, 1997; Dobson et al., 1999). This finding is highly significant in terms of increasing the range of potential applications of *Wolbachia* to population modification.

Wolbachia was first associated with the cytoplasmic incompatibility phenotype (henceforth abbreviated to CI) in *Cx. pipiens* (Yen and Barr, 1971; 1973). Tetracycline curing was used to clear

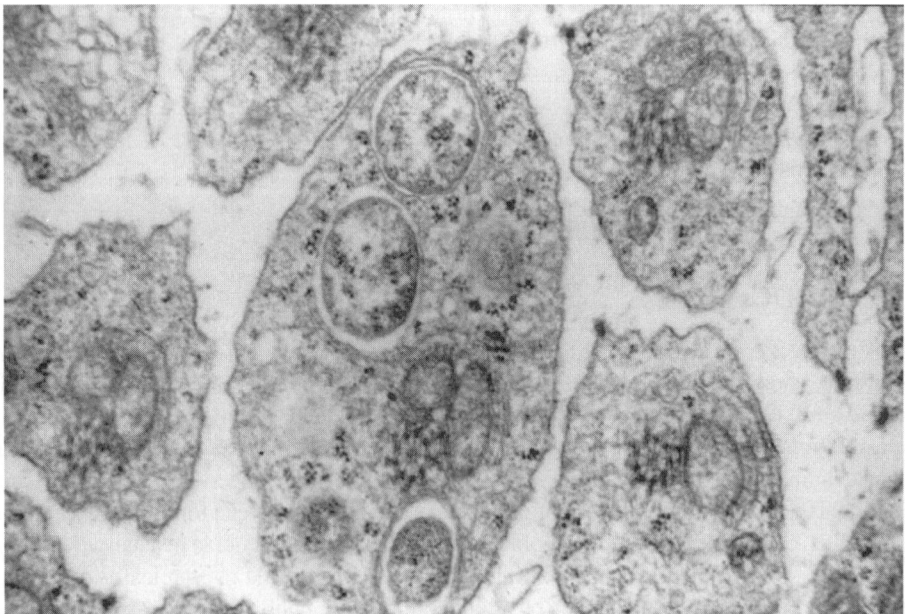

FIGURE 15.1 Transmission electron micrograph of *Wolbachia* within a developing spermatid of the moth *Ephestia cautella*. (From O'Neill, S. L. et al., *Influential Passengers: Inherited Microorganisms and Arthropod Reproduction*. Oxford University Press, Oxford, 1997. By permission of Oxford University Press.)

the infection, resulting in loss of the previously observed maternally inherited crossing sterilities between different mosquito strains. CI has been described in many other insects, and as far as is known is always induced by *Wolbachia* infections. In contrast, *Wolbachia* does not always induce CI in its host; its effects also include feminization in isopods (e.g., Bouchon et al., 1998) and parthenogenesis induction in wasps (e.g., Stouthamer et al., 1993).

15.2.2 HOST RANGE AND PHYLOGENY OF *WOLBACHIA*

The development of simple, reliable PCR assays for the presence of *Wolbachia*, amplifying the 16S rRNA gene and *ftsZ* gene (O'Neill et al., 1992; Holden et al., 1993; Werren et al., 1995b) has allowed the rapid accumulation of data on the host range of *Wolbachia*. It has proved remarkably widespread within phylum Arthropoda, across the Arachnida, Crustacea, and Insecta (more than 32 insect families are known to harbor infections; reviewed in O'Neill et al., 1997a). A survey of arthropods in Panama revealed its presence in 15 to 20% of the insect species assayed (Werren et al., 1995a). By extrapolation, that would imply that *Wolbachia* may infect the largest number of host species of any parasitic or mutualistic organism. It is not restricted to the Arthropoda, in fact, being also present in phylum Nematoda, although these *Wolbachia* form a separate sister clade to their arthropod counterparts and appear to be mutualists rather than reproductive parasites (Sironi et al., 1995; Bandi et al., 1998; Hoeraf et al., 1999; Taylor et al., 1999).

Phylogenetic analysis and tree reconstruction within the *Wolbachia* clade has proceeded in stages as more informative genes for finer phylogenetic levels have become available. The initial 16S rRNA gene trees (O'Neill et al., 1992; Rousset et al., 1992) were followed by much more resolved *ftsZ* gene trees (Werren et al., 1995b), which indicated a major subdivision into A and B subgroups. More recently, the highly variable *wsp* gene (cloned by Braig et al., 1998), coding for a *Wolbachia* surface protein, has allowed at least 12 clades of *Wolbachia* to be discriminated and reaffirmed the major A–B division (Zhou et al., 1998). The *wsp* trees appear to show sufficient resolution to permit some association between crossing type and *Wolbachia* strain groupings to be made (e.g., Bourtzis et al., 1998).

All the studies that have compared host and *Wolbachia* phylogeny have found high levels of incongruence between them (O'Neill et al., 1992; Rousset et al., 1992; Werren et al., 1995b). This shows clearly that *Wolbachia* has moved horizontally between species in its evolutionary history and that particular strains of *Wolbachia* are not restricted to one host species. It also implies that the mechanism of CI must affect a relatively conserved target. A second implication of this finding is that transfers across species boundaries should result in the expression of CI. This has been experimentally confirmed by artificial transfer of *Wolbachia* between *Drosophila* species (Boyle et al., 1993) and from *Aedes albopictus* mosquitoes into *D. simulans* (Braig et al., 1994). Cytoplasmic incompatibility was expressed when the transinfected male flies were crossed to uninfected females, and in fact new crossing types were generated.

Its wide host range and the clear evidence for interspecific transfer during its evolution have been major stimuli for the exploration of potential applications of *Wolbachia* as a gene drive mechanism of broad applicability.

15.3 UNIDIRECTIONAL CYTOPLASMIC INCOMPATIBILITY AND GENE DRIVE

15.3.1 POPULATION DYNAMICS OF UNIDIRECTIONAL CYTOPLASMIC INCOMPATIBILITY

The most important form of incompatibility in terms of the ability of *Wolbachia* to invade populations is unidirectional CI, which classically occurs when uninfected females mate with infected males (Figure 15.2). Complete or partial sterility is seen, whereas the reverse cross is

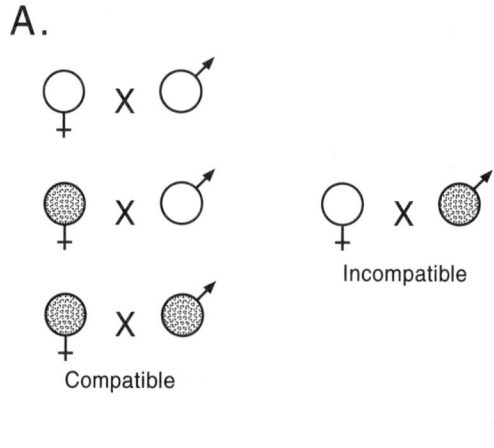

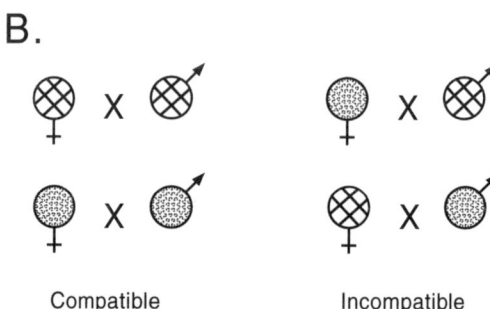

FIGURE 15.2 Cytoplasmic incompatibility. (A) Unidirectional incompatibility is expressed when *Wolbachia*-infected males (shading) mate with unifected females (no shading). These matings produce few viable offspring. All other crosses are compatible. (B) Bidirectional incompatibility is expressed when insects carrying different *Wolbachia* strains are mated together. In this case only crosses between individuals carrying the same *Wolbachia* strain are compatible.

fully fertile. The net result of this crossing pattern is that, in a mixed population of infected and uninfected individuals, females that do not carry the bacteria will lay eggs that die if they mate with an infected male, while infected females mated to infected males suffer no such loss. Therefore, infected females will produce a greater average number of offspring, resulting in an increase in the percentage of infected individuals in each generation. The rate of increase in the population will rise as the number of infected males increases until, if maternal transmission is perfect, all individuals in the population carry *Wolbachia*. The fertile mating of infected females with uninfected males means that a spreading *Wolbachia* infection would not carry nuclear genes into populations, but useful transgenes inserted into and expressed by *Wolbachia* could be spread by unidirectional CI.

Unidirectional incompatibility was first mathematically modeled prior to the discovery of its endosymbiotic causation by Caspari and Watson (1959) and generalized by Fine (1978) to include imperfect maternal transmission. Hoffmann et al. (1990) further modified the model so that the *Wolbachia* infection affects female fecundity but not male mating success, which is consistent with empirical data they collected in *D. simulans*. If p_t is the frequency of infected adults in generation t, F the fecundity of infected females relative to uninfected females, H the relative hatch rates from incompatible vs. compatible crosses, $s_h = 1 - H$ and $s_f = 1 - F$, and μ is the fraction of uninfected offspring produced by infected females, then

$$p_{t+1} = \frac{p_t(1-\mu)F}{1 - s_f p_t - s_h p_t(1 - p_t) - \mu s_h p_t^2 F}$$

The roots of the equation:

$$s_h(1 - \mu + \mu s_f)p^2 - (s_f + s_h)p + s_f + \mu - \mu s_f = 0$$

give an unstable low equilibrium frequency and also a high stable equilibrium at which *Wolbachia* will be maintained in the population under these conditions. Where $F(1-\mu) < 1$, at infection frequencies below the low unstable equilibrium in isolated populations, loss of infection ($p = 0$) is predicted. Thus, where *Wolbachia* reduce fecundity or show imperfect maternal inheritance, and if the infection is to spread through an isolated population, then the frequency of infected individuals must first exceed a threshold value (which will nevertheless be low under most described conditions, and could be overcome locally by drift).

The dynamics of the local and geographic spreading process have been well documented in the field in studies of the population invasion of uninfected *D. simulans* in California (Turelli and Hoffmann, 1991). The infection wave was reported to be spreading geographically at a rate of approximately 100 km/year, and at two localities increased from infection frequency below 30% to greater than 80% in approximately 15 generations. A probable second example of *Wolbachia* invading a natural population of plant hoppers has been reported in Japanese *Laodelphax striatellus* (Hoshizaki and Shimada, 1995).

There appears to be a female fecundity loss associated with harboring the bacteria in laboratory populations of *D. simulans* (Hoffmann et al., 1990), although this effect was not observed in field populations. Stably infected *D. simulans* populations in California (assumed to be at equilibrium) show approximately 6% uninfected individuals (Hoffmann et al., 1990). Small numbers of uninfected offspring are produced by some wild-caught infected females, which is probably sufficient to explain the population polymorphism, although immigration of uninfected flies could also have been a contributing factor. In contrast, under laboratory conditions there is 100% infection in the Riverside strain of *D. simulans*, perhaps because of reduced environmental stress. *Drosophila melanogaster* in the field show a higher frequency of uninfected individuals (Hoffmann et al., 1998), and CI in this species is only partial, although again, CI is weaker in the field than in the laboratory (Hoffmann et al., 1994; 1998; Karr et al., 1998). No fecundity loss associated with *Wolbachia* could be found in *D. melanogaster* in the laboratory (Hoffmann et al., 1994), perhaps because of lower bacterial densities than in *D. simulans* (Solignac et al., 1994).

Drosophila simulans represents a model system for the dynamics of *Wolbachia* in the field; *D. simulans* and *D. melanogaster* are the only species where transmission, relative fecundity, and incompatibility parameters have been measured in the field as well as in the laboratory (see Hoffmann and Turelli, 1988; 1997; Hoffmann et al., 1998). Cage studies of CI have been conducted for a number of other species, but care should be taken in extrapolating parameters observed under these conditions to what might be observed in the field; the differences observed for *D. simulans* between laboratory and nature can be significant. Although detailed field parameters are not available for mosquito species, percentage infection in populations at equilibrium is very high in *Aedes albopictus* and *Cx. pipiens*. The same is true of *L. striatellus* (Hoshizaki and Shimada, 1995). In the case of unidirectional CI between *Ae. albopictus* strains in the laboratory, 100% sterility is observed (Kambhampati et al., 1993; Sinkins et al., 1995b). Therefore, these species probably show even more favorable spread dynamics than the *D. simulans* model. *Wolbachia* strain differences within a species also influence the observed parameters. For example, in contrast to the *D. simulans* Riverside infection, another *Wolbachia* strain infecting *D. simulans* shows perfect maternal transmission (Hoffmann et al., 1996). The intraspecific variation in spread dynamics between different *Wolbachia* strains suggests that favorable host–*Wolbachia* combinations could be identified for many target species, with high levels of CI produced and frequencies of population infection at equilibrium close to 100%.

CI-associated sterility can decrease with male age, as observed in *Cx. pipiens* (Singh et al., 1976; Subbarao et al., 1977a). Again, this effect varies significantly between different *Wolbachia*–host associations, with rapid decreases observed in *D. melanogaster* (Karr et al., 1998). Interspecific transfer suggests that in this species the incomplete penetrance of CI is solely a host effect rather than a *Wolbachia* strain effect (Poinsot et al., 1998). Other complicating factors that vary between both species and populations are environmental effects. The presence of natural antibiotics will reduce bacterial density and hence may influence transmission and penetrance of CI. In addition, an influence of larval-rearing conditions on penetrance of CI in the male was observed for *D. simulans*, whereby males raised under crowded conditions showed reduced CI-associated sterility (Sinkins et al., 1995a). The population dynamics of CI are reviewed in detail by Hoffmann and Turelli (1997).

15.3.1.1 Population Structure

The rate of spread of *Wolbachia* within a species across wide geographic areas will be dependent on the natural migration of individuals and the degree to which structured local populations exist, which will obviously differ among species. If the species population structure consists of numerous fragmented subpopulations which are relatively isolated, with little intermigration and gene flow, this would impede the successful spread of the bacterium. Occasional infected migrants would never begin a local wave of *Wolbachia* spread if threshold frequencies are not exceeded. *Drosophila* are associated with human activity and are spread by the transportation of fruit, so it is possible that rates of migration and gene flow could be significantly lower in other species.

15.3.2 FITNESS EFFECTS OF TRANSGENES

If *Wolbachia* is transformed to spread useful genes, additional fitness costs associated with the expression of the transgene can be expected. A distinction between fitness costs to *Wolbachia* and fitness costs to the host can be made. Fitness costs to the host will be expressed in the model as a reduction in the relative fecundity parameter F, which will be additional to any host fecundity cost imposed by the *Wolbachia* itself. Transgenic *Wolbachia* will still be able to spread as long as the level of CI is greater than the combined fecundity deficit produced. However, the reduction in F will result in an increase in the value of the unstable equilibrium value of population infection frequency that must be exceeded to set the Bartonian wave of *Wolbachia* spread in motion. This translates to a higher initial release frequency required. Most release strategies would in any case aim to begin in seasons of lowest target population density to increase the ratio of initially released insects to wild insects, so small increases in spread threshold values may not represent a serious barrier to success. However, if populations were highly fragmented and migration rates between them low, the ability of the wave of *Wolbachia* infection to spread in a self-sustaining manner might be compromised by significantly higher threshold values. Large-scale concurrent releases over wide areas would not be possible in many cases, especially for disease vector species in many tropical countries where resources for disease control are limited. Transgenic *Wolbachia* that impose large fecundity reductions on their host would not represent a practical strategy in these circumstances.

Any fitness costs associated with transgene expression to the *Wolbachia* itself are most likely to be expressed as a slowing of *Wolbachia* replication, which may result in an increase in the percentage of uninfected ova produced by infected females, μ. An increase in μ will increase the threshold spread frequency and decrease the final equilibrium frequency reached by the *Wolbachia* when the spreading process is complete. However, potential effects on μ will depend on the transmission rates of the wild-type *Wolbachia* strain (before transformation). If μ in the untransformed strain–host combination is zero, probably associated with maintenance of high *Wolbachia* densities, then minor fitness costs of the transgene to transformed *Wolbachia* may not cause μ to rise.

Fitness cost to *Wolbachia* of the transgene is an especially important factor if the transgene may occasionally be lost during bacterial replication. If the loss of the transgene resulted in a significantly increased growth/multiplication rate for the "deleted" variant, the latter would gradually outcompete the transformed *Wolbachia* within the host reproductive tissues. Although the deleted variant would start from a low density relative to the transformed *Wolbachia*, and thus low transmission rate or high initial values of μ, over time μ would rise for the transformed *Wolbachia* and decrease for the deleted *Wolbachia*, as the mean relative densities of the two shifted. The final equilibrium would be the elimination of the transformed *Wolbachia* from the population. It is therefore extremely important that *Wolbachia* be transformed in a stable manner.

15.3.2.1 Transgenes Compatible with Cytoplasmic Drive

Those transgenes with the lowest negative fitness effects are the most desirable for spread through populations. Strategies that aim to achieve population suppression based on the spread of deleterious genes, imposing a genetic load on a population, are much less likely to be successful, creating high spread thresholds and strong selective pressure for transgene-deleted variants. Transgenes aiming to modify pest populations to reduce their impact would be much more favorable as long as negative fitness effects are not large. A transgene that interrupts the transmission of a pathogen by its insect vector is considered likely to fall into the category of imposing only a small negative fitness effect. It would be desirable to express two independently acting transgenes for the interruption of parasite transmission, such that a mutation in the parasite that allowed it to evade one of the mechanisms would still not be able to spread because of the vulnerability of the parasite to the other. The expression of more than one construct might be expected to increase fitness costs further.

A transgene introduced by transformed *Wolbachia* would be expressed in a wild-type background, and must therefore be dominant in its effect relative to any wild-type alleles. Natural examples of refractoriness to transmission of *Plasmodium*, for example, can be genetically dominant or partly dominant (e.g., Collins et al., 1986; Zheng et al., 1997) or recessive; only the dominant cases may provide cloned genes compatible with a *Wolbachia*-based expression/drive system. Transgenes blocking parasite transmission that are of nonhost origin would also fulfill the dominance requirement. An example would be a gene expressing a single-chain mammalian monoclonal antibody raised against pathogen antigens (e.g., Taviadoraki et al., 1993).

15.3.3 Cytoplasmic Incompatibility between *Wolbachia*-Infected Populations

CI can also occur between two insect strains that are both infected with *Wolbachia*. In fact, the first group in which CI was characterized, mosquito species of the *Cx. pipiens* complex, shows a particularly complex array of crossing types in the field (e.g., Laven, 1967b), yet all populations were later shown to harbor *Wolbachia* infections. Crossing incompatibilities between infected strains are commonly bidirectional (see Figure 15.2B). However, unidirectional CI also occurs between infected populations, and has also been reported in *Ae. albopictus* (Kambhampati et al., 1993), *D. simulans* (Nigro, 1991) and *Nasonia* wasps (Breeuwer and Werren, 1993).

The overall dynamics of unidirectional CI between infected populations are essentially the same as that seen between infected and uninfected insects. That is, one of the *Wolbachia* types will replace the other in a mixed population, just as the infected state will replace the uninfected. If this phenomenon could be suitably manipulated, species and populations that are naturally infected with *Wolbachia* would also become targets for population invasion, not just currently naive populations. In addition the occurrence of unidirectional CI between infected populations provides a mechanism for a second transgene spread into an already infected population. Repeat sweeps would be required if the first population invasion did not achieve complete success in rendering the pest population harmless, due, for example, to dissociation between transgene and *Wolbachia* or to the development of resistance to introduced refractoriness mechanisms by the transmitted parasites.

In *D. simulans*, microinjection techniques were used to combine, in one host, two strains of *Wolbachia* that produce bidirectional incompatibility when crossed with each other (Sinkins et al., 1995a). Single-infected females were incompatible with double-infected males, the female being unable to rescue the CI produced by the *Wolbachia* strain it lacks. However, double-infected females were fertile when crossed to both parental types of single-infected male, the strain required for CI rescue being carried in both cases. Hence, the presence in a host of two strains of *Wolbachia* that have mutually incompatible CI systems results in unidirectional CI when crossed to hosts containing only one of the strains. In population experiments the double infection introduced into a single-infected population spread to high frequencies, as would be expected from the unidirectional CI observed between them (Sinkins et al., 1995a).

Two-strain *Wolbachia* infections have been found in natural populations of *D. simulans* and appear to have a similar effect on crossing type (Rousset and Solignac, 1995; Mercot et al., 1995). *Aedes albopictus* mosquitoes also contain populations that are naturally superinfected with different *Wolbachia* strains, and the double-infected males are incompatible with single-infected females but not the reverse cross, as for *D. simulans* (Sinkins et al., 1995a). It is possible that many of the naturally occurring examples of unidirectional CI between *Wolbachia*-infected populations will prove to have a similar basis mechanistically. A number of other examples of two-strain infections have been uncovered that are yet to be associated with crossing type (Werren et al., 1995b).

It should also be noted that an explanation of unidirectional CI between naturally *Wolbachia* infected populations of *Nasonia* wasps has been proposed based on an association between crossing type and bacterial density (Breeuwer and Werren, 1993). Males of a strain showing higher infection densities were incompatible with females from a strain naturally showing low density, but the reciprocal cross was fertile. However, the *Nasonia* strain containing higher densities was also later shown to contain more than one *Wolbachia* strain, and the lower-density strain only one *Wolbachia* (Perrot-Minnot et al., 1996). When two-strain infections and higher bacterial density occur together, as was also observed in *Ae. albopictus* strains (Sinkins et al., 1995b), the contribution of each to an observed crossing pattern may be difficult to distinguish, but it seems likely that the effects of multiple *Wolbachia* strains are the more important.

The dynamics of two-strain population spread are more complex than that of single-strain infections if the transmission rates and fecundity effects of the two *Wolbachia* strains differ (see model by Hoffmann and Turelli, 1997). Some production of single-infected progeny from double-infected *D. simulans* females was seen in the laboratory (e.g., Sinkins et al., 1995a; Mercot et al., 1995) and would be expected in natural populations during the spread of the superinfection. However, if two-strain infections were used to spread transgenes and both of the strains carried the transgene, then individual strain transmission rates would have no influence on the population frequency reached by the transgene itself. Only the overall μ value, proportion of uninfected offspring, would be relevant in this regard. There is no evidence that μ is any higher for two-strain combinations, and in fact the reverse appears to hold. However, the speed of spread of a double infection into a single-infected population would decrease, and the threshold frequency for spread to begin would increase, if significant numbers of single-infected progeny were produced by double-infected females. Stable double-strain combinations that have high transmission fidelity in the field when forming a double infection would be preferable in terms of spread dynamics. For example, in the case of the *Ae. albopictus* natural double infection, no production of single-infected offspring was observed under laboratory conditions (unpublished data), in contrast to the artificial *D. simulans* double infections (Sinkins et al., 1995), although how this would translate to field conditions is not known.

Recent laboratory studies have shown that stable triple infections can be successfully established in *D. simulans*; these infections should spread into double, single, and uninfected populations (Rousset et al., 1999). No competition was observed between the different *Wolbachia* strains, suggesting that higher-order superinfections should also be possible. The upper limit of the number of different *Wolbachia* infections that *Drosophila* can stably maintain is not known.

15.3.4 USE OF OTHER MATERNALLY INHERITED ELEMENTS AS EXPRESSION SYSTEMS

An increase in the population frequency of a particular mitochondrial haplotype was observed in *D. simulans* in association with the spread of *Wolbachia* in California (Turelli et al., 1992). This increase was a consequence of the lack of assortment between cytoplasmic types, because of their uniparental inheritance. In fact, any element that shows perfect maternal inheritance would be spread into an uninfected population by *Wolbachia*. While the imperfectly transmitted *Wolbachia* itself reaches a stable population equilibrium frequency below 100%, a "hitch-hiking" element that is perfectly maternally transmitted will spread eventually to a population frequency of 100%; this was observed for the *D. simulans* mitochondrial haplotype in California. Uninfected females experience high mean levels of CI at equilibrium and thus leave few offspring. So, after many generations all uninfected individuals present in the population would have infected mothers or ancestors; all these uninfected individuals would therefore carry the *Wolbachia*-associated mitochondrial type.

The transformation of insect mitochondria for use as expression vectors would currently be technically demanding, but other maternally inherited elements exist that might be transformed and subsequently spread by CI.

15.3.4.1 Nutritive Symbionts

A number of nutritive bacterial symbionts are maternally inherited at high rates and offer more straightforward routes to transformation than *Wolbachia* itself. Tsetse flies (*Glossina*) have been shown to harbor three different bacterial endosymbionts (O'Neill et al., 1993; Aksoy et al., 1995; Aksoy, 1995). A large "primary symbiont," *Wigglesworthia glossinidia*, inhabits the midgut bacteriomes; a smaller secondary or S symbiont, *Sodalis glossinidius*, belongs to the gamma subdivision of the Proteobacteria, and is found both in the midgut and a variety of other tissues (Dale and Maudlin, 1999; Cheng and Aksoy, 1999); and *Wolbachia* is also present. The primary and secondary symbionts are not transovarially transmitted as is *Wolbachia*, but are apparently inherited through the milk gland secretions of the uterus. However they are still compatible with *Wolbachia* spread because their inheritance is solely maternal.

The tsetse secondary symbionts have been cultured and transformed with a plasmid containing an antibiotic resistance marker and using an origin of replication of broad host range (Beard et al., 1993). In this case the ultimate aim would be to express products that interfere with trypanosome transmission. The introduction of transformed symbionts into the fly would depend on cotransformation with an antibiotic-resistance gene as well as the transgene, followed by application of antibiotic to avoid mixtures of transformed and untransformed bacteria. However, this would also clear the primary mycetome symbionts that supply nutrients required for the survival of the fly. It would probably be necessary, therefore, to transform both symbionts with antibiotic-resistance genes, although the primary symbionts have thus far not been cultured or transformed.

Nutritive symbiotic bacteria are not uncommon in insects, but they do appear to be primarily associated with those species that have a restricted diet (tsetse feed exclusively on blood) and require nutritional supplements that are supplied by the symbionts. No nutritive endosymbionts of wide host range are known; in fact, most are confined to one host species or genus and are likely to have evolved a close obligate relationship with that host. Therefore, in contrast to the opportunity for a general arthropod gene expression system afforded by *Wolbachia*, a separate transformation effort would be required for the individual symbionts of each target species.

If the rate of maternal inheritance of a non-*Wolbachia* expression vector was lower than 100%, then the two elements will inevitably become dissociated during spread and, thus, the useful gene separated from its driving system. Obligate bacterial nutritive symbionts show, in effect, 100% maternal transmission, because any progeny that did not become infected would die, or be subjected to reproductive sterility (as is the case for tsetse). Therefore, there is no risk of the *Wolbachia* infection becoming dissociated from the expression vector by this means.

Wolbachia transmission through the male has been reported (at very low frequencies) in laboratory *Drosophila* populations (Hoffmann and Turelli, 1988). Therefore, if paternal transmission of *Wolbachia*, but not of the symbiont, took place, this would dissociate the two elements. However, from the evidence of mitochondrial DNA typing, paternal inheritance of *Wolbachia* appears not to occur in the field (Turelli et al., 1992), probably due to the lower densities of the bacterium that seem to occur in nature. Therefore, the risk of dissociation imposed by paternal inheritance of *Wolbachia* is probably very slight.

The most important consideration for the use of obligate nutritive symbionts as expression vectors is the same as for transformed *Wolbachia*, that is, the ease with which the transgene could be lost and the net fitness change to the symbiont produced by such loss. CI would not be able to carry the transgene to a high population frequency if there was a fitness cost imposed by expression of the transgene, and (1) the nutritive symbiont was unstably transformed or (2) if horizontal or paternal transmission of nutritive symbionts could occur to form mixed infections of transformed and untransformed bacteria.

15.3.4.2 Transovarially Transmitted Viruses

A second category of maternally inherited elements that might be used as expression vectors are transovarially transmitted viruses (discussed in Sinkins et al., 1997; modeled by Turelli and Hoffman, 1999). The best documented of these is the sigma virus of *D. melanogaster* (reviewed by Fleuriet, 1988). However, none are known that have complete maternal transmission; even the most stabilized sigma-infected lines in the laboratory give rise to small numbers of uninfected flies. Therefore, some degree of dissociation between *Wolbachia* and virus during the spreading process would be inevitable. The final equilibrium frequency of the virus would be dependent only on parameters associated with the virus itself, which in this case are maternal and paternal transmission frequency and negative fitness costs to the host. However, if maternal inheritance levels were high, CI would transiently increase the population frequency reached by the virus. The maximum population frequency reached and the number of generations before a frequency decline began would also be sensitive to the initial release level (for a numerical example, see Turelli and Hoffmann, 1999). The currently known imperfectly transmitted transovarially transmitted viruses are not promising expression vectors for *Wolbachia*-mediated population transformation.

15.4 OTHER APPLICATIONS OF *WOLBACHIA*

15.4.1 Bidirectional Cytoplasmic Incompatibility for Genome Replacement

If two bidirectionally incompatible insect strains were in contact or formed a mixed population, then females of the minority strain would be more likely to mate with males with which they would be incompatible and, thus, would be subject to a greater mean reproductive disadvantage than would the females of the strain in the majority. Therefore, the theoretical prediction is that the minority strain would be replaced by the majority strain in the absence of migration or other selective forces; a stable equilibrium between them within a population could not exist. This is a principle that applies to any genetic system of cross-sterility between insect strains in sympatry (Curtis, 1968). Random shifts in the crossing type over time in the regions where two bidirectionally incompatible crossing types are in contact may result, and such shifts have been reported in *Culex* populations (e.g., Irving-Bell, 1977; O'Neill and Paterson, 1992).

If both sexes of a bidirectionally incompatible strain were introduced to form a population majority with the aim of replacing the local wild-type (Laven and Aslamkhan, 1970), then the nuclear genome of the released strain would be spread in tandem. However, this would be heavily dependent on complete sterility being produced in all cross-matings. Genetic replacement using bidirectional CI would have the advantage of simplicity, in that it would not require the molecular isolation of the genes that render the pest species harmless. Classical selection procedures for useful traits could be employed, and genetic dominance would not be required.

Trials carried out on populations of the mosquito *Cx. quinquefasciatus* in laboratory cages (Curtis and Adak, 1974) and population field cages in India (Curtis, 1976) showed that, although the foreign cytoplasm did indeed replace the wild-type, linkage with a nuclear marker was not fully maintained. Reduced incompatibility with male aging (Singh et al., 1976; Subbarao et al., 1977a) and segregation of crossing types within populations (Subbarao et al., 1977b) had caused a breakdown in absolute sterility between the populations. The resulting loss of linkage between the nuclear genotype of interest and the cytoplasmic crossing type being released would be fatal to an aim of complete population replacement. Even if examples producing complete crossing sterility were found, there would still be severe practical constraints associated with this strategy. It would rely on large-scale release over wide areas to effect population majorities within local populations, which would not be economically viable for most species. Furthermore, large numbers of females would have to be released, and this could pose severe problems if the females bite humans or damage fruit. For these reasons, systems capable of spread from small release seedings, such as unidirectional CI, are a much more promising route to population replacement.

15.4.2 CYTOPLASMIC INCOMPATIBILITY AND POPULATION SUPPRESSION

CI also represents a natural sterility-producing system that has the capacity to augment the sterile insect technique (SIT), the mass release of males sterilized by irradiation to reduce the number of viable eggs being laid by wild females. One of the main factors associated with the cost and subsequent effectiveness of SIT programs is the competitiveness of released males. Radiation doses commonly used to sterilize males introduce secondary deleterious effects that reduce their ability to compete with wild males for mates. CI provides an alternative method for introducing "sterility" into a release strain that is independent of irradiation and, as such, could greatly reduce the cost of a given SIT program by reducing the numbers needed to be released for effective control.

In order for CI to be effective in this context it would be important that fertile females are not released together with males. Since none of the sexing mechanisms currently employed is 100% effective, it has been proposed that CI be used in conjunction with irradiation at lower doses. The males would be sterile on mating with wild females because of CI, while the small numbers of released females would be sterile due to irradiation (typically females require much lower doses of radiation than males to be sterilized). The lower levels of irradiation would produce males with significantly higher competitiveness than if male sterility were generated by irradiation alone. Such an approach has been experimentally tested in the mosquito *Cx. pipiens* (Curtis, 1976) and it has been shown that application of low radiation doses can generate sterile females and CI males that are as competitive as unirradiated males (Arunachalam and Curtis, 1984; 1985; Shahid and Curtis, 1987; Sharma et al., 1979).

The application of CI to population suppression of *Culex* mosquitoes was demonstrated in a World Health Organization field trial in Burma (Laven, 1967a). Although successful under the experimental field conditions used, the high population densities and reproductive rates of mosquitoes, coupled with their wide distribution over large areas, would render any population eradication scheme based on sterile release extremely expensive. The biology and life history of a species, especially its reproductive rate, is a crucial factor with respect to the feasibility of strategies based on the release of sterile males; the availability of necessary infrastructure and monetary resources for large-scale release programs is equally important. Insect pests of cash crops that have a suitably low reproductive rate would be the most realistic target for CI-based suppression. One tropical disease vector that does have a suitable biology, however, with very low reproductive rate, is the tsetse. CI could represent a complement to irradiation-based SIT programs that are already under way against tsetse in Africa. However, crossing patterns associated with *Wolbachia* in tsetse are so far poorly known.

15.4.3 WOLBACHIA AND MODIFICATION OF POPULATION AGE STRUCTURE

It has for some time been recognized that *Wolbachia* is responsible for various reproductive alterations in its arthropod hosts, but somatic effects of these infections were not considered significant. This viewpoint has now been overturned with the discovery of natural *Wolbachia* infections in *D. melanogaster* that display an intriguing virulence phenotype. Min and Benzer (1997) reported a strain of *Wolbachia* that in immature stages of the insect does not appear to affect fitness noticeably. However, in the adult insect this *Wolbachia* strain was shown to replicate rapidly within nervous tissue of the fly, leading to its early death. Min and Benzer were able to show that this effect was independent of host genotype, and appeared to be dependent on the particular *Wolbachia* strain involved, which they named *popcorn*. Another group has independently described a *Wolbachia* strain in isopods that produces an identical virulence phenotype, namely, early death of adults (Bouchon et al., 1998). These results suggest that the early death phenotype might be capable of being expressed in a range of different arthropod hosts.

Many studies have examined the relationship between age structure of a given insect population and disease transmission. Typically, the vast majority of transmission is attributable to a small fraction of the insect population that is the oldest. This is explained simply because an insect must first acquire a pathogen in a meal and then incubate that pathogen for a certain period of time before it is able to be transmitted to a new host. For many insects, the majority of reproduction in the population can be attributed to young individuals, while the majority of disease transmission is attributed to a small minority of old individuals.

Virulent *Wolbachia* forms that reduce the life span of the adult insect host, such as the *popcorn* strain in *D. melanogaster*, have the potential to affect disease transmission to humans by insects. These *Wolbachia* forms can persist in insect populations presumably because their reproductive fitness cost may not be too great, since the old individuals they kill do not contribute greatly to reproduction. However, if these *Wolbachia* were able to shorten life span in disease vectors, then they would effectively remove that fraction of the population which is responsible for the majority of pathogen transmission.

15.5 WOLBACHIA RESEARCH PRIORITIES

It has already been noted that much information still needs to be accumulated on the behavior of different *Wolbachia*–host combinations in the field; the collection of detailed parameters for stable infected populations would enable better models of spread into uninfected populations to be produced. Second, a molecular understanding of the mechanism of CI would be of great benefit to its utilization. Although there have been studies of cytological changes in sperm chromatin condensation/decondensation associated with karyogamy interruption in incompatible crosses (e.g., O'Neill and Karr, 1990; Reed and Werren, 1995; Callaini et al., 1997), the way in which *Wolbachia* induces such changes, or is able to rescue them, is unknown. This knowledge is not essential for the application of CI to pest control, much as, for example, insecticides can be used effectively without any biochemical understanding of their mode of action. Nevertheless, knowledge of the molecular mechanism of CI would significantly increase understanding of spread dynamics (such as when CI is produced between different *Wolbachia* infections) and potentially both facilitate and improve its manipulation. However, the two most immediate priorities that must be achieved are (1) more efficient techniques and a wider range of interspecific transfer and (2) the transformation of *Wolbachia*.

15.5.1 *WOLBACHIA* INTERSPECIFIC TRANSFER

The convincing evidence that horizontal transfer of *Wolbachia* over large phylogenetic distances has taken place naturally led to successful interspecific transfer among *Drosophila* species and from mosquitoes into *Drosophila*. Microinjection of preblastoderm embryos was performed to maximize the likelihood of forming pole-cell, and thus gonadal, infections. Purification of *Wolbachia* with sucrose gradients has been used, as well as direct cytoplasmic transfer from infected to uninfected embryos (Boyle et al., 1993; Braig et al., 1994; Chang and Wade, 1994; Clancy and Hoffmann, 1997).

It is routine in *Drosophila* species to obtain high hatching rates after microinjection of embryos, but when attempts were made to extend this technique to *Anopheles gambiae* and *An. stephensi*, sensitivity to desiccation meant that postinjection hatch rates were extremely low (Sinkins, 1996). The same problems are also experienced in attempts to inject these species with plasmids carrying transposable elements. However, the recent discovery that *Wolbachia* can have wide somatic distribution has opened exciting opportunities for moving *Wolbachia* between adult insects. Hemolymph transfer, which has been used successfully in other arthropods (Rigaud and Juchault, 1995; Bouchon et al., 1998; Grenier et al., 1998), may represent an approach that will yield better results. Experiments are currently under way to infect naive mosquito species by this means. It is likely that several generations of selection (using PCR assays) will be necessary to establish stably inherited germline infections from somatic tissue transfer. More extensive interspecific transfer in the laboratory is also needed to investigate such possibilities as the activation of the host immune system after *Wolbachia* transfer.

While *Wolbachia* can be moved between species that show natural infections, it is possible that naturally uninfected species do not harbor *Wolbachia* because they are resistant in some way to the action of CI or restrict its maternal transmission. The transinfection of the naturally uninfected species *Drosophila serrata* with *D. simulans Wolbachia* (Clancy and Hoffmann, 1997) produced strong CI in the new host–*Wolbachia* combination. However, significantly lower maternal transmission frequency was observed than for *D. simulans*, to the extent that this infection probably would not spread in this species (although other *Wolbachia* strains might produce significantly different transmission and incompatibility parameters in this host). An alternative hypothesis for the current phylogenetic distribution of *Wolbachia* is that the natural establishment of germline infections in new species is extremely rare, which given the obligately intracellular nature of the bacterium is perfectly feasible. If this is the case, then many currently uninfected species would be able to support the spread of *Wolbachia*. More examples of the behavior of *Wolbachia* in novel hosts are needed.

15.5.2 *WOLBACHIA* TRANSFORMATION

In order for *Wolbachia* to be used as a general system for expressing foreign genes in arthropods, a system for stably transforming the symbiont is required. The ability to develop transgenic *Wolbachia* will also be critical for understanding the molecular basis of the various reproductive phenotypes that *Wolbachia* induces in its hosts. The development of a transformation system for this obligately intracellular bacterium is not a trivial matter, however. The recent successful transformation of the close relative, *Rickettsia prowazekii* (Rachek et al., 1998), indicates that while technically challenging, it is not impossible to develop genetic transformation systems for fastidious bacteria.

In recent years a number of technical advances in the *Wolbachia* field suggest that transformation should be achievable in the near future. First, an *in vitro* culture system has been developed (O'Neill et al., 1997) in which it is possible to introduce and grow *Wolbachia*. This system will allow for the ready application of antibiotic resistance as a selectable marker for transformants. In addition, the recent cloning of the gene encoding the major surface protein of *Wolbachia* (*wsp*) has identified a strong endogenous promoter that can be used to drive foreign gene expression (Braig et al., 1998). This should allow for the construction of transposable element or homologous recombination-based transformation approaches that initially utilize antibiotic resistance driven by *wsp* promoter sequences in the *in vitro* culture system to select for transformants. Transformed *Wolbachia* can then be introduced back into insects using established microinjection technology.

15.6 CONCLUSIONS

CI is a powerful genetic system and offers considerable potential for the transformation of field populations, comparing favorably with nuclear methods of gene drive. Much research remains to be conducted before strategies based on driving transgenes into populations can be articulated in detail, without too much recourse to speculation. Nevertheless, it is clearly important to explore the implications of the experimental and theoretical work performed to date to evaluate what are likely to be the most productive avenues for further research. In a few cases the transformation of obligate nutritive endosymbionts, and subsequent population spread by *Wolbachia*, offers an attractive route to population modification. However, the transformation and spread of *Wolbachia* itself is considered to offer the most promise of the several CI-based possibilities that have been considered, offering fewest opportunities for breakdown in the field and affording the best prospects for a system of very wide utility.

REFERENCES

Aksoy S (1995) *Wigglesworthia* gen. nov. and *Wigglesworthia glossinidia* sp. nov., taxa consisting of the mycetocyte-associated, primary endosymbionts of tsetse flies. Int J Syst Bacteriol 45:848–851

Aksoy S, Pourhosseini AA, Chow A (1995) Mycetome endosymbionts of tsetse flies constitute a distinct lineage related to Enterobacteriaceae. Insect Mol Biol 4:15–22

Arunachalam N, Curtis CF (1985) Integration of radiation with cytoplasmic incompatibility for genetic control in the *Culex pipiens* complex (Diptera: Culicidae). J Med Entomol 22:648–653

Arunachalam N, Curtis CF (1984) Irradiation and cytoplasmic incompatibility for genetic control of the Culex pipiens complex. Progress Report, London School of Hygiene and Tropical Medicine, pp 45:40

Bandi C, Anderson TJ, Genchi C, Blaxter ML (1998) Phylogeny of *Wolbachia* in filarial nematodes. Proc R Soc London Ser B Biol Sci 265:2407–2413

Beard CB, O'Neill S, Mason P, Mandelco L, Woese CR, Tesh RB, Richards FF, Aksoy S (1993) Genetic transformation and phylogeny of bacterial symbionts from tsetse. Insect Mol Biol 1:123–131

Binnington KC, Hoffman AA (1989) *Wolbachia*-like organisms and cytoplasmic incompatibility in *Drosophila simulans*. J Invertebr Pathol 54:344–352

Bouchon D, Rigaud T, Juchault P (1998) Evidence for widespread *Wolbachia* infection in Isopod crustaceans: molecular identification and host feminization. Proc R Soc Lond B Biol Sci 265:1081–1090

Bourtzis K, Dobson SL, Braig HR, O'Neill SL (1998) Rescuing *Wolbachia* have been overlooked. Nature 391:852–853

Boyle L, O'Neill SL, Robertson HM, Karr TL (1993) Interspecific and intraspecific horizontal transfer of *Wolbachia* in *Drosophila*. Science 260:1796–1799

Braig HR, Guzman H, Tesh RB, O'Neill SL (1994) Replacement of the natural *Wolbachia* symbiont of *Drosophila* simulans with a mosquito counterpart. Nature 367:453–455

Braig HR, Zhou W, Dobson SL, O'Neill SL (1998) Cloning and characterization of a gene encoding the major surface protein of the bacterial endosymbiont *Wolbachia pipientis*. J Bacteriol 180:2373–2378

Breeuwer JA, Werren JH (1993) Cytoplasmic incompatibility and bacterial density in *Nasonia vitripennis*. Genetics 135:565–574

Callaini G, Dallai R, Riparbelli MG (1997) *Wolbachia*-induced delay of paternal chromatin condensation does not prevent maternal chromosomes from entering anaphase in incompatible crosses of *Drosophila simulans*. J Cell Sci 110:271–280

Caspari E, Watson GS (1959) On the evolutionary importance of cytoplasmic sterility in mosquitoes. Evolution 13:568–570

Chang NW, Wade MJ (1994) The transfer of *Wolbachia pipientis* and reproductive incompatibility between infected and uninfected strains of the flour beetle, *Tribolium confusum*, by microinjection. Can J Microbiol 40:978–981

Cheng Q, Aksoy S (1999) Tissue tropism, transmission and expression of foreign genes *in vivo* in midgut symbionts of tsetse flies. Insect Mol Biol 8:125–132

Clancy DJ, Hoffmann AA (1997) Behavior of *Wolbachia* endosymbionts from *Drosophila simulans* in *Drosophila serrata*, a novel host. Am Nat 149:975–988

Collins FH, Sakai RK, Vernick KD, Paskewitz S, Seeley DC, Miller LH, Collins WE, Campbell CC, Gwadz RW (1986) Genetic selection of a *Plasmodium*-refractory strain of the malaria vector *Anopheles gambiae*. Science 234:607–610

Curtis CF (1968) Possible use of translocations to fix desirable genes in insect pest populations. Nature 218:368–369

Curtis CF (1976) Population replacement in *Culex fatigans* by means of cytoplasmic incompatibility. 2. Field cage experiments with overlapping generations. Bull World Health Organ 53:107–119

Curtis CF (1976) Testing systems for the genetic control of mosquitoes. Proc Int Congr Entomol 15:106–116

Curtis CF, Adak Y (1974) Population replacement in *Culex fatigans* by means of cytoplasmic incompatibility 1. Laboratory experiments with non-overlapping generations. Bull World Health Organ 51:249–255

Dale C, Maudlin I (1999) *Sodalis* gen. Nov. and *Sodalis glossinidius* sp. Nov., a microaerophilic secondary endosymbiont of the tsetse fly *Glossina morsitans morsitans*. Int J Syst Bacteriol 49: 267–275

Dobson SL, Bourtzis K, Braig HR, Jones BF, Zhou W, Rousset F, O'Neill SL (1999) *Wolbachia* infections are distributed throughout insect somatic and germ line tissues. Insect Biochem Mol Biol 29:153–160

Fine PEM (1978) On the dynamics of symbiote-dependent cytoplasmic incompatibility in culicine mosquitoes. J Invertebr Pathol 30:10–18

Fleuriet A (1988) Maintenance of a hereditary virus: the sigma virus in populations of its host, *Drosophila melanogaster*. In Hecht MK, Wallace B (eds) Evolutionary Biology. Plenum, New York, pp 1–30

Grenier S, Pintureau B, Heddi A, Lassabliere F, Jager C, Louis C, Khatchadourian C (1998) Successful horizontal transfer of *Wolbachia* symbionts between *Trichogramma* wasps. Proc R Soc London Ser B Biol Sci 265:1441–1445

Hertig M (1936) The Rickettsia, *Wolbachia pipientis* and associated inclusions of the mosquito, *Culex pipiens*. Parasitology 28:453–490

Hertig M, Wolbach SB (1924) Studies on Rickettsia-like micro-organisms in insects. J Med Res 44:329–374

Hoerauf A, Nissen-Pahle K, Schmetz C, Henkle-Duhrsen K, Blaxter ML, Buttner DW, Gallin MY, Al-Qaoud KM, Lucius R, Fleischer B (1999) Tetracycline therapy targets intracellular bacteria in the filarial nematode *Litomosoides sigmodontis* and results in filarial infertility. J Clin Invest 103:11–18

Hoffmann AA, Clancy DJ, Duncan J (1996) Naturally occurring *Wolbachia* infection in *Drosophila simulans* that does not cause cytoplasmic incompatibility. Heredity 76:1–8

Hoffmann AA, Clancy DJ, Merton E (1994) Cytoplasmic incompatibility in Australian populations of *Drosophila melanogaster*. Genetics 136:993–999

Hoffmann AA, Turelli M (1988) Unidirectional incompatibility in *Drosophila simulans*: inheritance, geographic variation and fitness effects. Genetics 119:435–444

Hoffman AA, Turelli M (1997) Cytoplasmic incompatibility in insects. In O'Neill SL, Hoffmann AA, Werren JH (eds) Influential Passengers: Inherited Microorganisms and Arthropod Reproduction. Oxford University Press, Oxford, pp 42–80

Hoffmann AA, Hercus M, Dagher H (1998) Population dynamics of the *Wolbachia* infection causing cytoplasmic incompatibility in *Drosophila melanogaster*. Genetics 148:221–231

Hoffmann AA, Turelli M, Harshman LG (1990) Factors affecting the distribution of cytoplasmic incompatibility in *Drosophila simulans*. Genetics 126:933–948

Holden PR, Brookfield JF, Jones P (1993) Cloning and characterization of an *ftsZ* homologue from a bacterial symbiont of *Drosophila melanogaster*. Mol Gen Genet 240:213–220

Hoshizaki S, Shimada T (1995) PCR-based detection of *Wolbachia*, cytoplasmic incompatibility microorganisms, infected in natural populations of *Laodelphax striatellus* (Homoptera: Delphacidae) in central Japan: has the distribution of *Wolbachia* spread recently? Insect Mol Biol 4:237–243

Irving-Bell RJ (1977) Cytoplasmic incompatibility and rickettsial symbiont surveys in members of the *Culex pipiens* complex of mosquitoes. In Adiyodi KG, Adiyodi RG (eds) Advances in Invertebrate Reproduction, Peralam-Kenoth, Kerala, pp 36–48

Kambhampati S, Rai KS, Burgun SJ (1993) Unidirectional cytoplasmic incompatibility in the mosquito, *Aedes albopictus*. Evolution 47:673–677

Karr TL, Yang W, Feder ME (1998) Overcoming cytoplasmic incompatibility in *Drosophila*. Proc R Soc London Ser B Biol Sci 265:391–395

Laven H (1967a) Eradication of *Culex pipiens fatigans* through cytoplasmic incompatability. Nature 216:383–384

Laven H (1967b) Speciation and evolution in *Culex pipiens*. In Wright R, Pal R (eds) Genetics of Insect Vectors of Disease. Elsevier, Amsterdam, pp 251–275

Laven H, Aslamkhan M (1970) Control of *Culex pipiens pipiens* and *C. p. fatigans* with integrated genetical systems. Pak J Sci 22:303–312

Louis C, Nigro L (1989) Ultrastructural evidence of *Wolbachia* rickettsiales in *Drosophila simulans* and their relationship with unidirectional cross incompatibility. J Invertebr Pathol 54:39–44

Mercot H, Llorente B, Jacques M, Atlan A, Montchamp-Moreau C (1995) Variability within the Seychelles cytoplasmic incompatibility system in *Drosophila simulans*. Genetics 141:1015–1023

Min KT, Benzer S (1997) *Wolbachia*, normally a symbiont of *Drosophila*, can be virulent, causing degeneration and early death. Proc Natl Acad Sci USA 94:10792–10796

Nigro L (1991) The effect of heteroplasmy on cytoplasmic incompatibility in transplasmic lines of *Drosophila simulans* showing a complete replacement of the mitochondrial DNA. Heredity 66:41–45

O'Neill SL, Karr TL (1990) Bidirectional incompatibility between conspecific populations of *Drosophila simulans*. Nature 348:178–180

O'Neill SL, Paterson HE (1992) Crossing type variability associated with cytoplasmic incompatibility in Australian populations of the mosquito *Culex quinquefasciatus* Say. Med Vet Entomol 6:209–216

O'Neill SL, Giordano R, Colbert AM, Karr TL, Robertson HM (1992) 16S rRNA phylogenetic analysis of the bacterial endosymbionts associated with cytoplasmic incompatibility in insects. Proc Natl Acad Sci USA 89:2699–2702

O'Neill SL, Gooding RH, Aksoy S (1993) Phylogenetically distant symbiotic microorganisms reside in *Glossina* midgut and ovary tissues. Med Vet Entomol 7:377–383

O'Neill SL, Hoffmann AA, Werren JH (1997) Influential Passengers: Inherited Microorganisms and Arthropod Reproduction. Oxford University Press, Oxford

O'Neill SL, Pettigrew MM, Sinkins SP, Braig HR, Andreadis TG, Tesh RB (1997) In vitro cultivation of *Wolbachia pipientis* in *Aedes albopictus* cell line. Insect Mol Biol 6:33–39

Perrot-Minnot MJ, Guo LR, Werren JH (1996) Single and double infections with *Wolbachia* in the parasitic wasp *Nasonia vitripennis*: effects on compatibility. Genetics 143:961–972

Poinsot D, Bourtzis K, Markakis G, Savakis C, Mercot H (1998) *Wolbachia* transfer from *Drosophila melanogaster* into *D. simulans*: host effect and cytoplasmic incompatibility relationships. Genetics 150:227–237

Rachek LI, Tucker AM, Winkler HH, Wood DO (1998) Transformation of *Rickettsia prowazekii* to rifampin resistance. J Bacteriol 180:2118–2124

Reed KM, Werren JH (1995) Induction of paternal genome loss by the paternal-sex-ratio chromosome and cytoplasmic incompatibility bacteria (*Wolbachia*): a comparative study of early embryonic events. Mol Reprod Dev 40:408–418

Rigaud T, Juchault P (1995) Success and failure of horizontal transfers of feminizing *Wolbachia* endosymbionts in woodlice. J Evol Biol 8:249–255

Rousset F, Solignac M (1995) Evolution of single and double *Wolbachia* symbioses during speciation in the *Drosophila simulans* complex. Proc Natl Acad Sci USA 92:6389–6393

Rousset F, Bouchon D, Pintureau B, Juchault P, Solignac M (1992) *Wolbachia* endosymbionts responsible for various alterations of sexuality in arthropods. Proc R Soc London Ser B Biol Sci 250:91–98

Rousset F, Braig HR, O'Neill SL (1999) A stable triple *Wolbachia* infection in *Drosophila* with nearly additive incompatibility effects. Heredity 82:620–627

Shahid MA, Curtis CF (1987) Radiation sterilization and cytoplasmic incompatibility in a "tropicalized" strain of the *Culex pipiens* complex (Diptera: Culicidae). J Med Entomol 24:273–274

Sharma VP, Subbarao SK, Adak T, Razdan RK (1979) Integration of gamma irradiation and cytoplasmic incompatibility in *Culex pipiens fatigans*. J Med Entomol 15:155–156

Singh KR, Curtis CF, Krishnamurthy BS (1976) Partial loss of cytoplasmic incompatibility with age in males of *Culex fatigans*. Ann Trop Med Parasitol 70:463–466

Sinkins SP (1996) *Wolbachia* as a Potential Gene Drive System. PhD thesis, University of London

Sinkins SP, Braig HR, O'Neill SL (1995a) *Wolbachia* superinfections and the expression of cytoplasmic incompatibility. Proc R Soc London Ser B Biol Sci 261:325–330

Sinkins SP, Braig HR, O'Neill SL (1995b) *Wolbachia pipientis*: bacterial density and unidirectional cytoplasmic incompatibility between infected populations of *Aedes albopictus*. Exp Parasitol 81:284–291

Sinkins SP, Curtis CF, O'Neill SL (1997) The potential application of inherited symbiont systems to pest control. In O'Neill SL, Hoffmann AA, Werren JH (eds) Influential Passengers: Inherited Microorganisms and Arthropod Reproduction. Oxford University Press, Oxford, pp 155–175

Sironi M, Bandi C, Sacchi L, Di Sacco B, Damiani G, Genchi C (1995) Molecular evidence for a close relative of the arthropod endosymbiont *Wolbachia* in a filarial worm. Mol Biochem Parasitol 74:223–227

Solignac M, Vautrin D, Rousset F (1994) Widespread occurrence of the proteobacteria *Wolbachia* and partial cytoplasmic incompatibility in *Drosophila melanogaster*. C R Acad Sci Ser III 317:461–470

Stouthamer R, Breeuwer JAJ, Luck RF, Werren JH (1993) Molecular identification of microorganisms associated with parthenogenesis. Nature 361:66–68

Subbarao SK, Curtis CF, Krishnamurthy BS, Adak T, Chandrahas RK (1977a) Selection for partial compatibility with aged and previously mated males in *Culex pipiens* fatigans (Diptera: Culicidae). J Med Entomol 14:82–85

Subbarao SK, Krishnamurthy BS, Curtis CF, Adak T, Chandrahas RK (1977b) Segregation of cytoplasmic incompatibility properties in *Culex pipiens fatigans*. Genetics 87:381–390

Taviadoraki P, Benvenuto E, Trinca S, De Martinis D, Cattaneo A, Galeffi P (1993) Transgenic plants expressing a functional single-chain Fv antibody are specifically protected from virus attack. Nature 366:469–472

Taylor MJ, Bilo K, Cross HF, Archer JP, Underwood AP (1999) 16S rDNA phylogeny and ultrastructural characterization of *Wolbachia* intracellular bacteria of the filarial nematodes *Brugia malayi*, *B. pahangi*, and *Wuchereria bancrofti*. Exp Parasitol 91:356–361

Turelli M, Hoffmann AA (1991) Rapid spread of an inherited incompatibility factor in California *Drosophila*. Nature 353:440–442

Turelli M, Hoffmann AA (1999) Microbe-induced cytopasmic incompatibility as a mechanism for introducing genes into arthropod populations. Insect Mol Biol 8:243–255

Turelli M, Hoffmann AA, McKechnie SW (1992) Dynamics of cytoplasmic incompatibility and mtDNA variation in natural *Drosophila simulans* populations. Genetics 132:713–723

Werren JH, Windsor D, Guo LR (1995a) Distribution of *Wolbachia* among neotropical arthropods. Proc R Soc London Ser B Biol Sci 262:197–204

Werren JH, Zhang W, Guo LR (1995b) Evolution and phylogeny of *Wolbachia* — reproductive parasites of arthropods. Proc R Soc London Ser B Biol Sci 261:55–63

Yen JH, Barr AR (1971) New hypothesis of the cause of cytoplasmic incompatibility in *Culex pipiens* L. Nature 232:657–658

Yen JH, Barr AR (1973) The etiological agent of cytoplasmic incompatibility in *Culex pipiens*. J Invertebr Pathol 22:242–250

Zheng L, Cornel AJ, Wang R, Erfle H, Voss H, Ansorge W, Kafatos FC, Collins FH (1997) Quantitative trait loci for refractoriness of *Anopheles gambiae* to *Plasmodium cynomolgi* B. Science 276:425–428

Zhou W, Rousset F, O'Neill SL (1998) Phylogeny and PCR-based classification of *Wolbachia* strains using WSP gene sequences. Proc R Soc London Ser B Biol Sci 265:509–515

16 Bacterial Symbiont Transformation in Chagas Disease Vectors

Charles B. Beard, Ravi V. Durvasula, and Frank F. Richards

CONTENTS

16.1 Introduction 289
16.2 Chagas Disease Vector Control and the Potential Application
      of Transgenic Technology..289
16.3 The Biology and Ecology of Triatomine Symbionts 290
16.4 Transformation of Symbionts and Generation of Paratransgenic Insects 292
      16.4.1 Genetic Transformation of *Rhodococcus rhodnii* 293
      16.4.2 Cecropin A Expression and the Effects on *Trypanosoma cruzi* 293
      16.4.3 Single-Chain Antibody Expression in *Rhodnius prolixus*..........................295
      16.4.4 Use of Integrative Plasmids for Transformation of Actinomycetes............296
16.5 Strategy for Introducing Genetically Altered Symbionts..296
16.6 Applications of the Technology..298
16.7 Safety Concerns 298
References 302

16.1 INTRODUCTION

The vast majority of studies aimed at genetic modification of insects of medical and agricultural importance have focused on direct genome transformation of the target insect via integrative DNA elements (Warren and Crampton, 1994; Coates et al., 1998; Jasinskiene et al., 1998). There are, however, several alternative approaches that have the same ultimate aim of phenotypic alteration, but focus on different genetic targets and mechanisms. Specifically, these approaches involve the use of various viral and bacterial agents that are either native flora of, or can be introduced into the insect of interest (Beard et al., 1993; 1998; Higgs et al., 1993; Olson et al., 1996; see Chapter 8 by Carlson and Chapter 9 by Olson). The focus of this chapter is to review and summarize one group of microorganisms that has been used for this purpose, the actinomycete symbionts that live in the gut of triatomine vectors of Chagas disease. Specifically, the chapter will discuss the need and rationale for this approach, the general ecology of these microorganisms, the potential way that the methodology might be applied, and relevant safety concerns.

16.2 CHAGAS DISEASE VECTOR CONTROL AND THE POTENTIAL APPLICATION OF TRANSGENIC TECHNOLOGY

Chagas disease is caused by the parasitic protozoan, *Trypanosoma cruzi*, and transmitted by insects in the family Reduviidae and subfamily Triatominae, commonly referred to as kissing bugs, assassin bugs, or triatomines (Lent and Wygodzinsky, 1979). Transmission of the pathogen to humans most often occurs while the insect feeds at night, specifically via infected feces that are deposited by the insect on the skin near the bite site, and subsequently rubbed into the bite wound or a nearby membranous area, such as the eye. Chagas disease currently affects approximately 16 to 18 million people with an additional 100 million at risk who are living in Chagas endemic regions of South and Central America (World Health Organization, 1990; 1991). There is no treatment for chronic infections and no vaccine for prevention. Approximately 10 to 30% of infected individuals develop chronic life-threatening cardiac or digestive system disorders. At present, there are three multinational control programs aimed at elimination of domestic Chagas disease transmission. These programs, the Southern Cone, Andean Pact, and Central American initiatives, all focus on a two-pronged intervention that utilizes domiciliary insecticides and subsequent blood bank screening (additional information available from WHO at the following Web address: *http://www.who.int/ctd/html/chagcsstrat.html*). Since almost 90% of Chagas disease cases are thought to be insect transmitted, vector control is the key component of these control initiatives (Schofield and Dias, 1999); consequently, success rests almost exclusively on the effectiveness of insecticide-based vector control measures.

In the case of Chagas disease, the principal obstacle to insecticide control is reinfestation of treated homes by insects either from inadequately treated homes or from untreated peridomestic populations. Reinfestation poses a very serious threat to all three control programs that are currently under way due to the costly efforts associated with sustaining a long-term rural insecticide campaign that involves seasonal or annual retreatment of entire regions. In Guatemala, reinfestation due to the peridomestic vector *Triatoma dimidiata* already has been observed in insecticide-treated homes in less than 1 year following domiciliary application of synthetic pyrethroid insecticides (C. Cordon-Rosales, personal communication). In the Southern Cone region, where *T. infestans* is the principal domestic vector, sylvatic populations of these insects exist, and these are indistinguishable by isoenzymes and DNA sequence analysis from domestic populations (Dujardin et al., 1987; Monteiro et al., 1999). Consequently, these populations represent a potential source of reinfestation. Similarly, in regions of northern South America, peridomestic populations of *Rhodnius prolixus* pose a reinfestation threat. Consequently, Chagas disease control programs based solely on domiciliary insecticide use have inherent limitations that serve as the context for current efforts aimed at genetic manipulation of triatomine vectors using bacterial flora that are necessary inhabitants of the intestinal tract of the insect vectors that transmit this disease.

16.3 THE BIOLOGY AND ECOLOGY OF TRIATOMINE SYMBIONTS

Insects that feed on blood throughout their entire developmental cycle are known to harbor bacteria that produce nutritional factors that are limited or absent in the insect's restricted diet (Brooks, 1964). In blood-feeding insects, these factors are thought to include certain B vitamins. Triatomines are known to maintain a rich flora of various, mostly Gram positive, bacteria that have been shown in some cases to be involved in mutualistic symbiotic associations with these insects (Brecher and Wigglesworth, 1944; Nyirady, 1973; Dasch et al., 1984). *Rhodnius prolixus* is known to harbor the actinomycete symbiont *Rhodococcus rhodnii* (Figure 16.1). This bacterium lives in the digestive tract of the insect where it can reach numbers approaching 10^9, at 5 days following a blood meal (Dasch et al., 1984). Triatomines take large blood meals and shed copious amounts of feces that contain various blood breakdown products and large numbers of symbiotic bacteria. These bacteria are subsequently transmitted to progeny insects through the behavior of coprophagy, which is quite common in early instar nymphs (Figure 16.2). Triatomines can be made aposymbiotic (free of symbionts) by surface-sterilizing the eggs and rearing the progeny in clean conditions, separated from the waste or remains of other individuals of the same species. The resulting aposymbiotic individuals take a longer time to develop between molts, have higher stage-specific mortality rates, and fail to reach adulthood, dying off primarily as fourth-instar nymphs (Brecher and Wigglesworth,

Bacterial Symbiont Transformation in Chagas Disease Vectors

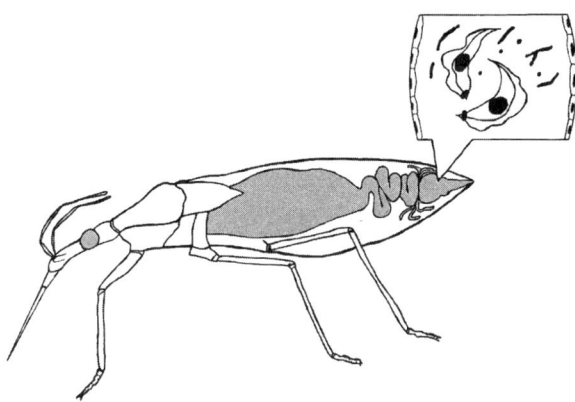

FIGURE 16.1 The actinomycete symbiont *Rh. rhodnii* lives in the gut of *R. prolixus*, in direct proximity to the Chagas disease agent *Trypanosoma cruzi*.

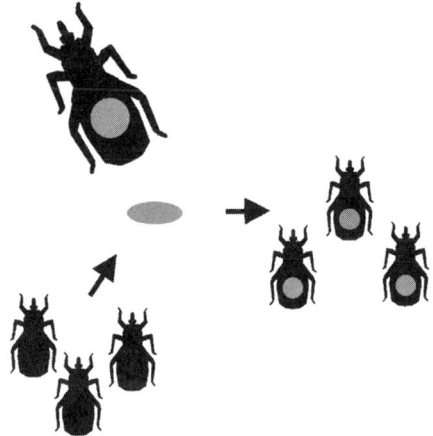

FIGURE 16.2 Essential symbionts are transmitted from adult to progeny through coprophagy — the ingestion of feces. Insects that do not acquire the symbiont die.

1944). In some cases, feeding the insects on alternative blood sources or on various dietary supplements has been reported to allow circumvention of the physiological limitations of aposymbiosis (Harington, 1960); consequently, the precise nutritional role of the bacteria in the symbiont–host relationship is not completely clear.

The specific association of *R. prolixus* with *Rh. rhodnii* has been observed on numerous occasions and is well established in the literature, leading to suggestions that this bacterial species may be unique in its association with *R. prolixus*. With other triatomine species, especially those of the genus *Triatoma*, however, it has been reported that a number of different bacterial species may function in a symbiotic role with the insect host (Dasch et al., 1984). We have conducted field studies that suggest this observation is valid for *T. dimidiata* in Guatemala and may in fact be valid for *R. prolixus*, as well. Table 16.1 shows a list of some of the microorganisms isolated from field-collected *T. dimidiata* and their tentative bacterial identification, based on Gram stain morphology and 16S rDNA analysis. Exact identification of these bacteria is often problematic because of the current state of taxonomy and systematics of coryneform bacteria (Goodfellow and Alderson, 1977;

Chun and Goodfellow, 1995). However, it is informative to note that the vast majority of microorganisms that have been isolated and identified from the digestive tract of triatomines fall into this broad group of Gram-positive actinomycete bacteria. In general, these microbes are common soil inhabitants. It is not clear at this point, however, how their niche in the intestine of triatomines relates to their overall ecology. While only a small number of these microbes have actually been shown experimentally in laboratory studies to function symbiotically with the insect, the fact that there is not one single species that is always present suggests that several, if not numerous, bacterial species may be able to function in a symbiotic capacity within the insect. Observations that we have made suggest that a certain group of bacterial species or strains may have metabolic characteristics that allow them to fulfill the symbiotic needs of the insect host and that a subset of these bacteria is actually able to survive and thrive in the intestinal tract of triatomines, thereby establishing a symbiotic association. It may be members of this subset that have actually been isolated from wild-caught bugs, as represented by the bacterial species listed in Table 16.1.

16.4 TRANSFORMATION OF SYMBIONTS AND GENERATION OF PARATRANSGENIC INSECTS

Several studies have been done demonstrating foreign gene expression in *R. prolixus*. Furthermore, methods for dispersal of foreign genes into field populations of this disease vector are currently being developed. To describe this work, the term *paratransgenic* has been used to emphasize that the insect is not actually transgenic, but harbors a second organism that is. Paratransgenesis of *R.*

TABLE 16.1
Examples of Microbial Flora Isolated from Wild-Caught *Triatoma dimidiata* in Guatemala[a]

Putative Bacterial Species	% Nucleotide Identity
Rhodococcus fascians	98
Gordona amarae	95.7
G. terrae	99.8
G. bronchialis	95
Tsukamurella paurometabolum	98
Streptomyces sp.	98
Staphylococcus xylosus	99.7
St. equorum	99
St. simulans	99
Micrococcus luteus	97
Enterococcus faecalis	100
E. malodoraturs	99
Saccharopolyspora spinosa	97
Bacillus cereus	99
Cellulomonas fermentans	95
Lactococcus lactis	99
Bordetella sp.	99
Pseudomonas sp.	99

[a] Identified based on sequence identity of the 16S rRNA gene.

Source: P. Pennington and J. Anderson, personal communication.

prolixus has relied primarily on genetic transformation of the actinomycete symbiont, *Rh. rhodnii* (Beard et al., 1992; Beard and Aksoy, 1997; Durvasula et al., 1997; 1999). This system has provided an excellent model to study arthropod paratransgenesis for a number of reasons. Both the symbiont and the parasite are extracellular organisms residing together in the same compartment within the lumen of the insect intestine. Therefore, transgene products released by genetically altered symbionts are in direct contact with the parasite. Multiplication of the parasite and the symbiont occur synchronously following a blood meal, thus permitting maximum release of recombinant gene products at the time of parasite replication. Methods for culture and genetic transformation of actinomycetes are well described, and the genetic similarity between rhodococci and mycobacteria permit the use of various selectable markers and regulatory elements that have been developed in *Mycobacterium* species that have been studied more thoroughly. Furthermore, aposymbiotic insect colonies can readily be established and maintained, and the adaptability of these insects to artificial membrane feeding allows introduction of genetically altered symbionts, antibodies, and/or trypanosomes without much difficulty. For these reasons, the greatest amount of work to date aimed at generating paratransgenic insects has been done in this insect–symbiont system.

16.4.1 GENETIC TRANSFORMATION OF *RHODOCOCCUS RHODNII*

The first studies of paratransgenesis in *R. prolixus* involved *Rh. rhodnii* genetically modified with the shuttle plasmid, pRr1.1 (Beard et al., 1992). This DNA vector was originally constructed from a low-copy-number endogenous plasmid resident in the *Rh. rhodnii* reference strain ATCC 35071. The native plasmid was digested into small DNA fragments with restriction enzymes and cloned into a pBR322-derived plasmid that contained a selectable genetic marker encoding resistance to the antibiotic thiostrepton. Following several rounds of passages back and forth between *Rh. rhodnii* and *Escherichia coli*, a stable shuttle plasmid was isolated that could be modified in *E. coli* then placed back into *Rh. rhodnii* for phenotypic analysis. Transformation was originally achieved by generation of protoplast forms and later improved by the use of electroporation. *Rhodococcus rhodnii* that had been successfully transformed with this shuttle plasmid were used to reconstitute aposymbiotic first-instar nymphs of *R. prolixus*. The bugs were maintained on blood meals supplemented with thiostrepton. Assays of adult bugs revealed persistence of the genetically altered bacteria without adverse effects on survival or fecundity of the host insects. Subsequently, it was shown that the resistant bacterial phenotype could be observed in adult insects that had not been maintained on blood containing thiostrepton, suggesting that pRr1.1 was stable through the development of the insect.

A series of DNA shuttle plasmids derived from the *E. coli* expression plasmids pUC19 and pBluescript (Figure 16.3), that contain origins of replication for both *E. coli* and *Rh. rhodnii*, have since been constructed. These plasmids reach high copy numbers in *E. coli* but maintain a low copy number in *Rh. rhodnii*. They have been modified in various ways to express genes that confer antibiotic-resistance markers in the host bacterium, as well as genes encoding antitrypanosomal molecules or single-chain antibodies (Durvasula et al., 1997; 1999).

16.4.2 CECROPIN A EXPRESSION AND THE EFFECTS ON *TRYPANOSOMA CRUZI*

Rhodnius prolixus refractory to infection with *Trypanosoma cruzi* have been generated using *Rh. rhodnii* genetically altered to express the pore-forming peptide, *L*-cecropin A (Figure 16.4) (Durvasula et al., 1997). Cecropin A, a 38 amino acid peptide, is an important part of inducible humoral immunity in the moth, *Hyalophora cecropia* (Lidholm et al., 1987). Cecropins and related peptides are distributed widely in both invertebrate and vertebrate species and exert a potent lytic effect against many types of bacteria. Cecropin A has significant activity against several strains of *T. cruzi*, but is relatively inactive against *Rh. rhodnii*. A shuttle plasmid, pRrThioCec, was constructed by modifying pRr1.1 to include a gene encoding the cecropin A mature peptide. Two colonies of aposymbiotic first-instar nymphs of *R. prolixus* were fed either wild-type *Rh. rhodnii* or *Rh. rhodnii*

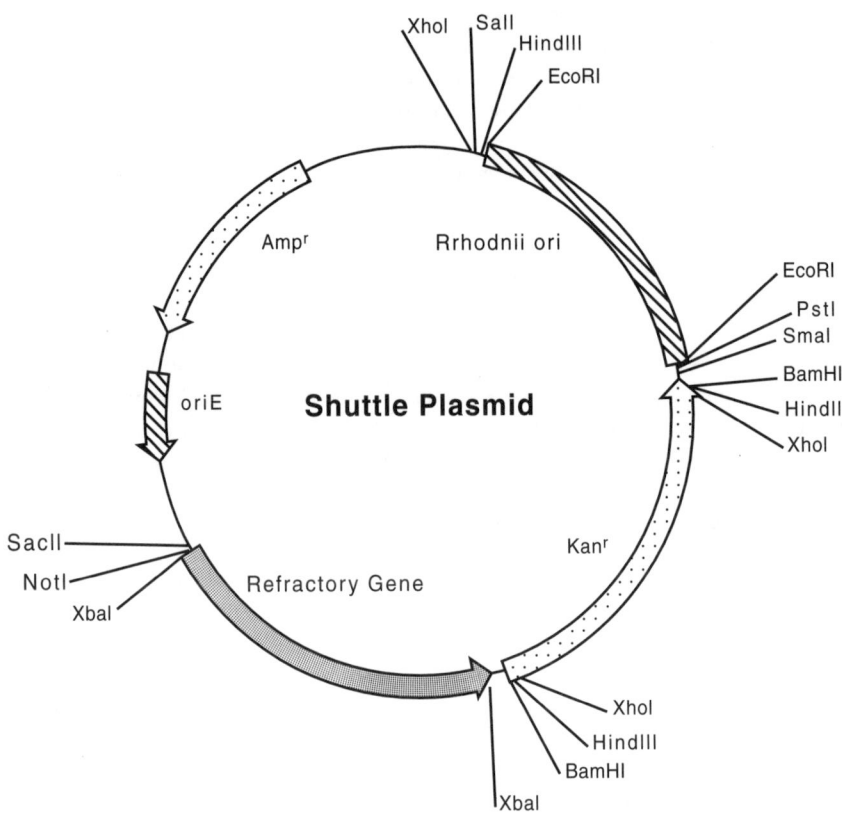

FIGURE 16.3 *Escherichi coli/Rh. rhodnii* shuttle plasmid containing ampicillin (Amp^r) and kanamycin (Kan^r) resistance markers, *E. coli* (oriE) and *Rh. rhodnii* (Rrhodnii ori) replication origins, potential refractory gene (Refractory Gene), and various restriction enzyme sites. (From Beard, C. B. et al., *Emerg. Infect. Dis.*, 4, 581–591, 1998.)

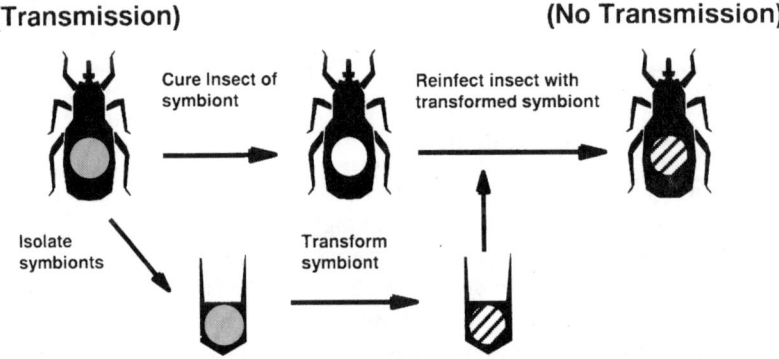

FIGURE 16.4 The symbiont *Rh. rhodnii* can be isolated from *R. prolixus*, genetically modified to express an anti–*T. cruzi* gene product, and placed back into *R. prolixus*, rendering the insect incapable of transmitting the Chagas disease agent *T. cruzi*. (From Beard, C. B. et al., *Emerg. Infect. Dis.*, 4, 581–591, 1998.)

transformed with pRrThioCec. Growth rates and fecundity of the bugs in the two colonies were similar. At the fourth-instar stage, bugs in each of the two groups were infected with the DM28 strain of *T. cruzi* by feeding on an infected blood meal via an artificial membrane. In 100 insects that harbored wild-type *Rh. rhodnii*, trypanosome counts ranged between 10^5 and 10^6 parasites/ml of gut contents at the adult stage. In 65 of 100 bugs that carried cecropin-producing bacteria, trypanosomes were absent completely; in the remaining 35 bugs, trypanosomes were detectable at concentrations of 10^3 to 10^4 parasites/ml of gut contents.

These studies are significant in that they demonstrated for the first time transmissible modulation (e.g., not simply somatic transmission, but allowing movement of the genetic factor from adults to progeny via coprophagy) of vector competence of a disease-transmitting arthropod via molecular genetic intervention. Constitutive expression of cecropin A by the engineered symbionts in the absence of antibiotic selection resulted in a refractory state in 65% of the experimental group insects and a 2 to 3 log reduction in parasite concentration in the remaining 35% of bugs. Trypanosomes that persisted despite cecropin expression were not resistant to normal lethal doses of cecropin A and appeared morphologically intact. Variation in expression of the cecropin gene in some of the experimental group insects perhaps explains these results.

16.4.3 Single-Chain Antibody Expression in *Rhodnius prolixus*

Emergence among target *T. cruzi* of resistance to foreign gene products is a likely outcome. Indeed, evolution of microbial resistance to chemotherapeutic agents and vector resistance to insecticides is a principal concern in any effort to control disease transmission. Therefore, development of multiexpression plasmids encoding products that target different regions of *T. cruzi* simultaneously is essential for success of the paratransgenic approach.

Advances in the field of antibody engineering and phage display permit generation of large libraries of single-chain antibodies through combinatorial arrangements of heavy- and light-chain genes derived from parent monoclonal IgG molecules (Huse et al., 1986). Successive "panning" of single-chain antibody fragments selects those fragments that bind target antigens with specificity comparable with the parent IgG molecule (Barbas et al., 1991). Therefore, single-chain antibodies that target selected *T. cruzi* epitopes could be expressed by engineered symbionts that are genetically transformed with multiexpression plasmids. Although inducible nonspecific immunity is well described in many insects, immunoglobulin molecules are absent. Furthermore, complement and effector cells are absent. Therefore, the ability to engineer bivalent antibody fragments that can block key receptor-mediated events or cross-link the parasite offers promise (Holliger et al., 1993).

Stable expression of an immunologically active single-chain antibody throughout the development of *R. prolixus* has been demonstrated recently (Durvasula et al., 1999). In this study, the murine single-chain (VH-Kappa) fragment, rDB3, that binds progesterone was used (He et al., 1995). A shuttle plasmid, pRrMDWK6, was constructed in which expression and secretion of rDB3 was under the control of a heterologous promoter/signal peptide complex derived from *Mycobacterium kansasii* (MKα). *Rhodococcus rhodnii* were genetically transformed with pRrMDWK6 and secretion of rDB3 was confirmed by immunoblot.

Two colonies of first-instar nymphs of *R. prolixus* were fed either wild-type *Rh. rhodnii* or rDB3-producing *Rh. rhodnii*. Gut contents of nine insects in each group were assayed at successive developmental stages for presence of genetically altered symbionts and progesterone–BSA binding activity by ELISA. Genetically altered *Rh. rhodnii* could not be detected in the control group insects. Likewise, gut extracts of these insects did not bind progesterone–BSA by ELISA. Experimental group insects carried a population of genetically altered bacteria that expressed the kanamycin-resistance phenotype throughout their development. Furthermore, progesterone–BSA binding was detected in gut extracts of these bugs at each developmental stage. Increase in OD 405-nm readings corresponded to the increase in number of genetically altered bacteria at successive stages. Addition of free progesterone–carboxymethyl-oxime to the ELISA reactions as a competitive inhibitor produced approximately a 40% decrease in binding.

The ability to express functional single-chain antibodies in the gut of *R. prolixus* via genetically engineered *Rh. rhodnii* is a significant advance in the strategy of arthropod paratransgenesis. Ongoing work is directed at expression of antitrypanosomal antibody fragments that target surface determinants of the parasite or critical attachment sites in the gut of the insect.

16.4.4 USE OF INTEGRATIVE PLASMIDS FOR TRANSFORMATION OF ACTINOMYCETES

One of the limitations of using plasmid expression vectors for transformation of insect symbionts is the instability sometimes associated with episomal DNA elements and their potential loss from the host cell. To circumvent this potential problem, we developed a series of chromosomal integration vectors that utilize a mycobacteriophage L1 integrase system (Lee et al., 1991). These plasmid constructs (Figure 16.5) comprise an *E. coli* replication origin, a selectable marker such as the kanamycin-resistance gene, β-galactosidase, or green fluorescent protein, the mycobacteriophage L1 integrase and *attP* chromosomal attachment site, and restriction sites for insertion of an anti-*T. cruzi* effector gene (B. Plikaytis and E. Dotson, personal communication). Using this family of constructs, we have been able to transform successfully bacterial symbionts from *R. prolixus*, *T. dimidiata*, *T. infestans*, and *T. sordida*. These elements appear to have a fairly broad host range for various actinomycetes and seem to be highly stable; consequently, they will probably be of great use for genetic applications involving symbiotic flora of various triatomines.

16.5 STRATEGY FOR INTRODUCING GENETICALLY ALTERED SYMBIONTS

Genetic manipulation of disease-transmitting arthropods can be of practical use only if an effective method exists for delivering foreign genes into field populations of insects. Any such approach should be able to deliver genes to the select target arthropods with a minimum risk for horizontal transfer of DNA to nontarget organisms. Reproductive fitness of the genetically altered insects should not be adversely affected to any significant degree. Furthermore, dispersal of foreign genes should pose a minimal risk for deleterious environmental effects and should not result in an increased population of potential disease-transmitting insects.

Triatomines transmit symbiotic bacteria to their progeny via fecal droplets. First-instar nymphs are transiently aposymbiotic after emerging from eggs. They rapidly establish symbiont infections in their gut through coprophagy. Actinomycete symbionts are hardy and resistant to desiccation. Thus, these bacteria can survive in dried fecal droplets for long periods of time, rendering this an effective method for dispersal of symbionts.

This naturally occurring phenomenon has been exploited as a mechanism for spreading transgenic symbionts in laboratory studies, using a simulated fecal paste we term CRUZIGARD. This material comprises guar gum as an inert substrate mixed in PBS, colored with sterile India ink, and supplemented with ammonium sulfate so that it approximates the color, texture, and attractancy of natural triatomine feces. Symbiotic bacteria can be impregnated into this mixture to a final concentration of 1×10^8 bacteria/ml. In preliminary assays, survival of the bacteria in CRUZIGARD has been found to be 6 to 8 weeks. Although this is much shorter than symbiotic bacteria in natural feces, it does permit application and transfer of symbionts to early instar nymphs.

Three studies have been done that demonstrate the efficacy of CRUZIGARD. In the first study, aposymbiotic first-instar nymphs of *R. prolixus* were exposed to CRUZIGARD that was impregnated with thiostrepton-resistant *Rh. rhodnii*. A control group of aposymbiotic *R. prolixus* were exposed to sterile CRUZIGARD. Bugs exposed to bacteria-impregnated CRUZIGARD had similar growth and survival as aposymbiotic bugs that were exposed to natural feces that contained wild-type symbionts. The thiostrepton-resistant phenotype was observed in bacteria isolated from adult bugs in the experimental group. None of the control group insects survived to the adult stage, with most of the mortality occurring before the second molt.

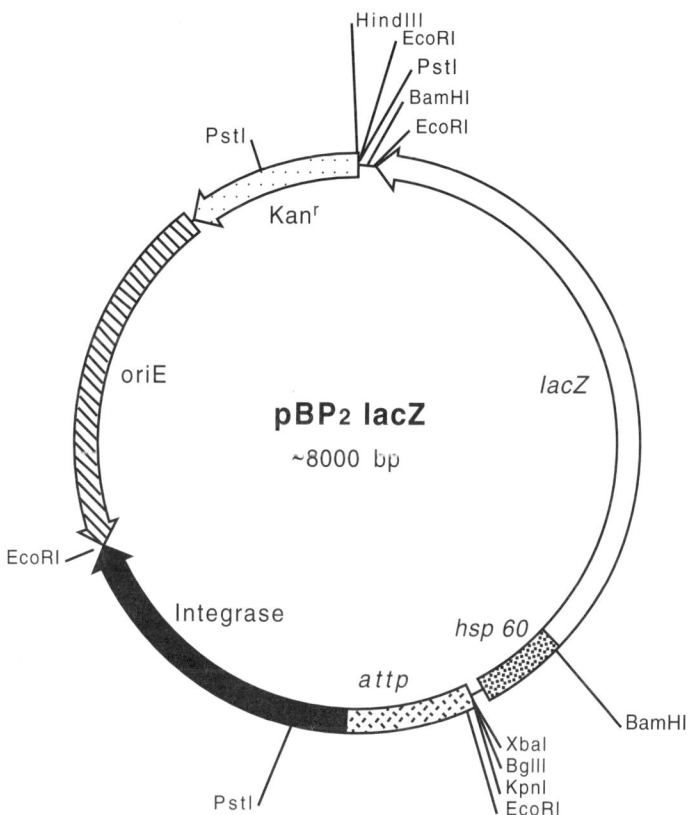

FIGURE 16.5 Actinomycete DNA integration element used to transfect triatomine symbionts. Specific features include an *E. coli* replication origin (oriE), mycobacteriophage L1 integrase (Integrase) and chromosomal attachment sequence (*attp*), the kanamycin-resistance marker (kan^r), the β-galactosidase (*lacZ*) selectable marker gene driven from the *M. tuberculosis* heatshock 60 (*hsp*60) promoter, and various restriction enzyme sites. (DNA element constructed by E. Dotson and B. Plikaytis.)

The second study assessed the efficacy of CRUZIGARD in first-instar nymphs that were already infected with wild-type *Rh. rhodnii*. Two groups of *R. prolixus* were established; one group was exposed to CRUZIGARD containing genetically transformed *Rh. rhodnii*, the other to sterile CRUZIGARD. In this study, the shuttle plasmid pRrMDWK6, conferring resistance to kanamycin and encoding the DB3 single-chain antibody, was used for genetic transformation of *Rh. rhodnii*. CRUZIGARD was applied to cages monthly. In the experimental group, all sampled insects from the third-instar to adult stages were coinfected with wild-type and kanamycin-resistant *Rh. rhodnii*. Control group insects carried only wild-type *Rh. rhodnii*. Genetically transformed symbionts comprised less than 1.0% of the total bacterial colony–forming units in the experimental bugs. This was an expected finding since these insects were infected with *Rh. rhodnii* before exposure to CRUZIGARD. Previous unpublished experiments with *R. prolixus* suggested that the initial coprophagic symbiont infection predominates throughout the life of the bug. Assays of gut contents of experimental group bugs were, however, positive by ELISA for progesterone-binding activity attributable to the DB3 single-chain antibody.

The third experiment assessed the efficacy of CRUZIGARD under simulated field conditions. Since a field release of genetically altered bacteria is still a future goal, this study attempted to recreate a field environment in closed containers. Panels of adobe and thatch were impregnated monthly with CRUZIGARD and placed in chambers containing dirt and building materials from a Chagas-endemic region of Guatemala (Figure 16.6). Adult *R. prolixus* were placed in the cages and removed after laying eggs. Progeny were assayed at third-instar, fifth-instar, and adult stages for coinfection with wild-type and genetically altered *Rh. rhodnii*. Between 50 to 60% of assayed insects were coinfected. In the coinfected insects, approximately 90% of bacterial colony–forming units were genetically altered *Rh. rhodnii*. The high concentration of genetically altered bacteria in these insects is likely the result of initial coprophagic infection of first-instar nymphs by the transgenic symbionts present in CRUZIGARD. CRUZIGARD was impregnated in the thatch and along cracks in the adobe panels, sites where eggs are usually laid. Therefore, emerging nymphs are more likely to be exposed to the genetically altered bacteria. Such an application could theoretically be achieved under real conditions since CRUZIGARD could be applied selectively to areas within a house. Thus, it is reasonable to think that CRUZIGARD might be used to deliver recombinant symbionts to bugs despite competition from wild-type symbionts and other endogenous flora.

16.6 APPLICATIONS OF THE TECHNOLOGY

The majority of work completed to date has been performed in the *R. prolixus/Rh. rhodnii* system; however, successful genetic transformation has also been achieved working with bacterial symbionts from *T. dimidiata*, *T. sordida*, and *T. infestans*. Studies are in progress to examine the potential efficacy that might be achieved in these other systems. Since the bacterial flora can vary among different triatomine species that invade the same homes, a CRUZIGARD formulation used for actual field application might contain a mixture of different bacterial species that have been similarly transformed to express the same antitrypanosomal agents.

A theoretical strategy for which CRUZIGARD might be used for Chagas disease control could entail applying CRUZIGARD either to new or to recently insecticide-treated homes in an effort to target triatomines that invade and colonize these homes. The goal would be to apply the bacteria in an attractant/bait formulation, in a density such that the progeny of colonizing individuals would be more likely to ingest CRUZIGARD than the native feces from the immigrant parental bugs, which would be present in relatively small numbers. The ultimate goal of this approach would not be to replace the use of insecticides, but to supplement insecticides as part of an integrated pest management program aimed at reducing transmission both by reducing numbers of bugs and by decreasing vector competence of any bugs that remained in homes or reinvaded homes. It can be envisaged that such an approach would probably be best suited for triatomine species that are particularly prone to colonize both domestic and peridomestic niches and for that reason are difficult to eliminate by domiciliary insecticide control measures. Currently, a greenhouse-based study is in being planned (Figure 16.7) that will allow testing of a number of different factors relating to treatment efficacy and potential safety concerns (discussed below). Some of the variables that will be tested include treatment regimens, formulation stability, supplemental attractants, competition with native flora in wild-caught bugs, and infectivity of the bacterial formulation to other insect species.

16.7 SAFETY CONCERNS

The safety concerns specifically associated with the use of bacterial symbionts for control of Chagas disease transmission can be placed broadly into two distinct categories. The first is the potential direct effect resulting from exposures of persons or domestic animals to bacteria that have been applied in large numbers in human habitations. The second concern is related to the potential long-term ecological effects resulting from horizontal movement of the introduced gene or genes. Both concerns are real, complex, and difficult to assess *a priori*. However, it is possible to minimize the potential risk associated with these concerns through a combination of thorough preliminary studies and careful monitoring of the progress of pilot field studies.

Bacterial Symbiont Transformation in Chagas Disease Vectors

FIGURE 16.6 Experimental design used to test stability and competition of genetically modified *Rh. rhodnii* in a laboratory study aimed at simulating possible field conditions. (A) The study used wooden frames that had been designed to approximate thatch and adobe building material and were treated with CRUZIGARD. (B) The frames were enclosed in Plexiglas containers to which were added field-collected *R. prolixus*.

With regard to potential human health risks resulting from direct exposure to bacterial formulations that have been applied to homes, it is important to realize that any candidate transformed bacterial symbiont will most certainly be a species that is naturally occurring (in a nontransformed state) in the same homes where it would be applied. If only a minimal health risk were associated directly with the transgene or its insertion into the bacteria (addressed below), then the primary concern is the concentration of applied bacteria and their potential importance either as allergens

FIGURE 16.7 (A) Greenhouse at the CDC Chamblee facility which contains a dual internal system of heavy-mesh enclosures that are sealed to the floor. (B) Inside these enclosures is a 6 × 6 × 6 ft hut constructed of thatch, which will be used for tests with genetically modified symbionts.

or as opportunistic pathogens in immunocompromised individuals. To minimize these risks, we would propose several measures aimed at ensuring safety. First, any unnecessary genetic material should be removed from the transgenic construct, including any antibiotic resistance markers or irrelevant DNA sequences. Additionally, a number of laboratory experiments should be conducted, including antibiotic sensitivity assessments of all candidate bacterial stocks and animal pathogenicity studies, both in immunocompetent and immunocompromised hosts. These studies would allow for a minimization of certain risks and more direct evaluation of the magnitude and probability of other potential risk factors.

The question of direct human health risk associated with introduced transgenes is more problematic. It is difficult to envision, for instance, any possible adverse affects that could be associated with expression by the bacteria of a mouse-derived single-chain antibody fragment that is specific for a unique antigenic epitope on the cell surface of *Trypanosoma cruzi*. It is possible that over time, trypanosomes might evolve resistance to immunogenetic control in the same way that resistance to drugs and/or insecticides evolves. These types of problems are inherent to almost any type of control efforts that could be devised. Of greater concern is the possibility for genetic disruption of the bacterial genome related to the potential movement of the genetic construct itself. Such mobilization could result in spontaneous mutations and subsequent changes in bacterial phenotypes, probably deleterious to bacteria and therefore self-eliminating, but nevertheless unpredictable. Consequently, it is important to understand the natural ecology and genetics of native insertion elements in populations of soil microorganisms. It will also be important to conduct laboratory studies to assess mobility of specific elements, both within the same genome and between different genomes (i.e., among different bacterial populations).

The other broad category of concern focuses on potential environmental or ecological risks that might be associated with the release of a GMO (genetically modified organism) into the environment. These risks are primarily related to movement of the transgenic bacteria and/or horizontal transfer of the introduced genetic material. To address some of the questions associated with movement of the bacteria, studies are currently in progress to determine what other insects might come in contact with CRUZIGARD if it were introduced into homes in Chagas-endemic regions, and if these insects are capable of harboring and/or transporting the bacteria. To evaluate the potential risk associated with horizontal transfer, work is in progress (described earlier) to examine the natural ecology of triatomine symbionts and other soil nocardiforms. Additionally, the mobility studies described above in the human safety section are essential for measuring the potential for horizontal transfer. If the probability of horizontal transfer is determined to be potentially significant, then one must look closely at the potential magnitude of the risk. For example, horizontal transfer of an antibiotic resistance gene among populations of different bacteria might be considered of great potential magnitude; however, the movement of a colorimetric marker such as the green fluorescent protein might not be. The question demands careful consideration and systematic experimental testing.

It is recognized that no matter how many preliminary laboratory studies are performed aimed at establishing safety in terms of potential human health or environmental hazards, pilot field trials will ultimately be required for final assessment of potential risks. If and when these types of limited releases are conducted, it is critical that the study design and protocol be evaluated carefully by a peer-review panel of recognized experts, be conducted in compliance with appropriate federal regulatory guidelines (National Institutes of Health (NIH), Centers for Disease Control and Prevention (CDC), United States Department of Agriculture (USDA/AHPIS), Environmental Protection Agency (EPA), United States Agency for International Development (USAID), Food and Drug Administration (FDA), etc.), and be carried out with appropriate surveillance measures in place. Such measures should include both careful monitoring for adverse outcomes and a contingency plan for halting the experiment and, if possible, eliminating any released GMOs, the latter of which would require specific containment measures designed *a priori* into the study protocol. A pilot field release also introduces other difficult questions associated broadly with specific research protocols and subsequent human subjects issues. Who is the population at risk? What is the specific magnitude and probability of that risk? How and from whom would informed consent be obtained? These are very complex questions that are just now beginning to be addressed in a generalized manner.

Ultimately, the risk associated with introducing a transgenic bacterial symbiont for control of Chagas disease must be weighed against the risk of doing nothing. Since there is neither a vaccine nor effective chemotherapy against this highly debilitating and sometimes fatal disease, prevention and control is currently based solely on the use of domiciliary insecticides. If insecticide campaigns prove successful and feasible, there may never be the need to embark on transgenic control. If, however, insecticide programs are significantly hampered by problems related to reinfestation of homes, insecticide resistance, or the high cost of repeated insecticide use, transgenic control may prove to be a valuable component of an integrated pest management program for control of Chagas disease.

REFERENCES

Barbas CF, Kang AS, Lerner RA, Benkovic SJ (1991) Assembly of combinatorial antibody libraries on phage surfaces: the gene III site. Proc Natl Acad Sci USA 88:7978–7982

Beard CB, Aksoy S (1997) Genetic manipulation of insect symbionts. In Crampton JM, Beard CR, Louis C (eds) The Molecular Biology of Insect Disease Vectors. Chapman & Hall, London. pp 555–566

Beard CB, Mason PW, Aksoy S, Tesh RB, Richards RR (1992) Transformation of an insect symbiont and expression of a foreign gene in the Chagas' disease vector *Rhodnius prolixus*. Am J Trop Med Hyg 46:195–200

Beard CB, O'Neill SL, Tesh RB, Richards FF, Aksoy S (1993) Modification of arthropod vector competence via symbiotic bacteria. Parasitol Today 9:179–183

Beard CB, Druvasula RV, Richards FF (1998) Bacterial symbiosis in arthropods and the control of disease transmission. Emerg Infect Dis 4:581–591

Brecher G, Wigglesworth VB (1944) The transmission of *Actinomyces rhodnii* Erikson in *Rhodnius prolixus* Ståhl (Hemiptera) and its influence on the growth of the host. Parasitology 35:220–224

Brooks MA (1964) Symbiotes and the nutrition of medically important insects. Bull Wold Health Organ 31:555–559

Chun J, Goodfellow M (1995) A phylogenetic analysis of the genus *Nocardia* with 16S rRNA gene sequences. Int J Syst Bacteriol 45:240–245

Coates CJ, Jasinskiene N, Miyashiro L, James AA (1998) *Mariner* transposition and transformation of the yellow fever mosquito, *Aedes aegypti*. Proc Natl Acad Sci USA 95:3748–3751

Dasch GA, Weiss E, Chang K (1984) Endosymbionts of insects. In Krieg NR (ed) Bergey's Manual of Systematic Bacteriology. Vol. 1. Williams & Wilkins, Baltimore, pp 811–833

Dujardin JP, Tibayrenc M, Venegas E, Maldonado L, Desjeux P, Ayala FJ (1987) Isoenzyme evidence of lack of speciation between wild and domestic *Triatoma infestans* (Heteroptera: Reduviidae) in Bolivia. J Med Entomol 34:544–552

Durvasula RV, Gumbs A, Panackal A, Kruglov O, Aksoy S, Merrifield RB, Richards FF, Beard CB (1997) Prevention of insect-borne disease: an approach using transgenic symbiotic bacteria. Proc Natl Acad Sci USA 94:3274–3278

Durvasula RV, Gumbs A, Panackal A, Kruglov O, Taneja J, Kang AS, Cordon-Rosales C, Richards FF, Whitham R, Beard CB (1999) Expression of a functional antibody fragment in the gut of *Rhodnius prolixus* via transgenic bacterial symbiont *Rhodococcus rhodnii*. Med Vet Entomol 13:1–5

Goodfellow M, Alderson G (1977) The Actinomycete-genus *Rhodococcus*: a home for the *"rhodochrous"* complex. J Gen Microbiol 100:99–122

Harington JS (1960) Studies on *Rhodnius prolixus*: growth and development of normal and sterile bugs, and the symbiotic relationship. Parasitology 50:273–277

He M, Hamon M, Liu H, Kang A, Tussig MJ (1995) Functional expression of a single-chain anti-progesterone antibody fragment in the cytoplasm of a mutant *Escherichia coli*. Nucl Acids Res 23:4009–4010

Higgs S, Powers AM, Olson KE (1993) Alphavirus expression systems: applications to mosquito vector studies. Parasitol Today 9:444–452

Holliger P, Prospero T, Winter G (1993) "Diabodies": small bivalent and bispecific antibody fragments. Proc Natl Acad Sci USA 90:6444–6448

Huse WD, Sastry L, Iverson SA, Kang AS, Atting-Mees M, Burton DR, Benkovic SJ, Lerner RA (1986) Generation of a large combinatorial library of the immunoglobulin repertoire in phage lambda. Science 246:1275–1281

Jasinskiene N, Coates CJ, Benedict MQ, Cornel AJ, Rafferty CS, James AA, Collins FH (1998) Stable transformation of the yellow fever mosquito, *Aedes aegypti*, with the *Hermes* element from the housefly. Proc Natl Acad Sci USA 95: 3743–3747

Lee MH, Pascopella L, Jocobs WR, and Hatfull GF (1991) Site-specific integration of mycobacteriophage L5: integration-proficient vectors for *Mycobacterium smegmatis*, *Mycobacterium tuberculosis*, and bacille Calmette-Guerin. Proc Natl Acad Sci USA 88:3111–3115

Lent H, Wygodzinsky P (1979) Revision of the Triatominae (Hemiptera, Reduviidae), and their significance as vectors of Chagas' disease. Bull Am Mus Natl Hist 163:125–520

Lidholm DA, Gudmundsson GH, Xanthopoulos KG, Boman HG (1987) Insect immunity: cDNA clones coding for the precursor forms of cecropins A and D, antibacterial proteins from *Hyalophora cecropia*. FEBS Lett 226:8–12

Monteiro FM, Perez R, Panzera F, Dujardin JP, Galvao C, Rocha D, Noireau F, Schofield C, Beard CB (1999) Mitochondrial DNA variation of *Triatoma infestans* populations and its implication on the specific status of *T. melanosoma*. Mem Inst Oswaldo Cruz, 94, suppl 1: 229–238

Nyirady SA (1973) The germfree culture of three species of Triatominae: *Triatoma protracta* (Uhler), *Triatoma rubida* (Uhler) and *Rhodnius prolixus* Stål. J Med Entomol 10:417–448

Olson KE, Higgs S, Gaines PJ, Powers AM, Davis BS, Kamrud KI, Carlson JO, Blair CD, Beaty BJ (1996) Genetically engineered resistance to dengue-2 virus transmission in mosquitoes. Sceince 272:884–886

Schofield CM, Dias JCP (1999) The Southern Cone Initiative against Chagas disease. Adv Parasitol 42:1–27

Warren AM, Crampton JM (1994) *Mariner*: Its prospects as a DNA vector for the genetic manipulation of medically important insects. Parasitol Today 10:58–63

World Health Organization (1990) Tropical Diseases (1990) World Health Organization, mimeographed document TDR/CTD/HH90, Geneva, p 1

World Health Organization (1991) Control of Chagas Disease. World Health Organization, Technical Report Series, no. 811, Geneva

Section VII

Strategies, Risk Assessment, and Regulation

17 The Application of Transgenic Insect Technology in the Sterile Insect Technique

Alan S. Robinson and Gerald Franz

CONTENTS

17.1 Introduction 307
17.2 Constraints on the Use of Transgenic Technology in SIT 308
17.3 Target Areas for Transgenic Technology 310
    17.3.1 Producing Only Males for Sterilization and Release 310
        17.3.1.1 Problems Associated with the Use of Conventional Genetic Sexing Strains 310
        17.3.1.2 Transgenic Approaches to the Development of Male-Only Strains 311
            17.3.1.2.1 Conditional Lethality 312
            17.3.1.2.2 Sex Determination 312
    17.3.2 Genetically Marked Insects for Release 313
        17.3.2.1 Problems with Current Techniques 313
        17.3.2.2 Use of Transgenes to Mark Released Insects 313
    17.3.3 Other Applications 314
        17.3.3.1 Stockpiling 314
        17.3.3.2 Vector Competence 314
        17.3.3.3 Radiation Protection 315
17.4 Conclusions 315
References 316

17.1 INTRODUCTION

From its beginnings almost 50 years ago the sterile insect technique (SIT) has evolved into an effective and widely accepted method for modern insect control and eradication (Krafsur, 1999). The area-wide concept, implicit in the SIT, is coming to be seen as the rational solution to many insect-related problems (FAO/IAEA, 1999). The power of the technique lies in the simplicity of the biological principles on which it is founded (Knipling, 1955) and the lack of any negative environmental effect following its application.

    The technique relies on the use of radiation to induce dominant lethal mutations and chromosome rearrangements in the sperm of adult male insects. Following release of the males and their subsequent mating with the wild females this sperm is transferred to wild females. If the required proportion of wild females are mated with sterile males at subsequent generations, then the population will collapse and eventually disappear. The proportion of wild females required to be mated by sterile males to initiate a population collapse will depend on the reproductive potential of the target species and the degree of density dependence which is operating. However, an overflooding ratio with sterile insects will always be needed, and this will require the release of large numbers of mass-reared sterile insects. It is essential that the ratio of wild to released insects is monitored during a program, and it is calculated by trapping insects in the field and identifying the released insects as they carry a fluorescent dye that is added to the pupae before fly emergence and release. The accurate and simple identification of released insects is critical to the evaluation of an SIT program.

In theory, all that is required to implement SIT is the release of large numbers of marked, sterile, male insects into a target population followed by a monitoring program to assess wild fly density and sterile male numbers. In practice, however, this simple scenario is complicated by both biological and logistical constraints that reduce the efficiency of most SIT programs. This chapter will identify some current constraints, suggest how they might be addressed using transgenic technology, and identify areas where transgenic technology can provide new opportunities for SIT.

The release of reproductively sterile insects is also an ideal system to evaluate the environmental impact of any transgenic insect, and thus can aid regulatory authorities in risk assessment. The use of sterility eliminates any possibility of vertical transmission of the transgene in the field and will enable many aspects of behavior of the transgenic insect to be studied in the field before the release of fertile insects is considered. For arthropod disease vectors where the long-term goal is to engineer strains that will not support the development of the pathogen, the use of sterile insect releases would be a safe way to initiate field evaluation of the strains. The current public perception of the potential risks of transgenic technology predicts that the first release of transgenic insects will be carried out using sterile individuals.

The development of transgenic strains of insects for use in the SIT has to take into account the fact that this technology is carried out over large areas and requires large numbers of insects to be reared, sterilized, and released over a fairly long time frame. This means that any transgenic strain that is proposed for SIT will have to be of good quality, robust, transferable, and amenable to industrial management.

17.2 CONSTRAINTS ON THE USE OF TRANSGENIC TECHNOLOGY IN SIT

A genetically manipulated strain of insects, developed using either Mendelian or molecular techniques, is generally established following the selection of a single event in a single individual. This ultimate genetic bottleneck creates concern in the minds of SIT managers regarding the overall fitness of the released flies. This concern is based on the general perception that a reduced level of genetic diversity will impact negatively on the fitness of the flies in the field. However, experience with genetic sexing strains that have now been used in the SIT program for the medfly, *Ceratitis capitata*, suggests that this type of inevitable genetic bottleneck does not impact negatively on final field fitness (De Longo et al., 1999; Rendon et al., 1999). In practical terms the effects of the bottleneck can be reversed by crossing new genetic material into the strain during the many generations between the initial isolation of the particular individual carrying the desired genotype and any field release.

The fitness of strains for use in SIT programs can also degrade following long-term colonization and mass rearing, and colonies are now generally replaced at regular intervals (Wyss, 1999). The replacement strain is usually established from the area where the releases will be made. This is done to try to ensure maximum genetic compatibility between the released and target population and to increase genetic variability in general. This matching of genetic backgrounds between the released and the target population will probably also be required for a transgenic strain. In medfly, before a genetic sexing strain was introduced into a rearing facility in Guatemala a series of backcrosses with a field strain was first required (Franz et al., 1996). Subsequent work in medfly has shown that the particular geographic origin of the material into which the sexing strain is crossed is not important, as wild populations of medfly show no mating incompatibilities (Cayol, 1999). The same might not be true for other species. It will, however, remain essential that backcrossing be carried out before a transgenic strain can be introduced into a facility for mass rearing. This backcrossing process is not always straightforward as the combination of different genomes can lead to genetic instability and other related phenomena (Louis and Yannopoulos, 1988; Torti et al., 1994). In the case of transgenic strains, the most obvious effect of this instability would be the possible remobilization of the transgene by closely related transposable elements in the different field populations. The remobilization could lead to changes in expression of the transgene and the loss of the specific phenotype for which the strain was being used. The magnitude of any destabilizing effect of strain backcrossing will depend on the families of elements present in the field population. To minimize this type of instability, transformation constructs should probably be developed using transposable elements that are not closely related to endogenous elements in the genome of the insect being transformed. However, here there is a paradox. On the one hand, a transposable element is required that induces a high frequency of transformants so that many transgenic strains can be evaluated. This requires that the recipient strain possess the host factors necessary for transformation, and the use of endogenous or related elements would seem to have an advantage. On the other hand, transformation using nonrelated elements might be difficult to achieve, but would lead to a more stable integration event as remobilization would be less likely to occur. Some thought should be given to the relative merits of these two strategies at the initiation of a transgenic program.

Mass rearing in operational SIT programs exerts significant pressure on many aspects of the biology of the cultured insect. During mass rearing of conventional strains, it is likely that genetic and molecular processes are operating that have little or no effect on the production of insects for release. However, if these same processes are superimposed on the mass rearing of a specifically constructed genetic or transgenic strain, then they can have major negative consequences for the specific characteristics of the strain. It is also conceivable that simply the number of insects that are required to be mass reared, up to several hundred million/week for an operational program, will expose new phenomena that are not observed under small-scale laboratory rearing and hence are to some extent unexpected. The background mutation rate assumes an unpredictable role under these conditions. Extremely rare events can have a major impact on the integrity of specially constructed strains. For example, in mass-rearing studies with current medfly genetic sexing strains based on Y-autosome translocations, an extremely rare intra-Y-chromosome recombination event led to the generation of a free Y chromosome and the rapid destabilization of the sexing system (G. Franz, unpublished). Although events such as this have a finite probability and can, to some extent, be predicted, their effects cannot be evaluated until the strains are tested under operational conditions of mass rearing.

Apart from these general considerations relating to biological fitness there are two other aspects of strain quality that have to be considered. First, recent data from transgenic *Drosophila* (Kaiser et al., 1997) have shown that both insert size and genetic background can impact negatively on important life traits of transgenic strains, e.g., longevity, irrespective of whether the transgene is expressed or not. Second, quality control procedures will have to be expanded to include a component that monitors the phenotype important for the functioning of the transgenic strain in the program. The phenotype has to be easily monitored on a regular basis, and if quality control reveals a change in the phenotype, then a procedure has to be in place to identify the cause of the problem and initiate remedial action.

Current successful transformation strategies routinely involve the use of constructs containing wild-type eye color coding sequences that are used to "rescue" altered eye color in the recipient mutant strain. This eye color selection system can be highly efficient, but it carries with it the problem that the recipient is mutant. This would probably rule out the use of a transgenic strain for an operational SIT program that originated from an eye color strain. In many cases the product provided by the transgene is unable to compensate fully for the mutation leading to a partial recovery of the phenotype. It will be essential to develop transformation technologies that introduce specific genes into a wild-type background. This refocuses attention on the types of marker that are most suitable, not just for the isolation of transformants, but also for their application in the SIT. What is required is a dominant marker that causes no deleterious effects on the transgenic insect. There is currently much interest in the use of green fluorescent protein (GFP) from the jellyfish *Aequorea victoria* acting as a vital marker (Chalfie et al., 1994). It has been used in *Drosophila* for this purpose (Yeh et al., 1995; Handler and Harrell, 1999) and attempts are under way to use it in different pest species both as a marker for transformation and as a marker for released flies. Details about this marker are given elsewhere in this volume (see Chapter 5 by Higgs and Lewis).

Given these important constraints, it will be essential that several independent transgenic strains are assessed for aspects of fitness and stability. This requires that a transformation system be used that generates many potential transgenic lines from which an appropriate one can be selected. Aspects relating to the management of strain fitness and replacement, backcrossing, and insertion effects require considerable attention for the design and construction of transgenic strains that are to be used in SIT programs. The ultimate hurdle for the introduction of a transgenic strain into an operational SIT program is efficiency. Any new strain, including a transgenic one, will have to demonstrate clearly that it is superior to the strain that is currently being used. In operational programs, superiority can be easily translated into economic terms and the question will be asked, "Will I get a cheaper, more effective product for the program if I use this strain?" The answer to this simple question is what must drive the integration of transgenic technology with the SIT.

17.3 TARGET AREAS FOR TRANSGENIC TECHNOLOGY

17.3.1 Producing Only Males for Sterilization and Release

SIT is generally carried out by rearing, sterilizing, and releasing both males and females even though the females, which need to be reared, irradiated, and released together with the males, do not contribute to population suppression. The removal of females from some or all of these procedures has obvious economic advantages to all SIT programs, as first recognized by Whitten (1969). Strains of insect that produce only males have so far been constructed using Mendelian principles of gene inheritance and chromosome mechanics. In essence, a gene expressing a chosen phenotype is linked by a translocation to the chromosome that determines maleness. Males and females can then be distinguished by the presence or absence of the phenotype. The phenotype can be either selectable, e.g., pupal color, or conditional lethal, e.g., insecticide susceptibility or temperature sensitivity.

For many diseases that are vectored by insects, females transmit the pathogen and even though in an SIT program they would be sterile they could still act as vectors and would need to be removed before the release was made. In several *Anopheles* species, genetic sexing strains were developed (Curtis et al., 1976; Seawright et al., 1978; Curtis, 1978; Robinson, 1986), and in one case a strain was used in an SIT field program (Dame et al., 1981). In an insect of agricultural importance, the medfly, the advantages of the use of genetic sexing strains for SIT have been well documented (Hendrichs et al., 1995; Franz and McInnis, 1995), and a recent review describes the development, mass rearing, and field application of these strains (Robinson et al., 1999).

17.3.1.1 Problems Associated with the Use of Conventional Genetic Sexing Strains

There are several disadvantages in using Mendelian genetics to construct genetic sexing strains and to mass-rear them. First, a large collection of mutations and chromosome rearrangements is required so that a basic genetic map of the species can be assembled. This includes the development of cytological techniques that enable both genes and chromosomal rearrangements to be accurately mapped. In most insect species of pest significance there are only rudimentary genetic tools available and assembling the appropriate mutations, genetic maps, and cytological techniques requires considerable resources. Second, genetic recombination can destabilize these strains when they are mass-reared. Recombination is an integral component of meiosis in most insect species, but even in the cases where it is drastically suppressed, e.g., meiosis in male Cyclorrapha Diptera, mitotic recombination is sufficient to cause problems of stability during mass rearing. In many mosquito species there are equivalent levels of recombination in both sexes, and stability of genetic sexing strains based on chromosomal translocations is very difficult to achieve. A solution to the problem can only be approached by positioning the translocation breakpoint very close to the gene in question or by using an inversion to suppress recombination (McDonald and Overland, 1973). Third, the components of the system are not transferable to other species, and the equivalent investment is required when a genetic sexing strain has to be developed in a closely related species.

17.3.1.2 Transgenic Approaches to the Development of Male-Only Strains

As in the current sexing strains for the medfly based on classical genetics, the choice of the appropriate sexing marker/method determines not only the feasibility and economy of the sexing system, but also its accuracy (Franz and Kerremans, 1994). The following general considerations have to be kept in mind:

1. A sexing system requiring the use of expensive or toxic chemicals would not be appropriate, especially for operational-scale mass rearing. Potential problems associated with worker safety and the disposal of the product would be considerable.

2. The elimination of females should occur as early as possible in the rearing process to save on rearing costs. If the sexing treatment can be initiated in the egg stage, then many millions of individuals can be easily handled. Once development proceeds to the larval stage, any treatment becomes more expensive and less practical.

3. The sexing procedure must be compatible with the normal rearing and release procedures. If chemicals are used to effect sexing by feeding them to the insect, then sexing (i.e., application of the chemical) must take place during the larval stages. The adult stage is usually not suitable, as in most cases this stage is held for only a short time before release. In many large programs, a chilled adult release system is used, and it is rather difficult to treat the insects at that stage. A similar argument applies for strategies where chemicals have to be absorbed by the insect (e.g., antibiotics). Also, here the earliest possible stage, the egg, probably cannot be used as it is unlikely that chemicals will penetrate the chorion. Consequently, promoters are required that are active during larval stages. Conversely, if temperature treatments are required to induce female-specific lethality, the larval stage is not particularly suitable. The volumes of larval diet that have to be handled are large at this stage, and it will be very difficult to maintain economical and accurate treatment conditions. This is especially true if a low temperature is required to eliminate the females as the rearing process already generates significant levels of metabolic heat, resulting in elevated temperatures in the rearing medium. Therefore, for temperature-induced female lethality, the treatment of eggs is optimal and, consequently, promoters are required that are active during that stage.

4. The sexing system must be very accurate. If flies are released at a ratio of 100:1 with the removal of 99% of the females, there will still be a doubling of the number of females in the field. In many cases this would be unacceptable. Accuracy has three aspects: first, how many females survive the sexing procedure and are released together with the males; second, how many males are killed together with the females thus reducing the number available for release; and, third, how many females are killed during the rearing process due to the leakiness of the sexing system leading to losses in colony production.

With existing temperature-sensitive lethal-based *(tsl)* sexing strains in the medfly, males can be produced with an accuracy of approximately 99.5% with virtually none being lost due to the temperature treatment (Fisher, 1998). However, for strain maintenance and colony production when no temperature treatment is applied, some of the temperature-sensitive females are lost because of a rise of temperature in the larval diet.

17.3.1.2.1 Conditional Lethality

It is obvious that a sexing system based on lethal genes can only be used if the lethality is both female specific and inducible. Female specificity is required to kill females, and inducibility is required for routine colony production where both sexes are required. Many strategies have been proposed and only a few generic examples are mentioned here. In this area there are many theoretical possibilities with as yet few systems being tested in practice.

1. *Sex-specific promoter in combination with a conditional lethal gene:* The enzyme alcohol dehydrogenase metabolizes environmental alcohol to aldehyde (Heinstra et al., 1983). If the *Adh* gene is placed under the control of a male-specific promoter and extra copies of this gene are introduced into the genome of the target species, then males will have a higher tolerance for alcohol in the diet. For colony rearing, a diet without additional alcohol would be used and, when males are required for release, alcohol would be added to the larval diet. The reverse strategy would also be possible. The *Adh* gene would be linked to a female-specific promoter and introduced as extra copies into the genome. For elimination of females, allyl alcohol would be added to the diet. This is converted by *Adh* to a lethal ketone, and the additional copies of the *Adh* gene would produce more ketones and would preferentially kill the females. However, for both strategies sex-specific promoters are needed that act during the larval stages, ideally as early as possible. At present, several female (chorion, Konsolaki et al., 1990; vitellogenin, Rina and Savakis, 1991; ceratotoxin, Rosetto et al., 1998) and male-specific genes (male-specific serum protein, MSSP; Thymiano et al., 1998) have been cloned from medfly, but none of them is useful for the above strategy as all are only active during the adult stage and some of them are also tissue specific.

2. *Constitutive promoter in combination with sex-specific splicing and a conditional lethal:* It is also possible to take advantage of sex-specific splicing to get specific gene products expressed in females. It has been proposed to link the medfly *doublesex* (*dsx*, Saccone et al., 1999) gene and the *Pseudomonas syringae ice-nucleation* gene (*InaZ*, Orser et al., 1985) in such a way that the ice-inducing protein is made only if the female-specific splicing of *dsx* occurs. A mild cold shock would lead to female-specific lethality. One big advantage of this strategy is, assuming that *dsx* expression starts early in embryogenesis, that it could be applied at the egg stage and would, therefore, minimize rearing costs. In principle, the *ice-nucleation* gene could also be linked to an early-acting female-specific promoter.

3. *Inducible promoter in combination with sex-specific splicing and a nonconditional lethal:* A system can also be envisaged where a nonconditional toxin is produced only in females because of sex-specific splicing. However, to rear such a strain, this construct must be controlled by an inducible promoter. Many promoters, inducible with a variety of different stimuli, are known from many different species, and they do function, at least to some degree, when transferred to other organisms. However, to be of use for a practical genetic sexing system, three criteria have to be met: (a) induction must lead to sufficient levels of expression — otherwise females will survive; (b) the system should not be leaky, which will make production difficult; and (c) induction must be economically/technically feasible as outlined above.

17.3.1.2.2 Sex Determination

An alternative approach to generate genetic sexing strains is to manipulate the sex determination mechanism directly. Sex determination systems in insects are extremely varied (Robinson, 1983), and each species will probably have to be approached individually. In medfly, and probably most other fruit flies of economic importance, male sex is determined by a dominant factor on the Y chromosome. As a first step to isolate this *Maleness* factor it was mapped on the Y chromosome by *in situ* hybridization and deletions (Willhoeft and Franz, 1996). Based on this information, the respective region of the Y was microdissected from mitotic chromosomes (Willhoeft et al., 1998), and this material is now being analyzed in an attempt to clone the *Maleness* factor.

Several groups are using different approaches to isolate genes involved in sex determination. Based on sequence homology, several genes have been cloned in pest insects using *Drosophila* genes as probes. These include *Sex lethal* from medfly (*Sxl*, Saccone et al., 1998), *Musca domestica* (Meise et al., 1998), *Chrysomya rufifacies* (Mueller-Holtkamp, 1995), and *Megaselia scalaris* (Sievert et al., 1997). However, in all these species *Sxl* is expressed equally in both sexes, and it is therefore believed that it is not the key control gene in the sex determination hierarchy as it is in *Drosophila*. In contrast, *dsx*, the final gene in the *Drosophila* sex determination cascade, shows sex-specific splicing in medfly (Saccone et al., 1999), *Bactrocera tryoni* (Shearman and Frommer, 1998), and *Anopheles gambiae* (Pannuti et al., 1999). In *Musca domestica* it has been recently shown that the male-determining activity on the Y chromosome consists of at least two elements (Hediger et al., 1998). Research is being actively pursued to identify other genes involved in sex determination.

The ultimate goal is to generate a strain where the inducible expression of a male-determining gene in genetically female flies would either kill them or would convert them into males. The latter scenario would be optimal, as no losses would occur due to the sexing procedure. Furthermore, it is envisaged that if such a mechanism is available for one species, it would be relatively straight-forward to transfer it to other species of interest.

17.3.2 GENETICALLY MARKED INSECTS FOR RELEASE

Evaluation of an SIT program through the monitoring of the field and released insects is essential, expensive, and difficult. It involves the placement of traps, the collection of flies, and their identification as either field or released. The only technology currently available that enables this discrimination to be made is by the use of a fluorescent dye that is added to the pupae before release. During fly emergence, dye particles are trapped within the head of the fly under the ptilinum, and they can be revealed by squashing the heads of the flies once they are returned from the traps to the laboratory. In an operational SIT program, large numbers of flies are trapped on a regular basis and each one has to be checked and identified as originating from the field or the mass-rearing facility.

17.3.2.1 Problems with Current Techniques

The use of fluorescent dye is the only technology available for monitoring the progress in an operational SIT program, and technical and logistical problems can arise during this important phase of a program. First, trapped flies are often in a very poor condition due to the type of trap used and the time that the fly has been in the trap. Sticky traps are often used and removal of flies in a condition in which they can be analyzed is difficult and laborious. There is also the problem that in a live trap released flies can contaminate field flies through contact. Second, the numbers of insects that have to be screened can be large, e.g., in California 100,000 flies/week have to be checked and accurately identified, and human error or uncertainty can become important. Third, the number of trapped released flies is far greater than the number of trapped field flies and when eradication approaches, virtually no field flies are trapped at all and any unmarked fly that is trapped has to be unequivocally identified. The misclassification of a single released fly as a wild fly can have major consequences for an SIT program.

17.3.2.2 Use of Transgenes to Mark Released Insects

In principle, any transgenic insect already carries a marker that can be used to discriminate released flies from nontransgenic, wild flies. PCR amplification of the integrated transformation vector would enable this to be achieved easily. Probably a certain degree of automation would be required even if one uses this molecular marker together with the traditional fluorescent dye-marking system.

The dye would be used to screen the major proportion of the trapped flies and PCR amplification would only be used for uncertain cases.

In an alternative approach, a specific gene would be inserted that encodes a visible product and allows screening of trapped flies without the need for time-consuming manipulations or expensive equipment. It has been suggested to use the autofluorescent properties of the GFP for this purpose. Several laboratories have already succeeded in generating such transgenics (P. Atkinson, personal communciation; A. Handler, personal communciation; C. Savakis, personal communciation). It will be required, however, to examine whether such visible markers have any negative effect on the mating behavior of the released flies, especially in species like the medfly where the females choose their mating partners.

One concern that applies to all transgenic marking systems is the stability of the marker under field conditions. Flies are monitored using different types of traps that may affect the efficacy of the marker over time, as servicing of traps can be several days apart and flies die. Therefore, it has to be shown that the marker can be unequivocally identified in all released flies found in the traps. This high degree of accuracy is essential in a scenario where unmarked wild flies are extremely rare and the identification of a single wild fly can trigger a very expensive eradication campaign. Program managers will need convincing evidence of the practicality of this approach.

17.3.3 OTHER APPLICATIONS

17.3.3.1 Stockpiling

The ability to stockpile insects has been a topic of much interest to SIT managers as it would enable them to manage insect production efficiently, when insects are required to be released in only part of the year. It could also provide a strategy that would be attractive to the commercialization of this technology. So far the only mechanism available has been the use of naturally occurring diapause, and this has been integrated into operational SIT programs for the onion fly, *Delia antiqua* (Loosjes, 1999), and the codling moth, *Cydia pomonella* (Bloem et al., 1999). However, in many insect species diapause is not a component of the life cycle and other mechanisms have to be explored. In *D. melanogaster,* Mazur et al. (1992) were able to use vitrification and a cryoprotectant to preserve embryos in liquid nitrogen. A recent review on the subject of cold storage of insects indicates the many ways that this problem can be approached (Leopold, 1998). With the advent of transformation, another option becomes available.

Many organisms express specific proteins that provide protection against freezing, the most well studied are the so-called antifreeze proteins from Arctic fish (De Vries, 1983). Many of these genes have now been cloned from different fish, and they have been used in transformation experiments with *Drosophila* (Walker et al., 1995). Using different antifreeze genes driven by the yolk protein promotor, Walker et al. were able to generate a degree of thermal hysteresis in transformed *Drosophila* females (Walker et al., 1995). Recently, the same group has cloned the gene for thermal hysteresis from the spruce budworm, *Choristoneura fumiferans* (Tyshenko et al., 1997). This protein is an order of magnitude more active than the fish proteins and would seem to be an good candidate for transformation studies. A word of caution is, however, essential when discussing the introduction of these genes into naive insects. The successful integration of a functional antifreeze gene could have serious consequences if individuals carrying the construct entered the environment. Theoretically, a species could dramatically expand its geographic range and so become a much more important pest.

17.3.3.2 Vector Competence

The use of the SIT to control and eradicate insect vectors of disease is often faced with the following technical and public relations problem. Radiation does not affect the ability of the released sterile insects to transmit the pathogen, and to achieve the goal of vector control, and hence disease control, many insects have to be released. If measures are not taken to address this problem, then release programs will not pass the required regulatory scrutiny. In some situations, e.g., malaria, where only the females transmit the disease and only the males are required for the SIT, then the answer is simply to remove the females. Examples of this have been given earlier. However, in other vector/parasite interactions such as those involving pathogenic trypanosomes both sexes can transmit the disease. In tsetse SIT (Msangi, 1999) this problem is currently taken care of by giving the sterile tsetse males two blood meals before release, a procedure that drastically reduces their ability to become subsequently infected. In addition, a trypanocide is added to the blood meal. Although these procedures do ensure that the released insects are not vectors, a strain of tsetse that is refractory to infection would be a great improvement.

The ability of many insects to vector diseases has a well-documented genetic basis and extensive efforts are under way to map the genes involved, but so far there has not been a convincing demonstration of such a refractory gene (Kidwell and Wattam, 1998). However, it has been possible to utilize symbionts in the gut of certain vectors as vehicles for the expression of antiparasitic genes (Durvasula et al., 1997; see Chapter 16 by Beard et al.). In general, SIT for insect vectors would become technically much simpler if transformation could be used to produce refractory strains. In a curious reversal of roles, sterile insects might well be an essential component for the eventual field evaluation of refractory insects before fertile individuals can be released with the aim to spread refractory genes in wild populations. Pettigrew and O'Neill (1997) provide a very balanced discussion of the control of vector-borne diseases using genetic manipulation of insects (also see Chapter 15 by Sinkins and O'Neill).

17.3.3.3 Radiation Protection

Ionizing radiation has shown itself to be the most suitable agent that will sterilize insects without negatively affecting their ability to survive and mate. Nevertheless, it does have a negative effect on the fitness of the treated insects. The enzyme Cu-Zn superoxide dismutase is very important for the inactivation of superoxide radicals that are produced by a variety of cellular and extracellular processes including radiation. An accumulation of these radicals leads to cell death. In *Drosophila* it has been shown that the gene coding for this enzyme can decrease adult somatic sensitivity to radiation damage (Parkes et al., 1998), but the study did not examine the effect of this gene on sterility induction. If there is a differential effect of this gene on reproductive as opposed to somatic cells, then it might be possible, by inserting an extra copy of this gene, that the negative somatic effects of radiation could be ameliorated.

17.4 CONCLUSIONS

There is no doubt that many of the approaches outlined above will be successfully developed and will make a major contribution to increasing the effectiveness of the SIT. The use of improved male-producing strains that are molecularly marked will simplify and facilitate the expanded use of this technology. There are no major technical hurdles to be overcome to produce transgenic strains of insects in the laboratory with some of the required properties outlined above. There remain, however, two critical areas, one operational and the other regulatory, either of which can have a major impact on the use of transgenic insects in operational SIT programs.

The operational concern revolves around the ability of a transgenic strain to withstand the rigors of mass rearing and maintain its required properties. This is an area in which experimentation is difficult, if not impossible, as there are no facilities where these kinds of "experiment" can be carried out. Managers of mass-rearing facilities will have to be convinced to replace a current strain with a new one with the attendant risk that "something might go wrong." The initial failure of a strain to meet the criteria of the facility both in terms of production and stability could be a serious setback to further development of this technology.

The regulatory problems for the use of transgenic insect strains in SIT are obvious. The current volatile public opinion on transgenic technology in general will inevitably impact on the integration of these strains in insect control programs (see Chapter 19 by Hoy). There are, of course, different levels of perceived risk for transgenic insects. Any strain carrying a transgene incorporating a toxin is likely to be viewed much more critically than one carrying an extra copy of a gene already present. The problems can also increase if a noninsect gene is used. The use of transgenic insects for SIT programs has the major advantage that the released insects are sterile and so vertical transmission of the transgene is impossible. Where the release of fertile transgenic insects is envisaged, regulatory bodies will be faced with much greater problems, especially if the transgene is designed to interact with vector competence. Whatever type of large-scale release is being planned will require that rearing facilities with fertile insects will probably be required to drastically improve quarantine precautions. Since the first demonstration of transgenic technology in insects almost 20 years ago (Rubin and Spradling, 1982), it has been heralded as opening the door to new approaches for insect control. The increasing use of the SIT for area-wide insect control programs provides an excellent vehicle for the realization of this goal.

REFERENCES

Bloem S, Bloem KA, Calkins CO Incorporation of diapause into codling moth mass-rearing: production advantages and insect quality issues. In Proceedings of FAO/IAEA Int. Conf. on Area-wide Control of Insect Pests, Malaysia, in press

Cayol JP (1999) World-wide sexual compatibility in Medfly, *Ceratitis capitata* Wied., and its implications for SIT. In Proceedings of FAO/IAEA Int. Conf. on Area-wide Control of Insect Pests, Malaysia, in press

Chalfie M, Tu Y, Euskirchen G, Ward WW, Prasher DC (1994) Green fluorescent protein as a marker for gene expression. Science 263:802–805

Curtis CF (1978) Genetic sex separation in *Anopheles arabiensis* and the production of sterile hybrids. Bull World Health Organ 45:453–456

Curtis CF, Akiyama J, Davidson G (1976) A genetic sexing system in *Anopheles gambiae* species A. Mosq News 36:492–498

Dame DA, Lowe RE, Williamson DL (1981) Assessment of released sterile *Anopheles albimanus* and *Glossina morsitans*. In Cytogenetics and Genetics of Vectors Proc. XVI Int. Congr. Entomol., Kyoto, Kitzmiller JB, Kanda T (eds) 1980 Elsevier Biological, Amsterdam, pp 231–241

De Longo O, Colombo A, Gomez Riera P (1999) Use of massive SIT for the control of medfly (*Ceratitis capitata* WEID.), using strain SEIB 6–96 in Mendoza, Argentina. In Proceedings of FAO/IAEA Int. Conf. on Area-wide Control of Insect Pests, Malaysia, June 1998

De Vries AL (1983) Antifreeze peptides and glycopeptides in cold-water fishes. Annu Rev Physiol 45:245–260

Durvasula RV, Gumbs A, Panackal A, Kruglov O, Aksoy S, Merrifield RB, Richards FF, Beard CB (1997) Prevention of insect-borne disease: an approach using transgenic symbiotic bacteria. Proc Natl Acad Sci USA 94:3274–3278

FAO/IAEA (1999) Proceedings of FAO/IAEA Int. Conf. on Area-wide Control of Insect Pests, Malaysia, June 1998

Fisher K (1998) Genetic sexing strains of Mediterranean fruit fly (Diptera: Tephritidae): Optimizing high temperature treatment of mass-reared temperature sensitive strains. J Econ Entomol 91:1406–1414

Franz G, Kerremans PH (1993) Requirements and strategies for the development of genetic sex separation systems with special reference to the Mediterranean fruit fly, *Ceratitis capitata*. In Fruit Flies and the Sterile Insect Technique Calkins CO, Klassen W, Liedo P (eds) CRC Press, Boca Raton, FL, pp 113–122

Franz G, McInnis DO (1995) A promising new twist — a genetic sexing strain based on a temperature sensitive lethal mutation. In The Mediterranean Fruit Fly in California: Defining Critical Research. The Regents of the University of California, Riverside pp 187–199

Franz GF, Kerremans Ph, Rendon P, Hendrichs J (1996) Development and application of genetic sexing systems for the Mediterranean fruit fly based on a temperature sensitive lethal. In McPheron BA, Steck GJ (eds) Fruit Fly Pests: A World Assessment of Their Biology and Management St. Lucie Press, Delray Beach, FL, pp 185–193

Handler AM, Harrell RA (1999) Germline transformation of *Drosophila melanogaster* with the *piggyBac* transposon vector. Insect Mol Biol 8:449–457

Hediger M, Minet AD, Niessen M, Schmidt R, Hilfiker-Kleiner D, Cakir S, Nöthiger R, Dubendorfer A (1998) The male-determining activity on the Y chromosome of the housefly (*Musca domestica* L.) consists of separable elements. Genetics 150:651–661

Heinstra PWH, Eisses KT, Schoonen WGEJ, Aben W, de Winter AG, van der Horst DJ, van Marrewijk WJA, Beenhakkers AMT, Scharloo W, Thorig GEW (1983) A dual function of alcohol dehydrogenase in *Drosophila*. Genetica 60:129–137

Hendrichs J, Franz G, Rendon P (1995) Increased effectiveness and applicability of the sterile insect technique through male-only releases for control of Mediterranean fruit flies during fruiting seasons. J Appl Entomol 119:371–377

Kaiser M, Gasser M, Ackermann R, Stearns SC (1997) *P*-element inserts in transgenic flies: a cautionary tale. Heredity 78:1–11

Kidwell MG, Wattam AR (1998) An important step forward in the genetic manipulation of mosquito vectors of human disease. Proc Natl Acad Sci USA 95:3349–3350

Knipling EF (1955) Possibilities of insect control or eradication through the use of sexually sterile males. J Econ Entomol 48:459–462

Konsolaki M, Komitopoulou K, Tolias PP, King DL, Swimmer C, Kafatos FC (1990) The chorion genes of the medfly, *Ceratitis capitata*, 1: structural and regulatory conservation of the s36 gene relative to two *Drosophila* species. Nucl Acids Res 18:1731–1738

Krafsur ES (1999) Sterile insect technique for suppressing and eradicating insect populations: 55 years and counting. J Agric Entomol, 15:303–317

Leopold RA (1998) Cold storage of insects for integrated pest management. In Hallman GY, Denlinger DL (eds) Temperature Sensitivity in Insects and Application in Integrated Pest Management Westview Press, Oxford, UK, pp 235–267

Loosjes M (1999) The sterile insect technique for commercial control of the onion fly. In Proceedings of FAO/IAEA Int Conf on Area-wide Control of Insect Pests, Malaysia, June 1998

Louis C, Yannopoulos G (1988) The transposable elements involved in hybrid dysgenesis in *Drosophila melanogaster*. In McLean M (ed) Oxford Surveys on Eukaryotic Genes, Vol 5. Oxford University Press, Oxford, pp 205–250

Mazur P, Cole KW, Hall JW, Schreuders, PD, Mahawold AP (1992) Cryobiological preservation of *Drosophila* embryos. Science 258:1932–1935

McDonald IC, Overland DE (1973) House fly genetics: 1. Use of an inversion to facilitate recovery of translocation homozygotes and to reduce genetic recombination on translocated third chromosomes. J Hered 64:247–252

Meise M, Hilfiker-Kliener D, Brunner C, Dubendorfer A, Nothiger R, Bopp D (1998) Sex-lethal, the master sex-determining gene in *Drosophila*, is not sex-specifically regulated in *Musca domestica*. Development 125:1487–1494

Msangi AM, Saleh KM, Kiwia N, Mussa WA, Mramba F, Juma KG, Dyck VA, Vreysen MJB, Parker AG, Feldmann U, Zhu ZR, Pan H (1999) Success in Zanzibar: eradication tsetse. In Proceedings of FAO/IAEA Int Conf on Area-wide Control of Insect Pests, Malaysia, June 1998

Muller-Holtkamp F (1995) The Sex-lethal gene homologue in *Chrysomya rufifacies* is highly conserved in sequence and exon-intron organization. J Mol Evol 41:467–477

Orser C, Staskawicz BJ, Panopoulos NJ, Dahlbeck G, Lindow SE (1985) Cloning and expression of bacterial ice nucleation genes in *Escherichia coli*. J Bacteriol 164:359–366

Pannuti A, Kocacitak T, Lucchesi JC (1999) A genetic sexing strategy for Anopheline and Aedine mosquitoes; abstr. In Proceedings of FAO/IAEA Int Conf on Area-wide Control of Insect Pests, Malaysia, June 1998

Parkes TL, Kirby K, Phillips JP, Hilliker AJ (1998) Transgenic analysis of the cSOD-null phenotype syndrome in *Drosophila*. Genome 41:642–651

Pettigrew WW, O'Neill SL (1997) Control of vector-borne disease by genetic manipulation of insect populations: technical requirements and research priorities. Aust J Entomol 36:309–317

Rendon P, McInnis D, Lance D, Stewart J (1999) Comparison of medfly male-only and bisex releases in large-scale field trials. In Proceedings of FAO/IAEA Int Conf on Area-wide Control of Insect Pests, Malaysia, June 1998

Rina M, Savakis C (1991) A cluster of vitellogenin genes in the Mediterranean fruit fly *Ceratitis capitata*: sequence and structural conservation in dipteran yolk proteins and their genes. Genetics 127:769–780

Robinson AS (1983) Sex ratio manipulation in relation to insect pest control. Annu Rev Genet 17:191–214

Robinson AS (1986) Genetic sexing in *Anopheles stephensi* using dieldrin resistance. J Am Mosq Control Assoc 2:93–95

Robinson AS, Franz G, Fisher K Genetic sexing strains in the medfly, *Ceratitis capitata:* development, mass rearing and field application. Trends in Entomology, submitted

Rosetto M, De Filippis T, Manetti AGO, Marchini D, Baldari CT, Dallai R (1998) The genes encoding the antibacterial sex-specific peptides ceratotoxins are clustered in the genome of the medfly *Ceratitis capitata.* Insect Biochem Mol Biol 27:1039–1046

Rubin GM, Spradling AC (1982) Genetic transformation of *Drosophila* with transposable element vectors. Science 218:348–353

Saccone G, Peluso I, Artiaco D, Giordano E, Bopp D, Polito LC (1998) The *Ceratitis capitata* homologue of the Drosophila sex-determining gene Sex-lethal is structurally conserved but not sex-specifically regulated. Development 125:1495–1500

Saccone G, Testa G, Pane A, DeMartino G, Polito LC (1999) Sex determination in the medfly:a molecular approach. In Proceedings of FAO/IAEA Int Conf on Area-wide Control of Insect Pests, Malaysia, June 1998

Seawright JA, Kaiser PE, Dame DA, Lofgren CS (1978) Genetic method for the preferential elimination of females of *Anopheles albimanus*. Science 200:1303–1314

Shearman DCA, Frommer M (1998) The *Bactrocera tryoni* homologue of the *Drosophila melanogaster* sex-determination gene doublesex. Insect Mol Biol 7:355–366

Sievert V, Kuhn S, Traut W (1997) Expression of the sex determining cascade genes *Sex-lethal* and *doublesex* in the phorid fly *Megaselia scalaris*. Genome 40:211–214

Thymianou S, Mavroidis M, Kokolakis G, Komitopoulou K, Zacharopoulou A, Mintzas AC (1998) Cloning and characterization of a cDNA encoding a male specific serum protein of the Mediterranean fruit fly, *Ceratitis capitata,* with sequence similarity to odourant binding proteins. Insect Mol Biol 7:345–353

Torti C, Malacrida AR, Yannopoulos G, Louis C, Gasperi G (1994) Hybrid dysgenesis-like phenomena in the medfly, *Ceratitis capitata* (Diptera, Tephritidae). J Hered 85:22–30.

Tyshenko MG, Doucet D, Davies PL, Walker VK (1997) The antifreeze potential of the spruce budworm thermal hysteresis protein. Nat Biotechnol 15:887–890

Walker VK, Rancourt DE, Dunker BP (1995) The transfer of fish antifreeze genes to *Drosophila*: a model for the generation of transgenic beneficial insects. Proc Entomol Soc Ont 126:3–13

Whitten MJ (1969) Automated sexing of pupae and its usefulness in control by sterile insects. J Econ Entomol 62:272–273

Willhoeft U, Franz G (1996) Identification of the sex-determining region of the *Ceratitis capitata* Y chromosome by deletion mapping. Genetics 144:737–745

Willhoeft U, Mueller-Navia J, Franz G (1998) Analysis of the sex chromosomes of the Mediterranean fruit fly by microdissected DNA probes. Genome 41:74–78

Wyss JH (1999) Screwworm eradication in the Americas: overview. In Proceedings of FAO/IAEA Int Conf on Area-wide Control of Insect Pests, Malaysia, June 1998

Yeh E, Gustafson K, Boulianne GL (1995) Green fluorescent protein as a vital marker and reporter of gene expression in *Drosophila*. Proc Natl Acad Sci USA 92:7036–7040

18 Control of Disease Transmission through Genetic Modification of Mosquitoes

Anthony A. James

CONTENTS

18.1 Introduction 319
18.2 Background 320
18.3 Laboratory Development of Pathogen Resistance in Mosquitoes 322
18.4 Moving Genes into Wild Populations 327
18.5 The Target Populations 328
18.6 Conclusions 329
Acknowledgments 329
References 329

18.1 INTRODUCTION

Vector-borne diseases remain one of the greatest menaces to humans throughout the world. Millions of people die every year from diseases such as malaria, and many more are incapacitated as a result of infections by a variety of vector-borne viral, protozoan, and metazoan pathogens. The exploitation of molecular biological techniques for research on arthropod vectors of disease has fostered the development of novel disease-control strategies based on attacking the vector. One strategy proposes to modulate vector competence genetically, reducing it to the point that it interrupts pathogen transmission (Collins and James, 1996; James et al., 1999). Vector competence as used here refers to the ability of an arthropod to serve as a suitable host for the transmission and propagation of a pathogen. We refer specifically to properties that result from the genetic background of the host arthropod, and therefore this is a more limited definition of vector competence than that provided originally by Hardy et al. (1983). The hypothesis of this strategy is that the introduction into a vector population of a gene that confers resistance to a pathogen will result in a decrease in transmission of that pathogen. To test this hypothesis, a gene or allele that interferes with pathogen development or propagation must be spread through a vector population. Once the gene or allele has become sufficiently abundant, measurable decreases in transmission and disease should be observed. Transgenesis technology plays an important role in the development and implementation of this strategy and is absolutely necessary for practical experiments that will test the hypothesis. This chapter highlights current research efforts to use transgenic technology to produce parasite-resistant mosquitoes. Mosquitoes are vectors of major diseases and most of the recent developments in vector transgenesis have taken place with them. The general background and research areas will be summarized, and this will be followed by a discussion of potential targets and molecular strategies for intervention.

18.2 BACKGROUND

The transmission of vector-borne diseases is, at minimum, a three-component system consisting of the pathogen, the vector arthropod, and a vertebrate host. Maintenance of the life cycle of the pathogen depends upon efficient transmission between the vector and vertebrate hosts. In a number of cases, pathogen life cycles are more complex and involve multiple vectors and vertebrate hosts. Some vertebrate hosts serve as reservoirs from which pathogens are transmitted to a second host that experiences disease as a result of infection. Furthermore, some pathogens, such as those that cause schistosomiasis, utilize a gastropod as the invertebrate vector. Each life stage of the parasite presents challenges and opportunities for interrupting transmission. Those pathogens with the least complex transmission cycles appear most vulnerable to genetic control because they provide a limited number of targets and therefore potential transmission "bottlenecks" for control.

Mosquito-borne diseases include malaria, filariasis, yellow fever, dengue fever, and a number of viral encephalites. The pathogens that cause malaria tend to be highly specific for their vertebrate hosts and vector mosquitoes. For example, only four species of *Plasmodium* cause malaria in humans, and these are transmitted solely by mosquitoes of the genus *Anopheles*. Many of the viral diseases have animal reservoirs and are transmitted to humans by a small number of mosquito species. Although the development of vaccines has provided a powerful tool to combat some of the viral pathogens, especially the yellow fever virus, the lack of effective vaccines for many of the other diseases has made mosquito control the principal approach to controlling disease transmission. The limited host range of pathogens for specific mosquitoes greatly facilitates an antivector approach to control disease transmission.

Mosquitoes belong to the family Culicidae, and two subfamilies, the Culicinae and Anophelinae, have hematophagic genera that transmit pathogens that cause disease in humans and animals. These two subfamilies are fairly old in evolutionary terms, perhaps last sharing a common ancestor more than 140 million years ago (Service, 1993). An awareness of the evolutionary relationships among mosquitoes is important, because although it is believed that certain techniques and approaches may be readily transferred and adapted to a wide range of mosquito species, the specific biology among the genera can be quite different, requiring novel developments within each group.

The Culicinae are a large and varied taxonomic group and transmit the widest variety of pathogens to humans. Perhaps the most notorious species is *Aedes aegypti*, which transmits yellow fever and dengue fever viruses (Clement, 1992; 1999; Gubler and Kuno, 1997). Effective control of *Ae. aegypti* is still a major objective, especially with the rapid spread of dengue viral serotypes throughout the tropical and subtropical zones. As mentioned above, mosquitoes of the genus *Anopheles* are the sole vectors of the pathogens that cause human malaria. This distinction has made them targets of efforts to control their population size or eliminate them entirely. Starting during World War II and continuing until the late 1970s, control of the mosquito was the principal objective in most of the major efforts to eradicate malaria (Collins and James, 1996). However, as recounted by Spielman et al. (1993), during the later years the primary objective of malaria programs changed from eradication of the disease to some measurable and tolerable level of control. Although the lack of a vaccine against the disease and the development of resistance in the pathogens to therapeutic drugs certainly contributed, it is likely that the failure to control the mosquito vectors was the major reason for the shift in the emphasis from eradication to control.

Current programs to control mosquitoes are designed around the use of insecticides and source-reduction methods. Insecticides are applied in different formulations to kill larval and/or adult mosquitoes. Source reduction involves altering or eliminating water sources that serve as sites for oviposition and larval breeding, and this can be an effective strategy for controlling both the Culicinae and Anophelinae. The well-documented emergence of insecticide resistance in mosquitoes threatens the continued use of insecticides for controlling the mosquitoes (Ferrari, 1996). However, source reduction remains a useful means of controlling local peridomestic, anthropophilic mosquitoes.

The failure to control mosquitoes by insecticides prompted research to find alternative means of control. A general effort was made to find new biological agents that would be highly specific to the vector and not be expected to select for resistance. Research in insect growth and molting hormones flourished, as did efforts to identify insect pathogens. The discovery of insecticidal toxins in bacteria, specifically *Bacillus thurengiensis* (BTI), has stimulated control methods that have been successfully applied to a number of different insects (Woodring and Davidson, 1996). BTI formulations are still being used for mosquito control and can be effective in programs focused on limited geographic areas (Skovmand and Sanogo, 1999). Part of the wave of new research included efforts in genetic control of insects. The genetic control approach is fundamentally different from that of chemical control. The latter provides the impetus for continuing research to find physiological targets that are vulnerable to exogenous chemical intervention, while genetic control research attempts to exploit intrinsic properties of the insect. However, it is possible to imagine how the two approaches can be combined for an effective control strategy. The discovery of lethal mutations and incompatible genotypes in hybrid insects fueled a burst of research activity on the genetics of mosquitoes in an attempt to utilize these genetic properties as tools for control. However, limitations on the speed at which new genotypes could be spread through populations, unachievable release ratios of modified to wild insects, and other constraints imposed by Mendelian transmission of useful traits curtailed research and field trials (Rai, 1996). Lately, the application of recombinant DNA technology and the discovery of transposable elements have opened up significant new areas of investigation and have rekindled interest in genetic control methods.

One objective of a strategy for genetic control of insects is to reduce the population size or to try to eliminate entirely the target species in a geographic area. Exploitation of specific genetic circumstances that lead to reproductive failure are the most likely means to achieve these kinds of objectives. Successful exploitation of these control methods is the principal goal of efforts focused on insects that have an agricultural impact (see Chapter 17 by Robinson and Franz). However, genetic control efforts involving insect vectors of disease can have different objectives. Here, the primary goal is to reduce or eliminate the transmission of disease-causing pathogens. Although in theory this can be achieved by elimination of the insect vector, total elimination of vector populations will be difficult, if not impossible to achieve. Therefore, alternative approaches that involve population replacement strategies have been proposed (Curtis and Graves, 1989). These approaches emphasize the introduction of genes into existing populations of vector insects that result in reduced or no capability for transmitting disease pathogens (Collins and James, 1996; James et al., 1999). This approach illustrates the shift in the general premise of genetic control of vector insects from causing reproductive failure (population reduction) to altering vector competence (population replacement).

The genetic control hypothesis is debated by a number of workers who argue, for example, that in the case of malaria, reducing transmission without achieving eradication may lead to worse disease than no intervention at all (Collins, 1994; Spielman et al., 1993). Data from a number of villages in Africa show that deaths due to severe malaria are lower in areas of high transmission than in areas of moderate transmission (Snow and Marsh, 1995). These conclusions stand in contrast to those from the study of the use of insecticide-impregnated bed nets, which prevent access of the mosquito to its human host and thus prevent transmission (Curtis, 1996). Use of nets can reduce by one half the incidence of clinical malaria cases in parts of Africa and Asia (Choi et al., 1995). The different conclusions on the effects of reducing transmission most likely will be reconciled following analyses of the transmission dynamics in the specific geographic areas examined in each study. However, the success of the bed net programs bolsters the argument for the development of additional methods for preventing transmission of malaria parasites with the expectation that deployment of these methods will result in less disease and death. Furthermore, it is likely that more than one strategy will be needed for the successful control of malaria. An integrated approach that includes bed nets, parasite-resistant mosquitoes, and vaccines (Miller and Hoffman, 1998), will ultimately control or eliminate the disease if the role of each component intervention method is optimized for a specific region of transmission.

Research in three general areas needs to be advanced to test the hypothesis of genetic control of vectors and disease (Collins and James, 1996; James et al., 1999). In laboratory-based work, insects must be developed that are resistant or refractory to parasites. The production of virus-resistant *Ae. aegypti* (Olson et al., 1996; Powers et al., 1996; Higgs et al., 1998; Chapter 9 by Olson), provides proof-of-principle for the development of refractory mosquitoes.

The second research area requires the development of techniques for moving antiparasite genes tested in the laboratory into wild insect populations. Mendelian transmission of a gene through a population will take too long for the gene to reach the critical threshold frequency necessary for interrupting pathogen transmission, and also requires a fitness advantage for the introduced gene. Therefore, alternate strategies to Mendelian transmission are being investigated for moving genes into populations. For example, models have been developed of genes spreading in an infection-like wave through mosquito populations, and there is enthusiasm for using transposable elements, other mobile nucleic acid vectors such as viruses, or symbionts to drive genes through populations (Kidwell and Ribeiro 1992; Turelli and Hoffmann, 1999; Chapter 15 by Sinkins and O'Neill).

Finally, one must have sufficient information about the target vector population so that how the gene will behave in the population can be modeled and predicted. This is important for both the introduction of the gene and establishing parameters by which the success of the introduction will be measured. The genetic structure, migration indices, gene flow, and other population genetic factors will have a major effect on the outcome of a genetic strategy, and one must be able to anticipate and account for these effects on the stability of the introduced gene in the target population.

18.3 LABORATORY DEVELOPMENT OF PATHOGEN RESISTANCE IN MOSQUITOES

The first area of research addresses the question of how to engineer genetically pathogen resistance in mosquitoes. Early researchers looking for a starting point turned to studying natural pathogen resistance existing in wild populations. A thorough review of the work of Ronald Ross, William Macdonald, George Craig, and other mosquito biologists and geneticists is outside the scope of this chapter. However, the aggregate work of these investigators revealed some general properties of mosquito-borne diseases. Not every species of hematophagic mosquito is capable of transmitting every mosquito-borne pathogen. There are clearly some species that can transmit a specific pathogen, and those that cannot. Many aspects of mosquito biology may contribute to these differences including behavioral properties (host preferences) and physiological properties. Furthermore, not only were differences in vector competence observable among species, but in some cases differences were seen within species (Collins et al., 1986; Feldmann and Ponnudurai, 1989). Different mosquito populations can vary in their vector competence. Genetic analyses showed that in some cases differences in vector competence could be traced to multiple alleles of a single genetic locus (Macdonald, 1962; Kilama and Craig, 1969; Feldmann et al., 1998). Recent analyses have shown that the genetic basis of vector competence is likely to be more complex, involving multiple genes with quantitative contributions (Vernick et al., 1989; Thathy et al., 1994; Severson et al., 1995; Zheng et al., 1997). However, the basic conclusion that a single gene or a small number of genes could affect vector competence encouraged workers to consider how these genes might be exploited to control pathogen transmission.

Researchers were faced with a number of possible approaches to exploiting pathogen-resistant genes. One alternative would be to study naturally occurring resistance, attempt to determine the genetic and molecular basis for this resistance, and then recreate or induce this resistance in pathogen-transmitting populations. Another approach would be to identify novel mechanisms for interfering with pathogen development and propagation within the vectors, and then engineer a gene whose product performs the interfering function. These approaches are not mutually exclusive, and are the basis for most ongoing work in mosquitoes.

The ideal properties of a gene conferring a pathogen-resistance phenotype were discussed in Meredith and James (1990). A gene with a dominant phenotype is required so that the antiparasite activity is manifest in heterozygous animals. The gene should have complete penetrance, so that all animals possessing a single copy show the phenotype. Finally, the gene should contribute minimally to the genetic load and therefore not reduce the fitness of the animal. Some have argued that the gene must confer a positive selective advantage, but this is not absolutely essential and will be discussed in the following section. None of the naturally occurring resistance genes appears to have all of these attributes, and, therefore, a major focus of research is developing novel, synthetic antiparasite genes.

The specific mechanism of action of an antiparasite gene will depend on the target pathogen. For example, interrupting viral propagation was achieved by expression of antisense viral RNA in the mosquito. The Sindbis virus is the basis of a transient expression system that allows the production of a large amount of antisense RNA (Chapter 9 by Olson). In experiments targeting the dengue virus, a modified Sindbis virus was used to express an antisense RNA derived from the premembrane coding region of the dengue virus (Olson et al., 1996). Mosquitoes were infected with the engineered Sindbis virus and subsequently inoculated with an infectious dengue virus. Transmission of the dengue virus was blocked in those animals expressing antisense RNA, but not in the controls. Similar results were achieved in separate experiments using antisense RNAs to the yellow fever (Higgs et al., 1998) and LaCrosse viruses (Powers et al., 1996). This approach is essentially an "intracellular immunization" (Olson et al., 1996) and is dependent on the replication properties of the virus *in vivo*. This approach may be a general mechanism for disabling viruses with RNA genomes.

Sindbis virus constructs do not integrate into the mosquito genome, and, therefore, the effects of the engineered gene are limited only to the generation that was infected. Clearly, a system for integrating high-expression constructs into the genome is required. Some of the transposable elements discussed below and in other chapters of this book may be adapted for this purpose.

Other pathogens targeted for genetic intervention include the protozoans and metazoans responsible for malaria and filariasis, respectively. Although it is judgmental to state that their development in the mosquito is more complex than that of the viruses, they do present a different set of problems for designing a gene that interferes with their development.

Mosquito-borne pathogens first interact with their insect host when they are ingested along with the blood meal (Figure 18.1). The number of infectious forms in the initial inoculum varies depending on the specific pathogen and the status of the infection in the vertebrate host from which the blood meal was obtained. The ingested pathogens must adapt to an environment, the midgut, that is significantly different from the vertebrate tissues that they just left. Temperature, pH, and other factors are much different in the insect midgut than in the vertebrate host. Therefore, the first challenge of the parasite is to adapt physiologically to these changes. These adaptations may involve immediate developmental changes, as seen with malaria parasites, or rapid penetration or infection of midgut cells, which in effect removes the pathogen from the extracellular midgut environment. Ultimately, all mosquito-borne pathogens must exit the midgut to continue their transmission cycle. Two potential targets present themselves for interfering with pathogens at this stage. First, disruption of development will prevent pathogens from progressing to the stage that is capable of penetrating the midgut. With malaria parasites this would involve blocking gametogenesis, fertilization, and ookinete mobility. The second target would be to prevent pathogen interactions with the midgut cells, thus preventing an exit from the midgut. Many of the naturally occurring barriers to vector competence result from an inability of the pathogen to survive successfully in the midgut. For example, many viruses when ingested by an incompetent host are unable to infect midgut cells (Hardy et al., 1983). A number of different strategies have been proposed for interrupting malaria parasite transmission by targeting the midgut developmental stages (Shahabuddin et al., 1998a).

Once pathogens have exited the midgut, they are in the open circulatory system of the insect, the hemolymph. The principal problems facing protozoan and metazoan pathogens during this stage

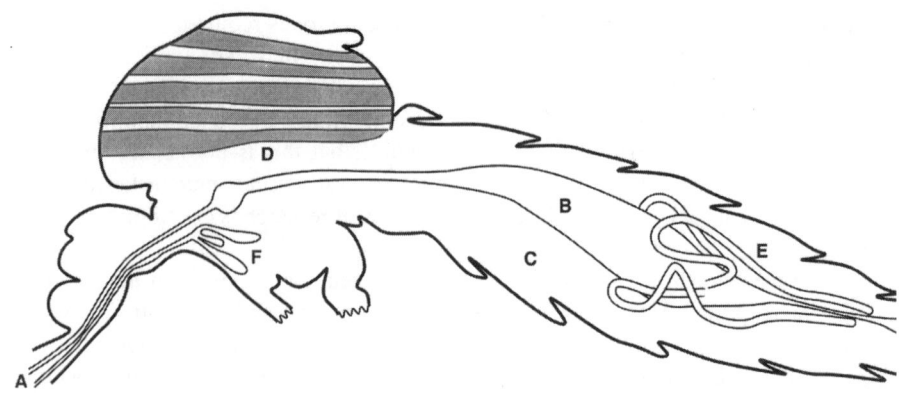

FIGURE 18.1 Pathogen development and movement in a vector host. The figure represents schematic of a cross section of a mosquito. Pathogens enter the host (A) following ingestion of an infectious blood meal and first encounter the midgut (B). Pathogens can complete a portion of their development here before crossing the midgut epithelium. Pathogens may continue their development on the basal surface of the midgut epithelium or transit the hemolymph (C) and develop in association with other tissues, such as the dorsal musculature (D) or Malpighian tubules (E). Most pathogens must make their way to the salivary glands (F) prior to transmission to a new vertebrate host.

are navigating the humoral environment to the appropriate tissues for their further development while evading components of the insect immune system. For example, filarial worms continue their development in association with flight muscles or the Malpighian tubules. The worms then make their way to the proboscis, where they extrude through the cuticle into the vertebrate host while the mosquito is feeding.

Malaria parasites develop in oocysts on the basal surface of the midgut and then burst out into the hemolymph as haploid sporozoites. They must make their way to the salivary glands in anticipation of transmission to a new vertebrate host. Invasion of the salivary glands is essential if they are to be secreted into a vertebrate host during the next feeding cycle. All of the destination tissues represent potential sites for interrupting transmission of pathogens and are therefore targets for the action of antiparasite genes.

Transgenesis technology plays an important role in validating the action and efficacy of engineered antiparasite genes. For example, a hybrid resistance gene could consist of a promoter region driving the expression of the antiparasite effector region. Such promoters will have to be tested *in vivo* to demonstrate that they express the desired product at sufficient levels in the appropriate tissues and at the correct developmental stage. Some of the analyses of the promoters may be done in viral expression systems (see Chapter 8 by Carlson et al.), in heterologous species, for example, *Drosophila melanogaster* (Skavdis et al., 1996; Xiong and Jacobs-Lorena, 1995) or in cultured cells (Fallon, 1991; Coates et al., 1999), but ultimately proper validation of the function of a hybrid gene requires testing it in the target vector species (Coates et al., 1999). Obviously, stable transformation technology must be developed for each vector species for which a genetic control strategy is planned.

The requirements and approaches for developing transgenesis technology are explored in much more detail in other chapters of this book. Although viruses have been used to integrate DNA into insect cells (Matsubara et al., 1996; Jordan et al., 1998; Chapter 7 by Burns), the most versatile technology most likely will use one of a number of Class II transposable elements (Finnegan, 1989). These elements are simple, consisting of inverted terminal repeat DNA flanking a self-encoded transposase gene. Class II elements transpose via a DNA intermediate and are readily adaptable to simple cloning strategies. These elements usually are developed for stable integration of DNA into chromosomes, and, as such, the transposase-encoding sequences are removed from the element and the tranposase is supplied *in trans* from another plasmid construct. Autonomously mobilizing elements keep the transposase gene intact between the inverted terminal repeats. Examples of elements that show great promise as general transformation vectors are *Hermes* (O'Brochta et al., 1996; Chapter 12 by Atkinson and O'Brochta), *mariner* (Medhora et al., 1991; Chapter 13 by Lampe et al.), *Minos* (Loukeris et al., 1995) and *piggyBac* (Fraser et al., 1996; Chapter 14 by Fraser). *Hermes* and *mariner* have been shown to integrate DNA into *Ae. aegypti* (Jasinskiene et al., 1998; Coates et al., 1998) and *Hermes* can integrate into chromosomes in *An. gambiae* cultured cells (Zhao and Eggleston, 1998; Chapter 2 by Eggleston and Zhao). *Minos* has been used to transform the medfly, *Ceratitis capitata* (Loukeris et al., 1995). The *piggyBac* element is noteworthy because it is derived from lepidopteran cells and can integrate into *D. melanogaster* (Handler and Harrell, 1999) and *C. capitata* (Handler et al., 1998). As discussed in other chapters of this book, other features of elements, such as autonomy, remobilization, and use of purified transposases, would greatly facilitate their use.

Key to most transgenesis technologies are appropriate marker genes. Most successful transformation experiments have relied on the partial restoration of eye color in mutant recipient strains with white eyes (*D. melanogaster,* Rubin and Spradling, 1982; *C. capitata,* Loukeris et al., 1995; *Ae. aegypti,* Coates et al., 1998; Jasinskiene et al., 1998; Chapter 4 by Sarkar and Collins). This complementation can be highly sensitive and partial rescues are easily scored. However, the use of other marker genes, for example green fluorescent protein (Higgs et al., 1996; Pinkerton et al., 2000; Chapter 5 by Higgs and Lewis), will allow the introduction of DNA into strains or species that do not have the appropriate eye color mutations.

The various phenotypes that prevent pathogen development or propagation in a vector can be classified as either "refractory" or "resistant" (Christensen and Severson, 1993). Refractory phenotypes are passive, usually involving a lack of something that is necessary for the continued viability of the pathogen, and it is reasonable to expect that these phenotypes will be recessive. Pathogens introduced into refractory hosts fail to develop without any obvious signs of activity by the host. On the other hand, resistant phenotypes are active and most likely dominant or semidominant. Molecules present in the host or host physiological responses to the infection actively kill or disable the parasite. Most of the phenotypes envisioned by those working to engineer antiparasite vectors belong to the active resistance class.

From the previous discussion, it is clear that specific recognition properties must exist that allow a pathogen to infect or invade the appropriate tissues. The generally accepted mechanistic model is that receptor/ligand-like interactions form the basis of the recognition properties. Ligands on the surface of the pathogen interact with specific molecules that function as receptors on the surface of the host insect tissues. A number of experiments with malaria parasites support the general model. For example, *P. knowlesi* is able to develop in the midgut, infect the salivary glands and be transmitted by *An. dirus*, a major vector in Southeast Asia. However, although *P. knowlesi* can develop in midgut of *An. freeborni*, they are not able to infect the salivary glands. Rosenberg (1985) transplanted salivary glands from *An. freeborni* into infected *An. dirus*, and the glands remained uninfected. Conversely, *An. dirus* salivary glands transplanted into infected *An. freeborni* were infected by the malarial sporozoites. This ability of the parasites to infect competent glands in the background of an incompetent host was interpreted to indicate that some intrinsic property of the salivary glands was responsible for the differences between competent and incompetent hosts. Molecules that would function as a receptor for sporozoites could be present on the *An. dirus* glands and missing from those of *An. freeborni*.

In other experiments, mouse monoclonal antibodies raised to salivary glands block the invasion of malaria parasites when they are injected into infected hosts (Barreau et al., 1995). It was suggested that the antibodies bind a tissue receptor preventing parasite recognition and invasion. Similar results were achieved using lectins, which bind carbohydrates, suggesting that glycoproteins may function as the receptor.

Studies using antibodies raised against pathogens also argue for the presence of ligands on the surface of pathogens. Specific monoclonal antibodies raised against a surface glycoprotein block La Crosse virus invasion of the mosquito midgut cells (Sundin et al., 1987). Furthermore, antibodies raised to the major surface protein, the circumsporozoite protein (CSP), of *P. gallinaceum*, prevent sporozoite invasion of salivary glands following their infection with infected mosquitoes (Warburg et al., 1992). Recently, it was shown that peptides derived from the CSP of *P. falciparum* bind salivary glands of *An. stephensi* (Sidjanski et al., 1997). These are but a few examples in support of the likelihood that receptor/ligand interactions form the basis of many vector–parasite interactions.

Specific molecular genetic interventions with receptor/ligand interactions can involve overexpression of either the receptor or ligand, and establishing circumstances for competitive inhibition as the basis of the resistance phenotype. Overexpression of the insect receptor in a soluble form should result in binding of this molecule to the ligand on the surface of the parasite and prevent the ligand from interacting with the bound receptor on the surface of the target tissue. Antibodies or other molecules that would bind the ligand and compete with the endogenous receptor also would provide the same type of inhibition. This latter approach has been argued for because it should not interfere with the insect ligands that may interact with the tissue receptors (James et al., 1999). In another approach, expression of soluble ligands would compete with the parasite ligand for binding to the target tissue and prevent parasite recognition of the target tissue. Thus, it is possible to devise strategies for blocking parasites in the mosquito by overexpression of either receptors or ligands.

Another approach involves expressing a gene product that would interfere with a specific developmental transition of the pathogen that is triggered by molecules in the mosquito. For example, gametogenesis of malaria parasites in the mosquito midgut is activated by xanthenuric acid (XA) (Billker et al., 1998; Garcia et al., 1998). Expression of a molecule that would selectively interfere with XA may prevent this important developmental transition. Both malaria parasites (Shahabbuding et al., 1993) and filarial worms (Fuhrman et al., 1992) secrete chitinases that may facilitate midgut penetration. Agents that block the functioning of these enzymes reduce infectivity and could form the basis of an antiparasite gene (Shahabuddin et al., 1998a).

Some of the most exciting ongoing work involves mosquito immune responses to pathogens. Indeed, many of the naturally occurring resistance phenotypes appear to result from immune responses that disable or kill pathogens. A number of heterologous immune peptides, for example, cercropins and magainins, when injected into mosquitoes, will kill protozoan parasites (Gwadz et al., 1989). The search for endogenous immune peptides has led to the description of a number of defensins from *Ae. aegypti* and *An. gambiae* (Richman et al., 1996; Lowenberger et al., 1999), as well as a number of novel gene products (Barillas-Mury et al., 1996; Dimopoulos et al., 1997). Furthermore, organs such as the salivary glands and midgut have been shown to respond to malaria parasite infections by the production of immune peptides (Dimopoulos et al., 1998). This latter observation is intriguing because it suggests that a number of tissues may be exploited for an immune-based antiparasite response.

Immune-based antiparasite approaches fall into the two general categories. The first involves an activation or elevation of a naturally occurring immune response that will kill the pathogens. Molecular studies are in progress to dissect such natural processes (Richman and Kafatos, 1996; Richman et al., 1997) and to discover the initiator events that trigger an immune cascade (Meister et al., 1997). An appropriate molecular intervention would stimulate the immune response in a competent host and thus disable the parasite.

An alternative approach is the overexpression of a specific component of an immune response and a subsequent killing of the parasite. For example, *Aedes* defensins interfere with filarial worm and malaria parasite development (Lowenberger et al., 1996; Shahabuddin et al., 1998b). Overexpression of such a defensin at the appropriate stages would produce a functional resistance phenotype.

The above-described approaches are all areas of active research. Undoubtedly, some will be more practical than others when it comes time to implement these in vector control. However, even those methods that do not lead directly to a control strategy may result in other discoveries that ultimately produce a useful approach to control.

18.4 MOVING GENES INTO WILD POPULATIONS

Development of laboratory strains that are resistant to parasites is necessary before the appropriate genes can be moved into the wild population. However, a number of problems must be solved before such genes can be introduced into a wild population. The first challenge is to determine how genes will be spread through the populations. As discussed previously (Collins and James, 1996; O'Brochta and Atkinson, 1998) standard Mendelian inheritance of an antiparasite gene is expected to take too long to establish the introduced gene. In addition, there would always be a chance that heterozygous animals would mate with each other and produce homozygous offspring that are competent to transmit pathogens. These individuals, if even a small percentage of the total population, may be sufficient to maintain the transmission cycle. Therefore, other mechanisms for moving genes within populations are required.

Meiotic drive (MD) is a genetic phenomenon in which only one of the two homologous chromosomes that pair during meiosis is represented in the gametes that contribute to the next generation. Meiotic drive loci have been described in *D. melanogaster* (Robbins et al., 1996) and *Ae. aegypti* (Wood and Ouda, 1987). In principle, if an antiparasite gene could be linked to an MD locus, its frequency in populations would increase rapidly. Meiotic drive genes were extensively evaluated in *Ae. aegypti* and shown to be capable of altering coupled gene frequencies in cage experiments (Wood et al., 1977). However, it was discovered in field experiments that wild-type populations have unlinked suppressor genes of MD loci that work to maintain recovery of both homologous chromosomes (Suguna et al., 1977). This presents a difficult problem because it is not possible to remove such loci from target populations before using an MD locus. However, the discovery of multiple MD loci may still make it possible to utilize this approach.

Two classes of active drive elements have been proposed for moving genes into wild populations. The first are the aforementioned transposable elements. These elements are spread through mating during which time they move from chromosomes that have elements to those that do not. Replicative transposition events are required to increase the copy number of elements in a genome. This essentially results in circumstances where either all or the majority of progeny arising from a mating of an animal carrying elements and an animal lacking elements have the potential to contain integrated elements in their genome. This mode of transmission circumvents the problem of heterozygosity and should rapidly increase transposon frequencies in populations. In addition, this type of transmission does not require a fitness advantage of the transposing element to see an increase in frequency in the population. Mobility and a positive reproductive rate are all that are required to get an increase in the frequency of the element (Ribeiro and Kidwell, 1992). Cage experiments with *P* elements in *D. melanogaster* have shown that it is possible to increase the frequency of marker genes linked to the transposing element (Meister and Grigliatti, 1993; Carareto et al., 1997). The worldwide dissemination of the *P* element through *D. melanogaster* populations may represent the best example of the rapid spread of a transposable element occurring in insects in nature. Indeed, the *P* element has spread throughout global populations of this fruit fly within the last 70 to 90 years (Daniels et al., 1990). A similar spread of an engineered element through a vector population could be expedited by targeted releases of mosquitoes carrying a linked antiparasite gene.

An alternative approach would be to spread the antiparasite gene using a symbiont or other infectious agent. Antiparasite genes are inserted into the genomes or plasmids of these microorganisms, and the transformed strains are used to infect the arthropod host. B. Beard and his colleagues have transformed the gut bacterium, *Rhodococcus rhodnii*, which colonizes the midguts of vector bugs of the genus *Triatoma*, with a plasmid that expresses and secretes cecropins (Durvasula et al., 1997; Chapter 16 by Beard et al.). Bugs infected with this transformed endosymbiont are resistant to infection by *Trypanosoma cruzi*, the etiological agent of Chagas disease. Similar approaches have been proposed to exploit the intracellular symbionts, *Wolbachia*, which are present in many Dipteran species, including mosquitoes (Pettigrew and O'Neill, 1997; Turelli and Hoffmann, 1999; Chapter 15 by Sinkins and O'Neill). Movements of *Wolbachia* through natural populations of *D. melanogaster* have been documented (Turelli and Hoffmann, 1991), providing support for the hypothesis that these microorganisms can be exploited for moving genes rapidly through populations.

Regardless of the specific element that forms the basis for the genetic drive, a number of important features must be included in the overall design. For example, it is important that linkage be maintained as long as possible between the antiparasite gene and the element or agent that is used to spread it through the target population. A loss of coupling of the antiparasite gene with the mobilizing vector obviously will reduce the effectiveness of the strategy. To circumvent this problem, elements or microorganisms containing antiparasite genes should be designed so that they are disabled or self-distruct should they lose the antiparasite gene. In the long term, it should be anticipated that elements or other drive agents will eventually lose the antiparasite gene. However, the key here is that the coupling of the two remain in effect long enough to have the desired effect on the target pathogen.

In most disease transmission circumstances, it is possible to determine which vector species are transmitting the infective pathogen. However, the major vector species may differ among geographic regions, or may change seasonally (Curtis et al., 1999). This means that there may be multiple targets into which the element must be introduced. Furthermore, it will be necessary to restrict as much as possible specific genetic drive agents to a limited range of target species. Release programs with indiscriminate vector targets risk unanticipated consequences of introducing novel genes into untargeted species. Therefore, a thorough evaluation of the host specificity of the drive mechanism must be done.

It is possible that pathogens will be selected for resistance to the antiparasite gene. An effective measure to counteract this is to put more than one resistance gene into each drive construct. This will make it less likely that resistant parasites will emerge and thus prolong the efficacy of the construct.

It is not anticipated that complete transformation of the wild population will be necessary to interrupt the transmission of the pathogen. Modeling of parasite transmission has indicated that there are theoretical threshold levels below which the pathogen transmission fails. The frequency of an antiparasite gene must achieve this threshold level before transmission will cease. This is an interesting area of research that requires additional work.

18.5 THE TARGET POPULATIONS

Much work has to be done to define the target populations in the wild that are the objects of genetic intervention. As discussed in the previous section, transmission of a specific pathogen may occur with different vector species in different geographic regions, or the vector species may vary seasonally. In addition, the local distribution of vector animals within a transmission zone may affect gene dispersal (Kiszewski and Spielman, 1998). Therefore, it is important to know the size and the distribution of genetically distinct vector populations in an area being considered for genetic intervention. Each target could require a specific intervention to effect control.

It will be important to have some idea of the size of the target population. The speed at which an introduced gene will move through a population will depend on the number of animals in the population, as well as the specific properties of the drive mechanism and the number of the initially released animals carrying antiparasite genes (Ribeiro and Kidwell, 1994). Knowing the size of the population also will be important in monitoring the frequency of the introduced gene in the population. It will be important to measure the rate of change of the frequency, and sampling sizes will be based on the maximum size of the target population.

The genetic structure of the population will be important. Since most of the anticipated drive mechanisms rely upon mating to spread the antiparasite genes, the breeding structure of the population must be determined (Favia et al., 1997). Species complexes have been defined for many of the vector species. Although there may be limited gene flow among the members of the complex, a certain level of inbreeding will allow an introduced element to move into another member of the complex.

Migration among vector populations also is important. Gene flow studies have shown that even the migration of a small number of individuals can have an effect on subsequent gene frequency in populations (Tabachnick, 1992). Migration can work in the favor of genetic control strategy by moving the antiparasite genes to new vector populations that are removed from the original release site.

Finally, one must have relevant measurements of the effectiveness of a genetic release program. Although it will be important to measure the change in gene frequency in target populations as the antiparasite gene is spread, the ultimate measure will be a reduction in disease and deaths in the human populations. This impact may be complicated by the concommitant use of vaccines, improved medical care, and a general improvement in the health and well-being of people living in disease-endemic areas. Careful analysis of each component may be difficult in this multicompent strategy, but it will be essential for maintaining the support and enthusiasm for these approaches.

18.6 CONCLUSIONS

Much has been accomplished since molecular techniques were first applied to studying vectors of parasitic diseases. Genes whose products are involved in hematophagy, digestion, and other important metabolic processes have been cloned, transformation has been developed for *Ae. aegypti*, and extensive new knowledge of the insect immune system has been acquired. In addition, many molecular approaches for interfering with pathogens in the vectors have been researched. What remains to be done is to broaden the number of vector species in which transformation can be done, as well as further studies on how genes will be moved into populations. The successful harnessing of genetic drive mechanisms remains the next biggest achievement for this field.

ACKNOWLEDGMENTS

The author thanks his many colleagues who support the concept and practice the use of genetic engineering to control parasite transmission. Original work in the author's laboratory is supported by grants from the John D. and Catherine T. MacArthur Foundation, the Burroughs-Wellcome Fund, the UNDP/World Bank/WHO Special Programme for Research and Training in Tropical Diseases (TDR), and the National Institutes of Health.

REFERENCES

Barillas-Mury C, Charlesworth A, Gross I, Richman A, Hoffman JA, Kafatos FC (1996) Immune factor Gambif1, a new rel family member from the human vector, *Anopheles gambiae*. EMBO J 15:4961–4701

Barreau C, Touray M, Pimenta PF, Miller LH, Vernick KD (1995) *Plasmodium gallinaceum*: sporozoite invasion of *Aedes aegypti* salivary glands is inhibited by anti-glandantibodies and by lectins. Exp Parasitol 81:332–343

Billker O, Lindo V, Panico M, Etienne AE, Paxton T, Dell A, Rogers M, Sinden RE, Morris HR (1998) Identification of xanthurenic acid as the putative inducer of malaria development in the mosquito. Nature 392:289–292

Carareto CM, Kim W, Wojciechowski MF, O'Grady P, Prokchorova, AV, Silva JC, Kidwell MG (1997) Testing transposable elements as genetic drive mechanisms using *Drosophila* P element constructs as a model system. Genetica 101:13–33

Choi HW, Bremen, JG Teutsch SM, Liu S, Hightower AW, Sexton JD (1995) The effectiveness of insecticide-impregnated bed net in reducing cases of malaria infection: a meta-analysis of published results. Am J Trop Med Hyg 52:377–382

Christensen BM, Severson DW (1993) Biochemical and molecular basis of mosquito susceptibility to *Plasmodium* and filarioid nematodes. In Beckage NE (ed) Parasites and Pathogens of Insects, Vol I, Thompson & Federici, Academic Press, Orlando, FL, pp 245–226

Clement AN (1992) The Biology of Mosquitoes, Vol 1, Development, Nutrition, and Reproduction. Chapman and Hall, London

Clement AN (1999) The Biology of Mosquitoes, Vol 2, Sensory Reception and Behavior CABI Publishing, Oxford

Coates CJ, Jasinskiene N, Miyashiro L, James AA (1998) *Mariner* transposition and transformation of the yellow fever mosquito, *Aedes aegypti*. Proc Natl Acad Sci USA 95:3748–3751

Coates CJ, Jasinskiene N, Pott GB, James AA (1999) Promoter-directed expression of recombinant fire-fly luciferase in the salivary glands of *Hermes*-transformed *Aedes aegypti*. Gene 226:317–325

Collins FH (1994) Prospects for malaria control through the genetic manipulation of its vectors. Parasitol Today 10:370–371

Collins FH, James AA (1996) Genetic modification of mosquitoes. Sci Med 3:52–61

Collins FH, Sakai RK, Vernick KD, Paskewitz S, Seeley DC, Miller LH, Collins WE, Campbell CC, Gwadz RW (1986) Genetic selection of a *Plasmodium*-refractory strain of the malaria vector *Anopheles gambiae*. Science 234:607–610

Curtis CF (1996) Impregnated bednets, malaria control and child mortality in Africa. Trop Med Int Health 1:137–138

Curtis CF, Graves PM (1988) Methods for replacement of malaria vector populations. J Trop Med Hyg 91:43–48

Curtis CF, Pates HV, Takken W, Maxwell CA, Myamba J, Priestman A, Akinpelu O, Yayo AM, Hu JT (1999) Biological problems with the replacement of a vector population by *Plasmodium*-refractory mosquitoes. Parassitologia 41:479–481

Daniels SB, Peterson KR, Strausbaugh LD, Kidwell MG, Chovnick A (1990) Evidence for horizontal transmission of the *P* transposable element between *Drosophila* species. Genetics 124:339–355

Dimopoulos G, Richman A, Müller HM, Kafatos FC (1997) Molecular immune responses of the mosquito *Anopheles gambiae* to bacteria and malaria parasites. Proc Natl Acad Sci USA 94:11508–11513

Dimopoulos G, Seeley D, Wolf A, Kafatos FC (1998) Malaria infection of the mosquito *Anopheles gambiae* activates immune-responsive genes during critical transition stages of the parasite life cycle. EMBO J 17:6115–6123

Durvasula RV, Gumbs A, Panackal A, Kruglov O, Aksoy S, Merrifield RB, Richards FF, Beard CB (1997) Prevention of insect-borne disease: an approach using transgenic symbiotic bacteria. Proc Natl Acad Sci USA 94:3274–3278

Fallon AM (1991) DNA-mediated gene transfer: applications to mosquitoes. Nature 352:828–829

Favia G, della Torre A, Bagayoko M, Lanfrancotti A, Sagnon N, Toure YT, Coluzzi M (1997) Molecular identification of sympatric chromosomal forms of *Anopheles gambiae* and further evidence of their reproductive isolation. Insect Mol Biol 6:377–383

Feldmann AM, Ponnudurai T (1989) Selection of *Anopheles stephensi* for refractoriness and susceptibility to *Plasmodium falciparum*. Med Vet Entomol 3:41–52

Ferrari JA (1996) Insecticide Resistance. In Beaty BJ, Marquardt WC (eds) The Biology of Disease Vectors. University of Colorado Press, Niwot, pp 512–529

Feldmann AM, van Gemert GJ, van de Vegte-Bolmer MG, Jansen RC (1998) Genetics of refractoriness to *Plasmodium falciparum* in the mosquito *Anopheles stephensi*. Med Vet Entomol 12:302–312

Finnegan DJ (1989) Eukaryotic transposable elements and genome evolution. Trends Genet 5:103–107

Fraser MJ, Ciszczon T, Elick T, Bauser C (1996) Precise excision of TTAA-specific lepidopteran transposons *piggyBac* (IFP2) and *tagalong* (TFP3) from the baculovirus genome in cell lines from two species of *Lepidoptera*. Insect Mol Biol 5:141–151

Fuhrman JA, Lane WS, Smith RF, Piessens WF, Perler FB (1992) Transmission-blocking antibodies recognize microfilarial chitinase in brugian lymphatic filariasis. Proc Natl Acad Sci USA 89:1548–1552

Garcia GE, Wirtz RA, Barr JR, Woolfitt A, Rosenberg R (1998) Xanthurenic acid induces gametogenesis in *Plasmodium*, the malaria parasite. J Biol Chem 273:12003–12005

Gubler DJ, Kuno G (eds) (1997) Dengue and Dengue Hemmorrhagic Fever. CABI Publishing, New York

Gwadz RW, Kaslow D, Lee JY, Maloy WL, Zasloff M, Miller LH (1989) Effects of magainins and cecropins on the sporogonic development of malaria parasites in mosquitoes. Infect Immunol 57:2628–2633

Handler AM, Harrell RA (1999) Germline transformation of *Drosophila melanogaster* with the *piggyBac* transposon vector. Insect Mol Biol 8:449–457

Handler AM, McCombs SD, Fraser MJ, Saul SH (1998) The lepidopteran transposon vector, *piggyBac*, mediates germ-line transformation in the Mediterranean fruit fly. Proc Natl Acad Sci USA 95:7520–7525

Hardy JL, Houk EJ, Kramer LD, Reeves WC (1983) Intrinsic factors affecting vector competence of mosquitoes for arboviruses. Annu Rev Entomol 28:229–262

Higgs S, Rayner JO, Olson KE, Davis BS, Beaty BJ Blair CD (1998) Engineered resistance in *Aedes aegypti* to a West African and a South American strain of yellow fever virus. Am J Trop Med Hyg 58:663–670

James AA, Beerntsen BT, Capurro M de L, Coates CJ, Coleman Jasinskiene NJ, Krettli AU (1999) Controlling malaria transmission with genetically engineered, *Plasmodium*-resistant mosquitoes: milestones in a model system. Parassitologia 41:461–471

Jasinskiene N, Coates CJ, Benedict MQ, Cornel AJ, Salazar-Rafferty C, James AA, Collins FH (1998) Stable, transposon-mediated transformation of the yellow fever mosquito, *Aedes aegypti*, using the *Hermes* element from the housefly. Proc Natl Acad Sci USA 95:3743–3747

Jordan TV, Shike H, Boulo V, Cedeno V, Fang Q, Davis BS, Jacobs Lorena M, Higgs S, Fryxell KJ, Burns JC (1998) Pantropic retroviral vectors mediate somatic cell transformation and expression of foreign genes in dipteran insects. Insect Mol Biol 7:215–222

Kidwell MG, Ribeiro JMC (1992) Can transposable elements be used to drive disease refractoriness genes into vector populations? Parasitol Today 8:325–329

Kilama WL, Craig GB Jr (1969) Monofactorial inheritance of susceptibility to *Plasmodium gallinaceum* in *Aedes aegypti*. Ann Trop Med Parasitol 63:419–432

Kiszewski AE, Spielman A (1998) Spatially explicit model of transposon-based genetic drive mechanisms for displacing fluctuating populations of anopheline vector mosquitoes. J Med Entomol 35:584–590

Loukeris TG, Livadaras I, Arca B, Zabalou S, Savakis C (1995) Gene transfer into the medfly, *Ceratitis capitata*, with a *Drosophila hydei* transposable element. Science 270:2002–2005

Lowenberger CA, Ferdig MT, Bulet P, Khalili S, Hoffmann JA, Christensen BM (1996) *Aedes aegypti*: induced antibacterial proteins reduce the establishment and development of *Brugia malayi*. Exp Parasitol 83:191–201

Lowenberger CA, Smartt CT, Bulet P, Ferdig MT, Severson DW, Hoffmann JA, Christensen BM (1999) Insect immunity: molecular cloning, expression, and characterization of cDNAs and genomic DNA encoding three isoforms of insect defensin in *Aedes aegypti*. Insect Mol Biol 8:107–118

Macdonald WW (1962) The genetic basis of susceptibility to infection with semi-periodic *Burgia malayi* in *Aedes aegypti*. Ann Trop Med Parasitol 56:373–382

Matsubara T, Beeman RW, Shike H, Besansky NJ, Mukabayire O, Higgs S, James AA, Burns JC (1996) Pantropic retroviral vectors integrate and express in cells of the malaria mosquito, *Anopheles gambiae*. Proc Natl Acad Sci USA 93:6181–6185

Medhora M, Maruyama K, Hartl DL (1991) Molecular and functional analysis of the *mariner* mutator element *Mos*1 in *Drosophila*. Genetics 128:311–318

Meister GA, Grigliatti TA (1993) Rapid spread of a *P* element/*Adh* gene construct through experimental populations of *Drosophila melanogaster*. Genome 36:1169–1175

Meister M, Lemaitre B, Hoffmann JA (1997) Antimicrobial peptide defense in *Drosophila*. Bioessays 19:1019–1026

Meredith SEO, James AA (1990) Biotechnology as applied to vectors and vector control. Annu Parasitol Hum Comp 65:113–118

Miller LH, Hoffman SL (1998) Research toward vaccines against malaria. Nat Med 4:520–524

O'Brochta DA, Atkinson PW (1998) Building the better bug. Sci Am 279:90–95

O'Brochta DA, Warren WD, Saville KJ, Atkinson PW (1996) *Hermes*, a functional non-drosophilid insect gene vector from *Musca domestica*. Genetics 142:907–914

Olson KE, Higgs S, Gaines PJ, Powers AM, Davis BS, Kamrud KI, Carlson JO, Blair CD, Beaty BJ (1996) Genetically engineered resistance to dengue-2 virus transmission in mosquitoes. Science 272:884–886

Pettigrew MM, O'Neill SL (1997) Control of vector-borne disease by genetic manipulation of insect populations: technological requirements and research priorities. Austr J Entomol 36:309–317

Pinkerton AC, Michel K, O'Brochta DA, Atkinson PW (2000) Green fluorescent protein as a genetic marker in transgenic *Aedes aegypti*. Insect Mol Biol 9:1–10

Powers AM, Kamrud KI, Olson KE, Higgs S, Carlson JO, Beaty BJ (1996) Molecularly engineered resistance to California serogroup virus replication in mosquito cells and mosquitoes. Proc Natl Acad Sci USA 93:4187–4191

Rai KS (1996) Genetic control vectors. In the Biology of Disease Vectors. Beaty BJ and Marquardt WC (eds) University of Colorado Press, Niwot, pp 564–574

Ribeiro JM, Kidwell MG (1994) Transposable elements as population drive mechanisms: specification of critical parameter values. J Med Entomol 31:10–16

Richman A, Kafatos FC (1996) Immunity to eukaryotic parasites in vector insects. Curr Op Immunol 8:14–19

Richman AM, Bulet P, Hetru C, Barillas-Mury C, Hoffmann JA, Kafatos FC (1996) Inducible immune factors of the vector mosquito *Anopheles gambiae*: Biochemical purification of a defensin antibacterial peptide and molecular cloning of preprodefensin cDNA. Insect Mol Biol 5:203–210

Richman AM, Dimopoulos G, Steeley D, Kafatos FC (1997) *Plasmodium* activates the innate immune response of *Anopheles gambiae* mosquitoes. EMBO J 20:6114–6119

Robbins LG, Palumbo G, Bonaccorsi S, Pimpinelli S (1996) Measuring meiotic drive [letter]. Genetics 142:645–647

Rosenberg R (1985) Inability of *Plasmodium knowlesi* sporozoites to invade *Anopheles freeborni* salivary glands. Am J Trop Med Hyg 34:687–691

Rubin GM, Spradling AC (1982) Genetic transformation of *Drosophila* with transposable element vectors. Science 218:348–353

Service MW (1993) Medical Insects and Arachnids. Chapman & Hall, London

Severson DW, Thathy V, Mori A, Zhang Y, Christensen BM (1995) Restriction fragment length polymorphism mapping of quantitative trait loci for malaria parasite susceptibility in the mosquito *Aedes aegypti*. Genetics 139:1711–1717

Shahabuddin M, Toyoshima T, Aikawa M, Kaslow DC (1993) Transmission-blocking activity of a chitinase inhibitor and activation of malarial parasite chitinase by mosquito protease. Proc Natl Acad Sci USA 90:4266–4270

Shahabuddin M, Cociancich S, Zieler H (1998a) The search for novel malaria transmission-blocking targets in the mosquito midgut. Parasitol Today 14:493–497

Shahabuddin M, Fields I, Bulet P, Hoffmann JA, Miller LH (1998b) *Plasmodium gallinaceum*: differential killing of some mosquitoes stages of the parasite by insect defensin. Exp Parasitol 89:103–112

Sidjanski SP, Vanderberg JP, Sinnis P (1997) *Anopheles stephensi* salivary glands bear receptors for region I of the circumsporozoite protein of *Plasmodium falciparum*. Mol Biochem Parasitol 90:33–41

Skavdis G, Sidén-Kiamos I, Müller HM, Crisanti A, Louis C (1996) Conserved function of *Anopheles gambiae* midgut-specific promoters in the fruitfly. EMBO J 15:344–350

Skovmand O, Sanogo E (1999) Experimental formulations of *Bacillus sphaericus* and *B. thuringiensis istraelensis* against *Culex quinquefasciatus* and *Anopheles gambiae* (Diptera: Culicidae) in Burkina Faso. J Med Entomol 36:62–67

Snow RW, Marsh K (1995) Will reducing *Plasmodium falciparum* alter malaria mortality among African children? Parasitol Today 11:188–190

Spielman A, Kitron U, Pollack RJ (1993) Time limitation and the role of research in the worldwide attempt to eradicate malaria. J Med Entomol 30:6–19

Suguna SG, Wood RJ, Curtis CF, Whitelaw A, Kazmi SJ (1977) Resistance to meiotic drive at the M^D locus in an Indian wild population of *Aedes aegypti*. Genet Res Camb 29:123–132

Sundin DR, Beaty BJ, Nathanson N, Gonzalez-Scarano F (1987) A G_1 glycoprotein epitope of La Crosse virus: a determinant of infection of *Aedes triseriatus*. Science 235(4788):591–593

Tabachnick WJ (1992) Microgeographic and temporal genetic variation in populations of the bluetongue virus vector *Culicoides variipennis* (Diptera: Ceratopogonidae). J Med Entomol 29:384–394

Thathy V, Severson DW, Christensen BM (1994) Reinterpretation of the genetics of susceptibility of *Aedes aegypti* to *Plasmodium gallinaceum*. J Parasitol 80:705–712

Turelli M, Hoffmann AA (1991) Rapid spread of an inherited incompatibility factor in California *Drosophila*. Nature 353:440–442

Turelli M, Hoffmann AA (1999) Microbe-induced cytoplasmic incompatability as a mechanism for introducing transgenes into arthropod populations. Insect Mol Biol 8:243–255

Vernick KD, Collins FH, Gwadz RW (1989) A general system of resistance to malaria infection in *Anopheles gambiae* controlled by two main genetic loci. Am J Trop Med Hyg 40:585–592

Warburg A, Touray M, Krettli AU, Miller LH (1992) *Plasmodium gallinaceum* — antibodies to circumsporozoite protein prevent sporozoites from invading the salivary glands of *Aedes aegypti*. Exp Parasitol 75:303–307

Wood RJ, Ouda NA (1987) The genetic basis of resistance and sensitivity to the meiotic drive gene *D* in the mosquito *Aedes aegypti* L. Genetica 72:69–79

Wood RJ, Cook LM, Hamilton A, Whitelaw A (1977) Transporting the marker gene *re* (red eye) into a laboratory cage population of *Aedes aegypti* (Diptera: Culicidae) using meiotic drive at the M^D locus. J Med Entomol 14:461–464

Woodring J, Davidson EW (1996) Biological control of mosquitoes. In Beaty BJ, Marquardt WC (eds) The Biology of Disease Vectors. University of Colorado Press, Niwot, pp 530–563

Xiong B, Jacobs-Lorena M (1995) Gut specific transcriptional regulatory elements of the carboxypeptidase gene are conserved between black flies and *Drosophila*. Proc Natl Acad Sci USA 91:9313–9317

Zhao Y-G, Eggleston P (1998) Stable transformation of an *Anopheles gambiae* cell line mediated by the *Hermes* mobile genetic element. Insect Biochem Mol Biol 28:213–210

Zheng L, Cornel AJ, Wang R, Erfle H, Voss H, Ansorge W, Kafatos FC, Collins FH (1997) Quantitative trait loci for refractoriness of *Anopheles gambiae* to *Plasmodium cynomolgi* B. Science 276:425–428

19 Deploying Transgenic Arthropods in Pest Management Programs: Risks and Realities

Marjorie A. Hoy

CONTENTS

19.1 Introduction 336
19.2 Genetic Manipulation of Pest Arthropods 337
    19.2.1 The SIRM Is the Only Genetic Control Tactic That Has Been Deployed 337
    19.2.2 Quality Control Problems Limit Efficacy and Increase Costs
        of Genetic Control Programs 338
    19.2.3 Complex Methods of Genetic Pest Control Have Not Been Deployed 339
19.3 Genetic Manipulation of Beneficial Arthropods 339
    19.3.1 Two Main Implementation Strategies Are Used with Genetically
        Modified Natural Enemies 340
    19.3.2 Genetic Manipulation and Deployment of Natural Enemies 340
        19.3.2.1 Strains with Complex Traits: A Nondiapausing Strain
            of Gypsy Moth 340
        19.3.2.2 Quality Control Issues: Inadvertent Selection for Nondiapause 341
        19.3.2.3 Genetic Improvement of *Metaseiulus occidentalis* 341
        19.3.2.4 Appropriate Deployment Models 344
        19.3.2.5 Driving Genes into Populations with *Wolbachia* 344
19.4 Transgenic Arthropods for Pest Management Programs 345
    19.4.1 A Transgenic Predatory Mite as a Model System for Assessing Risks 346
    19.4.2 Regulatory Agencies and Reviews of a Transgenic Strain
        of *Metaseiulus occidentalis* 348
    19.4.3 Results of Releasing a Transgenic Arthropod in Florida in 1996 350
    19.4.4 Some Responses from the Media 351
19.5 Potential Risks, Real Risks, and Public Perceptions 358
    19.5.1 Horizontal Gene Transfer 352
        19.5.1.1 Estimates of Frequency 353
        19.5.1.2 Transfer of Antibiotic- or Pesticide-Resistance Genes 353
        19.5.1.3 Horizontal Gene Transfer between Microorganisms
            within the Insect Gut 354
        19.5.1.4 Horizontal Gene Transfer by Transposons 354

19.5.1.5 Horizontal Gene Transfer by Feeding on Exogenous DNA 355
19.5.1.6 Movement of Host Genes into Symbionts and Parasites......................... 356
19.5.2 Can We Predict the Outcomes of Releasing Transgenic Insects with Models?......356
19.6 Laboratory Containment of Transgenic Arthropods..358
19.6.1 Uniform Containment Facilities and Procedures 358
19.6.2 Variable Risks of Accidental Releases of Transgenic Arthropods..........................359
19.7 Conclusions 359
Acknowledgments 361
References 362

19.1 INTRODUCTION

The development of molecular methods for genetic manipulation of arthropods has created exciting new opportunities for altering the genomes of both pest and beneficial arthropods (for example, Hoy, 1994; Ashburner et al., 1998). Authors in this book have developed elegant molecular genetic methods for inserting exogenous DNA by gene targeting or the use of viral or transposable element vectors. Others have identified interesting genes and regulatory elements that have the potential for conferring useful traits upon specific transgenic arthropods in a tissue-specific manner.

Unfortunately, the breakthroughs in the technology of transforming arthropods by recombinant DNA technology have not been matched with significant breakthroughs in our understanding of how to deploy these genetically manipulated arthropods in practical pest management programs or how to evaluate the potential risks associated with their permanent release into the environment (Spielman, 1994; Hoy, 1995; Ashburner et al., 1998). Significant efforts will have to be made to develop the data and resources to deploy the new transgenic strains that are being, and will be, developed.

The new transgenic technology has created both excitement about these new opportunities and concerns about the lack of knowledge (NABC, 1994; 1997; Abbott, 1996; Hoy et al., 1997). The possibility of controlling pests without the use of toxic pesticides, allowing pest managers to reduce crop losses in a world with a burgeoning population or to eliminate dreaded diseases that deplete the energy and productivity of millions of people around the world, is exciting. Yet there are concerns that biotechnology will yield a "bitter harvest," creating social and environmental problems (Goldberg et al., 1990; Abbott, 1996). Concerns about environmental risks are not unique to transgenic arthropods; concerns extend even to the potential risks of releasing nontransgenic natural enemies into the environment for classical biological control programs or nontransgenic pest arthropods for genetic control programs (Howarth, 1991; Simberloff, 1992; Goodman, 1993; Gubler, 1993; Ruesink et al., 1995; Simberloff and Stiling, 1996; Thomas and Willis, 1998; Ewell et al., 1999; Malakoff, 1999). The concern about invading species and biodiversity has heightened in recent years and many people do not want any type of nonindigenous genomes released because of concerns about maintaining biodiversity (Kaiser, 1999). For example, Sir Robert May noted that agriculture historically has reduced biodiversity, and that "The thrust of GM [genetically modified] crops is to accelerate this trend" (Anon., 1999a). Similar concerns have been raised about the release of arthropod biological control agents for classical biological control, and "...we're being challenged to meet tougher standards" for both safety and effectiveness (Malakoff, 1999).

Many questions need to be answered before we can safely release transgenic arthropods into the environment in practical pest management programs: What is the probability that the transgenic insects (released permanently into the environment) will create future environmental problems? Will transgenes inserted into insects somehow be transferred horizontally through known or currently unknown mechanisms to other species to create new pests? Can we develop mitigation methods or techniques for retrieving transgenic insects from the environment after their release should they perform in unexpected ways? The issues surrounding potential risks will require both researchers and regulatory agencies to accept new responsibilities and conduct creative research if the transgenic arthropods are to be deployed safely and appropriately in pest management programs (Hoy, 1995; Hoy et al., 1997).

One objective of this chapter is to review a small portion of the history involving genetic manipulation of pest and beneficial arthropods using traditional genetic methods because those who are unfamiliar with history may be forced to repeat it. Second, the chapter will review some of the deployment models and the empirical data obtained to support or reject them while working with genetically manipulated parasitoids and predators in integrated pest management (IPM) programs. These models or issues may be relevant to the deployment of transgenic arthropods. Third, the chapter will speculate on some of the logistical and risk issues that may be unique or new to the deployment of transgenic arthropods in pest management programs, using as examples some questions that arose during the development and release of a transgenic strain of the predatory mite *Metaseiulus occidentalis* (Acari: Phytoseiidae). Finally, the chapter reviews a recommendation regarding the facilities and procedures for containing transgenic arthropods prior to their release into the environment.

19.2 GENETIC MANIPULATION OF PEST ARTHROPODS

Genetic manipulation of pest arthropods is not a new concept (Laven, 1967; 1969; Curtis and Hill, 1971; Pal and Whitten, 1974; Whitten, 1979). One of the first attempts to control a pest by genetic methods involved crossing *Culex pipiens* mosquitoes from different geographic areas, which often produced eggs that did not hatch due to what we now know was the presence of different strains of *Wolbachia* endosymbionts (reviewed by Yen and Barr, 1974). The sterile insect release method (SIRM; also known as sterile insect technique, or SIT), first applied to control of the screwworm in the United States, was highly successful and provided a strong additional impetus to the use of genetic methods for suppressing pest arthropods (LaChance, 1979; see Chapter 17 by Robinson and Franz).

19.2.1 THE SIRM IS THE ONLY GENETIC CONTROL TACTIC THAT HAS BEEN DEPLOYED

In the SIRM, pest arthropods are genetically modified, mass-reared, and released to control agricultural or medical–veterinary pests. Males are sterilized (usually by γ irradiation), released repeatedly, and expected to compete with wild males in mating with wild females. Ideally, no progeny are produced from the matings between sterile males and wild females, or the progeny exhibit reduced fertility. Ideally, females mate only once. Because irradiated males have extensive somatic damage from irradiation, they are known to be less vigorous than wild males (LaChance, 1979). To compensate for this reduced fitness, program managers attempt to release more sterile males than wild males, usually by a factor of 100 to 1. Operational SIRM programs therefore are large, expensive, and carried out with public sector funding. One of the most successful programs is exemplified by the screwworm *(Cochliomya hominivorax)* eradication program, which involved the U.S. Department of Agriculture, several states, and countries in Central America. Early estimates of benefits accruing from the program ranged from $39 to $113 for each U.S. dollar spent (LaChance, 1979).

The SIRM concept has been applied to approximately 30 pest insects. Some SIRM projects have involved large-scale pest management programs, while others have consisted only of experimental field trials. SIRM program managers have struggled with immense behavioral, genetic, and logistical problems to achieve their successes, as summarized by Graham (1985):

> The concept that pest insects could be reared, sexually sterilized, and released to control or even to eradicate entire populations of their own species has been said to be the only truly original innovation

in insect control in this century. On first examination the method appears so simple and so sure that one might suppose that its application would be problem-free, almost magical in its ease of use. It is true ... that sterile insect releases have been used to eradicate the screwworm from a large portion, probably 90%, of its former distribution in North America, but this success has been achieved only because a long succession of problems has been confronted and solved....Those close to the screwworm eradication program know that it has been a program of crises. Time and time again program managers have had to cope with one crisis after another, each one of which seemed to doom the program to fail. Many of these crises have been nerve-racking, soul-trying, ulcer-producing crises.

Because very detailed knowledge of the target pest species is required, genetic control is likely to be successful for pest species with very significant health, economic, or ecological effects. The SIRM is a highly sophisticated technique that requires a thorough understanding of the dynamics, behavior, and ecology of the target pest population. For example, it is most economical when the female mates only once. It demands precision in its practical application. For maximum effect, it should be applied when a pest population exists naturally at a low level or after the population has been suppressed by other means because the efficiency of the method is inversely correlated with the density of the natural population.

19.2.2 Quality Control Problems Limit Efficacy and Increase Costs of Genetic Control Programs

One of the enduring problems of the SIRM (or any other program that involves mass rearing of arthropods) has been quality control (Boller, 1979; Bush, 1979). For example, Bush et al. (1975) identified a serious problem in the screwworm project in 1972 after screwworm *(C. hominivorax)* infestations, which had dropped from 50,000 in 1962 to less than 1000 in 1971 in the southeastern United States, suddenly jumped to more than 55,000 in 1972. This increase occurred despite the fact that the number of sterile flies released was increased from less than 2 billion per year in 1962 to about 10 billion in 1974. Bush et al. (1975) found a number of differences between the wild and factory-reared screwworm flies, including an enzyme (GDH) which is important in controlling flight activity. The GDH enzyme in the factory flies was of one type (GDH_2) that performed well at high temperatures, while the wild population retained the alternative form (GDH_1). The GDH_2 protein is less active in the temperature range experienced in nature. The GDH_2 form had apparently been selected for by the high temperatures used to speed up development in the factory. Because mating occurs in the air, Bush (1979) suggested that the factory-reared males would be at a disadvantage in competing for mates. Wild females were active from early morning to late afternoon but factory-reared females were not active until early afternoon.

Common quality problems include producing individuals that are diseased, do not respond to the same sexual stimuli, respond differently to daily light/dark cycles, or are more or less active than wild insects (King and Leppla, 1984). Other quality problems are due to inadequate artificial diets, insufficiently controlled environments, poorly trained personnel, and contaminants. Genetic changes have been found repeatedly in mass-reared insects due to their adaptation to the rearing environment; changes may include loss of genetic variability or the development of undesirable attributes due to inadvertent selection.

SIRM programs could fail if wild females are able to detect differences between wild and released males (LaChance, 1979). Several examples of apparent field-selected resistance to sterile males have been reported. For example, Hibino and Iwahashi (1991) reported that wild melon fly females became unreceptive to sterilized males in Okinawa. LaChance (1979) pointed out that there is a major difference in the risk of developing resistance between eradication and suppression programs. Eradication programs typically are of relatively short duration, whereas suppression programs may be carried out over many years. Resistance to SIRM programs could occur, especially in suppression programs, due to changes in the mating time, periodicity, behavior, host plants, or other behavioral differences between the colonized insect and the natural population (LaChance, 1979).

SIRM programs are not initiated without extensive analysis of the potential for success and the economics. Furthermore, such programs have been most successful where the native populations have occupied rather small, well-defined areas, perhaps because of logistical difficulties in mounting truly large-scale release programs (Wallace, 1985).

19.2.3 COMPLEX METHODS OF GENETIC PEST CONTROL HAVE NOT BEEN DEPLOYED

The induction of chromosomal aberrations (translocations), hybrid sterility, genetic incompatibility, and meiotic drive to suppress pest arthropods are genetically more complex manipulation methods than the SIRM (Curtis, 1979; Whitten, 1979). None have been used successfully in pest management programs, although some manipulated strains have been tested experimentally in the field. Wallace (1985) reviewed these proposed genetic control tactics and concluded:

> ... as I view the many problems facing the elaborate genetic and cytogenetic control methods, I am struck by the simplicity 1) of the sterile male release program where one merely overwhelms the target organism and 2) of the use of predators and parasites whose numbers grow with their success without requiring further extensive intervention by man. The problems facing the more sophisticated control programs are of the sort that can be of overwhelming importance in the field, but which often are not obvious in preliminary "pencil and paper" analyses.

Difficulties in implementing genetic control projects often are due to a lack of knowledge of the population biology and genetics of the target species, as noted by Curtis (1979):

> It must be anticipated that in many cases exotic populations would not be physiologically adapted to local conditions and/or that there would be behavioral barriers to cross-mating in the field with the local population, even where there are no apparent behavioral barriers in the laboratory. Furthermore, slight differences in mate recognition systems are likely to be enhanced by natural selection if those wild females which mate with released males produce no progeny or sterile progeny. Finally, one must consider the possibility that if an exotic population became established, it could be more noxious than the local one. For all these reasons, it should be a general rule to incorporate as much as possible of the genome of the local population into the release material, leaving only the necessary amount of exotic genome to produce sterility.

The diverse logistical and biological difficulties encountered in the SIRM and other genetic control programs no doubt will be found in genetic control programs involving transgenic arthropods. However, by learning from these past examples, it should be possible to resolve them.

19.3 GENETIC MANIPULATION OF BENEFICIAL ARTHROPODS

Genetic improvement of silkworms and honeybees by artificial selection or heterosis has been practiced for thousands of years and continues today with both traditional and transgenic methods. Genetic manipulation of natural enemies (parasitoids or predators) also has a long history (Beckendorf and Hoy, 1985), but the early projects primarily focused on determining whether or not it was possible to select natural enemies under laboratory conditions. The "improved" strains were not evaluated in the field or employed in practical pest management programs (Hoy, 1976). As a result, the feasibility of using genetically modified natural enemies was doubted by most pest management specialists. As with genetic control programs, *maintaining* quality is considered a serious issue in rearing natural enemies — so much so that it was considered unlikely that a natural enemy strain could be *improved*.

After 1973, several projects began to focus more on learning how to deploy genetically modified natural enemies in pest management programs than on determining if it is possible to select new strains (Hoy, 1985b; 1992c). The majority of these projects involved selecting a predator or parasitoid of a secondary pest for resistance to pesticides using traditional selection methods (Hoy, 1990). The resistant natural enemy strain was then released so that it could establish and survive the pesticides that were applied to control a primary pest that could not be controlled by any other method.

19.3.1 Two Main Implementation Strategies Are Used with Genetically Modified Natural Enemies

Implementation or deployment typically has used one of two strategies: the improved natural enemy can be deployed by *inoculation*, which means the new strain is released one or more times into the environment where it is expected to establish and persist. This strategy has the advantages of reducing rearing costs by requiring fewer individuals to release and the likelihood that fitness of the individuals reared in the field will be high.

The inoculation strategy has at least three possible population genetic mechanisms by which it can be achieved:

1. The new strain is released into a new environment where no native populations occur *(open niche)*, and it establishes and provides long-term control, even possibly undergoing postrelease genetic adaptations to the local environment.
2. The new strain is released into the environment after the native population is greatly reduced (usually after pesticide applications or severe winter conditions) and the new strain *replaces* the old, providing long-term control, especially if the wild population is unable to recolonize the release sites.
3. The new strain is released into the environment and, through *introgression and selection* in the field (perhaps with a pesticide), a new adapted strain is produced that is fit, vigorous, and able to persist.

The second strategy, *augmentation*, involves mass rearing the new strain and releasing it periodically because it is not expected to persist in the environment. Augmentation is particularly appropriate for greenhouse systems or other crops of high value and short persistence where the high costs of periodic releases can be justified. However, it is difficult to imagine that natural enemy augmentation would be cost-effective or technically feasible over vast acreages of relatively low-value crops, such as wheat or corn in the United States. Costs of present-day rearing and release technologies need to be reduced to approximately 1/100 of current costs to be competitive under current social policies (Cohen et al., 1999).

19.3.2 Genetic Manipulation and Deployment of Natural Enemies

Genetic manipulation projects involving parasitoids and predators using traditional selection and hybridization methods illustrate some of the issues that must be considered if a transgenic natural enemy is to be deployed in practical pest management programs. The complexities of implementing a pest management program, especially the importance of a detailed understanding of the population dynamics, population genetic structure, and behavior of arthropods in the field, becomes obvious from a review of several such projects and will be relevant when transgenic natural enemies are developed.

19.3.2.1 Strains with Complex Traits: A Nondiapausing Strain of Gypsy Moth

Modifying traits determined by multiple genes can have unexpected outcomes. Efforts to improve the effectiveness of parasitoids of the gypsy moth, *Lymantria dispar* (Hoy, 1975a,b) required that we have a method to rear the parasitoids throughout the year under laboratory conditions. This was made difficult because gypsy moth larvae were unavailable during several months due to an obligatory egg diapause. That problem was solved by selecting a nondiapausing strain (Hoy, 1977). The deployment of this nondiapausing gypsy moth strain in a genetic control program was suggested by Dr. E. F. Knipling before we realized that the nondiapause trait was not dominant (Lynch and Hoy, 1978). Furthermore, the "nondiapause trait" had no detectable fitness costs on overwintering eggs; eggs of the wild type, the nondiapausing strain, and their reciprocal F_1 hybrids survived equally well in field cages throughout the winter in Connecticut (Hoy and Knop, 1978). Thus, genetic manipulation yielded a strain with a shortened diapause interval, but cold hardiness was unaffected, presumably because it is determined by different genes.

19.3.2.2 Quality Control Issues: Inadvertent Selection for Nondiapause

Heterosis has been proposed as a mechanism to enhance the effectiveness of augmentative releases of natural enemies, and to evaluate this three different populations of the gypsy moth parasitoid *Apanteles melanoscelus* were crossed. As expected, the hybrid strain was easier to rear in the laboratory because it had a higher fecundity and improved sex ratio (Hoy, 1975a,b). Unexpectedly, we discovered a quality control problem in the colonies from France and Yugoslavia, which had been reared by another laboratory for inoculative releases into the leading edge of the gypsy moth infestation. Both apparently had been subjected to selection for nondiapause while being reared under a short daylength, making them unlikely to be able to establish permanently.

19.3.2.3 Genetic Improvement of *Metaseiulus occidentalis*

Genetic improvement of the predatory mite *M. occidentalis* was relatively easy to plan. We knew it was an effective predator of spider mites in deciduous orchards and vineyards in the western United States (Hoyt, 1969), so the development of additional pesticide-resistant strains might be expected to result in improved pest management. A strain of *M. occidentalis* had developed a useful level of resistance to organophosphorus (OP) insecticides under field selection and was more effective than the OP-susceptible strain in an IPM program in Washington State apple orchards (Hoyt, 1969). The OP-resistant strain had been transferred successfully to southern California apples (Croft and Barnes, 1971), indicating that this new strain could be released and established in new environments.

Laboratory selections with carbaryl and permethrin were both successful (Roush and Hoy, 1981a; Hoy and Knop, 1981), and laboratory tests failed to demonstrate biological differences between the carbaryl-resistant strain and the wild population (Roush and Hoy, 1981b). Even after crossing the carbaryl-resistant strain with a sulfur-OP resistant strain collected from the field and selecting for all three resistances (COS strain), detectable fitness costs were never identified (Hoy, 1984). The COS strain was subsequently deployed in an IPM program in almonds.

The large-scale deployment of the COS strain of *M. occidentalis* in California almond orchards required that several problems be solved during the 3 years of small-scale field trials. Mass production methods had to be improved (Hoy et al., 1982), and issues about release rates and release patterns had to be resolved (Hoy, 1982). Typically, the COS predators were released early in the growing season when native *M. occidentalis* populations were low and after an application of carbaryl had reduced the native population to a very low level. We assumed that the COS strain would replace the wild strain, and field results appeared to confirm that assumption.

The releases could not have been successful if additional research had not been conducted: predator and spider mite populations had to be monitored to ensure the spider mites did not cause excessive damage before the progeny of the released predators could control them. A simple, inexpensive, and rapid monitoring system was required that was grower-friendly. A remedy was required if the predator populations lagged too far behind the prey population because growers would not tolerate defoliation of almond trees by spider mites (Wilson et al., 1984; Zalom et al., 1984; Hoy, 1985a). The remedy was to apply a lower-than-label rate of a pesticide that was relatively

TABLE 19.1
Important Questions to Answer When Developing a Genetic Manipulation Project if It Is To Be Deployed Successfully

PHASE I. Defining the Problem and Planning the Project
- What genetic trait(s) limit effectiveness of beneficial species or might reduce damages caused by the pest?
  - Do we know enough about the biology, genetics, behavior, ecology of the target species to answer this question?
  - Is the trait to be modified determined by one or multiple genes?
- Can alternative control tactics be made to work effectively and inexpensively, and are they environmentally friendly?
- Can agencies be found to support the high costs and long duration of genetic manipulation projects?
- How will the genetically manipulated strain be deployed?
  - Will it be released inoculatively or augmentatively?
  - Which inoculative model will be employed: replacement, introgression with selection, new environment?
- What risk issues, especially of transgenic strains, should be considered in planning?
  - Can genes other than pesticide-resistance genes be used as selectable markers?
  - What is known about the potential for horizontal gene transfer in the target species?
  - If TEs or viral vectors or transgenic symbionts are used, what risks might they pose if the transgenic strain is released?
  - What health or other hazards might be imposed on human or animal subjects if the transgenic strain were released?
- What advice do the relevant regulatory authorities give regarding your plans to develop a transgenic strain?
  - Which agencies are relevant for your project?
  - Have you consulted these regulatory agencies early in the planning phase?

PHASE II. Developing the Genetically Manipulated Strain and Evaluating It in the Laboratory
- Where will you get your gene(s)?
  - Should the transgene(s) sequence be modified to optimize expression in the target species if it is from a species with a different codon bias?
- Is it important to obtain a high level of expression in particular tissues or life stages?
  - Where can you get the appropriate regulatory sequences to achieve appropriate expression levels?
- How can you maintain or restore genetic variability in your selection or transgenesis program after obtaining the pure lines?
  - Will you outcross the manipulated strain with a field population to improve its adaptation to the field or increase genetic variability by some other method?
- What methods can you use to evaluate fitness of the modified strains in artificial laboratory conditions that will best predict effectiveness in the field?
  - Can you compare life table attributes, analyze stability of the trait under no selection, and in competitive population cage studies?
- Do you have adequate containment methods to prevent premature release of the transgenic strains into the environment?
- Do you have adequate rearing methods developed for carrying out field tests?
  - Are high-quality artificial diets available to reduce rearing costs?
- What release rate will be required to obtain the goals you have set?
  - Do you have an estimate of the absolute population density of the target species?
  - What release model are you applying: inundative, inoculative, introgression, population replacement?
- Have you tested for mating biases, partial reproductive incompatibilities, or other population genetic problems?
- If the strain is transgenic, have you obtained approval from the appropriate regulatory authorities to release the strain in the environment or greenhouse?

TABLE 19.1 (continued)
Important Questions to Answer When Developing a Genetic Manipulation
Project if It Is To Be Deployed Successfully

- Can you contain it in the release site?
- Can you retrieve it from the release site at the end of the experiment?
- Can you mitigate if unexpected risks appear?
• How will you measure effectiveness of the modified strain in the field trials?

PHASE III. Field Evaluation and Eventual Deployment in Practical Pest Management Projects
• If the small-scale field trials were promising, what questions remain to be asked prior to deploying the manipulated strain?
    - Are true mass-rearing methods adequate?
    - Is the quality control program adequate?
    Is the release model adequate?
    - Were there unexpected reproductive incompatibilities or other mating biases between the released and wild populations?
• If permanent releases are planned, have all the risk issues been evaluated?
• How will the program be evaluated for effectiveness and potential environmental or other risks?
• Will the program be implemented by the public or private sector?
• What did the program cost and what are the benefits?
• What inputs will be required to maintain the effectiveness of the program over time?

nontoxic to the predators but reduced the spider mite density by about 50%. Application of low rates (about 1/10 the lowest label rate) had two benefits: low rates reduced costs and, because *M. occidentalis* is an obligatory predator, they allowed prey to be retained in the orchard to maintain predator populations. One or two additional applications of carbaryl were applied during the first growing season after release to ensure that the predator population remained resistant.

Studies on dispersal were essential to the successful deployment of the resistant strains. Aerial dispersal studies showed that *M. occidentalis* does not move rapidly into or out of almond orchards (Hoy, 1982; Hoy et al., 1985), so additional applications of carbaryl were required only rarely (Hoy, 1985a). Replacement of the susceptible population with the resistant strain probably was effective because this species has an unusually low dispersal rate (compared with other arthropods). The dispersal data were useful later when planning releases of a transgenic strain in Florida. This low mobility was a disadvantage, however, to growers hoping to get free resistant predators; they had to purchase the COS population from commercial producers if they wanted to use carbaryl in their pest management program.

To facilitate implementation by growers, the COS strain was provided free to commercial producers who wanted to sell it and we provided information on rearing methods and a list of the pesticides that could be used safely with the COS strain — it was not resistant to all pesticides! Adoption of the integrated mite management program in California almond orchards required that we work closely with growers, cooperative extension agents, pest control advisors, and commercial producers during an additional 3 years of large-scale field trials. An economic analysis conducted by Headley and Hoy (1987) found the project was successful in reducing production costs in almonds because the program allowed growers to reduce the number of pesticide applications for spider mite control, and, despite the costs of monitoring and purchasing resistant predators, economic benefits were estimated to be at least $20 million per year. The potential health benefit to workers and the environment of not applying these pesticides was not calculated.

This project showed that executing a genetic manipulation project can be divided into roughly three phases and all must succeed if a practical pest management program is to be deployed. First, the problem is identified and the project planned. Quality control issues must be dealt with throughout the project. The modified strain must be evaluated in the field to confirm it provides "improved" biological control. Ideally, an economic and environmental analysis should be conducted at the end of the project. The third phase actually required as much, or more, time than the first two phases and cost substantially more. Finally, additional research and development may be required to maintain the program. For example, a new pest may arrive that makes it necessary to use a different pesticide, pesticide resistance may develop in the pest, or cultural or other management practices may change.

19.3.2.4 Appropriate Deployment Models

Selection of *Trioxys pallidus*, a parasitoid of the walnut aphid, for resistance to azinphosmethyl was successful in the laboratory (Hoy and Cave, 1991). Because the resistant strain appeared to be relatively fit in laboratory tests, inoculative releases were made into several walnut orchards in California (Hoy et al., 1990). The deployment model (replacement model of implementation) was based on data from previous releases of a different biotype of this species; a population of this parasitoid from Iran had colonized the entire Central Valley of California within a year or two after its initial release. As a result, we hypothesized that the new strain would establish just as rapidly because we could provide strong selection by applying azinphosmethyl.

A simulation model examining this implementation strategy indicated, however, that some initial assumptions might be incorrect (Caprio et al., 1991). The model confirmed that regional establishment of the resistant strain of *T. pallidus* was promoted by movement of *T. pallidus* between orchards, low survival of the susceptible strain after treatment with insecticides, and minimal refuges from the pesticide treatment. Contrary to expectations, the model suggested that establishment of the resistant strain (i.e., 50% of the individuals are resistant in at least 90% of the orchards) would require at least 5 to 7 years! Several deployment problems were discovered, including the fact that growers often failed to apply pesticides to the entire (very tall) walnut tree, thus reducing selection intensity against the susceptible population. The rapid dispersal of *T. pallidus* was a disadvantage under these circumstances because susceptible individuals in the unsprayed refugia could promptly recolonize the pesticide-treated trees once the residues decayed sufficiently.

In an attempt to evaluate further the deployment model with *T. pallidus*, we used random amplified polymorphic DNA (RAPD) polymerase chain reaction (PCR) methods to discriminate between the resistant and susceptible strains (Edwards and Hoy, 1993; 1995a). Discriminate analysis of six RAPD fragments allowed separation of the resistant and field populations and tests in population cages indicated that the pure resistant and susceptible colonies and their hybrid progeny could be differentiated reliably over eight generations (Edwards and Hoy, 1995a).

RAPD-PCR analysis of *T. pallidus* individuals recovered from three walnut orchards where the resistant strain had been released 1 to 3 years previously suggested that no single deployment model (replacement or introgression) was appropriate (Edwards and Hoy, 1995b). In one site, where releases had been made 3 years previously, some individuals appeared to be genetically similar to the resistant strain, but others appeared to be the result of matings between resistant and susceptible wasps; some apparently pure susceptible individuals also were found. Thus, during the 3 years subsequent to initial releases (which is equivalent to about 30 generations), both replacement and introgression appeared to be occurring. Without the RAPD analysis, this would not have been evident because the phenotype of *T. pallidus* collected from the orchards did not provide evidence for hybridization. Phenotypically, over 80% of the individuals in this orchard were resistant to azinphosmethyl based on discriminating-dose bioassays (Edwards and Hoy, 1995b). In this case, molecular tools allowed us to unveil some of the complexity of the colonization process. Molecular tools such as RAPD-PCR will be important in empirical analyses of other deployment models.

19.3.2.5 Driving Genes into Populations with *Wolbachia*

Releases of transgenic organisms into the environment, especially where native populations exist, may require that some type of mechanism be employed to "drive" the gene into the population.

One proposed drive mechanism is based on the endosymbiont *Wolbachia*, which can confer reproductive advantages on individuals carrying it (Werren 1997; see Chapter 15 by Sinkins and O'Neill). Despite extensive discussion, little field research has been conducted on the role of *Wolbachia* as a drive mechanism for species of insects other than *Drosophila* (Turelli et al., 1992). The potential role of *Wolbachia* as a drive mechanism for *M. occidentalis* was investigated by Johanowicz and Hoy (1999). Like many arthropods, populations of *M. occidentalis* are genetically diverse and appear to have a different genetic architecture in different geographic sites. *Metaseiulus occidentalis* is native to western North America and is widely distributed in deciduous orchards and vineyards. Surveys in California for pesticide resistance in apples, pears, grapes, and almonds indicated that local populations often had distinctive differences in resistances to pesticides in close geographic proximity, apparently due to different local selection pressures (Hoy, 1985b). Crosses between different populations to obtain multiresistant strains uncovered partial reproductive isolation between some of these populations (Hoy and Cave, 1985; 1988) that appeared similar to the reproductive isolation observed in insect populations containing the rickettsial-like endosymbiont *Wolbachia*. Electron microscopic analysis indicated rickettsia-like microorganisms occurred in the ovaries and eggs of *M. occidentalis* (Hess and Hoy, 1982). Once molecular probes for *Wolbachia* became available, we found that some laboratory and field populations of *M. occidentalis* contained *Wolbachia,* while others did not (Johanowicz and Hoy, 1996; 1998).

Replicated population experiments were conducted using an inbred strain of *M. occidentalis* that had *Wolbachia* and a strain derived from it that lacked *Wolbachia* after heat treatment (Johanowicz and Hoy, 1999). Mixed populations were initiated with 10% of the individuals containing *Wolbachia* and 90% of the individuals lacking it (due to heat curing). The frequency of *Wolbachia* was monitored over eight generations, but this strain of *Wolbachia* did not "sweep" through the *M. occidentalis* populations. This *Wolbachia* might be effective if it were introduced in much higher frequency to counteract the significant fitness costs observed from the infection (Johanowicz and Hoy, 1999). However, compensating for these fitness costs by releasing larger proportions (at least 45%) of *Wolbachia*-containing individuals is unlikely to be economical because predator densities may exceed 10 million per acre in orchards or vineyards during the growing season. Additional studies will have to be conducted to determine whether other strains of *Wolbachia* could be used to drive genes into wild populations of *M. occidentalis*.

19.4 TRANSGENIC ARTHROPODS FOR PEST MANAGEMENT PROGRAMS

A benefit of transgenic technology is the ability to insert cloned genes from any prokaryotic or eukaryotic species so that we are no longer limited by the intrinsic genetic variability within a species. There also may be some disadvantages to recombinant DNA methods at present; it is possible to manipulate only traits that are determined by single genes and, because transgenic insect strains typically go through severe bottlenecks during their production (because transgenic strains typically are initiated with single females), the effects of inbreeding and loss of genetic variability will have to be mitigated, especially if the released strain must be adapted to the environmental conditions of the release site. It is straightforward, albeit time-consuming, to perform reciprocal single-pair crosses and subject a number of subsequent generations to appropriate selection for the desired traits to obtain a population that is homozygous for the transgene. However, the stability and expression of the transgene will have to be monitored continuously because the presence of even a few revertants could result in the loss of the transgenic strain over time during the massrearing phase of the program. Instability of transgenes and gene silencing in transgenic plants and mammals have been ongoing problems that could be observed in some transgenic insect strains. Finally, unless the insect species has a high reproductive rate and short generation time, multiplying the transgenic strain to the level that it can be used in a "rearing factory" could require substantial additional time.

Finally, if the transgenic arthropod is to be implemented in a practical pest management program, the potential risks associated with its permanent release into the environment must be assessed; yet guidelines for such an assessment do not at present exist (Hoy, 1992 a,b; 1995; see Tables 19.1 and 19.2). A more general discussion of risk issues will be conducted in Section 19.5, and this section will focus on a model project carried out with *M. occidentalis* that provides insights into some of the issues that will be relevant to other genetic manipulation projects.

19.4.1 A Transgenic Predatory Mite as a Model System for Assessing Risks

After observing the problems that arose during the first releases of transgenic plants and microbes, I considered it desirable that the first release of a transgenic arthropod raise as few concerns about risk as possible. Thus, the organism, gene inserted, and insertion method were chosen to serve as a potential model for beginning the process of assessing potential risks of releasing transgenic arthropods into the environment.

Furthermore, I assumed that a beneficial species would elicit less concern about potential risks than a pest species. After discovering that *P*-element transformation could not be extended to arthropod species outside the genus *Drosophila* (O'Brochta and Handler, 1988), we chose to attempt transformation of the predatory mite *M. occidentalis* without using a transposable element (TE) or viral vector. The construct injected contained the microbial *lacZ* open reading frame with a *Drosophila* heat shock protein 70 promoter and was intended to serve only as a molecular marker. Maternal microinjection was developed as a method for introducing plasmid DNA after we discovered that injection of plasmid DNA into *M. occidentalis* eggs caused very high rates of mortality (Presnail and Hoy, 1992).

Maternal microinjection involves injecting plasmid DNA directly into or near the ovaries of adult females and has the advantage of being a simple and inexpensive method that may be useful for many arthropod species (Figure 19.1). Injected plasmid DNA is transmitted to multiple eggs of *M. occidentalis* females (Presnail and Hoy, 1994). In some transformant lines of *M. occidentalis*, the injected plasmid DNA is inserted into the nuclear genome and stably transmitted to multiple generations in the laboratory (Presnail et al., 1997), but in others the DNA is transmitted in large extrachromosomal arrays (Jeyaprakash et al., 1998). Maternal microinjection

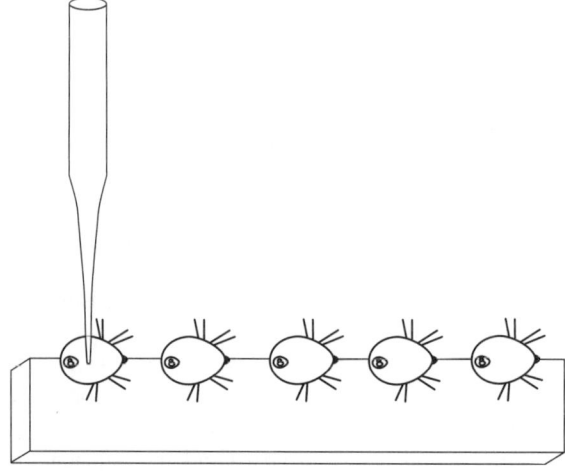

FIGURE 19.1 Maternal microinjection of *M. occidentalis* involves inserting a needle into the body of the adult gravid female. Plasmid DNA is injected into or near the ovaries and is transmitted to multiple eggs by unknown mechanisms.

also has been used to transform a parasitoid wasp (Presnail and Hoy, 1996) and may prove useful for delivering transposable element or viral vectors to diverse arthropod species without causing high rates of mortality.

No native TE vector was identified for inserting plasmid DNA into the chromosomes of *M. occidentalis* because we considered such vectors might pose a small, but unnecessary, risk for horizontal gene transfer. Concerns about horizontal transfer were based on the assumption that *M. occidentalis* may contain active native TEs that could provide the necessary transposase for transposition of disabled vectors. An inactive *mariner* element has been found in *M. occidentalis* (Jeyaprakash and Hoy, 1995) so it is possible that an active form of this TE, or others, could be present in one or more populations. While the plasmid contained *P* element ends, no *P* elements were known to occur in this species (M. Hoy and A. Jeyaprakash, unpublished), and we assumed that illegitimate recombination could insert the plasmid DNA into the genome of *M. occidentalis*.

Once the transgenic strain of *M. occidentalis* was obtained, it was time to consider the risks posed by experimental releases into the environment (Hoy, 1992a,b; 1995). The considerations relied heavily on consideration of potential risks discussed by Tiedje et al. (1989; Table 19.2). I focused on evaluating whether unintended changes in the prey specificity of this natural enemy had occurred as a result of the genetic modification and whether changes had occurred in their climatic tolerance. Tiedje et al. (1989) proposed that risks of transgenic organisms be assigned to four categories: attributes of the unmodified organism, attributes of the genetic alteration, phenotype of the modified organism compared with the unmodified organism, and attributes of the accessible environment (Table 19.2). Their analysis of potential risks is very helpful in identifying potential concerns but, unfortunately, does not provide any mechanism for quantifying the risks.

As noted above, the unmodified organism, *M. occidentalis*, was a beneficial predator of pest spider mites, it had only a molecular marker inserted, the modified strain did not express the inserted marker, and releases into Florida would occur in an environment in which no native populations of *M. occidentalis* existed. *Metaseiulus occidentalis* had been released in very large numbers over many years for biological control of two-spotted spider mites in strawberries, yet had never established permanently in Florida (McDermott and Hoy, 1997).

To confirm that the transgenic strain of *M. occidentalis* was unlikely to establish permanently, we released nontransgenic *M. occidentalis* in an experimental plot on the campus of the University of Florida and confirmed that they failed to survive the very humid summers (McDermott and Hoy, 1997). This species is native to western North America and thrives where summer rains are rare and relative humidities low. Eggs of *M. occidentalis* failed to hatch when relative humidities were greater than 97%, which is common in the summer in Florida. Thus, unless the transgenic strain of *M. occidentalis* had a new and unexpected ability to survive the summers in Florida, it was unlikely to establish permanently. Laboratory tests of relative humidity tolerance later confirmed there were no differences between the transgenic and nontransgenic populations with regard to this attribute (Li and Hoy 1996).

In addition, we evaluated the ability of the transgenic populations to feed on plant materials, including pollen, in the laboratory to confirm that its status as an obligatory predator remained intact. *Metaseiulus occidentalis* preys on spider mites and other small arthropods on plants but normally is unable to reproduce on alternative foods; the transgenic populations likewise were unable to develop or reproduce on pollens or leaf tissues (Li and Hoy, 1996).

An experimental analysis of the potential risks of horizontal gene transfer was not carried out with *M. occidentalis* because such risks were considered to be low and, even if transfer occurred, the consequences of a transfer of the *lacZ* construct were considered minimal. Risk assessments typically consider both the probability of the event happening *and* the potential harm that would ensue if the event happened.

TABLE 19.2
Some Risk Issues to Resolve before Releasing Transgenic Arthropods into Experimental Field Plots

Attributes of the unmodified organism
- What is the origin of the transgenic organism (indigenous or nonindigenous) in the accessible environment?
- What is the arthropod's trophic level and host range?
- What other ecological relationships does it have?
- How easy is it to monitor and control?
- How does it survive during periods of environmental stress?
- What is the potential for gene exchange with other populations?
- Is the arthropod involved in basic ecosystem processes?

Attributes of the genetic alteration
- What is the intent of the genetic alteration?
- What is the nature and function of the genetic alteration?
- How well characterized is the genetic modification?
- How stable is the genetic alteration?

Phenotype of modified organism compared with unmodified organism
- What is the host/prey range?
- How fit and effective is the transgenic strain?
- What is the expression level of the trait?
- Has the alteration changed the organism's susceptibility to control by natural or artificial means?
- What are the environmental limits to growth or reproduction (habitat, microhabitat)?
- How similar is the transgenic strain being tested to phenotypes previously evaluated in field tests?

Attributes of the accessible environment
- Describe the accessible environment, whether there are alternate hosts or prey, wild-type populations within dispersal capability of the transgenic organism, and the relationship of the site to the potential geographic range of the transgenic arthropod strain.
- Are there endangered/threatened species present that could be affected?
- Are there vectors or agents of indirect dissemination present in the environment?
- Do the test conditions provide a realistic simulation to nature?
- How effective are the monitoring and mitigation plans?

Source: Modified from discussions at the workshop on Evaluating Risks of Releasing Transgenic Arthropod Natural Enemies, 1993, Gainesville, FL. The original outline was based on concepts originally identified by Tiedje et al. (1989).

19.4.2 REGULATORY AGENCIES AND REVIEWS OF A TRANSGENIC STRAIN OF *METASEIULUS OCCIDENTALIS*

The first releases of transgenic microbes and plants elicited considerable concern from the public and the press about possible environmental and human health risks, and there is a continuing fear, especially in Europe and Japan, that transgenic organisms and foods may not be safe, suggesting that many people do not trust the risk analyses that have been conducted (Anon., 1999b; Butler et al., 1999; Macilwain, 1999).

The project with *M. occidentalis* was reviewed by regulatory agencies in the United States throughout the program. The first review occurred when I obtained permission from the institutional biosafety committee at the University of California at Berkeley to develop the transgenic strain of *M. occidentalis* containing the *lacZ* marker gene. At the same time, in consultation with the biosafety officer, I developed containment methods to prevent the accidental release of potential transgenic strains.

Before moving the transgenic strain to the University of Florida in 1992, I obtained permission from the state of Florida to import the strain into another containment facility on the campus. Once in Florida, I consulted with the University of Florida institutional biosafety committee, the Florida Department of Agriculture and Consumer Services (Division of Plant Industry), and the U.S. Department of Agriculture Animal and Plant Health Inspection Service (USDA–APHIS) regarding potential risks of releasing the transgenic strain.

Early on, I was unsure which federal agency(ies) I should consult and considered consulting with the Environmental Protection Agency (EPA). However, staff at the USDA–APHIS indicated that they were charged with this responsibility. Officials in the USDA–APHIS were very helpful and, with partial support from the USDA–APHIS National Biological Control Institute, I organized a small workshop in Gainesville in November 1993 to gain additional perspectives on the potential risks of releasing a transgenic natural enemy into the environment (Hoy, 1995). The workshop included people with a variety of viewpoints, including state and federal regulatory officials, sociologists, environmentalists, cooperative extension agents, community ecologists, molecular biologists, and biological control specialists. We considered guidelines for field-testing transgenic plants and microorganisms (Purchase and MacKenzie, 1990) to determine if they were suitable for arthropods.

A major issue raised was whether transgenic arthropods are different from transgenic plants or microorganisms in terms of risk assessment? The majority of the workshop participants concluded that insects probably have more complex interactions in their environment than plants and microbes because natural enemies interact with each other, with their hosts/prey, and with plants, although this viewpoint certainly can be challenged. The participants also discussed whether there were any scientific principles learned from the many releases of transgenic plants and microorganisms that could make risk assessments of transgenic arthropods "generic" rather than on a case-by-case basis. After discussion, the group agreed that each transgenic arthropod strain should be considered individually until a considerable body of information is built up about the effects of releasing such organisms into the environment (see Table 19.2).

Another issue that emerged in the discussions at the workshop was whether it is possible to make temporary releases of transgenic arthropods. One participant argued that risks of releasing transgenic arthropods should be assessed *as though all releases were permanent releases into the environment* because insects typically disperse rapidly and it might be impossible to contain them or retrieve them from the release site. (Exceptions, of course, would include the release of sterile transgenic arthropods that could not establish permanently, or, possibly, insects that contain a "suicide" gene, such as inability to diapause, lack of high or low temperature tolerance, that limits them from surviving or colonizing new environments.) That proposal was discarded by the group because no one wanted to give up the premise that all initial releases of transgenic organisms should be considered short-term experiments that could be terminated should unexpected results occur. At the time this chapter is being completed (April 2000), there still are no guidelines regarding what risk issues must be resolved before permanent releases of transgenic insects can be approved.

After the workshop in November 1993, the USDA–APHIS recognized the need for risk assessments of transgenic arthropods and developed a team charged with this duty in 1995 (see Chapter 20 by Young et al.). All release applications were to be made "transparent" by making them available for comment on the World Wide web at *www.aphis.usda.gov:80/bbep/bp/arthropod/#tgenadoc*. After discussions with the USDA–APHIS scientist assigned to lead this team, Dr. Orrey Young, I filled out an application to release a strain of *M. occidentalis* containing a *lacZ* gene and submitted it to the University of Florida institutional biosafety committee on October 10, 1995. This committee evaluated the application and, contingent upon approval of the application by the USDA–APHIS, approved the release on November 2, 1995. I then submitted the application (APHIS Form 2000) to the USDA–APHIS on November 20, 1995.

The USDA–APHIS consulted with the Division of Plant Industry about potential risks to agriculture, and with the U.S. Department of Fish and Wildlife regarding concerns about threatened and endangered species. Permission (Permit No. 95–326–02r) was obtained from USDA–APHIS on February 23, 1996 to make releases of a transgenic strain of *M. occidentalis*. Releases were approved for a 1-year interval in a single fenced field site on the campus of the University of Florida at Gainesville. As expected, I was required to retrieve and dispose of the released predators, or their progeny, at the end of the test.

One participant at the 1993 workshop noted that public acceptance of transgenic organisms requires considerable efforts to explain their benefits and risks (Asner, 1990). Because the press and public were expected to be interested in the release, the University of Florida public relations office was asked to develop press releases. To assist the public relations office I wrote up a brief statement describing what we had done, why we were doing it, and what the potential risks and benefits might be, and made an effort to avoid scientific jargon. I showed this draft to several people to determine if it was clear and concise; yet when I presented it to the public relations office it indicated the draft was inappropriate because it had been written at a level suitable for high school graduates. Writers in the public relations office indicated the news release should be written at a level suitable for people reading at the eighth-grade level. This requirement created a serious challenge when attempting to explain material that is technically complex.

Because some previous releases of transgenic organisms have elicited injunctions and other legal actions, the university administration and legal office were informed of the impending release. I also considered obtaining personal liability insurance should I be required to obtain legal counsel as a result of a legal challenge to the release. Fortunately, no personal or institutional legal costs were associated with the release of *M. occidentalis*, but other releases of transgenic organisms have engendered substantial legal costs; for example, the University of California at Berkeley had high legal expenses when the ice-minus bacterium was released.

19.4.3 Results of Releasing a Transgenic Arthropod in Florida in 1996

The first field release of a transgenic arthropod, a strain of *M. occidentalis* carrying a molecular marker, was designed to document the persistence of the molecular marker in a field population and the ability to contain the transgenic strain in the release site. The transgenic strain was released in Alachua County, FL for the first time on 10 April 1996, with regulatory authorities from the USDA–APHIS, Florida Division of Plant Industry, and University of Florida institutional biosafety committee present. The fenced release site contained potted bean plants infested with prey (two-spotted spider mites) surrounded by "trap plants" (potted plants lacking prey and treated with a pesticide toxic to *M. occidentalis*). In addition, sticky panels were present around the plot to monitor aerial dispersal of the predators from the plot. Low rates of aerial dispersal were expected because *M. occidentalis* remains on plants with abundant prey if the plant quality remains high. To reduce the likelihood that predators would be carried out of the plot by people, access to the site was limited and workers wore lab coats while working in the plot. The lab coats were treated with alcohol and stored in a container near the release block between sampling intervals.

The colony of mites released, line 18, appeared fit and genetically stable in laboratory tests (Li and Hoy, 1996). The field experiment was terminated after 3 weeks, however, because both predatory mite and prey spider mite populations declined rapidly. This decline occurred in part due to unsuitable climatic conditions. Several days after the initial release, many mites were killed when heavy rains washed mites off the plants. Subsequently, an unexpected late freeze reduced plant quality and further reduced mite populations. Samples of predators recovered from the field after these events were subjected to allele-specific PCR assays to monitor the presence of the transgene in the released line. By the end of this experiment (equivalent to approximately three generations), few individuals in the population contained the transgene, indicating that the transgene in line 18 was unstable under field conditions. This was surprising because line 18 had remained stable in the laboratory for over 150 generations.

Another field release was conducted at the same site on September 27, 1996 and involved six additional transgenic lines of *M. occidentalis* produced by maternal microinjection. Approximately 200 females from each of six colonies were released initially and an additional 200 predators from each line each were released into the plot on October 9, 1996. Predator–prey dynamics and stability of the transgene were monitored as before, with the experiment terminating on November 20, 1996. The results indicated that the transgene again was lost rapidly in all six colonies. We speculated that loss was due to the insertion of tandem multiple-copy arrays, which often results in poor expression, or loss (Henikoff, 1998). As required, the potted plants were placed in large plastic bags at the end of the experiment and carried into the laboratory where they were frozen or autoclaved to ensure that no transgenic predators escaped.

The field trials did set a precedent and demonstrated that the transgenic mites released at the single site could be physically contained by the combination of pesticide residues on trap plants, barren zones, and management of host plant quality and predator:prey ratios. Aerial dispersal of the predators was not detected on the sticky panels surrounding the plot. As predicted by McDermott and Hoy (1997), there was no evidence that the transgenic strains established at the test site. Monitoring during the field trial did not detect the presence of threatened or endangered species.

19.4.4 SOME RESPONSES FROM THE MEDIA

We attempted to follow the advice of people who had released transgenic plants and microbes; they indicated it was a good idea to tell people what you were going to do before doing it. Thus, the University of Florida press office prepared a news release about the application submitted to the USDA–APHIS for permission to release the transgenic strain of *M. occidentalis*. After that first press release and immediately after the actual release of the transgenic *M. occidentalis*, I was interviewed by reporters from newspapers, television, and radio. Each time I tried to explain the goals and the potential benefits transgenic natural enemies might contribute to improving pest management in agriculture and in reducing the use of toxic pesticides in the environment. Some reporters commented that this natural enemy strain appeared to be harmless and was unlikely to create any problems. However, I was surprised by the number of reporters who asked very critical questions in tones that implied that these predators really might be monsters. I also was impressed that, while some reporters were not very knowledgeable about molecular genetics, many did have a fairly sophisticated and detailed understanding of the genes inserted and the methods we employed. Responding to the media became tedious when I had to answer many of the same questions already discussed in the press releases. The media interest was sustained, and some reporters called back weeks or months later to ask for updates on the release. Unfortunately, I have no suggestions on how to make the effort to communicate with the public and the press more effective.

There were more negative responses to the predator releases than I expected. Some negative connotations given to the story are exemplified by newspaper headlines such as "Mutant Bugs: Genetically Altered Heroes or Spineless Menaces?"; "Mite Fakes [sic] Fight Storm of Environmental Controversy"; "UF Entomologist Opens a Can of Mites"; and "Genetically Altered Heroes or Ecological Dynamite?" Some reporters gave me quotes that were attributed to others, including Jeremy Rifkin, perhaps hoping I would respond with a controversial comment. A lawsuit was filed against the USDA by Jeremy Rifkin, president of the Foundation on Economic Trends, to prevent this release (article dated 12/18/95 by Rick Weiss, *Washington Post*). Jeremy Rifkin was quoted in that same news article as stating, "We are playing with ecological dynamite here. These are alive and can reproduce and can mutate from generation to generation very quickly. They can proliferate over large territories and they cannot be recalled after they are released." The Union of Concerned Scientists criticized the review carried out by the USDA–APHIS, especially its use of the world Wide Web Pages as a means of communicating with the public. They wanted the announcement to be published in the *Federal Register* (J. Rissler, personal communication, 1996).

19.5 POTENTIAL RISKS, REAL RISKS, AND PUBLIC PERCEPTIONS

Many people, especially some ecologists, are concerned about potential ecological damage or negative health effects if exotic or genetically engineered organisms are released into the environment. There is a growing concern that releases of any organism, even those intended to be beneficial, could have unexpected negative attributes once they are permanently established in the environment. The potential ecological risks of introducing foreign arthropods, arthropods with novel genomes, classical biological control agents, or transgenic arthropods into the environment have been discussed from different points of view (for example, Lundholm and Stackerud, 1980; Palca, 1988; Tiedje et al., 1989; Ehler, 1991; Howarth, 1991; Hoy, 1992; Simberloff, 1992; Goodman, 1993; Gubler, 1993; Ruesink et al., 1995; Simberloff and Stiling, 1996; Hoy et al., 1997).

The release of pest arthropods, especially mosquitoes, has elicited concern by the public (Palca, 1988). In 1988, an experimental release of marked mosquitoes was blocked by the public in California despite the project goal of improving control of the vector of encephalitis by learning more about the mosquito's dispersal range. Spielman (1994) argued that the use of transgenic mosquitoes for control of malaria has biological, social, and ethical problems and further argued that a transgenic intervention is unlikely to succeed for these reasons. Spielman (1994) noted that one ethical issue includes the fact that the agency responsible for releases may acquire the burden of nurturing a pest and also argued that the proposed drive mechanisms are nonrenewable and thus limited in their usefulness. Naturally, there are very different views on these issues presented elsewhere in this book.

It is urgent that we learn how to evaluate the potential risks of releasing transgenic arthropods into the environment. The possibility of damaging the long-term sustainability of agroecosystems (critical to the world food supply for the estimated 8 to 11 billion people who are projected to inhabit the Earth in 2050) is a serious concern. Such evaluations must occur despite the fact that some critics argue that we cannot yet predict ecological trends 30 to 40 years in the future based on experiments conducted over a few years. For example, Rasmussen et al. (1998) pointed out, "Conclusions based on 10 to 20 years of data can be very different than those based on 50-plus years of data.... Current technology is continually expanding our capability to measure and monitor chemical or biological components that were not possible to measure two or three decades ago."

Because we face uncertainties in ecological evaluations, it seems prudent to exhibit great care in our initial releases of transgenic arthropods. We are just now beginning to understand how genomes evolve and how genetic diversity is developed and maintained. Only a few years ago, we believed that genomes were relatively stable and we had little knowledge of the potential role played by TEs, retroelements, introns, noncoding DNA, and horizontal gene transfer in remodeling the genomes of organisms. Much remains to be learned about genome organization and evolution, but it is becoming clear that foreign DNA sequences inserted into the genome can sometimes serve as a source of variation that is selected on and capable of surviving over a long evolutionary period (Jeltsch and Pingoud, 1996; Britten, 1997; Miller et al., 1997).

19.5.1 HORIZONTAL GENE TRANSFER

The sequencing and analysis of diverse genomes, including the ~13,600 genes in the *D. melanogaster* genome, soon will lead to an increased understanding of the role and rate of horizontal gene transfer in genome evolution. *Drosophila* genome analysis also may serve as a source of new genes for use in genetic manipulation projects (Smaglik, 1999). We already know that horizontal gene transfer has a significant effect on genomes: on the basis of codon usage in *Escherichia coli*, Medigue et al. (1991) predicted that at least 10 to 15% of its genome consists of genes that were horizontally transferred recently and Lawrence and Ochman (1998) confirmed this. Lawrence and Ochman (1998) examined the complete sequence of *E. coli* strain MG1655 and found 755 of 4288 open reading frames (547.8 kb) have been introduced into the *E. coli* genome in at least 234 horizontal gene transfers since the species diverged from the *Salmonella* lineage 100 million years ago. The average age of the introduced genes in *E. coli* was 14.4 million years, yielding a rate of transfer of 16 kb/million years/lineage since divergence. Apparently many of the acquired genes subsequently were deleted, but the sequences that have persisted (approximately 18% of the current genome) "...have conferred properties permitting *E. coli* to explore otherwise unreachable ecological niches" (Lawrence and Ochman, 1998).

Horizontal gene transfer is a natural phenomenon that has occurred in many genomes, including those of arthropods, and can have significant effects on the evolution of species (Droge et al., 1998). Some horizontal transfers can be extreme: the complete genome of *Chlamydia trachomatis*, the agent of trachoma and the most common bacterial sexually transmitted disease in the United States, contains 20 or more *eukaryotic* genes, many of which more closely resemble genes of plants than of animals (Hatch, 1998).

It is likely that additional evolutionary mechanisms will be discovered and new routes of horizontal gene transfer discovered. Will that be relevant or important to risk assessments of transgenic arthropods? In many cases, it will not be. If the gene that is moved horizontally is lost, inactivated, or benign, then harm will be minimized. If, however, the gene confers some type of increased fitness (such as antibiotic or pesticide resistance, or ability to extend the organism's ecological range), then the harm could be greater. If the foreign gene inserts into genes in native nontarget populations that affect fitness of individuals and populations, a subsequent loss of biodiversity could be important if the organism has an essential role to play in the ecosystem.

"Exogenous DNA incorporation into a host genome is of great interest not only in the field of biotechnologies, but also for the understanding of evolutionary mechanisms, since horizontal transfers could lead to the emergence of new species" (Arnault and Dufournel, 1994). Understanding the mechanisms of horizontal gene transfer will help us design appropriate experiments to assess the potential risks of releasing transgenic arthropods into the environment. The following data suggest several mechanisms may exist, but horizontal gene transfer probably involves close contact between the recipient and the donor DNA, and requires uptake by the recipient, and incorporation into the recipient's genome in a stable fashion.

19.5.1.1 Estimates of Frequency

Nielsen et al. (1998) reviewed evidence for horizontal gene transfer from transgenic plants to terrestrial bacteria after 12 years of field releases and 15,000 field trials at different locations. They concluded that experimental approaches in both field and laboratory studies "have not been able to confirm the occurrence of such HGT [horizontal gene transfer] to naturally occurring bacteria although ... recently two studies have shown transfer of marker genes from plants to bacteria based on homologous recombination. The few examples of HGT indicated by DNA sequence comparisons suggest that the frequencies of evolutionarily successful HGT from plants to bacteria may be extremely low." Nielsen et al. (1998) cautioned, however, "this inference is based on a small number of experimental studies and indications found in the literature. Transfer frequencies should not be confounded with the likelihood of environmental implications, since the frequency of HGT is probably only marginally important compared with the selective force acting on the outcome." They argue that "attention should therefore be focused on enhancing the understanding of selection processes in natural environments. Only an accurate understanding of these selective events will allow the prediction of possible consequences of novel genes following their introduction into open environments."

19.5.1.2 Transfer of Antibiotic- or Pesticide-Resistance Genes

The use of a dominant selectable marker facilitates the detection of putative transgenic arthropods, but such genes are difficult to identify for many arthropods for which relatively little genetic information is available. The use of pesticide- or antibiotic-resistance genes as a dominant selectable marker could pose an environmental risk if the transgenic arthropod strain containing such genes is released into the environment and the resistance gene can be taken up by organisms in the environment. Fortunately, the green fluorescent protein (GFP) gene appears to be a useful selectable marker for many arthropods.

While antibiotic-resistance genes are useful for plasmid production in *E. coli,* excising the antibiotic-resistance gene before the transgenic insect is released into the environment would reduce potential risks (see Chapter 3 by Rong and Golic for mechanisms to delete markers). As noted by Pennisi (1998), "Bacteria are promiscuous gene swappers. Their ability to pass genes for antibiotic resistance from one strain to another is legendary." There is great concern that the control of infectious diseases is threatened by the ever-increasing number of bacteria that are resistant to multiple antibiotics (Williams and Heymann, 1998; Witte, 1998). Transfer of antibiotic-resistance genes to some pathogens could be facilitated by a newly discovered acquisition system called the "integron," which is able to capture antibiotic-resistance genes, adhesion protein genes, and toxin genes (Mazel et al., 1998). However, as suggested by Droge et al. (1998), if specific antibiotic-resistance genes already are widespread in natural bacterial communities, the transfer of the corresponding gene would not add something new to the gene pool although it could increase the frequency of horizontal transfer.

19.5.1.3 Horizontal Gene Transfer between Microorganisms within the Insect Gut

Bacteria may be able to take up genes from transgenic microorganisms within insect guts. Genetic manipulation of symbiotic bacteria in the guts of insects has been proposed by several authors, including Beard et al. (1992; see Chapter 16 by Beard et al.). If genes inserted into gut symbionts can be transferred to other microorganisms found in the insect gut, including bacteria normally found on the surfaces of plants, then genes could escape into the environment.

Watanabe and Sato (1998) showed that gene transfer occurred between different bacteria in the guts of insects at a relatively high frequency; five strains of the bacterium *Enterobacter cloacae,* isolated from several species of plants and insects, grew in the guts of silkworm larvae. When insects were fed artificial diet containing the epiphytic bacterium *Erwinia herbicola,* a high frequency of transconjugations occurred between *Ent. cloacae* and *Er. herbicola,* with plasmids being transferred between the insect-associated and plant-associated bacteria. Watanabe et al. (1998) suggested that "gene exchange among epiphytic bacteria occurs commonly in insects as well as on/in plants. Therefore, *in insecta* gene transfer may play an important role in the evolution of plant epiphytic bacteria or insect-resident bacteria." Armstrong et al. (1990) previously found transconjugation between insect-associated bacteria in the digestive tract of the lepidopteran *Peridroma saucia.*

The frequency of horizontal gene transfer will be difficult to quantify, although the frequency of horizontal transfer of antibiotic-resistance genes from transgenic plants to plant-associated or indigenous soil bacteria was estimated prior to the release of the FLAVR SAVR tomato (reviewed by Droge et al., 1998). Using a "worst-case scenario" (all soil microorganisms have a natural transformation system, every fragment of the plant DNA in the soil contains the resistance gene, the transformed gene is expressed after integration, and the gene product is active and stable), horizontal transfer was expected to occur at a frequency of 9×10^5 transformants per acre. In a less risky scenario (in which only 10% of the soil bacteria are transformable, the resistance gene constitutes 10^{-5} part of the plant genome, the gene is not always integrated and expressed), only two transformants per acre would be expected. Droge et al. (1998) concluded that, "Even very rare events may have an ecological impact if the transferred gene increased the ecological fitness of the recipient organism. Hence, the genes encoded by the recombinant DNA pose a potential risk which should be the focus of the biosafety considerations, rather than a natural phenomenon — horizontal gene transfer — itself."

19.5.1.4 Horizontal Gene Transfer by Transposons

Large-scale genomic analyses have shown that active and inactive TEs comprise a large fraction of the genomes of most organisms (Kidwell and Lisch, 1997). TEs have been proposed as a potential "drive mechanism" for inserting genes into pest populations (Carareto et al., 1997), but the use of TEs to drive specific genes into populations raises two concerns: the potential risk of developing resistance to the TE (which would reduce the effectiveness of the program) and the risk of horizontal transfer of the TEs containing transgenes to nontarget arthropods or other organisms.

TEs typically are found both as intact and deleted elements within a species, and the deleted forms may act as repressors of transposition by active forms. The introduction of a novel TE would probably be required to achieve a rapid sweep of the introduced gene because there is strong selection for a mechanism to regulate transposition (Brookfield, 1996; Brookfield and Badge, 1997). TEs are regulated in complex (and usually poorly understood) ways, but it appears that they sometimes can be activated when the organism is challenged or stressed by ultraviolet light, x rays, γ irradiation, or other stressors (Petrov et al., 1995). The efficacy of using a novel TE to insert a specific gene into a wild population would be reduced if the introduction unleashed endemic (suppressed) TEs that would lead to such high levels of genetic damage that the population becomes extinct. Petrov et al. (1995) found that at least four unrelated TEs were mobilized following a dysgenic cross in *D. virilis*.

Many TEs, such as *mariner*, are transmitted both vertically and horizontally. The *mariner* elements are found in planaria, nematodes, centipedes, mites, insects (Robertson, 1997), and, even, in humans (Robertson and Zumpano, 1997; see Chapter 13 by Lampe et al.). Gueiros-Filho and Beverley (1997) introduced the *D. mauritiana mariner* element into the human protozoan parasite *Leishmania major*, thereby demonstrating it could transpose to different kingdoms separated by an evolutionary distance of more than 1 billion years. Handler et al. (1998) used a lepidopteran TE vector to transform the Mediterranean fruit fly, a dipteran. The *mariner* (and other TEs) thus potentially are very powerful transformation vectors for introducing genes into diverse organisms, and the movement of such vectors from the genome of the target species will be deterred if the vector has been genetically disabled. However, if native TEs are present in the genome in the native population with which the transgenic population can interbreed, then transposase could be supplied by the indigenous TEs, leading to instability and the possibility of transspecies and, even, transkingdom movements of genes.

TEs are ubiquitous, present in all genomes, and horizontal movement is a rare event, occurring on a evolutionary timescale (Kidwell and Lisch, 1997). Thus, the frequency of horizontal TE movement from transgenic insects to other organisms may be low. Despite this, consideration should be given to the potential harm that could arise if the specific transgene moved horizontally.

19.5.1.5 Horizontal Gene Transfer by Feeding on Exogenous DNA

Is horizontal gene transfer by feeding a general phenomenon and does it increase the rate of mutation? Arnault and Dufournel (1994) noted that the first assays to measure the effects of feeding calf thymus DNA to *Drosophila* took place in 1937. Numerous mutations and sex-linked lethality were observed following ingestion or injection of foreign DNA from calf thymus, fish sperm, or viruses; the interactions between the genome and foreign DNA are "supposed to occur during chromosome replication by inhibition of the enzymes which participate in the DNA synthesis ... so that a few nucleotides or a chromosome segment are lost, leading to visible or lethal mutations."

Schlimme et al. (1997) detected horizontal gene transfer between bacteria within digestive vacuoles and fecal pellets of the protozoan *Tetrahymena pyriformis*. More than 90% of the fecal pellets contained viable bacteria and conjugational gene transfer increased by three orders of magnitude. Schlimme et al. (1997) conclude that this "micro-biotope provides a selective pressure which might enhance the acquisition of virulence genes in cases of mutual interactions between genetically modified micro-organisms and wild-type pathogens. This finding is important for biosafety considerations." Most insects contain multiple microorganisms, some of which are symbionts, transients, or entomopathogens, and gene transfer between them could result in organisms with new attributes.

Predatory arthropods also may move DNA horizontally while feeding. Houck et al. (1991) suggested that a predatory mite *(Proctolaelaps regalis)* could serve as a mechanical vector of *Drosophila* genes by moving P elements from eggs of one *Drosophila* species to another because the eggs are not always killed by feeding.

19.5.1.6 Movement of Host Genes into Symbionts and Parasites

More than 10% of insect species contain maternally inherited symbionts (bacteria, fungi, or viruses), which either live inside cells and are transovarially transmitted or live outside cells and are transmitted transovum (Hurst et al., 1997). Many symbionts contribute in a positive manner to the physiology and metabolism of the host, although others disrupt survival or fertility of their hosts. There is a possibility that transgenes in arthropods could be picked up by these symbionts or parasites and transmitted to other hosts (see Chapter 16 by Beard et al.).

Jehle et al. (1998) found an insertion mutant of the granulovirus of the codling moth *(Cydia pomonella)* contained a *Tc1*-like transposable element 3.2 kb long. The transposon most likely was inserted into the viral genome during infection of host insects, and Jehle et al. (1998) speculated that baculoviruses could serve as an interspecies vector in the horizontal transmission of insect transposons (see Chapter 14 by Fraser). During an abortive infection of a nonsusceptible host by a baculovirus, transcription and expression of a few viral genes can occur, which makes it possible for a transposon to be mobilized and transferred into the new host genome (Jehle et al., 1998).

Viruses may serve as vectors of foreign DNA. Chiura (1997) quantified the rate at which five marine bacteria could produce virus-like particles and transfer genes to a strain of *Escherichia coli*. Overall average efficiency of virus-like particle-mediated gene transfer was estimated to be between 2.62×10^{-3} and 3.58×10^{-5}. Chiura concluded that the virus-like particles produced by these marine bacteria might be an "important element for ... non-specific generalized horizontal gene transfer towards a broad range of bacterial hosts." Jordan et al. (1998) reported that pantropic retroviral vectors may serve as a general and efficient transformation system for insects. They demonstrated the vectors could transform somatic cells and allow expression of foreign genes in mosquitoes and *Drosophila*. The pseudotyped retroviral vectors are able to enter a wide variety of cells because cellular entry is facilitated through lipid binding and all of the necessary components (viral genome, integrase and reverse transcriptase) enter the cell together. The broad host range of the pantropic retroviruses raises the question of whether there is a risk of an accidental infection of nontarget species.

Iwamura et al. (1991) found diverse sequences in the mouse genome were present in the DNA of their schistosome parasites and they speculated that the "transposition of these mouse repetitive sequences may have occurred in the DNAs of the schistosomes during the intimate contact between parasite and host."

19.5.2 CAN WE PREDICT THE OUTCOMES OF RELEASING TRANSGENIC INSECTS WITH MODELS?

Many types of population and genetic models could be used in attempts to predict what will happen when genetically modified insects are released into the environment in pest management programs. We do not know, however, which model types are most likely to be predictive of the actual outcome of field releases because few models have been evaluated with empirical data.

An example of the difficulties in predicting field results from a mathematical model include three models developed to predict the success or failure of a biological control program involving applications of fungi for control of grasshoppers and locusts (Wood and Thomas, 1999). All three models fit the empirical data well, but one predicted sustained control at low levels after a single pathogen application. The other two models predicted that repeated pathogen applications would be necessary. The results demonstrated that two assumptions made by ecologists and modelers are suspect. First, quantitatively similar models need not give even qualitatively similar predictions (contrary to expectations). Second, the sensitivity analysis of model predictions to parameter variation is not always sufficient to ensure the accuracy of the predictions (Wood and Thomas, 1999).

Some current population models may lack key ingredients, such as partial reproductive isolation. For example, Caprio and Hoy (1995) developed a stochastic simulation model that varied the degree of mating bias between resistant and susceptible strains, diploidy state (diplo- or haplo-diploid), degree of dominance of the resistance allele, and degree to which mating biases extended to the hybrid progeny. The results obtained were somewhat counterintuitive and illustrated the point that models can offer insights into the complexities of population genetics and dynamics that might be overlooked (Figure 19.2). The amount of mating bias significantly affected the rate of establishment of the resistant strain in this model and also interacted significantly with the other three factors (Caprio and Hoy, 1995). As expected, dominance of the resistance allele significantly affected the rate of resistance development; when dominance varied, rates of resistance for diplo-diploid simulations varied more than with haplo-diploid populations. Thus, mating biases could have a large effect on establishing resistant (or other) strains and, in some cases, could prevent their establishment. The common assumption made in models is that all genotypes of a species mate at random, but this assumption may mask a considerable number of important interactions. The efficacy of transgenic insect release programs could be jeopardized if mating biases exist between released and wild populations.

Empirical data generally are lacking to compare the relative usefulness of different model types in predicting insect population dynamics. Theoretical ecologists usually assume homogeneous and continuous populations. Metapopulation models, by contrast, assume that populations exist in patches varying in area, degree of isolation, and quality. Metapopulation biology increasingly is being recognized as relevant to our understanding of population ecology, genetics, and evolution (Hanski, 1998). Recent data, and a variety of metapopulation models, indicate that spatial structure affects populations as much as birth and death rates, competition, and predation.

Metapopulation dynamics may be important in understanding how to deploy transgenic arthropods. For example, a stochastic metapopulation model examined the implementation of a pesticide-

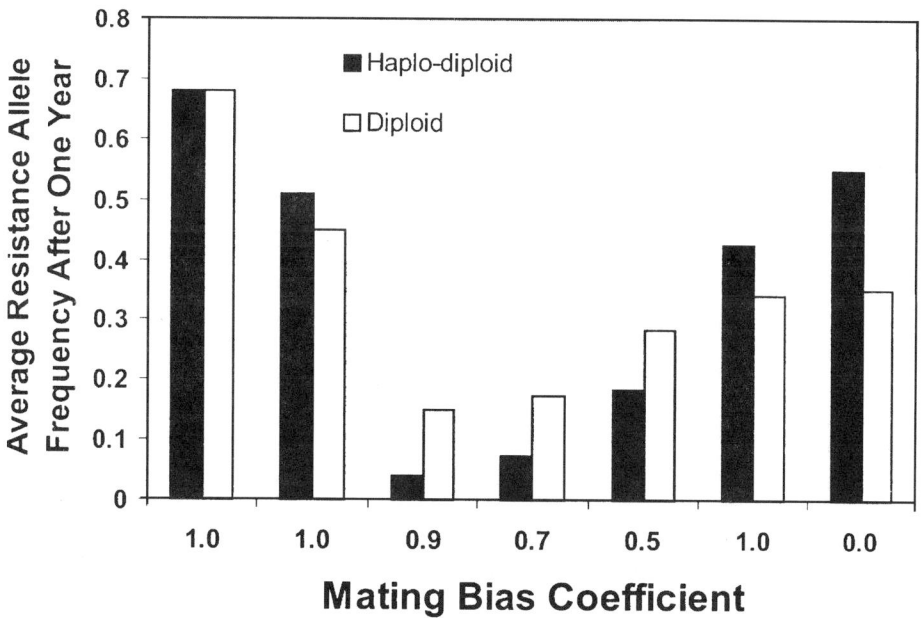

FIGURE 19.2 Mating biases could affect the establishment of transgenic arthropod strains released inoculatively. The mating bias in this simulation model generated frequency-dependent selection and altered establishment rates of pesticide-resistant strains. The mating bias rates varied with the genetic system (haplo-diploid or diplo-diploid). (Redrawn from Caprio and Hoy, 1995.)

resistant strain of the predator *M. occidentalis* and the model results matched the observed prior outcomes of releasing the COS-resistant strain (Caprio and Hoy, 1994). Sensitivity analyses of this model indicated metapopulation dynamics could affect establishment of resistant strains by increasing local homozygosity within patches, and that overwintering may represent a genetic weak point in many life cycles, which could be exploited.

19.6 LABORATORY CONTAINMENT OF TRANSGENIC ARTHROPODS

19.6.1 Uniform Containment Facilities and Procedures

If we assume that every field release of a transgenic arthropod should be conducted only after thorough peer review by scientists and regulatory agencies, efforts to contain transgenic insects and mites in the laboratory prior to their purposeful release into the environment should be effective (Hoy et al., 1997). At present, there are no U.S. or international guidelines on appropriate methods to contain transgenic arthropods in the laboratory prior to their purposeful release into the environment, although participants at an international conference on "Safe Utilization of New Organisms in Biological Control" organized by the Organization for Economic Cooperation and Development (OECD) recommended that such uniform guidelines should be developed (OECD, 1998).

Hoy et al. (1997) suggested that entomologists developing transgenic arthropods should voluntarily adopt the following principles:

1. No releases of transgenic arthropods should be made into the environment without prior evaluation by regulatory authorities and peer reviews by knowledgeable scientists.
2. To reduce the likelihood that accidental releases will occur, transgenic arthropods and putatively transgenic arthropods should be contained in the laboratory by appropriate facilities and procedures.

What facilities and procedures are appropriate and effective? Greenhouses or other general laboratory facilities usually are inadequate to prevent the accidental release of arthropods. The containment facilities for classical biological control projects are certified by the USDA–APHIS and state departments of agriculture and have been designed to contain the beneficial arthropod species until permission has been granted to release them into the environment. The facilities and procedures were developed to prevent the escape of undesired organisms, including pest arthropods, plant pathogens, and hyperparasitoids (Fisher, 1973; Ertle, 1993). Personnel working in these facilities adhere to specific handling and disposal procedures designed to prevent accidental releases. If guidelines for transgenic arthropods are adopted that are equivalent to the containment of beneficial (nontransgenic) natural enemies and their hosts or prey for classical biological control programs, the containment facilities would include the following features:

1. The laboratory for handling transgenic arthropods would be insect-tight and self-contained.
2. Personnel working in the room would be trained in specific handling procedures to minimize escape of transgenic arthropods.
3. There would be minimal movement of equipment into and out of the room to reduce the accidental transport of transgenic arthropods.
4. Air conditioning and heating systems would be made insect-proof with screening or filters to avoid accidental escape.
5. The containment room would have a method to kill the transgenic arthropods (such as an autoclave, incinerator, or freezer), unless permission has been obtained to release them into the environment.

Several *Drosophila* geneticists have argued that it is not necessary to contain transgenic strains of insects (especially *Drosophila*) because such strains are less fit than wild strains and they could not establish in the environment or, if they did, would not be harmful. They noted that the costs associated with providing adequate containment facilities would exclude some scientists from conducting transgenic research and that the containment facilities and procedures are too complex or burdensome. I believe that these arguments are inappropriate until we can convince the public that transgenic arthropods are useful in pest management programs and that the risks associated with their release are minimal.

19.6.2 Variable Risks of Accidental Releases of Transgenic Arthropods

The level of risk associated with accidental (or purposeful) releases of transgenic arthropods will depend upon the species involved, the transgenes and their regulatory sequences utilized, the geographic location of the release site, the presence or absence of conspecific populations, the method by which the transgene was inserted into the arthropod, and the weather conditions at the time of the release (Tiedje et al., 1989; Hoy, 1995).

Accidental releases of transgenic subtropical or tropical insects would not be risky during the winter in temperate climates because the population probably could not establish. Accidental releases of a beneficial arthropod containing a molecular marker should be less risky than release of a pest species containing a functioning transgene. Releases of transgenic arthropods with active TE vectors could be more risky than releases of transgenic arthropods with inactivated TE vectors.

Accidental releases of laboratory populations frequently are considered harmless because the populations are unfit due to laboratory adaptation, genetic drift, inbreeding, inadequate diets, and diseases. However, if the transgenic laboratory strain could interbreed with wild individuals in the environment and transfer the transgene to a wild population, risk could be increased. For example, accidental releases of transgenic *Drosophila* could be risky if the flies could interbreed with wild populations in tropical or subtropical areas and transfer the transgene to the wild population. The accidental release of a transgenic fruit fly carrying an eye color gene is unlikely to elicit much concern, but if it carries a pesticide-resistance gene it might elicit substantially more concern, especially in areas where *D. melanogaster* is an agricultural pest.

Releases of sterilized transgenic insects used in sterile insect release programs should pose a minimal risk as long as the released insects were effectively sterilized and unable to establish permanently in the environment. Containment of the transgenic population during factory rearing and prior to sterilization may be more important, depending on what transgene(s) are present.

The public's concern about transgenic technology requires that we maintain the highest standards of care in containing transgenic insect strains until the issues surrounding their safe and effective deployment in practical pest management programs are resolved. The adoption of uniform procedures and facilities for containing transgenic arthropods in the laboratory should reassure the public that we are responsible and responsive to their concerns.

19.7 CONCLUSIONS

Recombinant DNA techniques may offer exciting new opportunities for managing arthropod populations, but they currently make genetic manipulation more complex and expensive than programs that use traditional genetic methods. The increased cost and developmental time is due to the need to assess potential risks associated with releases of transgenic arthropods into the environment. To date, relatively little critical research has been conducted on the risks and benefits of releasing transgenic arthropods into the environment. Quantification of potential risks is expected to be difficult. However, even if the risk is low, the potential harm that could ensue, should the risk be realized, needs to be considered.

The issue is not *if* transgenic arthropods should be released, but when and how? The debate over evaluation methods and interpretations should include a variety of viewpoints. Ewel et al. (1999) reported the conclusions of a workshop on deliberate introductions of species. The workshop participants did not discriminate between the potential risks of genetically-modified organisms and unmodified organisms introduced into new enviroments. Ewel et al. (1999) discussed the complexities of assessing risks and concluded that the

> Benefits and costs of introductions [of new organisms] are unevenly distributed among ecosystems, within and across regions, among sectors of society, and across generations. Although an introduction may meet a desired objective in one area, at one time, or for some sectors of society, unwanted and unplanned effects may also occur. Introduced organisms can, therefore, simultaneously have both beneficial and costly effects. Furthermore, the relative magnitudes of costs and benefits vary both in space and over time.

Ewel et al. (1999) listed research questions under four headings: guarding against risks without sacrificing benefits, alternatives to introductions, purposeful introductions, and reducing negative impacts. They recommended a single framework for evaluating all types of introductions; a need for retrospective analyses of introductions; a holistic view of the invasion process; and fewer, more effective introductions. Ewel et al. (1999) concluded

> At the extremes, these views [of risks] range from a handful of advocates of no introductions, or of such rigorous pre-introduction proof of benignness that all introductions are effectively prohibited, to an equally small group that advocates a freewheeling global eco-mix of species... [M]ost proponents of purposeful introductions understand the risks (but believe that technology can deal with them), and most conservation biologists recognize the potential benefits to be derived from carefully controlled introductions. Clearly, there is a need to bring all parties together on common ground that can lead to objective, science-based decisions by policymakers.

I wish to echo this recommendation, especially if the debate includes transgenic arthropods, in addition to other transgenic organisms.

Enhanced funding and effort should be devoted to fundamental research on risk assessment methods for transgenic arthropods. Despite my perception that considerable effort has been put into assessing the risks of transgenic plants and microorganisms, Butler et al. (1999) disagreed and indicated that, "The apparent low level of studies into the long-term risks of plant biotechnology seems to result partly from the relatively small sums devoted to risk assessment research in comparison to overall budgets for biotechnology." Butler et al. concluded that a public backlash by consumers has forced action on the labeling of transgenic foods and "could now influence research goals." Butler et al. reported that the head of food safety at the World Health Organization commented that, "It is clear [at least] in Europe that the consumer backlash means that more investment in research [and monitoring] is needed to clarify the scientific uncertainties. Society needs to begin a broad discussion on the risks and benefits of GMOs." By contrast, almost no funding has been allocated to fundamental research on risk assessments of transgenic arthropods.

The potential value of transgenic arthropods to practical pest management problems is exemplified by the social, public health, and economic costs associated with malaria (Curtis and Townson, 1998). Despite enormous efforts, malaria is an increasingly important health problem with at least 500 million people infected and approximately 3 million deaths annually (Crampton, 1994). Will deployment of transgenic mosquitoes unable to vector malaria contribute to a solution? Miller (1989) reviewed malaria control strategies and pointed out that there is unlikely to be "a magic bullet that will eliminate malaria." He noted that "even DDT could not be called such a weapon, at least in retrospect." The strategy of relying on single tactics in pest management, whether it be control of mosquitoes or of agricultural pests, has always failed. The complexity of genetic structure in *Anopheles gambiae* populations in west Africa (Lanzaro et al., 1998), which may be reflective of the complex genetic architecture of other arthropod species, suggests that release programs involving a single transgenic strain are unlikely to be successful. The integration of several compatible tactics has been found to be more sustainable than reliance upon a single management tactic; multitactic management of medically important disease vectors is more likely to be sustainable, as well (Miller, 1989).

Past experience with nontransgenic, genetically manipulated natural enemies (Section 19.3) suggests that the most readily implemented pest management projects employing transgenic natural enemies in biological control will be those where augmentative releases can be conducted and the transgenic beneficial insect used in relatively small areas such as temporary cropping systems, or where the natural enemy has a low dispersal rate and can be established in individual orchards, or where the natural enemy is released into a geographic region where the wild strain does not occur. The most difficult projects to implement are those in which the new biotype is expected to replace the endemic arthropod population. It is likely that projects that require the transgenic strain to replace a wild strain will require whole teams of experts to develop the mass-rearing technology and quality control methods and to provide the necessary information on population structure and hidden partial reproductive isolation mechanisms that are likely to occur.

One field release of a transgenic arthropod has occurred, and it is logical to assume that releases of sterile transgenic insects (such as sterile Mediterranean fruit flies that contain a marker transgene) could occur soon because the sterile flies will be unable to establish permanently in the environment. Unfortunately, there are no guidelines yet for evaluating the risks of permanent releases of nonsterile transgenic arthropods into the environment and it is difficult to predict what issues will be considered. Based on the experiences of companies that developed transgenic plants and microorganisms, it could take 5 to 10 years of evaluating short-term releases of transgenic arthropods before permanent releases are permitted.

Significant and rapid advances are being achieved in transformation of arthropods and the identification of potentially useful genes to insert into pests such as mosquitoes. Knowledge about where to obtain useful genes for improving the effectiveness of parasitoids and predatory insects remains more limited at this time. New opportunities should arise over the next few years, as methods for achieving genetic manipulation of arthropods by recombinant DNA methods are improved and the *Drosophila* genome project is completed. Still, deploying a transgenic arthropod in a pest management program will be an awesome challenge, requiring risk assessments, detailed knowledge of the population genetics, biology, and behavior of the target species under field conditions, as well as coordinated efforts among molecular and population geneticists, ecologists, regulatory agencies, pest management specialists, and sustained efforts to educate the public about the benefits and potential risks of releasing transgenic insects into the environment.

ACKNOWLEDGMENTS

I thank Al Handler and Tony James for inviting me to contribute to this book and the graduate students, postdoctoral scientists, technical support staff, and colleagues who contributed to this research. Al and Tony provided thoughtful comments on an earlier version of the manuscript. Although I accepted most of their suggestions, a few points remain that I believe are subjects deserving discussion by people with differing points of view. Public debate on these issues is both desirable and appropriate. This work was supported in part by the Davies, Fischer and Eckes Endowment in Biological Control and is University of Florida Agricultural Experiment Station Journal Series T-00474.

REFERENCES

Abbott A (1996) Greens attack transgenic plant trials. Nature 382:746

Anon. (1999a) Briefing GM Crops. Assessing the threat to biodiversity on the farm. Nature 398:654

Anon. (1999b) GM foods debate needs a recipe for restoring trust. Nature 398:639

Arnault C, Dufournel I (1994) Genome and stresses: reactions against aggressions, behavior of transposable elements. Genetica 93:149–160

Armstrong JL, Wood ND, Porteous LA (1990) Transconjugation between bacteria in the digestive tract of the cutworm *Peridroma saucia*. Appl Environ Microbiol 56:1492–1493

Ashburner M, Hoy MA, Peloquin J (1998) Transformation of arthropods–research needs and long-term prospects. Insect Mol Biol 7(3):201–213

Asner M (1990) Public relations: the scientist and the public, the government, and the media. In Purchase HG, MacKenzie DR (eds) Agricultural Biotechnology. Introduction to Field Testing. Office of Agricultural Biotechnology, U.S. Department of Agriculture, p 35.

Beard CB, Mason PW, Aksoy S, Tesh RB, Richards FF (1992) Transformation of an insect symbiont and expression of a foreign gene in the Chagas' disease vector *Rhodnius prolixus*. Am J Trop Med Hyg 46:195–200

Beckendorf SK, Hoy MA (1985) Genetic improvement of arthropod natural enemies through selection, hybridization or genetic engineering techniques. In Hoy MA, Herzog DC (eds) Biological Control in Agricultural IPM Systems. Academic Press, Orlando, pp 167–187

Boller EF (1979) Behavioral aspects of quality in insectary production. In Hoy MA, McKelvey JJ (eds) Genetics in Relation to Insect Management. Working Papers, Rockefeller Foundation Press, New York pp 153–160

Britten RJ (1997) Mobile elements inserted in the distant past have taken on important functions. Gene 205:177–182

Brookfield JFY (1996) Models of the spread of non-autonomous selfish transposable elements when transposition and fitness are coupled. Genet Res Camb 67:199–209

Brookfield JFY, Badge RM (1997) Population genetics models of transposable elements. Genetica 100:281–294

Bush GL (1979) Ecological genetics and quality control. In Hoy MA, McKelvey JJ (eds) Genetics in Relation to Insect Management. Working Papers, Rockefeller Foundation Press, New York, pp 145–152

Bush GL, Neck RW, Kitto GB (1975) Screwworm eradication: inadvertent selection for noncompetitive ecotypes during mass rearing. Science 193:491–493

Butler D, Reichhardt T, Abbott A, Dickson D, Saegusa A (1999) Long-term effect of GM crops serves up food for thought. Nature 398:651–653

Carareto CMA, Kim W, Wojciechowski MF, O'Grady P, Prokchorova AV, Silva JC, Kidwell MG. (1997) Testing transposable elements as genetic drive mechanisms using *Drosophila* P element constructs as a model system. Genetica 101:13–33

Caprio MA, Hoy MA (1994) Metapopulation dynamics affect resistance development in a predatory mite. J Econ Entomol 87:525–534

Caprio M, Hoy MA (1995) Premating isolation in a simulation model generates frequency-dependent selection and alters establishment rates of resistant natural enemies. J Econ Entomol 88(2):205–212

Caprio MA, Hoy MA, Tabashnik BE (1991) Model for implementing a genetically improved strain of a parasitoid. Am Entomol 37(4):232–239

Chiura HX (1997) Generalized gene transfer by virus-like particles from marine bacteria. Aquatic Microbial Ecol 13:75–83

Cohen AC, Nordlund DA, Smith RA (1999) Mass rearing of entomophagous insects and predaceous mites: are the bottlenecks biological, engineering, economic, or cultural? Biocontrol News Inf 20(3):85N–90N

Crampton JM (1994) Molecular studies of insect vectors of malaria. Adv Parasitol 34:1–31

Croft BA, Barnes MM (1971) Comparative studies on four strains of *Typhlodromus occidentalis*. III. Evaluation of releases of insecticide resistant strains into an apple orchard ecosystem. J Econ Entomol 64:845–850

Curtis CF (1979) Translocations, hybrid sterility and the introduction into pest populations of genes favorable to man. In Hoy MA, McKelvey JJ (eds) Genetics in Relation to Insect Management. Working Papers, Rockefeller Foundation Press, New York, pp 19–20

Curtis CF, Hill WG (1971) Theoretical studies on the use of translocations for the control of tsetse flies and other disease vectors. Theor Pop Biol 2:71–90

Curtis CF, Townson H (1998) Malaria: existing methods of vector control and molecular entomology. Br Med Bull 54:311–215

Droge M, Puhler A, Selbitschka W (1998) Horizontal gene transfer as a biosafety issue: a natural phenomenon of public concern. J Biotechnol 64:75–90

Edwards OR, Hoy MA (1993) Polymorphism in two parasitoids detected using random amplified polymorphic DNA polymerase chain reaction. Biol Control 3:243–257

Edwards OR, Hoy MA (1995a) Monitoring laboratory and field biotypes of the walnut aphid parasite, *Trioxys pallidus*, in population cages using RAPD-PCR. Biocontrol Sci Technol 5:313–327

Edwards OR, Hoy MA (1995b) Random amplified polymorphic DNA markers to monitor laboratory-selected, pesticide-resistant *Trioxys pallidus* (Hymenoptera: Aphidiidae) after release into three California walnut orchards. Environ Entomol 24(3):487–496

Ehler LE (1991) Planned introductions in biological control. In Ginzburg LR (ed) Assessing Ecological Risks of Biotechnology. Butterworth-Heinemann, Boston, pp 231–239

Ertle LR (1993) What quarantine does and what the collector needs to know. van Driesche RG, Bellows TS Jr (eds) Steps in Classical Biological Control. Thomas Say Publ Entomol, Entomological Society of America, Lanham, MD, pp 53–65.

Ewel JJ, O'Dowd DJ, Bergelson J, Daehler CC, D'Antonio CM, Gomez LD, Gordon DR, Hobbs RJ, Holt A, Hopper KR, Hughes CE, LaHart M, Leakey RRB, Lee WG, Loope LL, Lorence DH, Louda SM, Lugo AE, McEvoy PB, Richardson DM, Vitousek, PM (1999) Deliberate introductions of species: research needs. Benefits can be reaped, but risks are high. BioScience 49:619–630

Fisher TW (1973) Quarantine handling of entomophagous insects, In DeBach P (ed) Biological Control of Insect Pests and Weeds. Chapman and Hall, London, pp 53–65.

Goldberg R, Rissler J, Shand H, Hassebrook C (1990) Biotechnology's Bitter Harvest Herbicide-Tolerant Crops and the Threat to Sustainable Agriculture. Rept Biotech Working Group, 73 pp

Goodman B (1993) Research community swats grasshopper control trial. Science (Washington, DC) 260:887

Graham OH (ed) (1985) Introduction. In Symposium on Eradication of the Screwworm from the United States and Mexico. Misc Publ 62, Entomological Society of America, College Park, MD, pp 1–3

Gubler DJ (1993) Release of exotic genomes. J Am Mosq Control Assoc 9(1):104

Gueiros-Filho FJ, Beverley SM (1997) Trans-kingdom transposition of the *Drosophila* element *mariner* within the protozoan *Leishmania*. Science 276:1716–1719

Handler AM, McCombs SD, Fraser MJ, Saul SH (1998) The lepidopteran transposon vector, *piggyBac*, mediates germ-line transformation in the Mediterranean fruit fly. Proc Natl Acad Sci USA 95:7520–7525

Hanski I (1998) Metapopulation dynamics. Nature 396:41–49

Hatch T (1998) Chlamydia: old ideas crushed, new mysteries bared. Science 282:638–639

Headley JC, Hoy MA (1987) Benefit/cost analysis of an integrated mite management program for almonds. J Econ Entomol 80:555–559

Henikoff S (1998) Conspiracy of silence among repeated transgenes. BioEssays 20(7):532–535

Hess RT, Hoy MA (1982) Microorganisms associated with the spider mite predator *Metaseiulus (= Typhlodromus) occidentalis*: electron microscope observations. J Invertibr Pathol 40:98–106

Hibino Y, Iwahashi O (1991) Appearance of wild females unreceptive to sterilized males on Okinawa Is. in the eradication program of the melon fly, *Dacus cucurbitae* Coquillett (Diptera: Tephritidae). Appl Entomol Zool 26:265–270

Houck, MA, Clark JB, Peterson KR, Kidwell MG (1991) Possible horizontal transfer of *Drosophila* genes by the mite *Proctolaelaps regalis*. Science 253:1125–1129

Howarth FG (1991) Environmental impacts of classical biological control. Annu Rev Entomol 36:485–509

Hoy MA (1975a) Forest and laboratory evaluations of hybridized *Apanteles melanoscelus* [Hym.: Braconidae], a parasitoid of *Porthetria dispar* [Lep.: Lymantriidae]. Entomophaga 20:261–268

Hoy MA (1975b) Hybridization of strains of the gypsy moth parasitoid, *Apanteles melanoscelus*, and its influence upon diapause. Ann Entomol Soc Am 68:261–264

Hoy MA (1976) Genetic improvement of insects: fact or fantasy. Environ Entomol 5:833–839

Hoy M (1977) Rapid response to selection for a non-diapausing gypsy moth. Science 196:1462–1463

Hoy MA (1982) Aerial dispersal and field efficacy of a genetically improved strain of the spider mite predator *Metaseiulus occidentalis*. Entomol Exp Appl 32:205–212

Hoy MA (1984) Genetic improvement of a biological control agent: multiple pesticide resistance and nondiapause in *Metaseiulus occidentalis* (Nesbitt) (Phytoseiidae). In Griffiths DA, Bowman CE (eds) Acarology VI, Vol 2, Ellis Horwood, Chichester, pp 673–679

Hoy MA (1985a) Almonds: integrated mite management for California almond orchards, In Helle W, Sabelis MW (eds) Spider Mites, Their Biology, Natural Enemies, and Control, Vol 1B, Elsevier, Amsterdam, pp 299–310

Hoy MA (1985b) Recent advances in genetics and genetic improvement of the Phytoseiidae. Annu Rev Entomol 30:345–370

Hoy MA (1990) Pesticide resistance in arthropod natural enemies: variability and selection responses. In Roush RT, Tabashnik BE (eds) Pesticide Resistance in Arthropods. Chapman & Hall, New York, pp 203–236

Hoy MA (1992a) Commentary: biological control of arthropods: genetic engineering and environmental risks. Biol Control 2:166–170

Hoy MA (1992b) Criteria for release of genetically improved phytoseiids: an examination of the risks associated with release of biological control agents. Exp Appl Acarol 14:393–416

Hoy MA (1992c) Genetic engineering of predators and parasitoids for pesticide resistance In Denholm I, Devonshire AL, Hollomon DW (eds) Resistance 91, Achievements and Developments in Combating Pesticide Resistance. Elsevier Applied Science, London

Hoy MA (1994) Insect Molecular Genetics. An Introduction to Principles and Applications. Academic Press, San Diego, 540 pp

Hoy MA (1995) Impact of risk analyses on pest management programs employing transgenic arthropods. Parasitol Today 11(6):229–232

Hoy MA, Cave FE (1985) Mating behavior in four strains of *Metaseiulus occidentalis* (Acari: Phytoseiidae). Ann Entomol Soc Am 78: 588–593

Hoy MA, Cave FE (1988) Premating and postmating isolation among populations of *Metaseiulus occidentalis* (Nesbitt) (Acarina: Phytoseiidae). Hilgardia 56(6):1–20

Hoy MA, Cave FE (1991) Genetic improvement of a parasitoid: response of *Trioxys pallidus* to laboratory selection with azinphosmethyl. Biocontrol Sci Technol 1:31–41

Hoy MA, Knop NF (1978) Development, hatch dates, overwintering success, and spring emergence of a "non-diapausing" gypsy moth strain (Lepidoptera: Orgyiidae) in field cages. Can Entomol 110:1003–1008

Hoy MA, Knop NF (1981) Selection for and genetic analysis of permethrin resistance in *Metaseiulus occidentalis*: genetic improvement of a biological control agent. Entomol Exp Appl 30:10–18

Hoy MA, Groot JJR, van de Baan HE (1985) Influence of aerial dispersal on persistence and spread of pesticide-resistant *Metaseiulus occidentalis* in California almond orchards. Entomol Exp Appl 37:17–31

Hoy MA, Cave FE, Beede RH, Grant J, Krueger WH, Olson WH, Spollen KM, Barnett WW, Hendricks LC (1990) Release, dispersal, and recovery of a laboratory-selected strain of the walnut aphid parasite *Trioxys pallidus* (Hymenoptera: Aphidiidae) resistant to azinphosmethyl. J Econ Entomol 83:89–96

Hoy MA, Gaskalla RD, Capinera JL, Keierleber C (1997) Laboratory containment of transgenic arthropods. Am Entomol 43(4):206–209, 255–256.

Hoyt SC (1969) Integrated chemical control of insects and biological control of mites on apple in Washington. J Econ Entomol 62:74–86

Hurst GDD, Hammarton TC, Bandi C, Majerus TMO, Bertrand D, Majerus MEN (1997) The diversity of inherited parasites of insects: the male-killing agent of the ladybird beetle *Coleomegilla maculata* is a member of the Flavobacteria. Genet Res Camb 70:1–6

Iwamura Y, Irie Y, Kominami R, Nara T, Yasuraoka K (1991) Existence of host-related DNA sequences in the schistosome genome. Parasitology 102:397–403

Jehle JA, Nickel A, Vlak JM, Backhaus H (1998) Horizontal escape of the novel Tc1-like lepidopteran transposon TCp3.2 into *Cydia pomonella* granulovirus. J Mol Evol 46:215–224

Jeltsch A, Pingoud A (1996) Horizontal gene transfer contributes to the wide distribution and evolution of type II restriction-modification systems. J Mol Evol 42:91–96

Jeyaprakash A, Hoy MA (1995) Complete sequence of a *mariner* transposable element from the predatory mite *Metaseiulus occidentalis* isolated by an inverse PCR approach. Insect Mol Biol 4:31–39

Jeyaprakash A, Lopez G, Hoy MA (1998) Extrachromosomal plasmid DNA transmission and amplification in *Metaseiulus occidentalis* (Acari: Phytoseiidae) transformants generated by maternal microinjection. Ann Entomol Soc Am 91(5):730–736

Johanowicz DJ, Hoy MA (1996) *Wolbachia* in a predator-prey system: 16S ribosomal DNA analysis of two phytoseiids (Acari: Phytoseiidae) and their prey (Acari: Tetranychidae). Ann Entomol Soc Am 89(3):435–441

Johanowicz DL, Hoy MA (1998) Experimental induction and termination of non-reciprocal reproductive incompatibilities in a parahaploid mite. Entomol Exp Appl 87:51–58

Johanowicz DL, Hoy MA (1999) *Wolbachia* infection dynamics in experimental laboratory populations of *Metaseiulus occidentalis*. Entomol Exp Appl 93:259–268

Jordan TV, Shike H, Cedeno V, Fang Q, Davis BS, Jacobs-Lorena M, Higgs S, Fryxell KJ, Burns JC (1998) Pantropic retroviral vectors mediate somatic cell transformation and expression of foreign genes in dipteran insects. Insect Mol Biol 7:215–222

Kaiser J (1999) Stemming the tide of invading species. Science 285:1836–1839

Kidwell MG, Lisch DR (1997) Transposable elements as sources of variation in animals and plants. Proc Natl Acad Sci USA 94:7704–7711

King EG, Leppla, NC (eds) (1984) Advances and Challenges in Insect Rearing. U.S. Department of Agriculture, Agric Res Serv Publ, Washington, D.C.

LaChance, LE (1979) Genetic strategies affecting the success and economy of the sterile insect release method. In Hoy MA, McKelvey JJ (eds) Genetics in Relation to Insect Management. Working Papers, Rockefeller Foundation Press, New York, pp 8–18

Lanzaro GC, Toure YT, Carnahan J, Zheng L, Dolo G, Traore S, Petrarca V, Vernick KD, Taylor CE (1998) Complexities in the genetic structure of *Anopheles gambiae* populations in west Africa as revealed by microsatellite DNA analysis. Proc Natl Acad Sci USA 95:14260–14265

Laven H (1967) Eradication of *Culex pipiens fatigans* through cytoplasmic incompatibility. Nature 216:383–384

Laven H (1969) Eradicating mosquitoes using translocations. Nature 221:958–959

Lawrence JG, Ochman H (1998) Molecular archaeology of the *Escherichia coli* genome. Proc Natl Acad Sci USA 95:9413–9417

Li J, Hoy MA (1996) Adaptability and efficacy of transgenic and wild-type *Metaseiulus occidentalis* (Acari: Phytoseiidae) compared as part of a risk assessment. Exp Appl Acarol 20:563–574

Lundholm B, Stackerud M (eds) (1980) Environmental protection and biological forms of control of pest organisms. Ecol Bull (Stockholm) 31:1–171

Lynch CB, Hoy MA (1978) Diapause in the gypsy moth: environment-specific mode of inheritance. Genet Res Camb 32:129–133

Macilwain C (1999) US sets up "round-table" talks with scientists. Nature 398:641

Malakoff D (1999) Fighting fire with fire. Science 285:1841–1843

Mazel D, Dychinco B, Webb BA, Davies J (1998) A distinctive class of integron in the *Vibrio cholerae* genome. Science 280:605–608

McDermott GJ, Hoy MA (1997) Persistence and containment of *Metaseiulus occidentalis* (Acari: Phytoseiidae) in Florida: risk assessment for possible releases of transgenic strains. FL Entomol 80:42–53.

Medigue C, Rouxel T, Vigier P, Henaut A, Danchin A (1991) Evidence for horizontal gene-transfer in *Escherichia coli* speciation. J Mol Biol 222:851–856

Miller LH (1989) Strategies for malaria control: realities, magic, and science. In Biomedical Science and the Third World: Under the Volcano. Ann NY Acad Sci 569:118–126

Miller WJ, McDonald JF, Pinsker W (1997) Molecular domestication of mobile elements. Genetica 100:261–270

NABC (1994) Agricultural Biotechnology and the Public Good, National Agricultural Biotechnology Council Report 6, Ithaca, NY

NABC (1997) Resource Management in Challenged Environments, National Agricultural Biotechnology Council Report 9, Ithaca, NY

Nielsen KM, Bones AM, Smalla K, van Elsas JD (1998) Horizontal gene transfer from transgenic plants to terrestrial bacteria — a rare event? FEMS Microbiol Rev 22:79–103

O'Brochta DA, Handler AM (1988) Mobility of P elements in drosophilids and nondrosophilids. Proc Natl Acad Sci USA 85:1068–1080

OECD (1998) Issues and recommendations from a workshop on "Safe Utilization of New Organisms in Biological Control," Montreal, Canada, September 1998, Phytoprotection 79(Suppl):1–155

Pal R, Whitten MJ (eds) (1974) The Use of Genetics in Insect Control. Elsevier/North-Holland, Amsterdam

Palca J (1988) Mosquito release blocked by fearful California residents. Nature 335:7

Pennisi E (1998) Versatile gene uptake system found in cholera bacterium. Science 280:521–522

Petrov DA, Schutzman JL, Hartl DL, Lozovskaya ER (1995) Diverse transposable elements are mobilized in hybrid dysgenesis in *Drosophila virilis*. Proc Natl Acad Sci USA 92:8050–8054

Presnail JK, Hoy MA (1992) Stable genetic transformation of a beneficial arthropod, *Metaseiulus occidentalis* (Acari: Phytoseiidae), by a microinjection technique. Proc Natl Acad Sci USA 89:7732–7736

Presnail JK, Hoy MA (1994) Transmission of injected DNA sequences to multiple eggs of *Metaseiulus occidentalis* and *Amblyseius finlandicus* (Acari: Phytoseiidae) following maternal microinjection. Exp Appl Acarol 18:319–330

Presnail JK, Hoy MA (1996) Maternal microinjection of the endoparasitoid *Cardiochiles diaphaniae* (Hymenoptera: Braconidae). Ann Entomol Soc Am 89(4):576–580

Presnail JK, Jeyaprakash A, Li J, Hoy MA (1997) Genetic analysis of four lines of *Metaseiulus occidentalis* (Nesbitt) (Acari: Phytoseiidae) transformed by maternal microinjection. Ann Entomol Soc Am 90:237–245

Purchase HG, MacKenzie DR (1990) Agricultural Biotechnology Introduction to Field Testing, Office of Agricultural Biotechnology USDA, Washington, D.C., 58 pp

Rasmussen PE, Goulding KWT, Brown JR, Grace PR, Janzen HH, Korschens M (1998) Long-term agroecosystem experiments: assessing agricultural sustainability and global change. Science 282:893–896

Robertson HM (1997) Multiple mariner transposons in flatworms and hydras are related to those of insects. J Hered 88:195–201

Robertson HM, Lampe DJ (1995) Distribution of transposable elements in arthropods. Annu Rev Entomol 40:333–357

Robertson HM, Zumpano KL (1997) Molecular evolution of an ancient *mariner* transposon, *Hsmar1*, in the human genome. Gene 205:203–217

Roush RT, Hoy MA (1981a) Genetic improvement of *Metaseiulus occidentalis*: selection with methomyl, dimethoate, and carbaryl and genetic analysis of carbaryl resistance. J Econ Entomol 74:138–141

Roush RT, Hoy MA (1981b) Laboratory, glasshouse, and field studies of artificially selected carbaryl resistance in *Metaseiulus occidentalis*. J Econ Entomol 74:142–147

Ruesink JL, Parker IM, Groom MJ, Kareiva PM (1995) Reducing the risks of nonindigenous species introductions: guilty until proven innocent. BioScience 45:465–477

Schlimme W, Marchiani M, Hanselmann K, Jenni B (1997) Gene transfer between bacteria within digestive vacuoles of protozoa. FEMS Microbiol Ecol 23:239–247

Simberloff D (1992) Conservation of pristine habitats and unintended effects of biological control, In Kauffman WC, Nechols JE (eds) Selection Criteria and Ecological Consequences of Importing Natural Enemies. Proc Thomas Say Publ Entomol, Lanham, MD, pp 103–117

Simberloff D, Stiling P (1996) How risky is biological control? Ecology 77(7):1965–1974

Smaglik P (1999) *Drosophila* sequenced — now the tricky part. The Scientist 13(19):7

Spielman A (1994) Why entomological antimalaria research should not focus on transgenic mosquitoes. Parasitol Today 10:374–376

Thomas MB, Willis AJ (1998) Biocontrol — risky but necessary? TREE 13(8):325–328

Tiedje JM, Colwell RK, Grossman YL, Hodson RE, Lenski RE, Mack RN, Regal PJ (1989) The planned introduction of genetically engineered organisms: ecological considerations and recommendations. Ecology 70:298–315

Turelli M, Hoffman AA, McKechnie SW (1992) Dynamics of cytoplasmic incompatibility and mtDNA variation in natural *Drosophila simulans* populations. Genetics 132:713–723

Wallace B (1985) Reflections on some insect pest control procedures. Bull Entomol Soc Am Summer 8–13

Watanabe K, Sato M (1998) Plasmid-mediated gene transfer between insect-resident bacteria, *Enterobacter cloacae*, and plant-epiphytic bacteria, *Erwinia herbicola*, in guts of silkworm larvae. Curr Microbiol 37:352–355

Watanabe K, Hara W, Sato M (1998) Evidence for growth of strains of the plant epiphytic bacterium *Erwinia herbicola* and transconjugation among the bacterial strains in guts of the silkworm *Bombyx mori*. J Invertibr Pathol 72:104–111

Werren JH (1997) Biology of *Wolbachia*. Annu Rev Entomol 432:587–609

Whitten MJ (1979) The use of genetically selected strains for pest replacement or suppression. In Hoy MA, McKelvey JJ (eds) Genetics in Relation to Insect Management. Working Papers, Rockefeller Foundation Press, New York, pp 31–40

Williams RJ, Heymann DL (1998) Containment of antibiotic resistance. Science 279:1153–1154

Wilson LT, Hoy MA, Zalom FG, Smilanick JM (1984) Sampling mites in almonds: I. Within-tree distribution and clumping patterns of mites with comments on predator–prey interactions. Hilgardia 52(7):1–13

Witte W (1998) Medical consequences of antibiotic use in agriculture. Science 279:996–997

Wood SN, Thomas MB (1999) Super-sensitivity to structure in biological models. Proc R Soc Lond B266:575–570

Yen JH, Barr AR (1974) Incompatibility in *Culex pipiens*. In Pal R, Whitten MJ (eds) The Use of Genetics in Insect Control. Elsevier/North-Holland, Amsterdam, pp 97–118

Zalom FG, Hoy MA, Wilson LT, Barnett WW (1984) Sampling mites in almonds: II. Presence-absence sequential sampling for *Tetranychus* mite species. Hilgardia 52(7):14–24

20 Regulation of Transgenic Arthropods and Other Invertebrates in the United States

Orrey P. Young, Shirley P. Ingebritsen, and Arnold S. Foudin

CONTENTS

20.1 Historical Overview 369
20.2 USDA Regulation of Biotechnology 370
20.3 APHIS Review and Evaluation Process Associated with Proposed Introductions
   of Transgenic Arthropods 371
   20.3.1 Overview of the Process 371
   20.3.2 Determination of Jurisdiction 372
   20.3.3 Determination of Categorical Exclusion under NEPA 372
   20.3.4 Evaluation of the Nontransgenic Form Proposed for Introduction 373
   20.3.5 Evaluation of the Transgenic Form 374
   20.3.6 Summary of Requested Information for Introduction of a Transgenic Arthropod
      (in addition to that requested for introduction of the nontransgenic form) 375
20.4 Outline of Current APHIS Procedures for Permitting Transgenic Invertebrates
   under FPPA and PQA 375
20.5 Future Developments in the Regulatory Arena 376
Abbreviations 377
References 378

20.1 HISTORICAL OVERVIEW

During the 1970s in the United States, public concern about the safety of conducting experiments with recombinant DNA under conditions of adequate laboratory containment led to the issuance of the "National Institutes of Health Guidelines for Research Involving Recombinant DNA Molecules" (NIH Guidelines, 1979; current version, 1998). Public concern was again elicited in the 1980s, this time by proposals to test and use genetically engineered organisms in the environment. The U.S. Congress and the executive branch were then stimulated to reexamine the issues surrounding widespread commercial development of the products of biotechnology. The issues ranged from speculation about the long-term ecological effects of environmental releases of genetically engineered organisms, to concerns that unwarranted restraints on the biotechnology industry would deprive the American public of the benefits of the new technology. In addition, the NIH Guidelines, which were originally written for NIH grantees doing biomedical research in the laboratory, were proving inadequate for testing in the environment a broad spectrum of genetically engineered organisms and potential commercial products.

It was in a climate of renewed interest that an interagency working group was formed in April 1984 within the Executive Office of the President. The charge of the working group included identifying the existing laws and regulations applicable to biotechnology and determining their adequacy for regulating the products of the new technologies. The results of these efforts were first published for public review and comment in December 1984, and in final form in June 1986 as the "Coordinated Framework for Regulation of Biotechnology" (OSTP, 1986). The Coordinated Framework included an index of laws applicable to biotechnology products in various stages of research, development, marketing, shipment, use, and disposal. This index of laws was published in final form in November 1985 (OSTP, 1985).

The U.S. federal biotechnology policy has been based on several conclusions:

1. The products of biotechnology will not differ fundamentally from unmodified organisms or from conventional products.
2. The product, rather than the process by which the product was created, should be regulated.
3. Regulation should be based on the end use of the product and conducted on a case-by-case basis.
4. The existing laws provide adequate authority for regulating the products of biotechnology.

An important corollary to this policy is the federal commitment to promote the safe development of genetically engineered organisms and products.

Each Federal agency is committed to ensuring protection for public health and the environment from any potentially harmful effects of biotechnology. The Coordinated Framework contained final policy statements by the U.S. federal agencies that share a major responsibility for regulating the products of biotechnology. These agencies include the Food and Drug Administration (FDA), the Environmental Protection Agency (EPA), and the U.S. Department of Agriculture (USDA).

Because statutory responsibility is assigned by Congress, the Coordinated Framework could not confer exclusive authority where jurisdiction lay with more than one agency regarding the review of a single biotechnology product. One agency is usually designated as a lead agency to facilitate review. The Animal and Plant Health Inspection Service (APHIS) is designated the lead agency for plant and animal biotechnology. It is important to note that the authority of FDA, EPA, and USDA is based on statute, and that the implementing regulations for those statutes are published in the U.S. Code of Federal Regulations (CFR). In most instances, sanctions for noncompliance with these regulations are based on statutory authority and include administrative, civil, and/or criminal penalties. The NIH Guidelines by contrast have a contractual, rather than a statutory basis. In recognition of this fact and to avoid interagency duplication, the NIH proposed amendments to the Guidelines in December 1986 to provide that if certain experiments are submitted to another federal agency for approval or official clearance, NIH review is not required. This amendment, which applies only to experiments covered by Section III-A (Major Actions) of the Guidelines, was adopted in August 1987 (USDHHS, 1987).

20.2 USDA REGULATION OF BIOTECHNOLOGY

The USDA has broad statutory authority to protect U.S. agriculture against threats to animal health, to protect against the adulteration of food products made from livestock and poultry, and to prevent the introduction and dissemination of plant pests. This authority is also applicable to genetically engineered animals, plants, and microorganisms. The USDA has major responsibilities for research in agricultural biotechnology (Agricultural Research Service; Cooperative States Research, Education, and Extension Service) and also for regulating genetically engineered organisms and products. A delegation of authority by the Secretary of Agriculture (USDA, 1985) assigned responsibility for departmental regulation of biotechnology to the assistant secretary for Marketing and Inspection Services (now Marketing and Regulatory Programs), responsibilities subsequently conferred downward on the APHIS and the Food Safety and Inspection Service (FSIS).

Under the authority granted by the Federal Plant Pest Act (FPPA) of May 23, 1957, as amended, and the Plant Quarantine Act (PQA) of August 20, 1912, as amended, USDA–APHIS regulates the movement into and throughout the United States of plants, plant products, plant pests, and any product or article that may contain a plant pest at the time of movement. These articles are regulated to prevent the introduction, spread, or establishment of plant pests new to or not widely prevalent in the United States. The regulations that implement this statutory authority are found in 7 CFR parts 300 through 399.

Specifically, under regulations codified at 7 CFR part 330.200, APHIS Plant Protection and Quarantine (PPQ) administers a permit program which prohibits the movement of any plant pest from a foreign country into the United States or movement interstate unless authorized under a permit issued by APHIS. Remedial measures are also exercised by APHIS to prevent the spread intrastate of a plant pest that would constitute a threat to agriculture.

USDA–APHIS published a final rule on June 16, 1987, under the FPPA and PQA, 7 CFR part 340 (USDA, 1987), that established a permit requirement for the introduction of genetically engineered organisms that are plant pests or which APHIS has reason to believe are plant pests. Genetic engineering at that time was defined as "the genetic modification of organisms by recombinant DNA techniques." The regulations in 7 CFR part 340 can be seen as an expansion of the preexisting regulations of 7 CFR part 330.200 to cover the organisms and products of genetic engineering technology. For additional discussion of federal regulatory oversight for biotechnology, see Cordle et al., 1991; Shantharam and Foudin, 1991.

20.3 APHIS REVIEW AND EVALUATION PROCESS ASSOCIATED WITH PROPOSED INTRODUCTIONS OF TRANSGENIC ARTHROPODS

It is expected that arthropod researchers contemplating eventual introduction of transgenic arthropods will be unfamiliar with the regulatory procedures and guidelines already in place for the introduction of transgenic plants and microorganisms. The current federal regulations governing the permitting of transgenic organisms under FPPA and PQA are detailed in 7 CFR 340, "Introduction of Organisms and Products Altered or Produced through Genetic Engineering Which Are Plant Pests or Which There Is Reason to Believe Are Plant Pests" (USDA, 1987). The Scientific Services (SS) directorate of PPQ–APHIS–USDA currently distributes upon request a guidance document for potential applicants, "Questions and Answers on Biotechnology Permits for Genetically Engineered Plants and Microorganisms" (APHIS 21–35–001). The present discussion represents an ongoing process of developing guidance relative to transgenic arthropods. Further information can be obtained by viewing the Web site, "The Regulation of Transgenic Arthropods (and other transgenic invertebrates)" at *www.aphis.usda.gov/biotech/arthropod*.

20.3.1 Overview of the Process

APHIS–USDA is obligated by statute and regulation to evaluate the potential impact to plants and the environment of transgenic arthropods proposed for release that are or may become plant pests. This evaluation process starts with a determination of jurisdiction: does the proposed introduction involve a "regulated" article as defined under the FPPA? (see below). If it is determined that APHIS does not have authority to regulate the particular article, then the process is complete and the author of the proposal is so informed. However, if it is determined that APHIS does have jurisdiction, then the evaluation proceeds to an assessment of the risks of the proposed introduction. In some situations the risk of the organism or activity has already been assessed, and been determined to be of no risk to plants or the environment. This may be the case for previously permitted similar organisms, or those activities categorically excluded from National Environmental Policy Act (NEPA) analysis and which are listed in the NEPA Implementing Procedures for APHIS (see below). A Courtesy Permit, if requested, may be issued for the introduction of organisms that fall within these categories. If the activity is not within these categories, the assessment then is conducted in two phases: first, an examination of the risks associated with the introduction of a nontransgenic form of the proposed species; and, second, an examination of the potential additional risks associated with the introduction of the transgenic form. When risks are identified, consideration is given to the mechanisms proposed to manage that risk.

At this point in the evaluation process, it may be determined that the proposed introduction of the candidate organism represents an obvious and significant threat to agricultural crops and under the FPPA and PQA cannot be permitted, and the applicant is requested to withdraw and revise the application. When this is not the case, an Environmental Assessment (EA) document is then prepared under NEPA and Council on Environmental Quality (CEQ) guidelines. The EA document outlines for the APHIS decision maker the potential impact of the introduction on the environment (including the potential of the organism as a plant pest) and recommends either a Finding of No Significant Impact (FONSI) or the preparation of an Environmental Impact Statement (EIS). When the decision maker chooses the FONSI alternative, the EA is made available for public comment, after which a FONSI could be prepared and a permit issued. When the EIS alternative is chosen, the possible issuance of a permit is delayed until the EIS is prepared. Based on the identified and analyzed potential impacts to the environment documented in the EIS, and after the public has had an opportunity to contribute to the evaluation process, a decision can be made to permit or prohibit the introduction.

20.3.2 Determination of Jurisdiction

The FPPA and PQA statutes, as promulgated in 7 CFR 330 and 340, define a plant pest as "any living stage of: any insects, mites, nematodes, slugs, snails, protozoa, or other invertebrate animals, bacteria, fungi, other parasitic plants or reproductive parts thereof, viruses, or any organisms similar to or allied with any of the foregoing, or any infectious substances, which can directly or indirectly injure or cause disease or damage in any plants or parts thereof, or any processed, manufactured, or other products of plants." The current regulatory process concerning the introduction of plant pests involves the analysis of potential risk to the environment of a proposed introduction and the issuance of a permit by USDA–APHIS if the risk is deemed acceptable. For organisms that are not transgenic, the process is governed by 7 CFR 330 and implemented by the Plant Pest Permits unit of APHIS–PPQ–SS (www.aphis.usda.gov/ppq/ss/). In the case of transgenic organisms, 7 CFR 340 authorizes the Biotechnology Permits unit of APHIS–PPQ–SS (*www.aphis.usda.gov/bbep/bp/*) to regulate the "introduction of organisms and products altered or produced through genetic engineering which are plant pests or which there is reason to believe are plant pests." To import, move interstate, or release into the environment a genetically engineered organism or product, an individual must obtain a permit from USDA–APHIS if (1) the organism has been altered or produced through genetic engineering from a donor, vector, or recipient organism that can be classified as a plant pest or whose classification is unknown; (2) the product contains such an organism as described above; or (3) any other organism or product not included in (1) or (2) altered or produced through genetic engineering which APHIS determines is a plant pest or has reason to believe is a plant pest.

20.3.3 Determination of Categorical Exclusion under NEPA

The NEPA Implementing Procedures for APHIS (7 CFR 372) (USDA, 1995) define actions normally requiring environmental assessments as "(372.5) (b)(4) Approvals and issuance of permits for proposals involving genetically engineered or nonindigenous species, except for actions that are categorically excluded.... (b)(5) Research or testing that: (i) Will be conducted outside of a laboratory or other containment area (field trials, for example). ..." Categorically excluded actions are defined as "(372.5)(c)(2) Research and development activities. (i) activities that are carried out in laboratories, facilities, or other areas designed to eliminate the potential for harmful environmental effects — internal or external — and to provide for lawful waste disposal. (c)(3) Licensing and permitting. (ii) permitting, or acknowledgment of notifications for, confined field releases of genetically engineered organisms and products." Exceptions for categorically excluded actions are defined as "(372.5)(d)(4) When a confined field release of genetically engineered organisms or products involves new species or organisms or novel modifications that raise new issues." An activity that is categorically excluded by APHIS from NEPA analysis would have been previously determined to have no significant impact on the environment.

20.3.4 Evaluation of the Nontransgenic Form Proposed for Introduction

The transgenic organism prior to its transformation may have been indigenous or nonindigenous to the United States, and either (1) phytophagous and thus an actual or potential direct plant pest or (2) a nonphytophagous predator, parasite, or competitor of plant pests or beneficials and thus potentially an indirect plant pest. The information required in these various situations for APHIS–PPQ–SS to evaluate the permit request can be organized in a manner similar to the format utilized by SS for nontransgenic plant pests. For the release of actual or potential plant pests under FPPA and NEPA, SS requests from the applicant certain basic information about the proposed release in the context of USDA–APHIS–PPQ Form 526, "Application for Permit to Move Live Plant Pests and Noxious Weeds" (see below). These evaluations focus on potential impacts of the released organism on both the target organism and nontarget organisms, at the levels of populations, species, communities, and ecosystems. Impacts on humans and threatened/endangered species are considered, as are potential impacts on the physical, economic, and cultural environments. The general informational requirements for the introduction of a nontransgenic organism under FPPA and NEPA (summary of SS guidelines) are:

1. Statement of intent from applicant
2. Confirmed taxonomic identification of organism
3. Recent and historical geographic distribution
4. Review of biology and ecology of organism in area of origin, to include host specificity and sources of mortality
5. Review of target pest, to include economic impact, sources of mortality, and host range
6. Provisions for shipment and containment
7. Activities conducted in containment
8. Summary of biology and ecology of organism, as evidence of appropriateness for release

Environmental Assessments prepared by SS for a particular species as part of the FPPA, PQA, and NEPA permitting procedure would be incorporated into EAs prepared by SS when a transgenic form of the same species is being evaluated for introduction.

Specific information requested (if available) for the evaluation of a proposed introduction of a nontransgenic arthropod under FPPA, PQA, and NEPA (summary of SS guidelines) would include:

1. Introduced Organism Information:
    a. Taxonomic identity and relationships
    b. Documentation of efforts to ensure the organism is "pure"
    c. Geographic and habitat distribution

d. Food/host/prey range
   e. Dispersal characteristics
   f. Mortality factors, to include climate, natural enemies, and competitors
2. Target Pest Information:
   a. Economic importance
   b. Political, social, and cultural factors
   c. Benefits of pest control
   d. Geographic and habitat distribution
   e. Food/host/prey range
   f. Mortality factors, to include climate, natural enemies, and competitors
   g. Vulnerability to biocontrol organisms
3. Site-Specific Information (when appropriate):
   a. Location of releases, including proximity to significant sites
   b. General description (ecological, spatial, temporal)
   c. Specific habitat and community descriptions
   d. Influence of site characteristics on success of release
4. Composite Factors, i.e., potential impact of introduced organism on
   a. Ecological community
   b. Endangered and threatened species
   c. Pollinators
   d. Other biocontrol agents
   e. Other nontarget organisms
   f. Target pest relationship with rest of community
   g. Relationship between the introduced organism and currently employed pest management strategies

20.3.5 Evaluation of the Transgenic Form

The evaluation process next considers whether the genetic alterations associated with the proposed organism to be released have altered the risks associated with the unmodified organism. The fundamental risk assessment issue addressed is, Will the genetic alteration modify ecologically or environmentally relevant properties of the organism? Specific potential risks associated with the release of a transgenic arthropod can include the displacement of native populations, a change in host or prey utilization or ecological distribution, the transfer of exogenous DNA to other organisms, or, if one of the characteristics of the transgenic arthropod is increased resistance to herbicides or pesticides, subsequent increased usage of such chemicals. To address risk issues adequately, APHIS would consider such information as:

1. How the recipient organism was transformed through recombinant DNA technology, to include characteristics of the donor, vector, and recipient organisms and a description of the methods employed.
2. The characteristics of the modified organism, to include the stability of the new genotype and the probability of gene transfer to other organisms with resultant consequences.
3. Potential impact of the transgenic arthropod on native populations, communities, and ecosystems.
4. Methods for evaluation of the safety of the transgenic organism in field trials before unrestricted release.

This information is requested on the current USDA–BBEP Form 2000, "Application for Permit or Courtesy Permit under 7 CFR 340 (Genetically Engineered Organisms or Products)."

20.3.6 SUMMARY OF REQUESTED INFORMATION FOR INTRODUCTION OF A TRANSGENIC ARTHROPOD (IN ADDITION TO THAT REQUESTED FOR INTRODUCTION OF THE NONTRANSGENIC FORM)

Information that would be appropriate before a confined field trial is authorized can include:

1. History of introductions of the nontransgenic form
2. Life table/life history attributes of the transgenic form
3. Nature/function of the genetic alteration, e.g., mode of inheritance, stability, degree of expression
4. Behavior of the trait in caged or mesocosm situations
5. Mathematical modeling of released populations, to include probability of establishment
6. Consequences of inadvertent escape and establishment
7. Methods for monitoring and control.

A field trial is considered "confined" when the candidate arthropod cannot become established and subsequently spread. Confinement may be achieved by means of physical (e.g., dispersal barriers), chemical (e.g., insecticide spraying of dispersal areas), cultural (e.g., trap plants, bare zones), and/or biological (e.g., inability to overwinter) methods. Information that field trials can provide, before unrestricted release is authorized, includes:

1. Changes in biology/ecology/behavior of transgenic form relative to unmodified form, such as dispersal characteristics, habitat utilization, climatic tolerances, life history
2. Stability under environmental conditions of characteristics associated with transgenic material, e.g., between generations, movement of new gene from insertion site, transmittability of gene to other species, such as by means of endosymbionts, transposons, or viruses
3. Probability of increased fitness of transgenic arthropod after release
4. Interaction with native strains, such as the probability of new phenotype acting as an agent of natural selection in the community, or the consequences of adding a new species or genotype to the ecological community.

Further discussion of these topics can be found in Christensen et al., 1995; Gaugler et al., 1997; Hoy, 1991, 1992a,b, 1995, 1997; Levin and Strauss, 1991; Parsons, 1990; Ruesink et al., 1995; Tiedje et al., 1989; USDA 1987, 1993; Williamson, 1992.

20.4 OUTLINE OF CURRENT APHIS PROCEDURES FOR PERMITTING TRANSGENIC INVERTEBRATES UNDER FPPA AND PQA

1. Applicant submits materials — Form 2000 and supporting documents — to SS–PPQ–APHIS for review (electronic version encouraged). *Note:* The submission of an application may (should) be preceded by an informal consultation, in which the applicant can obtain (a) information concerning protocols, data requirements, and APHIS procedures, (b) a preliminary determination of regulatory status for the proposed introduction, and (c) application materials, some of which are also available at the APHIS Web site (*www.aphis.usda.gov*).
2. Initial processing of application.
    a. Application is reviewed for completeness; i.e., all requested information is present; missing material is requested or application returned for further development.

b. The mandated 120-day interval for completion of the review process is initiated upon determination of application completeness.
c. Application is made available to the public for comment (without Confidential Business Information, CBI) by placement on a APHIS Web site (*www.aphis.usda.gov/biotech/arthropod*) within 7 days of receipt of complete application.
3. Preliminary review initiated.
   a. State(s) informed of receipt and availability of application and start of review process.
   b. Regulatory status is determined; i.e., is the organism a "regulated article" under FPPA?
   c. Courtesy permit may be issued (if requested) if proposed introduction is not a regulated article.
   d. Relevant APHIS documents are reviewed; e.g. has the same or related organism been previously permitted?
   e. If organism previously permitted or similar to such, a permit extension may be issued.
   f. Preliminary review continues, considering:
      i. Risks associated with introduction of nontransgenic form of organism;
      ii. Risks associated with introduction of transgenic organism.
4. Preliminary review concluded within 30 days of initiation, with a preliminary determination that:
   a. The introduction of organism be allowed; state(s) informed of results of review; concurrence requested within 30 days.
      i. State(s) concurs (go to 5).
      ii. State(s) concurs, with conditions; consultation with state(s) and applicant (go to 5 if all parties agree to conditions).
      iii. State(s) does not concur; consultation with state(s) and applicant; may lead to modification or withdrawal of application.
   b. The introduction of organism not be allowed (permit denied). SS Director informs applicant and state(s) of basis of decision under FPPA and describes appeal process.
5. NEPA process is initiated.
   a. Determination of categorical exclusion under APHIS Implementing Procedures for NEPA.
   b. Environmental Assessment prepared if appropriate.
      i. Endangered Species Act requirements considered, consultation with USDI–FWS initiated if appropriate, and applicant informed of possible delay in permitting.
      ii. EA is completed, Director of SS reviews application and EA, and if approves, signs FONSI.
      iii. Availability of EA/FONSI is announced in Federal Register and at APHIS Web site.
      iv. After appropriate interval, permit may be issued.

20.5 FUTURE DEVELOPMENTS IN THE REGULATORY ARENA

The preceding discussion was restricted to transgenic arthropods and other invertebrates that are or may become plant pests and as such come under the perview of the several plant health statutes. Transgenic arthropods that may be vectors of animal (and human) diseases quite logically are not included in those plant health statutes. However, USDA–APHIS does have general statutory authority under 21 U.S.C. 111 to take whatever measures are necessary to prevent the introduction and/or dissemination of contagious/infectious/communicable diseases of animals, and specific regulatory authority under 9 CFR 122 to restrict the introduction of organisms and vectors of those diseases. USDA–APHIS is now (March 2000) in the process of developing specific regulatory authority and guidelines for transgenic arthropods and other invertebrates that are vectors of animal diseases. It is anticipated that these new regulations will contain many components similar to the existing regulations for transgenic plant pests.

Those same mechanisms that were employed in the development of guidelines and procedures for regulating transgenic arthropods impacting plants will be utilized in the development of APHIS oversight for transgenic arthropod vectors of animal disease. The Biotechnology Permits unit, USDA–APHIS–PPQ, announced in December 1995 the formation of an agency "virtual team" that would be involved with the regulation of transgenic arthropods. Members of that team remained in their current units, but were available to assist the team leader when appropriate. Activities of the team included reviewing applications, contributing to or reviewing EA, proposing modifications in existing procedures, and liaisoning with the user community. The membership of the team represented a broad spectrum both of agency activities and scientific disciplines relevant to the regulation of transgenic arthropods. It was hoped that the availability, and utilization, of this expertise would ensure that APHIS transgenic arthropod activities and decisions would be innovative, efficient, and well informed. A perusal of the APHIS Transgenic Arthropod Virtual Team Web site (*www.aphis.usda.gov/biotech/arthropod*) should indicate that the team was successful in meeting its objectives. This conclusion was reinforced in October 1997, when the team received Vice President Gore's "Hammer Award" from the National Performance Review, for "reducing costs, cutting paperwork, increasing efficiency, and improving customer service."

Effective team innovations that will be continued include:

1. Electronically connected APHIS "virtual team" of experts
2. Team Web site as focus of federal transgenic arthropod regulatory activities and as clearinghouse for information within the transgenic arthropod research community
3. Extensive dialogue with permit applicants before submission of formal applications, as a means of improving the quality of applications and reducing subsequent processing time
4. Placing all permit applications, and subsequent agency decision documents, on the team Web site in an on-screen version within 7 days of receipt or issuance
5. Involvement of team members in professional scientific society and international organization activities such as workshops, symposia, and conferences relating to transgenic arthropods
6. Use of the "Transgenic Arthropod Interest Group" to inform interested individuals and entities, both outside APHIS and outside the federal government, of permit applications, agency decisions, team activities, and as a mechanism to encourage comment on transgenic arthropod regulatory activities of APHIS.

The APHIS Transgenic Arthropod Virtual Team Web site will be *the* location for information concerning future development of federal regulations, guidelines, and procedures involving transgenic arthropods, as well as a source for information on recent and current activities within the regulatory and research communities.

ABBREVIATIONS

APHIS	Animal and Plant Health Inspection Service (USDA)
CEQ	Council on Environmental Quality
EA	Environmental Assessment
EIS	Environmental Impact Statement
EPA	Environmental Protection Agency
FDA	Food and Drug Administration
FONSI	Finding of No Significant Impact
FPPA	Federal Plant Pest Act
FSIS	Food Safety and Inspection Service (USDA)
FWS	Fish and Wildlife Service (USDI)

NEPA	National Environmental Policy Act
NIH	National Institutes of Health
OSTP	Office of Science and Technology Policy (Office of the President)
PPQ	Plant Protection and Quarantine (USDA–APHIS)
PQA	Plant Quarantine Act
SS	Scientific Services (USDA–APHIS–PPQ)
USDA	U.S. Department of Agriculture
USDHHS	U.S. Department of Health and Human Services
USDI	U.S. Department of Interior

REFERENCES

Christensen BM, Beckage NE, Raikhel AS, James AA, Fallon AM, ffrench-Constant R (1995) Toward the genetic manipulation of insects. Keystone Symposium, Tamarron, CO, J Cell Biochem Suppl 21A (abstr.)

Cordle MK, Payne JH, Young AL (1991) Regulation and oversight of biotechnology applications for agriculture and forestry. In Ginzburg LR (ed) Assessing Biological Risks of Biotechnology. Butterworth & Heineman, Boston, MA, pp 289–311

Gaugler R, Wilson M, Shearer P (1997) Field release and environmental fate of a transgenic entomopathogenic nematode. Biol Control 9:75–90

Hoy MA (1991) Use of parasites and predators in the biological control of arthropod pests: emerging technologies and challenges. In Vinson SB, Metcalf RL (eds) Entomology Serving Society: Emerging Technologies and Challenges. Entomology Society of America Lanham, MD, pp 272–297

Hoy MA (1992a) Biological control of arthropods: genetic engineering and environmental risks. Biol Control 2:166–170

Hoy MA (1992b) Criteria for release of genetically improved phytoseiids: an examination of the risks associated with release of biological control agents. Exp Appl Acarol 14:393–416

Hoy MA (1995) Impact of risk analyses on pest-management programs employing transgenic arthropods. Parasitol Today 11(6):229–232

Hoy MA (1997) Laboratory containment of transgenic arthropods. Am Entomol 43:206–209, 255–256

Levin MA, Strauss HS (1991) Introduction: overview of risk assessment and regulation of environmental biotechnology. In Levin MA, Strauss HS (eds) Risk Assessment in Genetic Engineering. McGraw-Hill, New York, pp 1–17

OSTP (Executive Office of the President, Office of Science and Technology Policy) (1985) Coordinated framework for regulation of biotechnology; establishment of the biotechnology science coordinating committee. Fed Regis 50 (220):47174–47195

OSTP (Executive Office of the President, Office of Science and Technology Policy) (1986) Coordinated framework for regulation of biotechnology; announcement of policy; notice for public comment. Fed Regis 51 (123):23302–23350

Parsons PA (1990) Risks from genetically engineered organisms: energetics and environmental stress. Funct Ecol 4:265–271

Ruesink JL, Parker IM, Groom MJ, Kareiva PM (1995) Reducing the risks of nonindigenous species introductions: guilty until proven innocent. BioScience 45:465–477

Shantharam S, Foudin A (1991) Federal regulation of biotechnology: jurisdiction of the U.S. Department of Agriculture. In Maramorosch K (ed) Biotechnology for Biological Control of Pests and Vectors. CRC Press, Boca Raton, FL, pp 239–250

Tiedje JM, Colwell RK, Grossman YL, Hodson RE, Lenski RE, Mack RN, Regal PJ (1989) The planned introduction of genetically engineered organisms: ecological considerations and recommendations. Ecology 70:298–315

U.S. Department of Agriculture (USDA), Animal and Plant Health Inspection Service (1987) Plant pests; introduction of genetically engineered organisms or products; final rule. 7 CFR Parts 330 and 340. Fed Regis 52 (115):22892–22915

U.S. Department of Agriculture (USDA), Animal and Plant Health Inspection Service (1993) Genetically engineered organisms and products: notification procedures for the introduction of certain regulated articles; and petition for nonregulated status; final rule. 7 CFR Part 340. Fed Regis 58 (60):17043–17059

U.S. Department of Agriculture (USDA), Animal and Plant Health Inspection Service (1995) National Environmental Policy Act implementing procedures; final rule. 7 CFR Part 372. Fed Regis 60(21):6000–6005

U.S. Department of Agriculture (USDA), Office of the Secretary (1985) Delegation of authority pertaining to biotechnology; final rule. 7 CFR Part 2. Fed Regis 50 (139):29367–29368

U.S. Department of Health and Human Services (USDHHS) National Institutes of Health (1987) Recombinant DNA research; actions under guidelines. Fed Regis 52:31848–31849

Williamson M (1992) Environmental risks from the release of genetically modified organisms (GMOs) the need for molecular ecology. Mol Ecol 1:3–8

Index

A

a, 85
AaPV (*Aedes albopictus* densovirus), 140, 142–143
AAV (adeno-associated viruses), 140, 150–152, 155
Ac, 9, 223, 232
Ace, 110–111
Acetylcholinesterase (AChE), 110–111, 114
AcFP6, 252
AcMNPV, 251
AcNPV (*Autographa californica* nuclear polyhedrosis virus), 250
Actinomycete symbionts, 293, 296
Activator, 4
Adeno-associated viruses (AAV), 140, 150–152, 155
Adh (alcohol dehydrogenase), 7, 312
Aedes aegypti (mosquito)
    AeDNV densovirus, 140, 141
        ability to establish persistent infections, 156
        gene expression, 143–145
        genomic organization, 142–143
        integration, 152–155
        Sindbis as helper, 149
        transducing systems development, 146–148
    antiviral RNA expression, 176–178
    assay *Rdl* promoter activity, 118
    AthDNV densovirus, 140
    as disease vector, 320
    Hermes vector development, 222
    infection with dsSIN virus, 101
    mariner transformation of, 238
    meiotic drive genes, 327
    protein expression in midgut, 174–176
    protein expression using dsSIN virus, 171–174
    SIN virus strains in, 165
    successful *piggyBac* transformations, 261
    targeted gene case study, 42–43
Aedes albopictus (mosquito)
    AaPV densovirus, 140, 142–143, 156
    AeDNV production and, 147–148
    C6/36 cells
        AeDNV production and, 147–148
        densovirus isolation, 140
        dsSIN infection, 171
        flavivirus replication, 176
        SIN replicon RNA into, 180
    dsSIN and protein expression, 171
    frequency of *Wolbachia* infection, 275
    SIN virus infection, 165
    unidirectional CI, 277–278
Aedes pseudoscutellaris virus, 140
Aedes triseriatus (mosquito)
    antiviral RNA expression, 176–178
    infection with dsSIN virus, 101
    SIN virus infection, 165
AeDNV (*Aedes aegypti* densovirus), 140, 141
    ability to establish persistent infections, 156
    gene expression, 143
        nonstructural gene promoter, 144–145
        structural gene promoter, 145
        transactivation of promoter, 145
    genomic organization, 142–143
    integration, 152–155
    Sindbis as helper, 149
    transducing systems development, 146–148
Aequorea sp. (jellyfish), 88, 96
Air-pulse injection systems, 20–21
alb, 86
Alcohol dehydrogenase (*Adh*), 7, 312
Alphaviruses, *see* Sindbis (SIN) virus
Alternative foods, 347
γ Aminobutyric acid (GABA), 115
Anal papillae and transduction evidence, 149–150
Anastrepha suspensa (Carribean fruit fly), 10, 261
Animal and Plant Health Inspection Service (APHIS), 370
    certification of facilities, 358
    procedures for release permission, 375–376
    review and evaluation process
        determination of categorical exclusion, 372–373
        determination of jurisdiction, 372
        evaluation of nontransgenic form, 373–374
        evaluation of transgenic form, 374
        information requested, 375
        process overview, 371–372
        Web site resources, 371
Anopheles (mosquito)
    albimanus, 111
    dirus, 325
    as disease vectors, 320
    eye pigment mutations in, 84
    freeborni, 325
    gambiae
        extrachromosomal homologous recombination, 31–34
        Hermes integration into, 220
        Hermes vector development, 222
        huni element discovery, 229
        infection efficiency in MOS-55, 128
        infection with dsSIN virus, 101
        sex-specific splicing and, 313
        transposon-mediated gap repair, 46, 48
    malaria control programs, 320
    stephensi, 101, 326
Anophelinae, *see* Mosquitoes
Antibiotic-resistant gene transfer, 353–354
Antifreeze proteins, 20, 314

Antisense RNA
    expression of, 176–179
    Sindbis virus and, 323
    for use in insects, 139–140
Antiviral RNA expression, 176–178
Apanteles melanoscelus, 341
APHIS, *see* Animal and Plant Health Inspection Service
Apis mellifera (honeybee), 81, 85, 339
Arboviruses, 139–140
Artemia sp., 102
Arthropoda, 273
AthDNV (*Aedes aegypti* densovirus), 140
Augmentation strategy, 340
Aurodrosopterin, 82
Autographa californica nuclear polyhedrosis virus (AcNPV), 250
Azinphosmethyl, 344

B

Bacillus thurengiensis (BTI), 321
Backcrosses and biofitness, 308–309
Bacterial genes as transgenesis candidates, 112
Bacterial symbiont transformation; see also *Wolbachia pipientis* symbiont vector
    Chagas disease vector
        biology and ecology, 290–292
        control efforts, 289–290
        CRUZIGARD use, 296–298
        need for control, 301–302
    genetically altered symbionts
        application in disease control, 298
        introduction strategy, 296–298
    paratransgenic insects generation
        cecropin A expression, 293–295
        genetic transformation of *Rh. rhodnii*, 293
        integrative plasmids for actinomycetes transformation, 296
        single-chain antibody expression, 295–296
        system description, 292–293
    safety concerns
        environmental/ecological risks, 301
        human health risks, 299–301
Bactrocera dorsalis (Oriental fruit fly)
    hopper discovery, 228
    and *piggyBac*, 5
    successful *piggyBac* transformations, 10, 261
Bactrocera tryoni (fruit fly)
    hobo excision assays, 222
    Homer discovery, 227
    sex-specific splicing and, 313
Baculoviruses
    FP plaque morphology mutation, 250–251
    horizontal gene transfer and, 263, 356
    transposon-induced mutations, 251–252
    for use in insects, 140
Balancer chromosomes, 71
Ballistic method, 17
Banchinae, 206; *see also* Polydnaviruses (CsIV)
Bed net programs, 321

Beneficial insects and genetic manipulation
    background of efforts, 339–340
    implementation strategies, 340
Biodiversity loss, 338
Biodiversity maintenance, 336
Biolistics, 17
Biological control programs, 336, 358
Biopterin, 82
Biotechnology Permits unit, 377
BLAST analysis, 19
BmDNV (*Bombyx mori* densovirus), 140
Bombyx mori (silkmoth), 4, 11
    BmDNV densovirus, 140
    successful *piggyBac* transformations, 261
Braconids, *see also* Polydnaviruses (CsIV)
    phylogeny of polydnaviruses and, 206
    segment integration sites, 207
Bracoviruses, 206; *see also* Polydnaviruses (CsIV)
Brevidensovirus, 140
Brown gene mutation effects, 84–85
Brown pigments, *see* Ommochrome pigments
BTI (*Bacillus thurengiensis*), 321
Buckeye butterfly (*Precis coenia*), 101

C

C6/36 cells (*Aedes albopictus*)
    AeDNV production and, 147–148
    densovirus isolation, 140
    dsSIN infection, 171
    flavivirus replication, 176
    SIN replicon RNA into, 180
Caenorhabditis elegans (nematode), 238
Campoletis sonorensis polydnaviruses (CsIV), *see* Polydnaviruses (CsIV)
Campoplegineae, 206; *see also* Polydnaviruses (CsIV)
Carbaryl, 341, 343
cardinal, 86
Cardiochilinae, 206; *see also* Polydnaviruses (CsIV)
carnation, 86
CAT (chloramphenicol acetyltransferase), 169, 171, 174
Cecropin A expression, 293–295
Cell autonomy, 86, 88
CEQ (Council on Environmental Quality guidelines), 372
Ceratitis capitata (Mediterranean fruit fly)
    Hermes vector development, 222
    Maleness factor isolation, 312–313
    piggyBac element in, 10
    piggyBac transposon using, 261
    pigment gene cloning, 85
    SIT program, 308
Cesium chloride, 12
Chagas disease vector, *see also* Bacterial symbiont transformation
    biology and ecology, 290–292
    control efforts, 289–290
    CRUZIGARD use
        application in disease control, 298
        description, 296
        efficacy experiments, 296–298

Index

need for control, 301–302
Trypanosomas cruzi
    cecropin A expression effects on, 293–294
    transmission of disease, 289
chartreuse, 86
Cheloninae, 206; *see also* Polydnaviruses (CsIV)
Chemical resistance, *see* Resistance genes
Chlamydia trachomatis, 353
Chloramphenicol acetyltransferase (CAT), 169, 171, 174
Choristoneura fumiferans (spruce budworm), 314
Chromosomal integration detection, 18
Chromosome rearrangements in SIT, 307
Chrysomya rufifacies, 313
Chrysoperla plorabunda (green lacewing), 241
CI, *see* Cytoplasmic incompatibility
cinnabar, 11, 86, 87, 95
Circular molecule integration, 62
Circumsporozoite protein (CSP), 326
cis-acting elements, 192, 197
Class I elements, 45
Class II transposons
    described, 4, 45
    hAT element, *see hAT* element gene vectors
    mariner, *see mariner* transposons
    transgenesis technology and, 324–325
    TTAA-specific, *see* TTAA-specific transposons
Cochliomya hominivorax (screwworm), 337–338
Codling moth (Cydia pomonella), 314, 356
Cold storage of insects, 314
Coleopteran and *piggyBac*, 10
Conditional lethal phenotypes in SIT programs, 310
Confocal microscope, 104
Constitutive promoters and SIT, 312
Containment facilities for release preparations, 358
"Coordinated Framework for Regulation of Biotechnology," 370
Copraphagy, 290, 295, 296
Cost of genetic programs
    equipment and supplies, 15–16
    genetic transformation, 20
    SIRM, 337
Council on Environmental Quality guidelines (CEQ), 372
Courtesy Permit, 372
Cre-*lox*P, 20
    FLP recombinase and, 56
    site-specific recombination system, 54, 64
Cross-mobilizing system, 18, 232
Crossover, 35
CRUZIGARD
    applications in disease control, 298
    description, 296
    efficacy experiments, 296–298
Cryopreservation, 20
CsIV (*Campoletis sonorensis* polydnaviruses), *see* Polydnaviruses (CsIV)
CSP (circumsporozoite protein), 326
Culex (mosquito)
    pipiens
        infection with dsSIN virus, 101
        protein expression in saliva, 174
        SIN virus infection, 165
        sterility techniques, 281
        Wolbachia symbiont vector, *see Wolbachia pipientis* symbiont vector
    quinquefasciatus
        bidirectional CI strains, 281
        Hermes vector development, 222
    theileri, 140, 156
Culicidae, *see* Mosquitoes
Cut-and-paste recombination, 4–5, 257–258
Cu-Zn superoxide dismutase, 315
Cydia pomonella (codling moth), 314, 356
cys-motif gene family, 210
Cytoplasmic incompatibility (CI)
    bidirectional, 280–281
    male aging and, 281
    population suppression and, 281–282
    unidirectional
        associated sterility, 276
        expression systems, 279–280
        fitness effects of transgenes, 276–277
        geographic spreading process, 275–276
        between infected populations, 277–279
        mathematical model, 274–275
        population dynamics, 273–275
        population structure, 276
        transgenes compatibility, 277
    Wolbachia association with, 272–273

D

Ddc (dopa decarboxylase), 7
DDT, 111
Dechorionation, 13, 21
Delia antiqua (onion fly), 314
Dengue fever virus
    antisense viral RNA and, 323
    expression by dsSIN, 176–178
    mosquito disease vectors, 320
Densonucleosis viruses, 11, 157
    biology of, 140–141
    compared to site-specific recombination, 155
    GFP expression and, 101, 102
    molecular biology
        AeDNV gene expression, 143–145
        genomic organization, 142–143
        JcDNV gene expression, 145–146
        replication, 141–142
    potential transformation vectors
        AAV integration, 150–152, 155
        AeDNV integration, 152–155
        episomal transformation, 156–157
    transducing systems development
        by cotransfection of C6/36, 147–148
        helper characteristics, 147
        of mosquito larvae, 149–150
        packaging cell lines construction, 148
        packaging with Sindbis helper, 148–149
        strategy for, 146
Densovirus, 140
Densoviruses, *see* Densonucleosis viruses

Diapause, 314, 341
Dicistronic vectors, 128
Dieldrin-resistance (*Rdl*), see *Rdl* as transgenesis candidate
Dihydroneopterin triphosphate, 82, 86
Dihydropbiopterin, 82
Dioptric apparatus, 80
Dipterans, see *Drosophila*; Mosquitoes
Dipteridine compounds, 82
Discosoma striata, 97
Disease transmission controls, *see* Chagas disease vector; Malaria; Mosquitoes; Vector-borne diseases
DNA, *see* Recombinant DNA
Dominant lethal mutations in SIT, 307
Dopa decarboxylase (Ddc), 7
doublesex gene and SIT, 312
Double-strand gap repair
    mechanism for, 44
    potential of, 44–45
    transposon-mediated, *Drosophila*, 45–46
    transposon-mediated, other insects, 46, 48
Double subgenomic SIN (dsSIN) expression system
    advantages/disadvantages, 178–179
    applications
        downregulation, 183
        expression of antiviral RNAs, 176–178
        protein expression in cultured cells, 171
        protein expression in midgut, 174–176
        protein expression in mosquitoes, 171, 174
        protein expression in saliva, 174
        specificity development, 183–184
        virus-specific interactions observation, 183
    description, 166–167
    GFP expression and, 101
    infection of mosquitoes with, 101
    infection of non-dipterans with, 101–102
    methodology
        gene subcloning, 168
        intrathoracic inoculation into mosquitoes, 170
        per os infection of mosquitoes, 170–171
        transfection, 170
        in vitro transcription, 168, 170
Drive mechanism using transposable elements, 344–345, 355; see also *Wolbachia pipientis* symbiont vector
Drosophila (fruitfly)
    double-strand gap repair, 45–46
    hawaiiensis, 8
    hydei, 10
    mauritiana
        mariner and, see *mariner* transposons
        site-specific recombination, 11
    melanogaster
        early DNA transfer attempts, 4
        eye color as markers, *see* Eye color genes for selection
        frequency of *Wolbachia* infection, 275
        Gypsy element, 193–195
        Himar1 marginal activity, 240–241
        meiotic drive, 327–328
        non-*P* transposons, 9–11
        resistance genes as markers, 110–111
        site-specific recombination method, *see* Site-specific recombination
        P-element transformation, 87, 115–117
            excision assays, 257
            historical perspective, 4–9
            hobo and, 223–226
            lack of excision in non-drosophilids, 220–222
    simulans, 8
        female fecundity loss from *Wolbachia*, 275
        unidirectional CI, 277–278
        target transformation, 29–30
    virilis, 232
Drosopterin, 82; *see also* Pteridine pigments
DsRed (red fluorescent protein), 97, 98
dsSIN expression system, *see* Double subgenomic SIN (dsSIN) expression system
dsx expression, 312, 313

E

EA (Environmental Assessment document), 372
Eastern equine encephalitis (EEE), 166
EIS (Environmental Impact Statement), 372
Electroporation, 17
Embryo injection, 13–14, 101
End-joining, 33–34
Endogenous retroviruses (EnRV), *see* Retroelement classes
Enterobacter cloacae, 354
Entomopoxviruses use in insects, 140
env, 194, 196
Environmental Assessment document (EA), 372
Environmental/ecological risks, *see* Safety concerns/factors
Environmental Impact Statement (EIS), 372
Environmental Protection Agency (EPA), 349, 370
Ephestia kühniella (meal moth), 85, 86
Ephestia transformation studies, 4
Equipment and supplies
    air-pulse injection systems, 20–21
    dechorionation, 13, 21
    DNA delivery, 15–16
    DNA preparation, 12–13
    for embryo injection, 13–14
    for GFP expression, 102
    for GFP visualization, 103
    micromanipulators, 20
    microscopes for microinjection, 14–15, 20
    needle preparation, 14, 21
Eradication programs, 338
Erlichia, 272
Erwinia herbicola, 354
Escherichia coli
    efficiency of gene transfers, 356
    Himar1 activity in, 243–244
    horizontal gene transfers history, 352–353
    stable shuttle plasmid isolation, 293
Eukaryotic retroelements classes, 192
Excision assays
    non-drosophilid *P*-element studies, 220–222
    P-element transformation, 257
    in *piggyBac*

Index

cut-and-paste recombination, 257–258
plasmid-based, 256
precise, 256
terminal duplications, 256–257
in *tagalong*, 252–253
in transposable element mobility assays, 220–222, 223
Exogenous retroviruses (ExRV), *see* Retroelement classes
Extrachromosomal arrays, 346
Eye color genes for selection
  eye pigments
    ommochrome, 81–82, 83
    pteridine, 82
  eye structure and organization, 80–81
  as markers, 87, 94–95
  marker types, 79–80
  mutations affecting pigmentation
    granule group, 85
    ommochrome biosynthesis pathway, 85–86, 88
    overview, 82–84
    pteridine biosynthesis pathway, 86, 88
    transport, 84–85
    uncharacterized/lost, 86–87
  summary, 88
  Web site resources, 88–89

F

FDA (Food and Drug Administration), 370
Federal Plant Pest Act (FFPA), 371
Few polyhedra (FP) mutations, 10, 250–251
Field-selective resistance, 338
Field trials
  criteria for confined, 375
  M. occidentalis case study, 350–351
Filariasis, 320, 323
Finding of No Significant Impact (FONSI), 372
flamenco, 196–197
Flavobacterium sp., 113
FLAVR SAVR tomato, 354
Flesh fly (*Sacrophaga barbata*), 86
Florida Department of Agriculture and Consumer Services, 349
FLP recombinase (*FRT*), 20
  circular molecules and, 62
  description and mechanism, 54–56
  DNA integration techniques, 62–63
  gene expression manipulation using, 59–60
  integration efficiency, 63–65
  marker gene manipulation using, 57–59
  recombination dependence on distance, 60–61
  uses for, 56–57
Fluorescence
  dye use in SIT, 313
  in wt GFP, 96
FONSI (Finding of No Significant Impact), 372
Food and Drug Administration (FDA), 370
Food Safety and Inspection Service (FSIS), 371
Formylkynurenine, 82
Foundation on Economic Trends, 351
FP (few polyhedra) mutations, 10, 250–251

FRT, see FLP recombinase
Fruit fly, see *Bactrocera dorsalis*; *Bactrocera tryoni*; *Ceratitis capitata*; and *Drosophila*
FSIS (Food Safety and Inspection Service), 371

G

G418 (neomycin)
  Drosophila resistance to, 8, 9
  in negative selection targeting study, 41–42
  as promoter trap, 42
  selection as reporter gene, 94
  transgenesis candidate, 112–113
GABA (γ-aminobutyric acid), 115
gag, 193, 195
GAL4/UAS system, 8, 56
β-Galactosidase, 7
Galleria mellonella densovirus (GmDNV), 140, 143
Galleria mellonella NPV, 250
Gap repair
  double-strand
    mechanism for, 44
    potential of, 44–45
    transposon-mediated, *Drosophila*, 45–46
    transposon-mediated, other insects, 46, 48
  P-element precise excision, 257–258
garnet, 85
GDH enzyme, 338
Gene conversion, 35
Gene targeting, *see also* Site-specific recombination
  double-strand gap repair
    mechanism for, 44
    potential of, 44–45
    transposon-mediated, *Drosophila*, 45–46
    transposon-mediated, other insects, 46, 48
  homologous recombination
    biological significance of, 30
    extrachromosomal, 30–31
    extrachromosomal case study, 31–34
    genome manipulation using, 34–38
    knockout case study, 42–43
    negative selection case study, 38–42
    in vivo prospects, 43–44
  target transformation overview, 20
  transgenesis overview, 29–30
  viral vectors use, 29–30
Genetic recombination use in SIT programs, 311
Genetic sexing strains and SIT
  disadvantages of Mendelian genetics, 310–311
  male-only production considerations, 310
  system design
    conditional lethality strategies, 312
    considerations for, 311–313
    sex determination manipulation, 312–313
Genetic transformation
  cost considerations, 20
  critical factors in, 5
  first reported success, 87
  goals of transgenesis, 271–272
  historical perspective

early DNA transfer attempts, 3–4
    P-element transformation, 4–8
identification using markers, 18–19
methodology
    dechorionation, 13
    DNA delivery, 16–18
    DNA preparation, 12–13
    embryo injection preparation, 13–14
    microinjection, 14–16, 17
    needle preparation, 14
new transformation vectors, 9–11
perspectives on use, 19–20
potential applications, 7–8
viral and symbiont vectors, 11–12
Genetic variability loss, 338
Genome manipulation, *see* Gene targeting
Germplasm storage, 20
Germline transformation, *see* Genetic transformation
GFP, *see* Green fluorescent protein
Giant silkmoth (*Hyalophora cecropia*), 238, 293
Glossina (tsetse flies), 279, 282
GmDNV (*Galleria mellonella* densovirus), 140, 143
GmFP3, 252
GmMNPV, 251
Granule group mutations, 85
green, 85, 86
Green fluorescent protein (GFP)
    advantages/disadvantages, 7, 99–100
    background, 96
    characteristics of wild type, 96–97
    dsSIN application, 171
    methods and applications
        expression from viral vectors, 100–102
        transformations using, 102–103
        visualization *in vitro*, 103
        visualization *in vivo*, 103–104
    mutants and related proteins, 97–99
    potential problems/solutions
        histology, 104–105
        in vitro, 104
        in vivo, 105–106
    as selectable marker, 88, 353–354
    SIT use of, 314
    transgenesis technology and, 325
    Web site resources, 106–107
Green lacewing (*Chrysoperla plorabunda*), 241
GTP cyclohydrolase, 86
gypsy, 11
Gypsy element of EnRV
    comparison to other EnRVs, 197
    control by *flamenco* gene, 196–197
    infectious properties, 196
    proviral structure, 195–196

H

Haemagogus equinus, 140, 156
Haematobia irritans (horn fly), 241
Halocarbon oil for embryo injection, 13
hAT element gene vectors

description, 10, 45
mobility properties
    Ac, 223
    hobo in *D. melanogaster*, 223–226
    non-drosophilid species, 226
    Tam3, 223
non-drosophilid discoveries
    Hermes, 226–227
    Hermes mobility properties, 229–230
    hermit, 227–228
    Homer, 227–228
    Homer mobility properties, 230
    hopper, 228–229
    huni, 229
non-drosophilid transformation success elements, 219–220
selection considerations
    cross-mobilization, 231–232
    transposable element interactions, 231–232
    vector selection decision, 230–231
transposable element mobility assays
    excision assays, 220–222, 223
    overview, 220
    transposition assays, 222–223
Helicoverpa armigera (noctuid), 222
Helicoverpa zea NPV, 250
Hemolymph, 323
Hermes
    as alternative transposon, 29
    attributes, 10
    discovery as non-drosophilid *hAT* element, 226–227
    eye color genes as markers and, 87
    integration into *An. gambiae*, 220
    mobility properties
        chromosomal integration sites, 230
        germline integration, 229–230
        host range, 229
    transposon-mediated gap repair, 46, 48
    vector development, 222
hermit, 227–228
Heterosis, 341
Himar1
    activity demonstration attempts, 242–243
    autonomous activity test of *mariner*, 240–241
    hyperactive mutants of, 243–244
Hin recombination sites, 204
Hitch-hiking element, 279
hobo
    as alternative transposon, 29
    attributes, 9–10
    excision assays, 222
    eye color genes as markers and, 87
    mobility in *D. melanogaster*
        horizontal transfer evidence, 224
        integration characteristics, 225–226
        vs. *P*-element hybrid dysgenesis, 223–224
        vs. *P*-element somatic activity, 225
        vs. *P*-element spreading abilities, 224–225
    new transposon systems in, 10
    relationship to *Hermes*, 226–227
Homer

discovery as non-drosophilid *hAT* element, 227–228
  mobility properties, 230
Homochromy, 84
Homologous pairing, 63
Homologous recombination in gene targeting
  biological significance of, 30
  extrachromosomal, 30–31
  extrachromosomal case study
    end-joining, 33–34
    luciferse activity determination, 31, 33
    single-strand-annealing model, 33, 34
    substrate design, 31, 32
  genome manipulation using
    enrichment mechanisms, 38
    gene targeting description, 34–35
    gene targeting uses, 35
    insertion vectors, 36–37
    replacement vectors, 36, 37
    vector design importance, 35–36
  knockout case study, 42–43
  negative selection case study
    cellular *tk* gene activity, 39–40
    efficacy test, 41–42
    methodology and results, 40–41
    overview, 38–39
  in vivo prospects, 43–44
Honeybee (*Apis mellifera*), 81, 85, 339
hopper, 228–229
Horizontal gene transfer
  effect on genomes, 352–353
  by feeding on exogenous DNA, 355–356
  frequency estimates, 353
  host gene movement, 356
  mariner and, 11
  between microorganisms within gut, 354
  piggyBac gene
    considerations for, 262
    evidence outside order, 262
    safety concerns, 263–264
    viruses as DNA vectors, 262–263
  resistant genes transfer, 353–354
  safety concerns, 19, 263–264, 355
  by transposons, 354–355
Horn fly (*Haematobia irritans*), 241
Housefly, see *Musca domestica*
hsp70
  as eukaryotic promotor, 112
  expression of FLP and, 56, 57, 58, 60–61
  wings-clipped helper creation, 6
HSV-*tk* gene targeting
  cellular *tk* gene activity, 39–40
  methodology and results, 40–41
  overview, 38–39
Human health risks of bacterial symbionts use, 299–301;
  see also Safety concerns/factors
huni, 229
Hyalophora cecropia (giant silkmoth), 238, 293
3-Hydroxykynurenine, 82
Hygromycin (*hyg*) resistance gene, 42, 94, 152
Hymenoptera and polydnavirus systems, 205–206
Hyperactive transposase mutants, 243

Hyposoter fugitivus, 211
Hypoxanthine, 82

I

ice-nucleation gene (*InaZ*), 312
Ichneumonids, 206; see also Polydnaviruses (CsIV)
Ichnoviruses, 206; see also Polydnaviruses (CsIV)
IFP2, see *piggyBac*
InaZ (*Pseudomonas syringae ice-nucleation gene*), 312
Inducible promoters and SIT, 312
In insecta gene transfer, 354
Inoculation, 170, 340
Insect control
  conventional gene targets, 110
  of mosquitoes, 320
  organophosphorus insecticides, 341
  pest management programs, see Transgenic arthropods
    in PMP
  potential applications of gene transfer, 7–8
  SIT, see Sterile insect technique
Insecticides, see Insect control
Insect immune system, 324
Insect transgenesis, see Genetic transformation
Insertion site sequencing, 18
Integron, 354
Internal ribosomal entry site (IRES), 128
Interplasmid transposition of *piggyBac*, 258–259
Intramolecular recombination of *FRT*s, 55–56
Intrathoracic inoculation of dsSIN into mosquitoes, 170
Introgression and selection strategy, 340
Inverse PCR, 18
Inverted terminal repeat sequences (ITRs), 45, 225, 226,
  242
Ionizing radiation and SIT, 315
IRES (internal ribosomal entry site), 128
Iridoviruses use in insects, 140
Irritans, see *Himar1*
Isodrosopterin, 82
Isogenicity, 35
Isoxanthopterin, 82
Iteravirus, 140
ITRs (inverted terminal repeat sequences), 45, 225, 226,
  242
ivory, 85–86

J

JcDNV (*Junonia coenia* densovirus), 140, 141
  gene expression, 145–146
  genomic organization, 143
  transducing systems development, 146
Jellyfish (*Aequorea* sp.), 88, 96

K

Kanamycin-resistance phenotype, 295
karmoisin, 86
Kissing bugs, see Chagas disease vector

Knockouts, 35, 36, 37, 42–43
Kynurenine formamidase, 82, 86
Kynurenine hydroxylase, 86, 88
Kynurenine monoxygenase, 86

L

La Crosse virus, 176, 326
lacZ, 243, 347
Lamina, 80
Leishmania major, 241, 355
Lepidopterans
  Bombyx mori, 4, 11
    BmDNV densovirus, 140
    successful *piggyBac* transformations, 261
  piggyBac element in, 10
  polydnavirus/wasp parasite/host system, 205
  TTAA-specific transposons, *see* TTAA-specific transposons
Lipofection, 16–17
Liposome fusion, 16
little isoxanthopterin (lix), 86
Long terminal repeat-retrotransposons (LTR-RT), *see* Retroelement classes
Long terminal repeats (LTRs), 192
LOOPER, 262
Luciferase
  in biolistics, 17
  restoration of activity case study
    activity determination, 31, 33
    end-joining, 33–34
    single-strand-annealing model, 33, 34
    substrate design, 32
Lucilia cuprina (blowfly), 85
  hermit element discovery, 228
  hobo excision assays, 222
Lucilia striatellus, 275
LuIII, 148–149
Lymantria dispar (gypsy moth), 250, 340–341

M

Malaria, 319, 323
  Anopheles gambiae
    extrachromosomal homologous recombination, 31–34
    Hermes integration into, 220
    Hermes vector development, 222
    huni element discovery, 229
    infection efficiency in MOS-55, 128
    infection with dsSIN virus, 101
    sex-specific splicing and, 313
    transposon-mediated gap repair, 46, 48
  mosquito control issues, 352
  mosquito disease vectors, 320
  parasite development, 324
  potential value of PMP, 360
  transmission reduction effects, 321
Maleness factor isolation, 312–313
Male-specific genes in SIT, 312
Malpighian tubules, 324
mariner transposons, 6
  advantages/disadvantages as genetic tools, 244–246
  as alternative transposon, 29
  attributes of, 11, 45
  autonomous nature, 240–241
  background, 237–238
  diversity of, 238–239
  excision assays, 222
  eye color genes as markers and, 87
  features facilitating genetic transfer, 237
  function in diverse hosts, 241–242
  Himar1
    activity demonstration attempts, 242–243
    hyperactive mutants of, 243–244
  horizontal transfers evidence, 239, 245–246
  Mos1
    FLP and, 56
    function in diverse hosts, 241
    mariner nomenclature and, 238
    transgenic line generation, 222
  phylogenetic relationships, 240
  subfamily divergence, 244
  transmission of, 355
  transposon-mediated gap repair, 46, 48
Markers/reporters
  for *Drosophila*, 7
  eye color genes used as, 87, 94–95
  genetic marking in SIT, 313–314
  green fluorescent protein
    advantages/disadvantages, 7, 99–100
    background, 96
    characteristics of wild type, 96–97
    dsSIN application, 171
    methods and applications, 100–104
    as selectable marker, 88, 353–354
  ideal reporter characteristics, 95–96
  in laboratory development of pathogens, 325
  manipulation using FLP-*FRT*, 57–59
  P-element and, 6–7
  in promoter-trap strategies, 38
  properties for interspecies use, 66
  pugD, 66–67
  resistance genes use
    insect genes/enzymes, 110–111
    noninsect genes, 111–114
    overview, 109–110
    Rdl, 115–118
  scorable, 79, 80, 87
  in selection of transformed insects, 94–95
  in SIT programs, 309–310
  table of dominant marker transgenes, 67–71
  for trangenesis identification, 18–19
  types, 79
Mass rearing and SIT, SIRM, 309, 340
Maternal microinjection, 346–347
Mating bias, 357
MD (meiotic drive), 327–328, 339
mdg4, see *Gypsy* element of EnRV
Meal moth (*Ephestia kühniella*), 85, 86

Medfly, see *Ceratitis capitata*
Mediterranean fruit fly, see *Ceratitis capitata*
Megaselia scalaris, 313
Meiotic drive (MD), 327–328, 339
Mendelian genetics
    in SIT programs, 310–311
    usefulness of, 322
Metabolic genes as transgenesis candidates, 111
Metapopulation models, 357
Metaseiulus occidentalis (mite)
    field release results, 350–351
    genetic improvement example
        dispersal studies, 343
        economic analysis, 343
        monitoring system, 341, 343
        OP-resistant strain use, 341
        program phases, 343–344
    regulatory review of release
        institutional biosafety committees involvement, 348–349
        public acceptance, 350
    risk model system
        ability to survive in wild, 347
        maternal microinjection, 346–347
        risk considerations, 347, 348
Metazoan pathogens, 323
Methylation of plasmids, 259
Methylenetetrahydrofolate dehydrogenase (MTHFD), 66–67, 86
Microgastrinae, 206; see also Polydnaviruses (CsIV)
Microinjection in DNA delivery, 14–16, 17, 346–347
Micromanipulators, 20
Microplitis croceipes, 211
Microplitis demolitor, 205, 211
Microscopes
    for GFP visualization, 103, 104
    for microinjection, 14–15, 20
Midgut tissue infection of *Aedes aegypti*, 174–176
Milkweed bug (*Oncopeltus fasciatus*), 101
Minos
    as alternative transposon, 29
    attributes, 10
    eye color genes as markers and, 87
    transposon-mediated gap repair, 46, 48
Minute virus of mice (MVM), 140, 152
Mite, see *Metaseiulus occidentalis*
MKα (*Mycobacterium kansasii*), 295
Mobility assays
    hAT element
        excision assays, 220–222, 223
        overview, 220
        transposition assays, 222–223
    for *piggyBac*
        cut-and-paste recombination, 257–258
        interplasmid transposition, 258–259
        nonreplicative transposition, 259
        ORF encoding, 254–256
        plasmid-based excision, 256
        precise excision, 256
        terminal duplications excisions, 256–257
Models for risk assessment, 346–348, 356–358

Moloney murine leukemia virus (MoMLV), 126
Monocistronic vectors, 127
Mos1, see also *mariner* transponsons
Mosaic factors, 237
Mosquitoes, 319; see also *Aedes*; *Anopheles*; *Culex*
    control programs, 320
    as disease vectors, 320
    dsSIN intrathoracic inoculation into, 170
    dsSIN *per os* infection of, 170
    dsSIN protein expression, 171, 174
    immune-based responses to pathogens, 326
    laboratory development of pathogen resistance
        gene expression to interfere with specific transmission, 326
        genetic engineering approaches, 322
        ideal properties of phenotype gene, 323
        immune-based antiparasite approaches, 326
        marker gene use, 325
        pathogen interference targets, 323–324
        receptor/ligand interactions, 326
        recognition properties identification, 325
        resistant or refractory phenotypes, 325
        transposable elements use, 324–325
        validation of engineered antiparasite gene, 324
    non-drosophilid gene transfer and, 8–9
    pathogen interaction with host, 323–324
    Sindbis infection of, 101
MRE16, 1001 SIN virus
    infection of mosquitoes, 165
    midgut tissue infection, 174–176
MRE/3′2J
    dsSIN and, 166–167
    intrathoracic inoculation into mosquitoes, 181
    midgut tissue infection, 174–176
mRNA injection of FLP, 56–57
M strains, 5
MTHFD (methylenetetrahydrofolate dehydrogenase), 66–67, 86
Multimeric concatamers, 18
Musca domestica (housefly), 10, 85, 86
    Hermes transposable element
        as alternative transposon, 29
        attributes, 10
        discovery as non-drosophilid *hAT* element, 226–227
        eye color genes as markers and, 87
        integration into *An. gambiae*, 220
        mobility properties, 229–230
        transposon-mediated gap repair, 46, 48
        vector development, 222
    Sex lethal gene and, 313
    sex-specific splicing and, 313
Mutant rescue, 6
Mutations affect on pigmentation
    granule group, 85
    ommochrome biosynthesis pathway, 85–86
    overview, 82–84
    pteridine biosynthesis pathway, 86
    transport, 84–85
    uncharacterized/lost, 86–87
MVM (minute virus of mice), 140, 152
Mycobacteriophage L1 integrase system, 296

Mycobacterium kansasii (MKα), 295

N

Nasonia (wasp), 277–278
National Environmental Policy Act (NEPA), 372–374
"National Institutes of Health Guidelines," 369
Natural enemies and genetic manipulation
    background, 339
    considerations for, 342–343
    deployment models, 344
    drive mechanisms, 344–345
    genetic improvement example, 341, 343–344
    quality control issues, 341
    strains with complex traits, 340–341
Needle preparation, 14, 21
Negative selection gene targeting with HSV-tk
    cellular *tk* gene activity, 39–40
    efficiency test, 41–42
    methodology and results, 40–41
    overview, 38–39
Nematoda, 273
neo
    in negative selection targeting study, 41–42
    as promoter trap, 42
    transgenesis candidate, 112–113
Neodrosopterin, 82
Neomycin (G418)
    Drosophila resistance to, 8, 9
    in negative selection targeting study, 41–42
    as promoter trap, 42
    selection as reporter gene, 94
    transgenesis candidate, 112–113
Neomycin phosphotransferase (NPT), 8, 112
NEPA (National Environmental Policy Act), 372–374
Neuroactive proteins, 171
nomad element, 197
Nondiapause, 341
Nonreplicative transposition in *piggyBac* mobility assays, 259
NPT (neomycin phosphotransferase), 8, 112
Nutritive symbionts, 279–280

O

ocra, 86
Ommatidia, 80
Ommochrome pigments
    biosynthesis pathway mutations, 85–86, 88
    described, 81–82, 83
    Drosophila eye color and, 66
Oncopeltus fasciatus (milkweed bug), 101
Onion fly (*Delia antiqua*), 314
opd (organophosphorus dehydrogenase), 7, 113–114, 260–261
OPD (paraoxon)
    resistance to, 113–114
    selection as marker/reporter, 94
Open niche strategy, 340

Optic chiasma, 80
ORF encoding in *piggyBac* mobility assays, 254–256
Organophosphorus dehydrogenase (*opd*), 7, 113–114, 260–261
Organophosphorus insecticides (OP), 341
Oriental fruit fly (*Bactrocera dorsalis*)
    hopper discovery, 228
    and *piggyBac*, 5
    successful *piggyBac* transformations, 10, 261

P

p3E1.2, 10, 255
p7hyg genome, 152
Pantropic vectors, *see* Retroviral vectors
Papilio glaucus (tiger swallowtail butterfly), 101
Papillation assays, 243
para, 111
Paraoxon (OPD)
    resistance to, 113–114
    selection as marker/reporter, 94
Parathion hydrolase, 113
Paratransgenic insects generation
    cecropin A expression, 293–295
    genetic transformation of *Rh. rhodnii*, 293
    integrative plasmids for actinomycetes transformation, 296
    single-chain antibody expression, 295–296
    system description, 292–293
Paris, 232
PARP (poly-ADP-ribose polymerase) gene, 43
Parvoviridae, 140
Parvoviruses, *see* Densonucleosis viruses
PBS (primer binding site), 192–193
PCR amplification, 18, 38, 134
PDA (pyrimidodiazepine), 82
Pectinophora gossypiella (pink bollworm), 258, 261
P-element transformation, 87
    excision assays, 257
    historical perspective
        cut-and-paste mechanism for, 4–5
        success factors, 5
        use as genetic tool, 7–8
        use for non-drosophilid gene transfer, 8–9
        vectors and markers overview, 6–7
    hobo and
        hybrid dysgenesis, 223–224
        integration characteristics, 225–226
        somatic activity, 225
        spreading abilities, 224–225
    lack of excision in non-drosophilids, 220–222
Penelope, 232
Peptides and cecropin A expression, 293
Peridroma saucia, 354
Peripheral retina, 80
Permethrin, 341
per os infection of mosquitoes with dsSIN, 170–171, 174
Pesticide resistance, 340, 345, 353–354
Pest management programs (PMP), *see* Transgenic arthropods in PMP
Pests and genetic manipulation

control project implementation difficulties, 339
quality control issues, 338–339
SIRM description and mechanism, 337–338
Phage display, 295
Phenoxazinone synthase, 86
Photoreceptor cells, 80
phspBac helper construct, 258
piggyBac
   as alternative transposon, 29
   attributes, 10–11
   background, 253
   excision assays, 222
   eye color genes as markers and, 87
   horizontal transmission
      considerations for, 262
      evidence outside order, 262
      safety concerns, 263–264
      viruses as DNA vectors, 262–263
   identification of, 220
   insertion at target sites, 254
   mobile host DNA insertions, 251
   mobility assays
      cut-and-paste recombination, 257–258
      interplasmid transposition, 258–259
      nonreplicative transposition, 259
      ORF encoding, 254–256
      plasmid-based excision, 256
      precise excision, 256
      terminal duplications excisions, 256–257
   structural features, 253–254
   transposon-mediated gap repair, 46, 48
   use in transformations
      attempts to find markers, 260–261
      medfly model system, 261
   vector systems in, 5
Pigments
   in compound eye
      ommochrome, 81–82, 83
      pteridine, 82
   description of cells, 80
   Drosophila eye color and, 66
   mutations affecting
      granule group, 85
      ommochrome biosynthesis pathway, 85–86, 88
      overview, 82–84
      pteridine biosynthesis pathway, 86, 88
      transport, 84–85
      uncharacterized/lost, 86–87
Pink bollworm (*Pectinophora gossypiella*), 258, 261
Plant pest definition, 372
Plant Quarantine Act (PQA), 371
Plasmid-based excision in *piggyBac* mobility assays, 256
Plasmid DNA, *see* Recombinant DNA
Plasmid rescue, 19
Plasmodium, 320, 325, 326
PMP (pest management programs), *see* Transgenic arthropods in PMP
pMRE/3′2J, 166–167
pol, 193, 194

Poly-ADP-ribose polymerase (PARP) gene, 43
Polydnaviruses (CsIV)
   biology, 204–205
   genome organization
      coding/noncoding sequences, 207–208
      genes encoding structural proteins, 208–209
      segment characteristics, 206–207
      segment integration sites, 207
   genome organization and viral gene function
      class 2 gene characteristics, 210–211
      gene copy number, 211
      infectious dose vs. host range, 211
      segment nesting, 209–210, 211
   parasitoid phylogeny, 205–206
   relation to transposable elements, 204
   relation to viruses, 203–204
   relevance to transgenic research, 140, 203
   resident genomes characteristics, 204
   summary and applications, 212–213
popcorn, 282
Population and genetic models, 356–358
Population replacement strategies, 321
Position effects, 7, 48, 103
Positive–negative selection, 38
PQA (Plant Quarantine Act), 371
Precis coenia (buckeye butterfly), 101
Precise excision, *see* Excision assays
   piggyBac gene mobility assays, 256
   tagalong element, 252–253
Predator–prey dynamics, 351
Primer binding site (PBS), 192–193
Promoters in SIT systems, 312
Promoter-trap strategies, 38, 42
Proteobacteria, 279
Protozoan pathogens, 323
pRr1.1, 293
prune, 86
Pseudomonas diminuta, 113
Pseudomonas syringae ice-nucleation gene (InaZ), 312
Pseudotyping, 11, 126–127
pTE3′2J, 166–167, 168
Pteridine pigments
   biosynthesis pathway mutations, 86, 88
   in compound eye, 82
   Drosophila eye color and, 66
pUC18, 255
pU*Chs*neo plasmid, 112, 143
pugD, 66–67, 86
Punch, 86
Purine *de novo* synthesis, 67
purple, 86
Pyrethroids, 111
Pyrimidodiazepine (PDA), 82
6-Pyruvoyltetrahydropterin (PTP), 82, 86

Q

Quality control issues in SIRM, 338–339

R

Radiation and SIT, 315
RAPD-PCR analysis, 344
raspberry, 86
Rdl as transgenesis selection candidate, 7
   considerations for, 115
   functional receptor cloning/expression, 115
   large mosquito gene transformation, 118
   P-element transformation, 115–117
   promoter definition, *Drosophila*, 117–118
   promoter definition, mosquitoes, 118
Recombinant DNA
   delivery methods
      biolistics, 17
      electroporation, 17
      lipofection, 16–17
   early attempts at use, 3–4
   integration techniques, 62–63
   preparation, 12–13
   public safety concerns about use, 369
   stabilization methods, 63–65
   viruses as vectors of, 140, 262–263
Red flour beetle (*Tribolium castaneum*), 11
   infection with dsSIN virus, 101
   successful *piggyBac* transformations, 261
Red fluorescent protein (DsRed), 97, 98
Red/red-yellow pigments, *see* Pteridine pigments
Reductive evolution, 207–208
Reduviidae, 289
Regulation of transgenic arthropods
   abbreviations, 377–378
   authority over biotechnology, 370–371
   current procedures, 375–376
   future developments, 376–377
   historical overview, 369–370
   review and evaluation process
      determination of categorical exclusion, 372–373
      determination of jurisdiction, 372
      evaluation of nontransgenic form, 373–374
      evaluation of transgenic form, 374
      information requested, 375
      process overview, 371–372
      Web site resources, 371
Relative humidity tolerance, 347
rep gene family, 210
Replacement vectors, 42–43
Replicon gene expression
   advantages/disadvantages, 181
   applications, 180–181
   methodology, 180
   overexpression of antipathogen genes, 184–185
   toxicity effects, 184
Reporters, *see* Markers/reporters
Resistance
   to antibiotics, 353–354
   field-selected, 338
   to pesticides, 340, 345
Resistance genes as selectable markers
   insect genes/enzymes
      Ace, 110–111

      metabolic, 111
      para, 111
   noninsect genes
      characteristics and uses, 111–112
      neo, 112–113
      opd, 113–114
   overview, 109–110
   Rdl
      considerations for, 115
      functional receptor cloning/expression, 115
      large mosquito gene transformation, 118
      P-element transformation, 115–117
      promoter definition, *Drosophila*, 117–118
      promoter definition, mosquitoes, 118
Resistance to dieldrin (*Rdl*), *see Rdl* as transgenesis selection candidate
Retroelement classes (LTR-RT, EnRV, ExRV)
   common structural features
      gene products, 193
      long terminal repeat, 192
      poly(A) tract, 194
      polypurine tract, 193
      primer binding site, 192–193
   EnRV *Gypsy* element
      comparison to other EnRVs, 197
      control by *flamenco* gene, 196–197
      infectious properties, 196
      proviral structure, 195–196
   functional similarities, 194
   overview, 191–192
   retroviral vectors uses and limitations, 197–199
   structural and functional differences
      env gene, 194
      pol gene, 194
      replication cycle, 194
Retrotransposons, *see* Retroelement classes
Retroviral vectors
   genetic organization
      dicistronic, 128
      monocistronic, 127
   horizontal gene transfer and, 356
   infection efficiency
      of embryos, 128
      of insect cell lines, 128, 129, 130
      of larvae, 128, 130
   methodology for pantropic generation, 127
      luciferase assays protocol, 134–135
      PCR amplification of provirus, 134
      reagents and solutions, 135–136
      by stable producer cell line creation, 132–133
      by transient transfection, 130–131
      ultracentrifugation, 133–134
   pseudotyping to create, 126–127
   retrovirus transformation into, 126
   uses and limitations, 197–199
Retroviruses, 11, 126; *see also* Retroelement classes
Rhodnius prolixus
   bacterial symbionts of, 12
   biology of, 290–291
   cecropin A expression and, 293–295
   Chagas disease control and, 290

Index

paratransgenesis of, 292–293
single-chain antibody expression, 295–296
Rhodococcus rhodnii
cecropin A expression and, 293–295
Chagas disease and, 290–291
paratransgenesis of *R. prolixus* and, 293
Rickettsia, 272
Rickettsia prowazekii, 283
Rifkin, Jeremy, 351
Risk assessment
in biological control programs, 336
M. occidentalis model system
ability to survive in wild, 347
considerations, 347, 348
maternal microinjection, 346–347
models for, 346–348, 356–358
in SIT, 308
transgenic arthropods in PMP
accidental release risk, 359
ecological risks, 336, 352
models for risk, 346–348
risk considerations, 336–337
RNA viruses, antisense
expression of, 176–179
Sindbis virus and, 323
for use in insects, 139–140
rosy, 6, 86, 87

S

Sacrophaga barbata (flesh fly), 86
Safety concerns/factors
bacterial symbiont transformation, 299–301
environmental/ecological risks, 301, 336, 352
for horizontal gene transfer, 263–264, 355
laboratory containment
accidental release risks, 359
facilities and procedures, 358–359
regulations
authority over biotechnology, 370–371
current procedures, 375–376
future developments, 376–377
historical overview, 369–370
review and evaluation process, 371–374
regulatory review of transgenic strains, 348
risk assessment, 349
risk model system for *M. occidentalis*
ability to survive in wild, 347
maternal microinjection, 346–347
risk considerations, 347, 348
of *Wolbachia*-infected females, 281
Salivary gland
parasite development in mosquitoes and, 324
protein expression and, 174
Salmonella, 204, 352
Scarlet, 84–85
Scorable markers, 79, 87
Scorpion toxin gene (scotox), 171, 174
Screwworm (*Cochliomya hominivorax*), 337–338
SDSA (synthesis-dependent strand annealing), 45–46, 47

Segmentation of polydnaviruses, 211
Segment nesting, 209–210, 211
Selectable markers, 79
Selection of transgenic insects using eye color genes, *see* Eye color genes for selection
Sepiapterin, 82
Sex lethal gene (Sxl), 313
Sex-specific promoters and SIT, 312
Sex-specific splicing and SIT, 312
Silkmoth, see *Bombyx mori*
Silkworms, 339
Sindbis (SIN) virus, 11
antisense viral RNA and, 323
background
molecular biology, 163, 164
virogenesis in cultured cells, 163–165
virus/mosquito interactions, 165–166
densovirus packaging with helper, 148–149
dsSIN system and GFP expression, 101
dsSIN advantages/disadvantages, 178–179
dsSIN expression of antiviral RNAs, 176–178
dsSIN expression system methodology
gene subcloning, 168
intrathoracic inoculation into mosquitoes, 170
per os infection of mosquitoes, 170–171
transfection, 170
in vitro transcription, 168, 170
dsSIN protein expression
in cultured cells, 171
downregulation potential, 183
as marker genes, 183
in midgut, 174–176
in mosquitoes, 171, 174
in saliva, 174
engineered gene effects, 323
expression system applications
DNA-based replicon use, 184–185
downregulation, 183
effector molecule use determination, 181, 183
specificity development, 183–184
virus-specific interactions observation, 183
expression systems
dsSIN description, 101, 166–167
infectious clone technology, 166
replicon, 168, 169
GFP expression, 100–101
infection of mosquitoes, 101
infection of non-dipteran species, 101–102
overview, 162
replicon gene expression
advantages/disadvantages, 181
applications, 180–181
methodology, 180
overexpression of antipathogen genes, 184–185
toxicity effects, 184
for use in insects, 140
Single-chain antibody expression, 295–296
Single-strand-annealing model, 33, 34
SINrep5 replicon, 167
SIRM (sterile insect release method), *see* Sterile insect technique

SIT, *see* Sterile insect technique
Site-specific recombination, *see also* Gene targeting
   for chromosome rearrangements, 60–61
   D. hydei, 10
   D. mauritiana, 11
   densonucleosis virus integration compared to, 155
   FLP recombinase, 20
      circular molecules and, 62
      description and mechanism, 54–56
      DNA integration techniques, 62–63
      gene expression manipulation using, 59–60
      integration efficiency, 63–65
      marker gene manipulation using, 57–59
      recombination dependence on distance, 60–61
      uses for, 56–57
   in gene expression manipulation, 59–60
   marker gene manipulation using, 57–59
   marker genes
      pugD, 66–67
      requirements for interspecies use, 66
      table of transgenes, 67–71
   piggyBac precise excision, 256, 257–258
   tagalong precise excision, 252–253
   as transformation tool
      advantages to, 61–62
      DNA integration techniques, 62–63
      integration efficiency, 63–65
   uses for, 54
Snow, 85
Sodalis glossinidius, 279
Somatic transformation, 12
Southern DNA hybridization, 18
Spruce budworm (*Choristoneura fumiferans*), 314
Sterile insect release method (SIRM), *see* Sterile insect technique
Sterile insect technique (SIT)
   CI uses with, 281–282
   description and mechanism, 307–308
   genetic marking
      fluorescent dye use, 313
      technique problems, 313
      transgenes use, 313–314
   genetic sexing strains
      disadvantages of Mendelian genetics, 310–311
      male-only production considerations, 310
      system design considerations, 311–313
   ionizing radiation use, 315
   operational concerns, 315
   regulatory problems, 316
   sterile insect release method
      description and mechanism, 337–338
      quality control issues, 338–339
      risk considerations, 359
   stockpiling applications, 314
   technological constraints on use
      biological fitness concerns, 308–309
      efficiency need, 310
      marker selection, 309–310
      mass rearing pressures, 309
      monitoring need, 309
   vector competence considerations, 314–315

Suicide vectors, 20
Sustainability of agroecosystems, 352
Sxl (*Sex lethal* gene), 313
Symbiont vectors, *see also* Vectors
   antiparasite gene spread using, 327–328
   bacterial, *see* Bacterial symbiont transformation
   horizontal gene transfer and, 356
   introduction of genes into wild populations, 327–328
   nutritive, 279–280
   triatomine, 290–292
   vector-borne disease control and, 322
   Wolbachia, *see Wolbachia pipientis* symbiont vector
Symbiotic paratransgenesis, 12
Synthesis-dependent strand annealing (SDSA), 45–46, 47

T

tagalong
   mobile host DNA insertions, 251
   TTAA-specific transposons
      background, 252
      precise excision, 252–253
      target site preference, 252
Tam3, 9, 223
tangerine, 86
Targeted transformation, *see* Gene targeting
Target transposition, 7
Tc family
   horizontal gene transfer and, 356
   -*mariner* superfamily activity, 241
   members of, 10
TE/3'2J
   dsSIN and, 147, 166–67
   GFP expression and, 101–102, 149
   infection of mosquitoes, 165
   intrathoracic inoculation into mosquitoes, 181
   midgut tissue infection, 174–176
   protein expression in saliva, 174
Temperature-sensitive lethal-based (*tsl*) sexing strains, 311
Terminal duplications excisions in *piggyBac*, 256–257
Terminal Repeat Binding Protein (TRBP), 259–260
Tetrahymena pyriformis, 355
TFP3, see *tagalong*
Thermal hysteresis, 314, 338
Tiger swallowtail butterfly (*Papilio glaucus*), 101
TIR (terminal inverted repeat), *see* ITR
TN-368 cell line, 250
tom, 197
Topaz, 85
Toxorhynchites amboinensis, 140, 156
Traffic ATPase, 84, 85
Transduction, 139–140
Transgenesis, *see* Genetic transformation
Transgenic arthropods in PMP
   background and risk overview, 336–337
   genetic manipulation and deployment of natural enemies
      considerations for, 342–343
      deployment models, 344
      drive mechanisms, 344–345

genetic improvement example, 341, 343–344
quality control issues, 341
strains with complex traits, 340–341
genetic manipulation of beneficials
background of efforts, 339–340
implementation strategies, 340
genetic manipulation of pests
control project implementation difficulties, 339
quality control issues, 338–339
SIRM description and mechanism, 337–338
laboratory containment
accidental release risks, 359
facilities and procedures, 358–359
for pest management programs
disadvantages to recombinant DNA methods, 345–346
field release results example, 350–351
media responses, 351
regulatory agencies review, 348–350
risk model system, 346–348
regulation of
authority over biotechnology, 370–371
current procedures, 375–376
future developments, 376–377
historical overview, 369–370
review and evaluation process, 371–374
risk considerations
ecological, 336, 352
from horizontal gene transfer, 352–356
population and genetic models, 356–358
summary and conclusions
benefits and costs of introductions, 360
guidelines need, 361
multitactic management need, 361
research priorities, 360
value to PMP, 360
Translocation breakpoint in SIT programs, 311
Transmissible modulation of vector competence, 295
Transovarially transmitted viruses, 280
Transposable elements, 4; see also *hAT* element gene vectors, *mariner* transposons, *Minos*, *P*-element, and TTAA-specific transposons
horizontal gene transfer and, 354–355
Transposition assays, 222–223
Transposon-tag genes, 7
Trap plants, 350
TRBP (Terminal Repeat Binding Protein), 259–260
Triatoma sp., 290, 291
Triatominae, 289
Triatomine symbionts, *see* Chagas disease vector
Tribolium castaneum (red flour beetle), 11
infection with dsSIN virus, 101
successful *piggyBac* transformations, 261
Trichoplusia ni
excision assays, 222
plaque assay application, 250
transposition efficiency, 258
Trioxys pallidus, 344
Trypanosomas cruzi, see also Chagas disease vector
cecropin A expression effects on, 293–294
transmission of Chagas disease, 289

Trypanosomes, 12, 295, 315
Tryptophan oxygenase, 82, 88
Tsetse flies (*Glossina*), 279, 282
tsl (temperature-sensitive lethal-based sexing strains), 311
TTAA-specific transposons
host-specific accessory factors possibility, 259–260
origin of
FP plaque morphology mutation, 250–251
transposon-induced mutations, 251–252
piggyBac element
background, 253
insertion at target sites, 254
structural features, 253–254
piggyBac horizontal transmission
considerations for, 262
evidence outside order, 262
safety concerns, 263–264
viruses as DNA vectors, 262–263
piggyBac mobility assays
cut-and-paste recombination, 257–258
interplasmid transposition, 258–259
nonreplicative transposition, 259
ORF encoding, 254–256
plasmid-based excision, 256
precise excision, 256
terminal duplications excisions, 256–257
piggyBac use in transformations
attempts to find markers, 260–261
medfly model system, 261
summary, 264–265
tagalong element
background, 252
precise excision, 252–253
target site preference, 252
$\beta 2$-*Tubulin* gene, 56
Type I, II granules, 81

U

U.S. Department of Agriculture (USDA), 349, 350, 370–371
Ubiquitous, 232
Ulysses, 232
Unequal sister-chromatid exchange, 58–59
Union of Concerned Scientists, 351
Upstream activating sequences (UAS), 8, 56
USDA–APHIS, *see* Animal and Plant Health Inspection Service (APHIS)

V

Vector-borne diseases
control strategies, 319
genetic control objective and programs, 321–322
introduction of genes into wild populations
design features, 328
meiotic drive, 327
symbionts, 327–328
transposable elements, 327

laboratory development of pathogen resistance in mosquitoes
   expression of selectively interfering molecules, 326
   genetic engineering approaches, 322
   ideal properties of phenotype gene, 323
   immune-based antiparasite approaches, 326
   marker gene use, 325
   pathogen interference targets, 323–324
   receptor/ligand interactions, 326
   recognition properties identification, 325
   resistant or refractory phenotypes, 325
   transposable elements use, 324–325
   validation of engineered antiparasite gene, 324
target populations characteristics, 328–329
transmission system, 320
Vector competence, 319
Vectors, 11–12; see also Viruses
   densovirus potential use
      AAV integration, 150–152, 155
      AeDNV integration, 152–155
      episomal transformation, 156–157
   design importance in gene targeting, 35–36
   disease vectors and SIT, 314–315
   gene targeting case study, 42–43
   GFP expression from, 100–102
   -helper ratios in DNA concentrations, 13
   helper systems development, 4–5
   horizontal gene transfer and, 262, 356
   insertion, 36–37
   P overview, 6–7
   replacement, 36, 37
   replacement of *P*
      Hermes, 10
      hobo, 9–10
      mariner, 11
      Minos, 10
      piggyBac, 10–11
      viral and symbiont, 11–12
   retroviral, see Retroviral vectors
   symbiont, see Symbiont vectors
      antiparasite gene spread using, 327–328
      bacterial, see Bacterial symbiont transformation
      horizontal gene transfer and, 356
      introduction of genes into wild populations, 327–328
      nutritive, 279–280
      triatomine, 290–292
      vector-borne disease control and, 322
      Wolbachia, see *Wolbachia pipientis* symbiont vector
   transient excision assays to assess, 9
   transposable elements, see Transposable elements
   viral vectors use in gene targeting, 29–30
   viruses as, 262–263
Venezuelan equine encephalitis (VEE), 163
Vermilion (v) mutant line, 4, 85–86
Vertical transmission elimination with SIT, 308
Vesicular stomatitis virus (VSV-G), 126
Viral encephalites, 320
Viruses, see also Vectors
   baculoviruses
      FP plaque morphology mutation, 250–251
      horizontal gene transfer and, 263, 356
      transposon-induced mutations, 251–252
      for use in insects, 140
   definition, 203
   densonucleosis, see Densonucleosis viruses
   horizontal gene transfer and, 356
   polydnaviruses
      biology, 204–205
      genome organization, 206–209
      genome organization and viral gene function, 209–211
      parasitoid phylogeny, 205–206
      relation to transposable elements, 204
      relation to viruses, 203–204
      relevance to transgenic research, 140, 203
      resident genomes characteristics, 204
      summary and applications, 212–213
   retroviral vectors use and limitations, 197–199
   SIN, see Sindbis (SIN) virus
   transovarially transmitted, 280
   for use in insects, 139–140
   as vectors, 262–263
Voltage-gated sodium channel, 111
VSV-G (vesicular stomatitis virus), 126

W

Wasps
   polydnavirus systems and, 205–206
   unidirectional CI and, 277–278
Web sites
   APHIS virtual team, 377
   Chagas disease, 290
   eye color mutations and strains, 88–89
   GFP development, 106–107
   plant pest permits, 372
   release applications, 349
   review and evaluation process for proposed introductions, 371
Western equine encephalitis (WEE), 166
white
   chromosome rearrangements in mutants, 61
   mutation effects, 84–85
   naming inaccuracies, 87
   selection as marker, 87, 95
   success factors, 6–7
Wigglesworthia glossinidia, 279
Wings-clipped helper creation, 6
Wolbachia pipientis symbiont vector, 12
   applications
      bidirectional CI, 280–281
      CI and population suppression, 281–282
      population age structure modification, 282
   CI association, 272–273
   as drive mechanism, 344–345
   host range and phylogeny, 273
   overview, 272–273
   research priorities, 282–284
   unidirectional CI and gene drive associated sterility, 276

expression systems, 279–280
fitness effects of transgenes, 276–277
geographic spreading process, 275–276
between infected populations, 277–279
population dynamics, 273–275
population structure, 276
transgenes compatibility, 277

X

Xanthenuric acid (XA), 326
Xanthine dehydrogenase, 17, 82, 86
Xanthommatin, *see* Ommochrome pigments

Y

Yeast, 45
Yeast transcriptional activator, *see* GAL4/UAS system
yellow, 86
Yellow fever mosquito, see *Aedes aegypti*
Yellow fever virus, 176–178, 320
yellowish, 85

Z

ZAM element, 197
Zea mays, 232